ENZYME TECHNOLOGIES

CHEMICAL BIOLOGY OF ENZYMES FOR BIOTECHNOLOGY AND PHARMACEUTICAL APPLICATIONS

(A Series Consisting of Three Volumes)

Volume I. *Enzyme Technologies: Metagenomics, Evolution, Biocatalysis, and Biosynthesis*
Editors: Wu-Kuang Yeh, Hsiu-Chiung Yang, and James R. McCarthy

Volume II. *Enzyme Technologies: Pluripotent Players in Discovering Therapeutic Agents*
Editors: Hsiu-Chiung Yang, Wu-Kuang Yeh, and James R. McCarthy

Volume III. *Enzyme Technologies: Two Facets of Modern Structure-Based Design*
Editors: James R. McCarthy, Hsiu-Chiung Yang, and Wu-Kuang Yeh

ENZYME TECHNOLOGIES

Pluripotent Players in Discovering Therapeutic Agents

Edited by

HSIU-CHIUNG YANG
Eli Lilly and Company
Indianapolis, Indiana

WU-KUANG YEH
Indiana University Purdue University Indianapolis
Indianapolis, Indiana

JAMES R. McCARTHY
Indiana University Purdue University Indianapolis
Indianapolis, Indiana

WILEY

Published by John Wiley & Sons, Inc., Hoboken, New Jersey.
Published simultaneously in Canada.

For general information on our other products and services or for technical support, please contact our Customer Care Department within the United States at (800) 762-2974, outside the United States at (317) 572-3993 or fax (317) 572-4002.

Wiley also publishes its books in a variety of electronic formats. Some content that appears in print may not be available in electronic formats. For more information about Wiley products, visit our web site at www.wiley.com.

***Library of Congress Cataloging-in-Publication Data*:**

Yang, Hsiu-Chiung.
Enzyme technologies : pluripotent players in discovering therapeutic agents / by Hsiu-Chiung Yang, Wu-Kuang Yeh, and J.R. McCarthy.
pages cm. – (Chemical biology of enzymes for biotechnology and pharmaceutical applications)
Includes index.
ISBN 978-0-470-28626-5 (hardback)
1. Enzyme inhibitors. 2. Enzymes–Pharmacokinetics 3. Drug design.
I. Yeh, Wu-Kuang, 1942– II. McCarthy, J. R. III. Title.
QP601.5.M33 2014
615.3–dc23

2013013968

Printed in the United States of America

10 9 8 7 6 5 4 3 2 1

CONTENTS

CONTRIBUTORS

Herve Aloysius, MS, Department of Medicinal Chemistry, Ernest Mario School of Pharmacy, Rutgers, The State University of New Jersey, Piscataway, New Jersey

Michael Beck, MD, University Medical Center, Children's Hospital, Mainz, Germany

Rathnam Chaguturu, PhD, Del Shankel Structural Biology Center, High Throughput Screening Laboratory, Lawrence, Kansas; SRI International, Harrisonburg, Virginia

Yu Chen, MS, Department of Medicinal Chemistry, Ernest Mario School of Pharmacy, Rutgers, The State University of New Jersey, Piscataway, New Jersey

Ozlem Goker-Alpan, PhD, LSD Research and Treatment Unit, Center for Clinical Trials, Fairfax, Virginia

Longqin Hu, PhD, Department of Medicinal Chemistry, Ernest Mario School of Pharmacy, Rutgers, The State University of New Jersey, Piscataway, New Jersey

Daigo Inoyama, BS, Department of Medicinal Chemistry, Ernest Mario School of Pharmacy, Rutgers, The State University of New Jersey, Piscataway, New Jersey

Jutta Keller, MD, Department of Internal Medicine, Israelitic Hospital in Hamburg, Hamburg, Germany

Ivan S. Krylov, PhD, Department of Chemistry, University of Southern California, Los Angeles, California

Ley Nadine Lacbawan, PhD, LSD Research and Treatment Unit, Center for Clinical Trials, Fairfax, Virginia

Peter Layer, MD, PhD, FACG, AGAF, Department of Internal Medicine, Israelitic Hospital in Hamburg, Hamburg, Germany

Gerald H. Lushington, PhD, Molecular Graphics and Modeling Laboratory, University of Kansas, Lawrence, Kansas; LiS Consulting, Lawrence, Kansas

James R. McCarthy, PhD, Department of Chemistry and Chemical Biology, Indiana University Purdue University Indianapolis, Indianapolis, Indiana

James McGee, PhD, Quantitative Biology, Eli Lilly and Company, Indianapolis, Indiana

Charles E. McKenna, PhD, Department of Chemistry, University of Southern California, Los Angeles, California

Shujaath Mehdi, PhD, Immunoinflammation Therapeutic Strategy Unit, Sanofi Pharmaceuticals, Bridgewater, New Jersey

Taichi Ohshiro, PhD, Department of Microbial Chemistry, Graduate School of Pharmaceutical Sciences, Kitasato University, Tokyo, Japan; Section on Lipid Sciences, Department of Pathology, Wake Forest University School of Medicine, Winston-Salem, North Carolina

Richard G. Peterson, PhD, PreClinOmics, Inc., Indianapolis, Indiana

Anuradha Roy, PhD, Del Shankel Structural Biology Center, High Throughput Screening Laboratory, Lawrence, Kansas

Henrike von Schassen, MD, Department of Internal Medicine, Israelitic Hospital in Hamburg, Hamburg, Germany

Hiroshi Tomoda, PhD, Department of Microbial Chemistry, Graduate School of Pharmaceutical Sciences, Kitasato University, Tokyo, Japan

Yanhui Yang, PhD, Department of Medicinal Chemistry, Ernest Mario School of Pharmacy, Rutgers, The State University of New Jersey, Piscataway, New Jersey

Wu-Kuang Yeh, PhD, Indiana University Purdue University Indianapolis, Indianapolis, Indiana

Wei Zheng, PhD, National Center for Advancing Translational Sciences, National Institutes of Health, Bethesda, Maryland

PREFACE

The human genome was predicted to contain approximately 2742 genes that encode enzymes, which corresponds to 9.5% of the genome (HumanCyc version 7.5). These predicted enzymes can be subdivided into 1653 metabolic enzymes and 1089 nonmetabolic enzymes (including enzymes whose substrates are macromolecules, such as protein kinases and DNA polymerases). Enzymes play an important role in human physiology and the pathophysiology of disease. Understanding the function of an enzyme presents a significant opportunity for finding therapeutic agents. Therefore, a comprehensive understanding of all aspects of enzyme technology is critical in discovering therapeutic agents targeting enzymes, therapeutic enzymes, and enzyme-based applications, enabling research in drug discovery, chemistry, material science, and a vast number of other fields in science and technology.

This series on Chemical Biology of Enzymes for Biotechnology and Pharmaceutical Applications consists of three volumes. Volume I, *Enzyme Technologies: Metagenomics, Evolution, Biocatalysis, and Biosynthesis*, was published in 2010. Volume II, *Enzyme Technologies in Drug Discovery*, as listed in Volume I, has now been changed to *Enzyme Technologies: Pluripotent Players in Discovering Therapeutic Agents*. This book is intended both for biotech and pharmaceutical scientists in academic research institutes and industry as a comprehensive reference material for all common applications of enzyme technology in drug discovery. This book will be useful for biotechnology, biochemistry, molecular biology, and medicinal chemistry faculty members for teaching or conducting research in the field of enzyme technology. It is also a practical handbook for industrial scientists to study various aspects of enzyme technology and discover new treatments for unmet medical needs.

This book is divided into three parts: Part A: Enzymes – Essential Workhorses in Pharmaceutical Research; Part B: Enzymes – Indispensable Tools for Improving Druggability; and Part C: Enzymes – Powerful Weapons for Correcting Nature's Errors. Part A consists of four chapters. Chapter 1, by Dr. A. Roy et al., discusses the principles of assay development and cutting-edge technologies available for protease assays, using proteases as a prototype. Chapter 2, by Drs. Ohshiro and Tomoda, provides a case study on the design and development of selective enzyme inhibitors, using lipid metabolizing enzymes as a prototype. There is a belief that covalent enzyme inhibitors (also called "irreversible inhibitors") are not desirable for drug candidates. Chapter 3, by Dr. Mehdi, describes methods for characterizing covalent inhibitors and their therapeutic applications and explains how enzyme kinetics has been applied in drug discovery. Chapter 4, by Drs. Yeh and Peterson, provides a comprehensive coverage on various technologies that have been applied for *in vitro* enzymatic assays, as well as on common *in vivo* models to assess preclinical drug discovery for metabolic diseases. After going through this part, readers will have a better understanding as to how to select the best enzyme targets for drug discovery, the steps involved in designing enzyme inhibitors for therapeutic agents, and methods for evaluating selective enzyme inhibitors.

Part B consists of three chapters. It explains the principles of improving druggability and provides examples on how to utilize the properties of enzymes for designing therapeutic agents, specifically prodrugs. Chapter 5, by Dr. Hu et al., provides a comprehensive review of enzymes that are being, or can be, used to design prodrugs to improve druggability of existing drug molecules. Chapters 6 and 7 summarize case studies on the design of two successful prodrugs. Chapter 6, by Dr. McCarthy, provides a detailed approach and explains the hypothesis and rationale for the design and synthesis of a Gemcitabine prodrug. Chapter 7, by Drs. McKenna and Krylov, presents several examples of successful prodrug approaches that explicitly depend on enzyme-mediated activation. After going through this part, readers will have a complete understanding on how to select best target molecules for the prodrug approach to improve druggability and how to design successful prodrugs.

Part C consists of three chapters and provides a different viewpoint as to how enzymes can be used in pharmaceutical applications. There are many types of genetic in-born error disorders. Some of these disorders are involved in either deficiency or malfunctioning of specific enzymes. In such cases, functional enzymes can be reintroduced into patients. With recent advances in protein technology, there are several successful examples, such as the production of high-quality proteins and optimal methods for the delivery of these large molecules to the body. However, further improvements on either manufacturing or delivery of enzymes for therapeutics are still required. Chapter 8, by Dr. Beck, deals with Hunter's syndrome, while Chapter 9, by Dr. Lacbawan et al., discusses enzyme replacement therapy for Fabry disease. Chapter 10, by Dr. von Schassen et al., analyzes the level and activity of pancreatic enzymes as a means of diagnosing patients with pancreatic dysfunction, for example, Hunter's syndrome. After going through this part, readers will have an insight into how enzymes can be applied as therapeutic agents or diagnostic tools. Furthermore, readers may be able to identify possible enzyme targets to treat genetic disorders that still do not have effective medication.

Eli Lilly and Company
Indiana University Purdue University Indianapolis
Indianapolis, Indiana

Hsiu-Chiung Yang
Wu-Kuang Yeh
James R. McCarthy

PART A

ENZYMES – ESSENTIAL WORKHORSES IN PHARMACEUTICAL RESEARCH

1

ASSAY TECHNOLOGIES FOR PROTEASES

ANURADHA ROY

Del Shankel Structural Biology Center, High Throughput Screening Laboratory, Lawrence, Kansas

GERALD H. LUSHINGTON

Molecular Graphics and Modeling Laboratory, University of Kansas, Lawrence, Kansas; LiS Consulting, Lawrence, Kansas

JAMES MCGEE

Quantitative Biology, Eli Lilly and Company, Indianapolis, Indiana

RATHNAM CHAGUTURU

Del Shankel Structural Biology Center, High Throughput Screening Laboratory, Lawrence, Kansas; SRI International, Harrisonburg, Virginia

I. INTRODUCTION

Proteases are ubiquitously expressed enzymes which catalyze hydrolysis of peptide bonds and work under a wide range of conditions using diverse catalytic mechanisms [1]. Proteases specifically cleave protein substrates either from the N- or C-termini (aminopeptidases and carboxypeptidases, respectively) or in the middle of the molecule (endopeptidases) [2]. Proteolytic enzymes modulate many physiological processes ranging from nonspecific hydrolysis of dietary proteins to highly specific and regulated proteolysis in cell cycle regulation, tissue remodeling,

Enzyme Technologies: Pluripotent Players in Discovering Therapeutic Agents, First Edition.
Edited by Hsiu-Chiung Yang, Wu-Kuang Yeh, and James R. McCarthy.

blood coagulation, blood pressure control, angiogenesis, apoptosis, inflammation, ovulation, fertilization, and embryonic development [3,4]. Over 500 proteases each from humans, rat, mouse, and chimpanzee have been annotated and compiled in the Degradome database (http://degradome.uniovi.es) [5,6]. Information on all known proteases and their substrates/inhibitors is listed in the MEROPS database [7]. Based on the amino acid or metal that catalyzes the nucleophilic attack on substrate peptide bonds, the proteases are classified into five major types: aspartic (Asp), metallo-, cysteine (Cys), serine (Ser), and threonine (Thr) proteases. Aspartic and metalloproteases use an activated water molecule as a nucleophile to attack the peptide bond of the substrate, whereas in Cys, Ser, and Thr proteases, a catalytic amino acid residue (Cys, Ser, or Thr, respectively) serves as a nucleophile (Fig. 1). As a result, acyl–enzyme intermediates are formed only in the reactions catalyzed by Ser/Thr and Cys peptidases. Within each class of protease type are several enzymes that may have overlapping or distinct substrate recognition sites. Rawlings and Barrett proposed a classification of proteases into families based on amino acid sequence similarity, and families with similar three-dimensional folding are assembled into clans, indicating common ancestry [7,8]. The focus of this article is mainly on mammalian proteases and retroviral proteases which are of significant therapeutic relevance.

While pepsin in gastric juices digests a variety of proteins with broad specificity, renin is an example of Asp protease that shows high substrate specificity. Most proteases bind their substrates in fairly similar manner, first elucidated for papain by Schechter and Berger [9–11]. The catalytic site is flanked on one or both sides by sites that confer specificity of substrate binding to the protease and accommodate a side chain of an amino acid residue of the substrate. The enzymatic binding sites toward the N-terminus of the substrate are the non-prime side designated as S1, S2, ..., S*n* from the catalytic site, and the residues C-terminal to the cleavage site are the prime side designated as S1′, S2′, ..., S*n*′ [8,12,13]. The amino acid residues in the protein substrate which correspond to their respective subsites are numbered P1, P2, ..., P*n* and P1′, P2′, ..., P*n*′ (Fig. 1). Only few of the substrate binding sites have stringent specificities. For instance, site S1 confers specificity for Ser proteases and caspases, whereas the site S2, a hydrophobic subsite, defines specificity for the papain family of Cys proteases. In addition to the sites close to the catalytic site of the enzyme, distant sites on the enzyme may also contribute to the binding of substrates to the protease [9]. The specificity and biological activity of caspases are also determined by S4, which is distant from the catalytic site [14]. Proteolytic processing is being recognized as a mechanism for regulation of enzymatic activities, localization, and fate of proteins that are activated by limited and specific hydrolysis of peptide bonds. Dysregulation of proteolytic activity, structure, or expression results in major pathologies in the areas of cardiovascular diseases, cancer, neurodegenerative disorders, osteoporosis, diabetes type II, pancreatitis, inflammation, arthritis, and infectious diseases [4]. A large number of marketed drugs target the proteolytic enzymes that are involved in pathogenesis of various diseases [15] (Table 1). Although only a relatively small number of proteases are currently targeted for

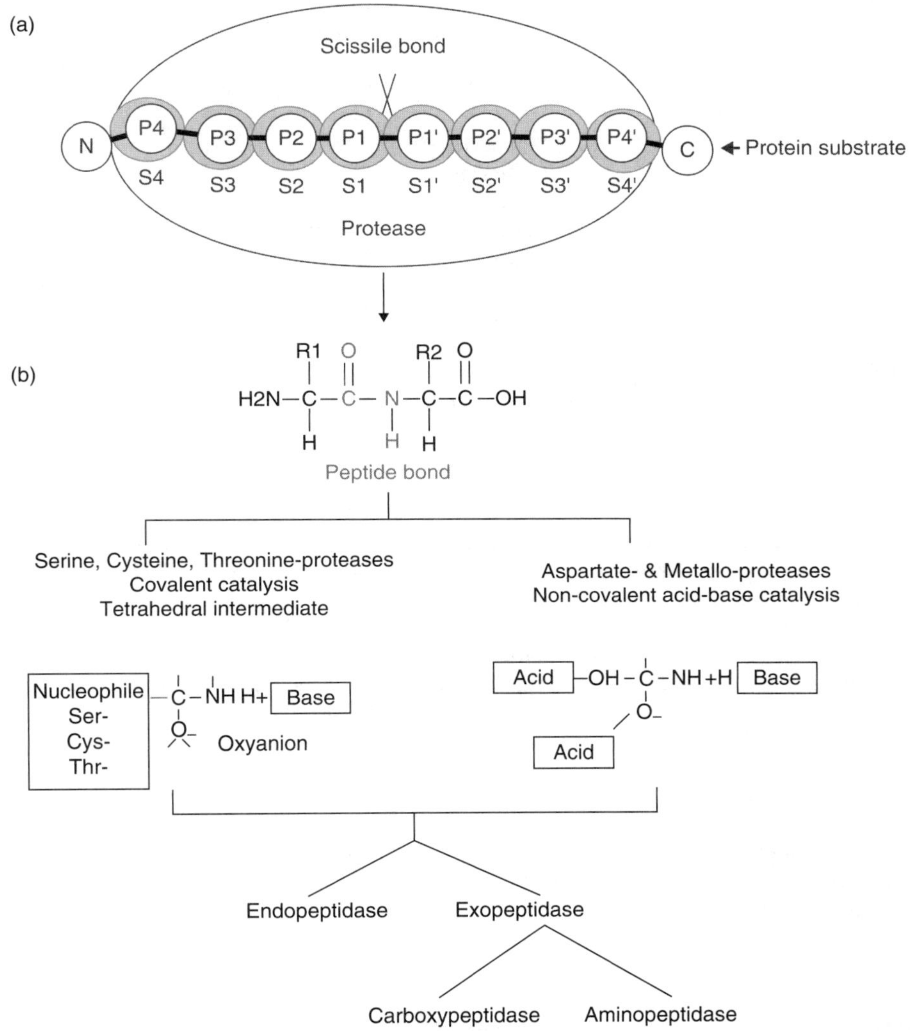

FIGURE 1 Schematic representation of binding of substrate to a protease site. (a) The binding sites of the protease are numbered on either side of the scissile bond, with the non-primed sites (S1, S2, …, S*n*) located toward the amino-terminus of the substrate and S1′ … S*n*′ or the primed subsites toward the carboxy-terminus. (b) Structure of the peptide bond which is hydrolyzed by proteases and the two basic catalytic mechanisms for all types of protease hydrolysis. In Ser, Cys, and Thr proteases, an amino acid at the active site serves as the nucleophile forming a transient covalent intermediate, whereas in metallo- and Asp proteases, an active water molecule functions as nucleophile (adapted from Reference [13]). The base in covalent catalysis is usually a His, and in non-covalent intermediate, Asp/Glu and zinc (metalloproteinases) serve as acids and bases. The proteases are also classified as endo- and/or exoproteinases based on their ability to cleave within or at the amino-/carboxy-terminus of the peptide chain.

TABLE 1 FDA-Approved Drugs for Select Proteases

Protease	Class	Compound	Company	Indication
HIV-1 protease	Asp	Atazanavir	Bristol-Myers Squibb	AIDS
		Darunavir	Prezista	
		Fosamprenavir	GlaxoSmithKline	
		Indinavir	Merck	
		Lopinavir	Abbott	
		Nelfinavir mesylate	Pfizer	
		Ritonavir	Abbott	
		Saquinavir	Hoffmann-La Roche	
		Tipranavir	Boehringer Mannheim	
Renin	Asp	Aliskiren (Tekturna)	Novartis	Hypertension
ACE	Metallo	Captopril	Bristol-Myers Squibb	Hypertension
		Enalapril	Merck	
		Lisinopril	AstraZeneca	
Carboxypeptidase A	Metallo	Penicillamine		Hypertension
MMP-1 and MMP-2 collagenases	Metallo	Periostat	Galderma Labs	Periodontitis
Enkephalinase	Metallo	Racecadotril		
Thrombin	Ser	Ximelagatran	AstraZeneca	Thrombosis
		Argatroban	Mitsubishi Pharma	
		Lepirudin	Aventis	
		Desirudin	Novartis	
Human FXa	Ser	Fondaparinux	Sanofi-Synthélabo	Thrombosis
Human neutrophil elastase	Ser	Sivelestat	Ono	Respiratory disease
Trypsin-like protease	Ser	Camostat mesilate	Ono	Pancreatitis I
Broad-spectrum protease	Ser	Nafamostat mesilate	Japan Tobacco	Pancreatitis inflammation
Plasminogen activator	Ser	Streptokinase		
Proteasome	Thr	Bortezomib (Velcade)	Millennium	Multiple myeloma
DPPIV	Ser	Pioglitazone	Takeda	Diabetes mellitus type II
		Saxagliptin	Bristol-Myers Squibb	
		Linagliptin	Boehringer Ingelheim	
Cathepsin K	Cys	Odanacatib	Merck	Osteoporosis/ bone cancer
		ONO5334	Ono	Osteoporosis

drug development, the commercial success of angiotensin-converting enzyme (ACE) inhibitors and human immunodeficiency virus (HIV) protease inhibitors makes the protease family a valuable target for disease treatment [16]. We have worked on a variety of protease assay formats using proprietary substrates for HTS and will present an overview of common protease assay technologies. We will end the review by discussing computational approaches to designing substrates for protease binding sites.

II. PROTEASE ACTIVITY ASSAYS

Identification of appropriate substrate is the first major step toward characterizing a protease and developing an assay to monitor its activity. A protein containing a short recognition sequence for a protease may work well in an *in vitro* assay, but may not be a physiologically relevant substrate for the proteolytic enzyme. In cases where the peptide sequences are unknown or ill defined, positional scanning experiments or phage display methods are used for screening combinatorial libraries of peptides for specificity determination. Bioinformatics-based prediction of cleavage sites and determination of protease preferences on synthetic substrates are important techniques in predicting natural protease substrates. Figure 2 summarizes the techniques employed in mapping substrates, a rapidly evolving field which is beyond the scope of this article and has been extensively reviewed [17–20]. The design of sensitive and selective synthetic peptide-substrate cleavage assays and the comprehensive mapping of active site specificity determinants are crucial for developing protease inhibitor drugs. The assays for proteolytic enzymes are generally continuous, homogeneous assays that can be performed in medium- to high-throughput formats. The assays are more relevant when the enzyme activity is measured with their specific native substrates, but assays with long native proteins have low hydrolytic rates and are time-consuming and costly. The substrate peptide sequences for most of the common proteases are very well established, and the fact that most proteases catalyze hydrolysis of small peptides has led to the development of technically simple and sensitive assays using fluorometric, colorimetric, and bioluminescent methods in which a single specific peptide bond is cleaved and the cleavage is monitored spectrometrically. A large number of kits are available from various vendors that serve to assay protease families using generic substrate peptides. In general, the minimal defined peptide substrate (average three to six amino acid residues) is synthesized based on the information on the binding fragments of natural substrates or inhibitors of proteases. In the case of many proteases like caspases, matrix metalloproteinases (MMPs), cathepsins, or HIV-1 proteases, the same substrate peptide or a minor variant of a substrate sequence is conjugated with either fluorophores or chromogenic groups or tagged with aminoluciferin to allow development of fluorescence-, absorbance-, or luminescence-based assay formats. Both cell-based and rapid mix-and-read biochemical methods have been developed for assaying the protease activity. The biochemical assays based on purified recombinant enzymes being

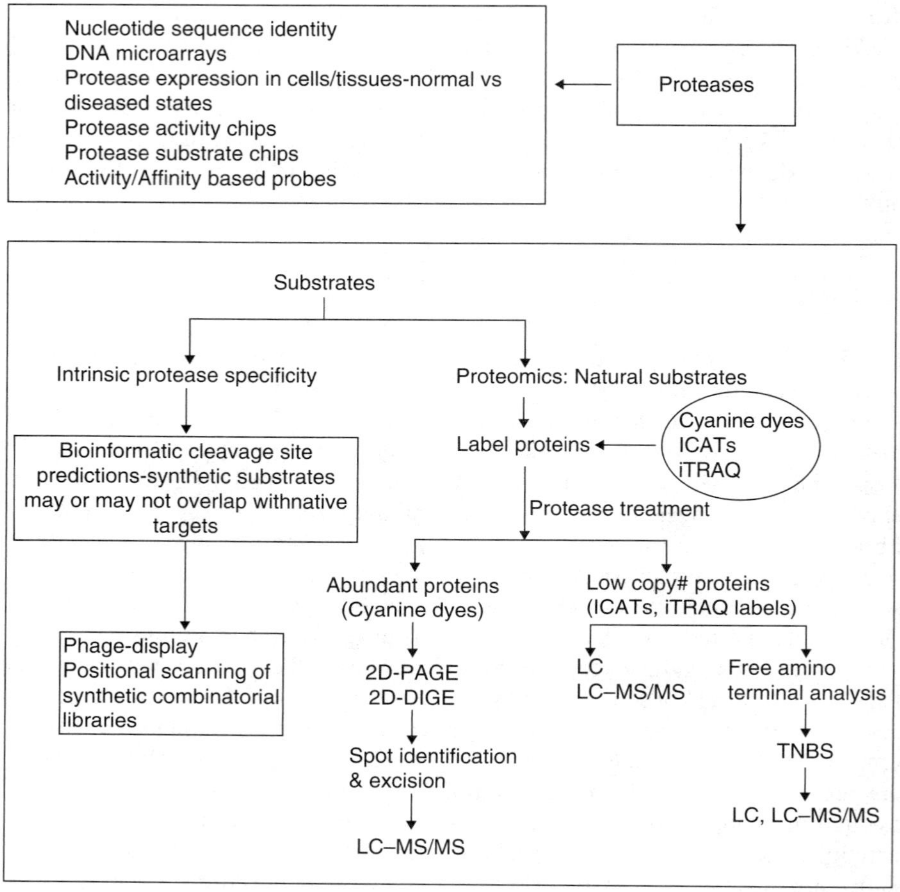

FIGURE 2 Overview of substrate and protease characterization. Protease identification/characterization may initiate from sequence identity/homology, from expression and activity chips, and from loss- or gain-of-function studies in cells or mouse models to establish their role in disease. *In vitro* biochemical hydrolysis and binding experiments based on short synthetic sequences of amino acids are often used to determine protease site binding and hydrolysis. Bioinformatics is useful for building theoretical predictions of substrate sequences, but may differ from real *in vivo* biological targets. Proteomics tools help in the identification of native substrates in cell environment. The protease-treated or protease-untreated cells are labeled with different cyanine dyes (Cy3, Cy5), and the proteins are resolved on two-dimensional (2D) gels (difference gel electrophoresis (DIGE) or 2D DIGE). The substrate and cleavage products are obtained from superimposition of the images from protease treated and controls. The proteins of interest are excised from gels and analyzed by liquid chromatography and mass spectrometry (LC–MS, multidimensional liquid chromatography, MuDPIT). Low-copy-number proteins can be labeled with more sensitive isotope mass tags (isotope-coded affinity tags (ICAT)), or label all primary amines in a trypsinized proteome (iTRAQ), followed by LC and LC–MS/MS. A comprehensive overview of protease substrate identification is available in Overall et al. [1].

more specific are useful for studying kinetics and for the screening of small-molecule libraries or peptidomimetic modulators. Figure 3 shows general aspects of activity assay optimization like defining optimal buffer components, pH, and stability; temperature determination for various substrates of interest; determining the concentration range for the purified protease; and determining the linearity of the reaction. The continuous assays are well suited for miniaturization and adaptation for HTS for identification of protease activity modulators. The modulators identified from the *in vitro* or cell-based assays are generally tested in orthogonal assays to ascertain specificity and selectivity (Fig. 3). The detailed guidelines for development and optimization of any enzyme including proteases for HTS are given in NIH assay guideline manual (http://assay.nih.gov/assay/), which, among others, is a useful reference manual. PubChem (http://pubchem.ncbi.nlm.nih.gov/assay) database lists over 3000 protease activity assays, which were used to identify modulators of protease activity. A few representative assays for proteases compiled from PubChem database are listed in Table 2. The following three major assay formats are widely employed for assaying the activity of various members of protease family:

A. Colorimetric Assays

The colorimetric assays, though less sensitive, have traditionally found more utility in active site titrations for estimating active protease concentrations or for detection of activity in biological fluids. Colorimetric assays based on substrates like amino acid–ethyl, amino acid–methyl, or amino acid–alpha-naphthyl esters were used earlier for quantification of activity of proteases such as trypsin, thrombin, chymotrypsin, pronase, plasmin, and urokinase [21]. These reactions could be used for kinetic analysis and for zymogram studies, but many of these substrates were unstable. The use of tripeptide or other polypeptide substrates afforded higher specificity and stability. In majority of direct colorimetric assays, a substrate peptide (three to five residues) is linked to a chromophore, usually *para*-nitroanilide (*p*-NA). Enzymatic hydrolysis of the substrate releases free *p*-NA, a yellow-colored compound which has high absorbance at 405 nm and is detected spectrophotometrically (Fig. 4a). In the presence of excess substrate concentration, an increase in the rate of absorbance of chromophore released is linearly related to enzyme concentration. A large number of chromogenic substrates are available from Spectrozyme and American Diagnostica for assaying general Ser proteases as well as coagulation factors and kallikreins. For example, ACTICHROME® factor X (fX) is a chromogenic assay for the measurement of fX activity in human plasma. The assay involves activation of fX in the plasma to factor Xa (FXa) by Russell's viper venom. Activated FXa hydrolyzes the Spectrozyme® chromogenic substrate and releases the chromophore, *p*-NA. The color of the reaction solution is read spectrophotometrically at 405 nm. In a large number of colorimetric caspase activity assays, short four-residue peptides linked to *p*-NA serve as substrates. For instance, the DEVD-*p*-NA chromogenic substrate is used to assay caspase-3 and caspase-8 activities in cell lysates. The colorimetric QuantiCleave Protease Assay

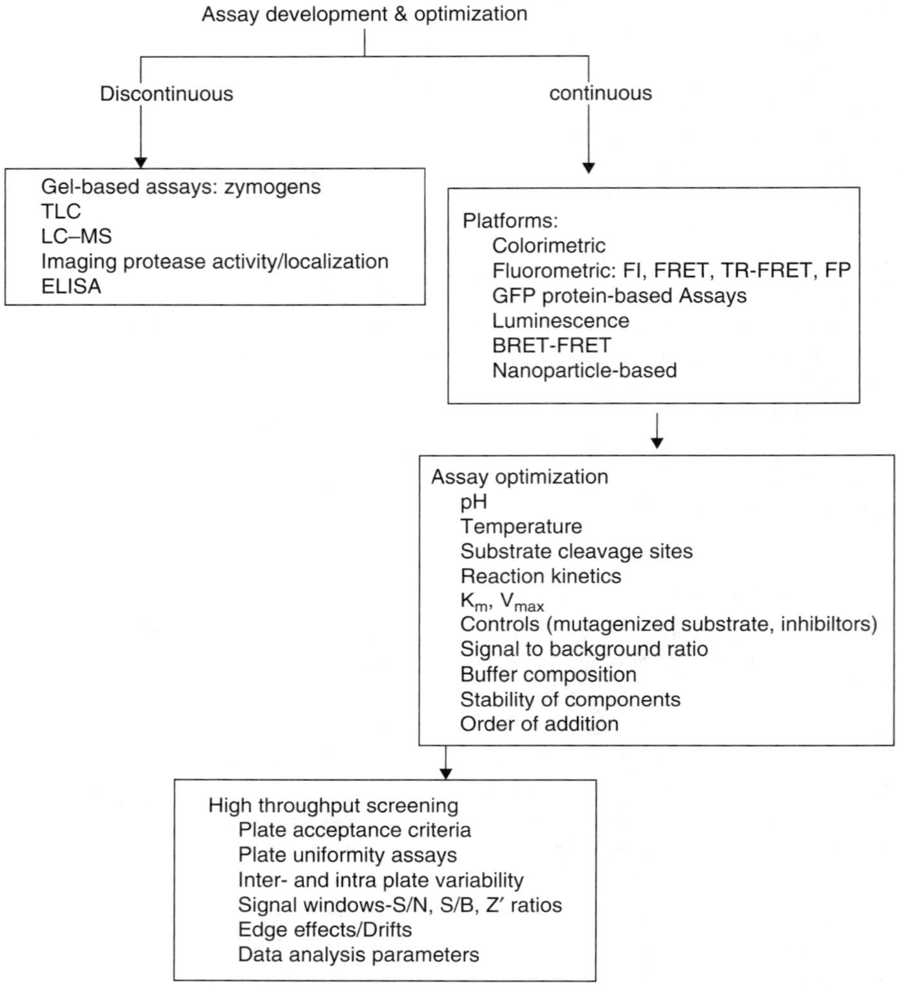

FIGURE 3 Protease activity assay development and optimization. Protease assays can be discontinuous or homogeneous. Examples of discontinuous, low throughput assays include protease activity determination in zymogens, analysis of peptide fragments by TLC or via LC–MS, and imaging of proteases in cell environment by using tagged peptides and ABPs. Biochemical or cell-based assays are continuous assays which are based on colorimetric, fluorometric, or luminescence platforms. The three basic formats can be used in conjunction with nanoparticles or used as such with substrates labeled with appropriate tags. The continuous assays can be further miniaturized and adapted for HTS by further optimizations to conform to stringent statistical acceptance criterion of HTS assays.

TABLE 2 Representative Protease Screens Compiled from PubChem Database

Protease	Assay format	Disease/significance
Cathepsin L	Fluorescence increase by hydrolysis of Z-Phe-Arg-AMC	Proteolysis by cathepsin L required for entry and replication of SARS and Ebola virus
High-temperature-responsive antigen (HTRA-1) Ser protease	FP with FP-TAM probe	Osteoarthritis, AD, and age-related macular degeneration
Insulin-degrading enzyme (IDE), an Abeta-degrading zinc metalloprotease	Fluorescence polarization: fluorescein-Abeta-(1–40)-Lys-biotin (FAbetaB) peptide incubated with IDE dissociates biotin moiety from fluorescein. Addition of avidin to the reaction increases the mass of intact substrate, slowing their rotation rate and decreasing depolarization of plane-polarized light. The low-molecular-weight cleaved substrate rotates rapidly and causes strong depolarization. Ratios determine the relative amounts of cleaved and intact forms of the FAbetaB substrate	AD is characterized by accumulation of amyloid beta-protein
Proteasome, ATP-dependent protease	Flow cytometry: FLAG-tagged, fluorescent proteasomes were captured on anti-FLAG beads, and disassembly was monitored by loss of bead fluorescence in the presence of ATP. The beads were sorted via high-throughput flow cytometer	Inhibitors of proteasome assembly and activity
Sentrin-specific protease-8 (SENP-8)	Luminescence: SENP8-dependent RLRGG-aminoluciferin peptide-substrate hydrolysis – releases aminoluciferin, measurable in a coupled luminescence detection assay	Involved in maturation of SUMO precursors (endopeptidase cleavage) and deconjugation of the targets (isopeptidase cleavage)
SARS coronavirus 3C-like protease	FRET assay: HiLyte Fluor TM 488 is attached at the N-terminus of a 3CLpro peptide substrate (HiLyte Fluor 488-ESATLQSGLRKAK (QXL520)-NH2 (AnaSpec) and is quenched by a QXLTM520 moiety at the C-terminus. Following cleavage, separation of the fluorescent compound and quencher leads to an increase in fluorescence	Cold, lower respiratory tract infections, and diarrhea

(*Continued*)

TABLE 2 ***(Continued)***

Protease	Assay format	Disease/significance
Ub-specific peptidase 2 isoform a	Ub-IsoPro1 kit (Progenera, LifeSensors, Inc.): a Ub or ubiquitin-like (UBL) protein conjugated to a reporter enzyme, phospholipase A2 (PLA2), which has an absolute requirement for a free amino-terminus. Fusion of a UBL to the N-terminus of PLA2 inactivates PLA2. When the UBL–PLA2 reporter enzyme is cleaved by USP2, the activated reporter can subsequently act on its substrate, available NBD C6-HPC (Invitrogen), with a fluorescent readout	Ub homeostasis

Source: http://pubchem.ncbi.nlm.nih.gov/assay.

Kit (Thermo Scientific) is based on the usage of modified casein as substrate, in which all surface primary amines have been succinylated. When succinylated casein is cleaved at peptide bonds by proteases, free alpha-amino-terminal groups of peptide fragments are released. The primary amines react with trinitrobenzene sulfonic acid (TNBSA) to form a yellow-colored TNB–peptide adduct [22] which is measured spectrophotometrically at 450 nm (Fig. 4b). The non-succinylated casein serves as a control substrate in these assays. Replacement of the peptide cleavage bond with a thioester bond results in creation of free sulfhydryl in an assay reported for MMPs [23,24]. The free sulfhydryl group reacts with 5,5′-dithiobis(2-nitrobenzoic acid) (DTNB, Ellman's reagent) to form 5-nitro-5-thiobenzoate, a colored compound with absorbance at 412 nm. A thiopeptide (Ac-PLG-[2-mercapto-4-methyl-pentanoyl]-LG-OC2H5)-based assay kit for MMP-9 is available from Enzo Life Sciences. Most colorimetric assays have been replaced by more sensitive fluorometric- or luminescence-based formats.

B. Fluorescence Assays

In recent years, homogeneous fluorescence- and luminescence-based assay formats have gained wider acceptance due to higher sensitivity, larger signal-to-noise windows, high adaptability to enzymes from various sources, and requirement for very low volumes of reagents. Fluorescence assays are most commonly employed for developing assay platforms for protease activity [25], and most of the assays are based on conjugating extrinsic fluorophores like fluorescein, rhodamine, and BODIPY dyes to a substrate of interest [26]. The fluorophores with distinct excitation (Ex) or emission (Em) properties can be multiplexed within the same reaction, enabling simultaneous measurements of multiple endpoints. Sensitivity of fluorescence-based assays is dependent on accumulation of cleaved product and may be limited by

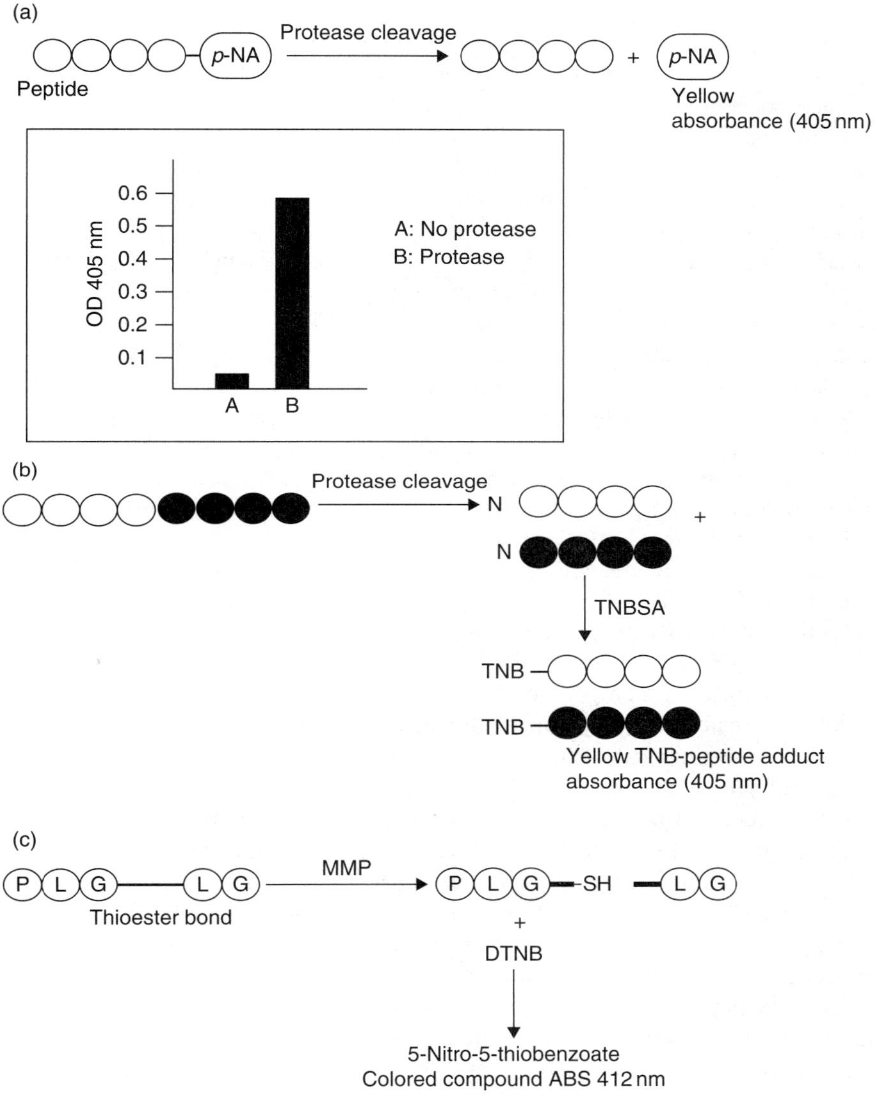

FIGURE 4 Colorimetric assays for proteases. Examples of some colorimetric methods in which protease cleavage is followed by increased absorbance in direct or in coupled assays. (a) Short peptides are conjugated with *p*-NA, which is released following hydrolysis of amino acid–*p*-NA bond, resulting in increased absorbance at 405 nm. (b) TNBSA reacts with the amino-termini of the protease hydrolyzed peptides to increase absorbance at 405 nm. (c) Thioester bond at the peptide cleavage site releases a peptide fragment with reduced sulfhydryl group which reacts with DTNB to form a colored compound with absorbance at 412 nm.

residual fluorescence of peptide-conjugated fluorophores or spectral overlap of cleaved fluorescent products. Fluorescence assays may involve either direct fluorescence intensity measurements [27], Förster resonance energy transfer (FRET), fluorescence polarization (FP) [28,29], or time-resolved measurements (homogeneous time-resolved fluorescence (HTRF), time-resolved Förster resonance energy transfer (TR-FRET)). The measurement method is dependent upon the basic design of the synthesized peptide substrate. In addition, FRET measurements are also performed using fluorescent protein-based substrates. In the simplest substrate type, a short peptide substrate is linked to a fluorophore, which is quenched when bound to the peptide, but fluorescence is detected upon peptide cleavage by a protease (Fig. 5). Many Ser and Cys protease assays are based on simple peptide-conjugated fluorophores. In these proteolytic reactions, the substrate specificity is defined by S subsites, and an aminoacyl intermediate is formed during proteolysis. Fluorogenic substrates for these proteases contain aromatic amines like 7-amino-4-methylcoumarin (AMC) or 4-methyl-7-coumarylaminde (NH-Mec), which are conjugated to short peptides that confers specificity. The cleavage of amide bond by the proteases increases fluorescence. The fluorescent group occupies the S1′ pocket of the proteases and undergoes a change in its fluorescence characteristics on acylation of the amino acid. Many of the caspase activity assays are based on such substrates in which a tetrapeptide is linked to the fluorophore like 7-amino-4-trimethylcoumarin. Cleavage of these substrates by purified caspases or apoptosis-induced cell lysates results in an increase in fluorescence. One of the disadvantages of such substrates is an associated high background signal and proper storage and intactness of the peptide substrate. This format may greatly reduce the sensitivity range for an assay. In other cases, the peptide substrates are synthesized such that the fluorophore Em is quenched by proximity of the second group, the quencher. The peptidase activity physically separates the quencher from the fluorophore and helps restore detectable Em of fluorescence. In FRET, FP, and HTRF/TR-FRET, the peptide substrate is designed to possess two fluorophores, one of which is the donor, which transfers energy to the second fluorophore, the acceptor. The Em spectrum of the donor overlaps with the absorption spectrum of the acceptor (Fig. 5). The acceptor either can emit energy (fluorescence increase) or may serve as a quencher. The same peptide substrate is utilized for FRET and FP assays, and many such peptides are commercially available for generic protease activities. In FRET, direct fluorescence intensity is measured at defined Ex and Em wavelengths and is sensitive to absorptive interference from colored compounds. This spectral interference can be bypassed using FP assay, in which the ratio of orthogonal fluorescence intensities is measured. An interesting FP assay was developed for assaying human cytomegalovirus protease activity using a peptide substrate which was biotinylated at amino-terminus and was conjugated to a fluorophore at the C-terminus [28,29]. After incubation of the substrate with recombinant cytomegalovirus protease, avidin was added. Binding of avidin to the biotin end of the probe produced a polarization signal which was a function of relative amounts of cleaved and uncleaved substrate. The higher-molecular-weight uncleaved substrate bound avidin to generate a high polarization value, whereas the cleaved, low-molecular-weight peptide produced a low polarization value.

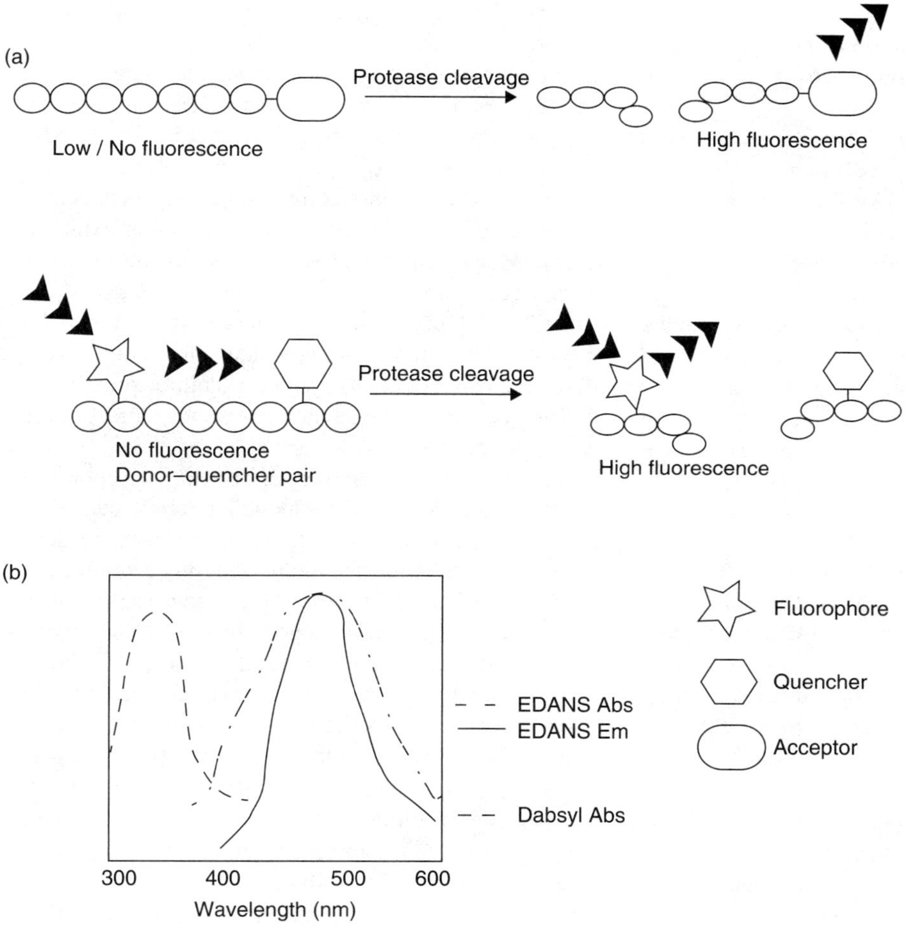

FIGURE 5 Fluorometric assays for proteases. Fluorescent groups are added to peptide substrates to generate substrates for fluorescence-based assays. The simplest substrate is synthesized by adding a coumarin class of fluorophores to linear peptide sequence. An increase in signal is registered with the release of fluorophore post-protease-mediated hydrolysis. (a) Donor–quencher combination in peptide substrate ensures higher signal above background. (b) The overlap of Em spectra of EDANS and Ex spectra of DABSYL characterizes a suitable FRET pair useful for many fluorescence-based platforms.

For the FRET/FP assays, the labeling moiety does not need to be placed at the catalytic site and permits design of substrates with optimal residues at the active site. Since FRET is based on electromagnetic energy transfer between the donor and acceptor pairs of fluorophores separated by very short distance of 10–100 Å, the substrates are designed such that fluorescence increases on cleavage of peptide as a result of loss of intramolecular fluorescence quenching. The selection of donor–acceptor fluorophores requires a high degree of overlap between the Em spectrum of the donor and absorption spectrum of the acceptor, preferably in the long wavelength region; the high quantum yield of the donor; and an effective quenching of donor fluorescence [27]. Several donor–acceptor pairs have been developed like EDANS/DABSYL (4-(4′-dimethylaminophenylazo) benzoic acid, Dansyl/Trp, and naphthalene/anthracene (5 (2′aminoethylamino) naphthalenesulfonic acid). In EDANS/DABSYL pair, quenching of EDNAS fluorescence in substrates with amino-terminal DABSL group is very efficient because of almost complete overlap between the excited state of EDANS and the DABSYL absorbance and the high molar extinction coefficient of DABSYL (Fig. 5). Many aspartyl proteases and metalloproteases require amino acids P′ to the cleavage site for enzyme recognition, and their peptide substrates are designed such that the fluorophore and quencher are separated by less than three or four residues, allowing short-range intramolecular interactions. Many proteasome-mediated events like conformational changes in protein or protein degradation, which change the distance between the acceptor and donor pairs, are efficiently assayed using FRET-based techniques.

The small stokes shifts and near-UV absorption/Em spectra of donor–acceptor pairs in FRET assays have significant drawbacks in HTS of compound libraries since FRET reactions are influenced by high backgrounds from compounds, microplates, reagents, low signal-to-background ratios, and short fluorescence lifetimes. Using a combination of fluorophores with widely separated wavelengths, for example, a protease assay using dual substrates labeled with fluorescent probes with nonoverlapping Ex/Em wavelengths like rhodamine 110- and coumarin-based fluorophores, in the same well allowed for identification of true inhibitors which were active at both fluorescence settings [30]. These problems of FRET assays are circumvented by TR-FRET assays or HTRF assays. The TR-FRET/HTRF assays, introduced by *Cisbio* International, are based on using rare earth lanthanides like dysprosium (Dy), samarium (Sm), terbium (Tb), and europium (Eu). The lanthanides have poor intrinsic fluorescence but can be complexed to rare earth chelates and cryptates to enhance their fluorescence and prevent decay (cisbio.com). The complexed lanthanides have large stokes shifts and long Em half-lives of 100–1000 μs and help avoid signal contamination by shorter-lived signals contributed by other components of the assay. The HTRF assays generally use complexed cryptates like EuK as an energy donor and XL665 protein as an acceptor (Em 665 nm) [31]. The Em spectra of donor and the Ex spectra of the acceptor overlap and the donor–acceptor pairs are placed at 10 nm or less from one another. A highly sensitive time-resolved fluorescence quench assay (TR-FQA) was developed for caspase-3 [32] using a hexapeptide substrate labeled with fluorescent Eu chelate at one end and a Dabacyl quencher at the other end. Cleavage with caspase-3 allowed measurement of time-resolved Eu signal to the acceptor allophycocyanin. Time resolution allowed the separation of the fluorescence signal by spectral and temporal filtering and resulted in very high signal-to-background ratios. In addition to the extrinsic fluorophores, many protease assays are based on the naturally fluorescent proteins, like the green fluorescent protein (GFP), which can be used for both *in vitro* and *in vivo* assay formats.

Protein-Based Fluorescence Reporter Substrates The reporter constructs in which the protease substrate sites are introduced between the coding sequences of the fluorescent proteins are used in many cell-based protease assays. The cell-based protease assays require transient or stable expression of the GFP which absorbs blue light and emits green light, in the absence of any cofactor requirement. The GFP chromophore arises from posttranslational modifications and comprises of *p*-hydroxybenzylideneimidazolinone formed by cyclization of Ser65, Tyr66, and Gly67 and 1,2-dehydrogenation of the tyrosine (Tyr) [33].The wild-type GFP has been extensively mutagenized (Fig. 6a, [33,34]) to improve spectral characteristics and quantum yields and generate more useful variants like eGFP, blue fluorescent protein (BFP), cyan fluorescent protein (CFP), and yellow fluorescent protein (YFP) [33,34]. Many protease assays use FRET-based reporters for monitoring protease activity in cellular environment in which a protease cleavage site is placed between two fluorescent variants of GFP or between a GFP and a fluorophore, with overlapping donor Em and acceptor Ex spectra [35]. Induction of protease activity in cells expressing the fusion protein results in peptide cleavage followed with the loss of the FRET signal between the donor and acceptor, when excited at the wavelength of donor absorbance. In general, both live cells expressing FRET constructs as well as cell lysates can be used for studying protease activity. Onuki et al. [36] reported a caspase-8 assay, in which its native substrate, BID, a BH3-interacting domain death agonist peptide, was flanked by CFP, the cyan fluorescent variant of GFP, and YFP, the yellow fluorescent variant. The absorbance of light at 433 nm by CFP results in energy transfer to YFP, which emits fluorescence at 527 nm (Fig. 6b). The hydrolysis of the protease site by caspase-8 results in loss of FRET Em at 527 nm, with a gain of Em at 475 nm from CFP [36]. The ratio of light Em at the two wavelengths is used for FRET quantification and is proportional to protease activity. Most assays are based on energy transfers between CFP/BFP and GFP or YFP separated by specific protease cleavage sites. Since most cells express a wide variety of proteases which have redundant activities, a direct correlation between the protease type and measured activity is obtained only when cells are transfected with highly specific peptidase sites in the FRET expression constructs. The FRET-based reporters can be used for imaging cell systems with detectable GFP expression. In addition to the reporter assays expressing fluorescent proteins, short fluorescence-based cell-permeable probes have been used for studying specific protease activities and their localization in live or fixed cells and in the small animal models.

Imaging of Protease Activities Fluorescence-based probes have been used in many *in vitro* and *in vivo* imaging techniques for monitoring and detecting active proteases within live cells and small animal models. For instance, the degradation of extracellular matrix (ECM) by collagenases, which are required for migration of cancer cells, was studied using BT549 breast cancer cells grown in the presence of fluorescent ECM protein substrate, FITC-labeled collagen IV. The areas containing cell clusters degraded the matrix and, after fixing, were imaged as nonfluorescent, cleared areas, showing that the ECM substrates have relevance in the cell migration through the tumor matrix. However, since high background fluorescence and cell fixation preclude the ability for real-time data imaging, the technical problems were circumvented by using the dye-quenched fluorescent protein substrates. The dye-quenched substrates contain high density of FITC molecules which are quenched due to high molecular proximity [37]. Protease-mediated cleavage of the substrate releases fluorescent degradation products. BioMol has developed cell-based assays which

(a)

Fluorescent Protein (FP)	Mutations	Excitation Max (nm)	Emission Max (nm)
Wt GFP	None	395, 495	511
Green (eGFP)	S65T, F64L	489	509
Blue (BFP)	Y66H, Y145F, F64L	385	445
Cyan (CFP)	Y66W, S65A, S72A, M153T, V163A	434	477
Yellow (YFP)	T203Y, S65G, S72A, V68L	514	527
T-Sapphire	T203I, Q69M, C70P, V163A, S17G	399	511

(b)

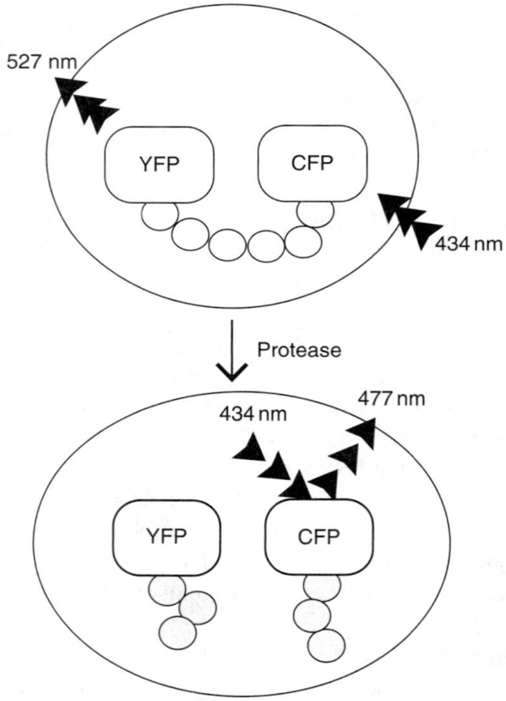

FIGURE 6 Fluorescent protein-based FRET assays. (a) Green fluorescent protein, an intrinsically fluorescent protein, and its variants derived from extensive mutagenesis of the wild-type GFP. (b) A schematic showing a FRET assay using YFP and CFP as acceptor/donor in a cell-based reporter assay. The peptide sequence between the two fluorescent proteins could represent any sequence specific for caspases or cathepsins or metalloproteinases.

utilize cell-permeable fluorescent substrate probes for *in situ* localization of active proteases. The cells with active Ser proteases stain green (FFCK-FAM) and cells with active caspases stain red (SR-VAD-FMK). Cells displaying either or both activities are easily detected using fluorescence microscopy and FACS. A number of polymer-based near-infrared fluorescence (NIRF) probes have been introduced for imaging protease activity [38] in whole organism models. The NIRF probes consist of a high-molecular-weight carrier like polyethylene glycol (PEG), a cleavable peptide spacer, and a near-infrared fluorophore, indocyanine green, which is an FDA-approved tricarbocyanine dye. Many such probes have been used to monitor activity of caspases [39], metalloproteases, cathepsin, and coagulation enzymes [40]. In one such example, an NIRF peptide substrate was developed to monitor general cathepsin activity *in vivo* [38]. Since most lysosomal cathepsins cleave the polylysine peptides, fluorophores were coupled at high density to an inert copolymer composed of polylysine backbone stabilized with PEG side chains, and the NIRF peptide was injected along with a dye in mice. The high density of fluorophores quenched fluorescence Em, but *in vivo* cathepsin-mediated proteolysis of the peptide decreased the fluorophore density and increased the Em signal in the near-infrared range, the long wavelength which was not subject to scattering or absorption by tissues [38].

FRET-based probes are not specific and can generally be hydrolyzed by many cellular proteases. The intracellular responses to a variety of physiological and pathological conditions can be monitored with the help of more specific and stable, fluorescently labeled activity-based probes (ABPs) [41]. Activity-based probes form covalent complex with the protease active site since they are often based on specific protease inhibitors that occupy protease active sites. The highly hydrophobic ABPs are membrane permeable and can detect specific protease activity in a cellular microenvironment. All ABPs consist of three parts: reactive group, linker, and tag. The reactive functional group, the warhead, is an electrophilic group, which covalently binds to the catalytic nucleophiles (Ser, Cys, or Thr in Ser, Cys, or Thr proteases, respectively) located at the active site of the enzymes [41,42]. The steric hindrance between the warheads from the tag is reduced by a variable linker, which may be a nonspecific spacer of alkyl or PEG groups or may be designed after a substrate. All ABPs contain a tag to enable detection and/or purification of labeled proteins. The tags may be radiolabels (I-125, H-3), fluorescent groups (cyanins, BODIPY, rhodamines, TAMRA), or biotin, which have been used for activity profiling of Cys proteases like cathepsins [43], caspases, and proteasomes. The design of specific ABPs requires that the target proteases have well-defined catalytic mechanisms and known inhibitors for which structural and kinetic data are available [41]. The ABPs have been used extensively for Cys and Ser proteases. DCG-04 was designed based on E-64, a natural product inhibitor of papain family of Cys proteases, and was used to determine the role of papain-like proteases in tumor progression, parasitic invasion, and cell cycle regulation [44]. The caspase-1 of Cys protease family was identified using ABPs containing a biotin-tagged specific peptide sequence coupled to acyloxymethyl ketone (AOMK) and aldehyde reactive groups [45]. The ABP based on specific Ser protease inhibitor, diisopropyl fluorophosphonate (DFP), was used to profile Ser hydrolases for biomarkers for cancer, and activities of at least seven Ser hydrolases were found to be differentially expressed in MDA-MB-231 breast cancer cells in cell culture and in xenograft model [46]. The metalloproteases, with their zinc-containing active sites, hydrolyze the substrate peptide bond without forming acyl–enzyme intermediates. An ABP against metalloproteinase, based on its inhibitor peptide hydoxamate [47], a zinc chelator, and a photocrosslinking group were used to profile MMP activity in melanoma cell lines. The study uncovered the role of neprilysin, a membrane protein, in progression of melanoma. The ABPs have proven useful in unraveling both previously unknown drug targets and novel players in various pathophysiological processes. In summary, fluorescence-based formats have extensive applications in development of many *in vitro* and *in vivo* imaging assays. At the same time there are several drawbacks in fluorescence-based approaches for biochemical and cell-based assays. Although fluorescence assays are very sensitive, they are susceptible to interference from inherent absorption characteristics of chemical compounds in the *in vitro* assays. In cell-based systems, the quantum yields of many GFP derivative proteins are weak, and many such fluorescent protein-based constructs have limited sensitivity, and the large size of fluorescent proteins limit the spatial resolution of the constructs. Many FRET-independent photochemical changes can result from external light source to initiate fluorescence transfer causing background noise or photobleaching [48,49]. In addition, the use of ultraviolet light for Ex of BFP in cell-based reporter assays has cytotoxic effects. The problems with fluorescent assays can often be circumvented using luminescence assay formats, which are also not affected by compound interference, photobleaching, and phototoxic effects.

C. Bioluminescence Assays

Bioluminescent assays are based on naturally occurring bioluminescence reactions catalyzed by the enzyme luciferase derived from *Photinus pyralis*. Bioluminescent assays are rapid, are sensitive, and are based on the reaction catalyzed by luciferase enzyme, which acts on its substrate luciferin or aminoluciferin in the presence of ATP and generates energy-rich peroxide intermediates, which spontaneously decompose and reach the ground state accompanied with Em of photons at 550–605 nm. Any modification of the amino or carboxy group of aminoluciferin makes the resultant aminoluciferin derivative insensitive to luciferase reactions [50]. Protease substrates are covalently conjugated via a peptide bond to the amino group or to the modified carboxy-terminal derivative of aminoluciferin. The protease-mediated cleavage of the peptide bond linking the substrate and aminoluciferin generates free aminoluciferin, which is available as a viable substrate for luciferase enzyme [51]. These assays are nonhomogeneous reactions that require compatibility between the protease assay conditions and the luciferase assay buffer especially with respect to pH requirements (Fig. 7a). The amount of aminoluciferin released is a direct measure of protease activity. A detailed comparison of the bioluminescent assays with corresponding fluorescent assays for caspase-3/7, caspase-8/9, DPPIV, and calpain showed that the luminescence assays were more sensitive (100–1000-fold) [50]. Ser and Cys proteases which do not require P′ sites for cleavage work well with aminoluciferin–peptides. A single-step homogeneous assay was reported for caspase-3, in which the aminoluciferin-tagged peptide substrate was premixed with luciferase enzyme before adding to wells containing caspase-3 enzyme [52]. Many FRET-based approaches have been replaced with BRET (bioluminescence resonance energy transfer) in which the fluorescent donor is replaced by a bioluminescent luciferase. The luciferase produces an initial photon Em, which is transferred to an acceptor (a fluorophore) that absorbs the donor energy and emits light at a longer

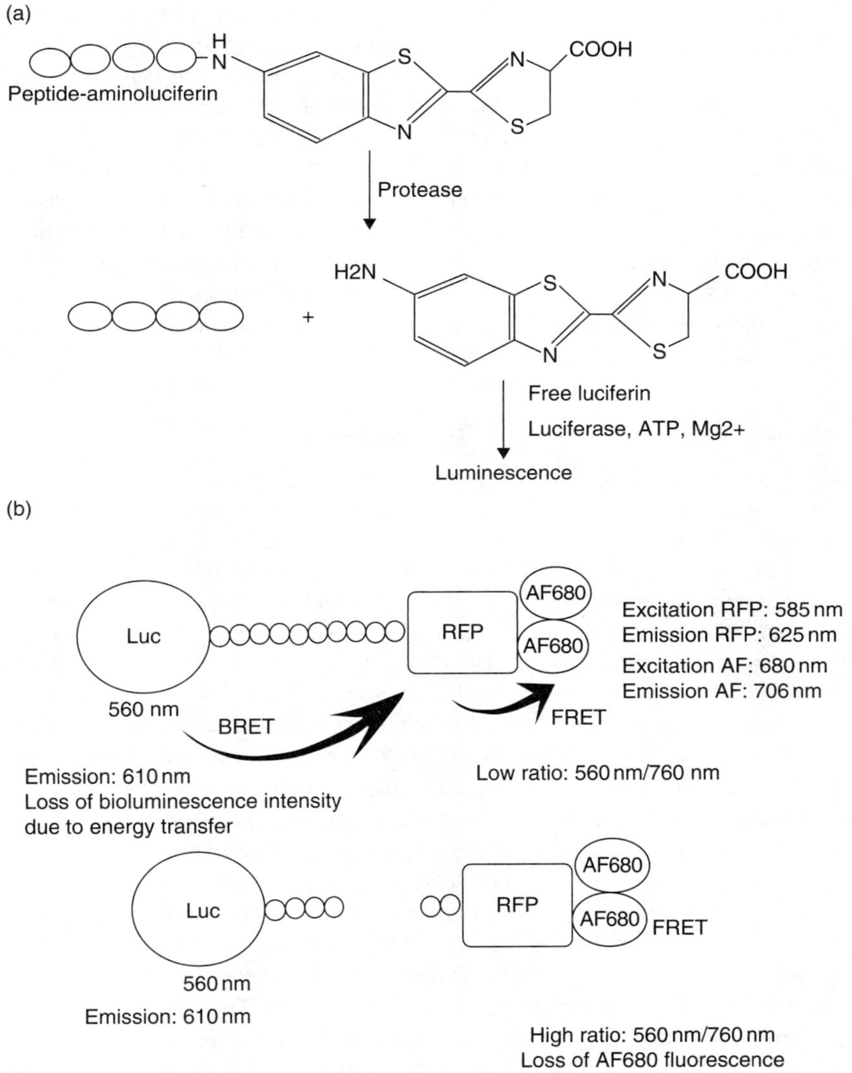

FIGURE 7 Luminescence-based protease assays. (a) Basic construction of the luciferase substrate conjugated to a protease-specific peptide sequence. The peptide–aminoluciferin has no detectable activity. Hydrolysis of the aminoluciferin–peptide bond by protease releases free aminoluciferin which is a usable substrate for luciferase in the presence of ATP and magnesium. (b) A luciferase-based bioluminescence transfer assay (BRET) in which luminescence is detected only after proteolysis.

wavelength. In one example of BRET format, the protease substrate decapeptides containing FXa, thrombin, or caspase-3 recognition sequences were flanked by the bioluminescent luciferase protein (Em 610 nm) and a red fluorescent protein (Ex 585 nm/Em 625 nm) covalently labeled with an NIRF dye, AF680 (Ex 680 nm/Em 705 nm). The energy transfers were possible due to overlap between the red bioluminescence Em at 617 nm of the Luc variant and the AF680 absorbance at 680 nm (Fig. 7b). The ratios of 560 to 760 nm, determined following protease cleavage of the protein substrates, were monitored by recording Em spectra and plotting the change in peak ratios over time [53]. The bioluminescent reporters have high sensitivity and have very high signal windows due to minimal autoluminescence of cells and tissues. The three basic assay formats described earlier are also being developed in conjunction with new technologies like nanoparticle-based assay systems.

D. Nanoparticle-Based Protease Activity Assays

Nanoparticles are nanometer-sized particles that have found wide utility in biology and other fields of study. Since nanoparticles are similar to subcellular components like proteins, they provide an interface to study events at nanoscale [54]. The nanoparticles are tagged with antibodies, fluorescent or chromogenic tags that enable detection and quantification of an enzyme activity, and in many cases the material composition of nanoparticles itself contributes to optical detection properties of the assays. Nanoparticle-based assays are very sensitive and can be used to monitor protease activity in real-time both *in vitro* and *in vivo*. The protease substrates immobilized onto a nanoparticle surface are acted upon by active enzymes resulting in a change of nanoparticle environment, which is measured via fluorescence, absorbance, and imaging or by biophysical methods. Gold nanoparticles (AuNP) exhibit characteristic size- and shape-dependent electronic and spectral properties, which have been exploited to develop protease activity assays. Colloidal AuNP tethered to peptides containing two Cys residues flanking protease cleavage site acquires blue color due to aggregation [55,56]. Protease-mediated cleavage of the activity on the immobilized substrate disperses the aggregates, changing the color to red [57]. In addition to colorimetric assays, AuNP and other metallic nanoparticles are known to quench fluorescence of transferred excited electrons [58]. The electrons transferred from fluorophore-containing peptides are quenched when attached to AuNP. After the cleavage of peptide by protease, the physical separation from the AuNP restores the previously quenched fluorescence [59]. An MMP-2 assay was established using Cy5.5-labeled substrate in which association with AuNP quenched the signals. With MMP-2 activity, the fluorescence signal recovered both *in vitro* and in mice experiments [60]. A new method to assay MMPs expressed in tumors was reported recently using composite gold–iron oxide (Au–Fe_3O_4) nanoparticles. An optical probe containing Cy5.5-GPLGVRG-TDOPA was immobilized on the iron oxide surface and SH-PEG (5000) for *in vivo* imaging on the gold surface. The methodology exploited the quenching properties of AuNP and stable surface chemistry of iron oxide nanoparticles. The MMP activity was assayed by fluorescence imaging of both *in vitro* and *in vivo* mouse tumor models [61]. In addition to AuNPs, magnetic [62,63] and polymeric- and silica-based nanoparticles [64,65] are also used in protease sensing assays. In addition to the fluorescence quenching properties, the iron oxide-based magnetic nanoparticles [64] are useful for tracing distribution of injected samples by magnetic resonance technology. The nanoparticle-based protease assay systems are quantitative, specific, and sensitive and will soon find wide applications in drug screening campaigns and in routine assay development.

III. ASSAYS FOR SOME CLINICALLY SIGNIFICANT PROTEASES

Because of their therapeutic relevance, members from each of the protease families have been targeted as significant targets for drug discovery. A successful identification of protease activity in disease and its development as a therapeutic target requires an understanding of the *in vitro* chemical mechanism and kinetics of proteolytic activity as well as the complexity of *in vivo* function and biological processes regulating proteolysis. Some of these proteases and the assays used for activity determination and for identification of small-molecule/peptidomimetic modulators are discussed briefly in the following sections.

A. Aspartic Proteases

Aspartyl proteases play an important role in several aspects of our overall health and physiology, including blood pressure (renin), digestion (pepsin and chymosin), and in the maturation of the HIV-1 protease [66]. Human Asp proteases are a small class of proteases with only 15 members, most of which are significant drug targets [67]. The aspartate proteases are monomeric enzymes, except for the HIV protease, which is a homodimer [68]. Most aspartyl proteases can accommodate up to 9-amino acid-long peptides into the active site binding cleft and specifically cleave dipeptide bonds that have hydrophobic residues [67]. The aspartate proteases consist of two domains arising from gene duplication, and the two halves of the enzyme are independent but similarly folded units that move relative to each other [67]. In contrast, the HIV protease consists of two identical subunits which are interconnected with six-stranded antiparallel β-sheets. Most of the amino acid sequences of aspartate proteases are divergent except for the highly conserved catalytic site motif comprising of Asp-Thr-Gly. The amino- and carboxy-terminal domains each contribute one catalytic Asp acid residue to the active site, each of which has different pK value. Peptide bond cleavage occurs by a general acid–base catalytic mechanism. One of the two catalytic Asp residues acts as a general acid and is protonated in the enzyme–substrate complex. The other Asp residue acts as a general base activating a water molecule which then attacks the carbonyl carbon of the amide bond, resulting in the formation of a tetrahedral diol intermediate. Subsequent deprotonation of the hydroxyl group by one of the catalytic aspartates and simultaneous activation of the leaving amine by the other protonated Asp residue ultimately lead to peptide bond cleavage. At least three types of Asp proteases are discussed in the succeeding text.

HIV Protease The HIV genome encodes a single Asp protease responsible for the cleavage of the viral polyprotein, which is an essential step for maturation of infectious viral particles [69]. Interference with the proteolytic activity of HIV protease renders viral particles immature and noninfectious. Consequently, inhibitors of HIV protease are important drug candidates as these compounds reduce the viral load in AIDS patients, and at least 15 anti-HIV drugs are known to act against the proteases. The rapid development of inhibitors of HIV proteases was guided by extensive structural information on the target peptide binding to its active site as well as from fluorescence-based assays [68,70]. The extensive modeling of the cleavage sites in the glycosaminoglycan (gag) and gag–polymerase (pol) precursor proteins in the active site of HIV protease led to the design of HIV protease inhibitor screens. All of the protease inhibitors like saquinavir, ritonavir, indinavir, nelfinavir, amprenavir, and lopinavir in the market for the treatment of HIV infection share a hydroxyethylene core, which mimics the peptide linkage in polyprotein gag–pol, and are uncleavable peptidomimetic competitive substrate analogs for the HIV protease [16]. However, drug–drug interactions, overlapping resistance patterns, and long-term side effects resulted in development of new, nonpeptidic inhibitors like tipranavir and atazanavir based on scaffolds such as 4-hydroxycoumarin, L-mannaric acid, or 4-hydroxy-5,6-dihydro-2-pyrone. In addition to hydrolyzing large precursor polyproteins (gag and gag–pol) *in vivo*, the HIV protease can also hydrolyze much smaller peptides, which allows convenient assay development for HTS campaigns. The naturally processed substrate site in the gag–pol polyprotein for HIV-1 protease is very well characterized and forms the basis for the design of small peptide substrates for homogeneous FRET assays. This substrate SQNYPIVWL or SQNYPIVQ is labeled at the N-terminus with DABSYL and at the C-terminus with EDANS. The cleavage of the peptide with HIV-1 protease resulted in a 40-fold increase in DABSYL fluorescence [71], allowing initial reaction velocities to be determined. The same or related peptides have been tagged with chromophores to develop various other assay formats. Many retroviral proteases cleave bonds adjacent to the prolyl residues (phenylalanine–proline (Phe-Pro) bonds). A colorimetric assay was developed, in which the amino-terminal Pro residue in the product was shown to react with isatin, in the presence of 2-(4-chlorobenzoyl) benzoic acid, an aromatic carboxylic acid catalyst to form a blue product that was quantitated spectrophotometrically [72]. The reaction is discontinuous but simple, sensitive, and specific for routine evaluation of potential protease inhibitors. The N- and C-termini of substrate peptides were protected to make them specific for HIV protease. The substrates Ac-ARALAEA-NH2 and Ac-ATIMMQR-NH2 are cleaved between Leu-Ala and Met-Met, exposing primary amino groups that can be detected in a reaction with 2,4,6-trinitrobenzene sulfonic acid, yielding yellow products.

Renin Renin, an Asp protease, catalyzes the first and the rate-limiting step in an enzymatic cascade controlling hypertension and, unlike other Asp proteases like pepsin and cathepsin D, has high specificity for its only substrate angiotensinogen. Renin is produced by kidney juxtaglomerular cells and released into the circulation following regulated and constitutive pathways [67]. Renin cleaves angiotensinogen to release the N-terminal decapeptide angiotensin I (Ang I), which is cleaved by ACE, a carboxypeptidase, to release a vasoconstrictor hormone, angiotensin II. A number of compounds based on the natural substrate of renin were found to be potent inhibitors (peptide-based, CGP29287; peptide-like, CPG 38560) *in vitro* but could not be developed as drugs due to poor pharmacokinetic properties. The use of homology models and docking studies using various conformers of CPG38560 allowed substitution of the peptide backbone with a tetrahydroquinoline and, eventually, a phenyl-based scaffold extended by a methoxy group. Derivatives of this phenoxy scaffold led to the development of aliskiren, the third generation of nonpeptidic inhibitors of renin [73]. Plasma Renin Activity (PRA) is a well-established biomarker for assessing the circulating renin levels and to test the activity of antihypertensive agents. Most of the PRA measurements utilize assays to detect Ang I, an unstable decapeptide released when plasma renin cleaves the circulating angiotensinogen. The Ang I peptide is captured and stabilized by specific antibodies. The amount of Ang I is quantified by either radioimmunoassay or enzyme immunoassay [74]. These nonhomogeneous methods, though very specific, are tedious and require large amounts of plasma. Many homogeneous formats have been reported in literature for detection of renin activity by quantifying its product, Ang I. A high-throughput immunoassay based on AlphaScreen (Amplified Luminescent Proximity Homogeneous Assay) technology (Perkin-Elmer) was reported by Cauchon et al. [75]. The AlphaScreen is based on the use of latex-based donor and acceptor beads coated with a hydrogel which provides reactive aldehyde groups for conjugating biomolecules to the bead surface. Donor beads contain a phthalocyanine, a photosensitizer, which converts ambient oxygen to singlet oxygen upon illumination at 680 nm. If acceptor bead is in close proximity, energy is transferred from singlet oxygen to thioxene derivatives within the acceptor beads. This generates chemiluminescence at 370 nm, which activates a fluorophore in the acceptor bead that subsequently emits light. Biotin–Ang I competes with unlabeled Ang I for binding to its antibody, which is immobilized on acceptor beads. The antibody-bound Biotin–Ang I is captured by streptavidin-coated donor beads. The assay was reported to be very sensitive (1–50 nM Ang I), robust, and fast and did not require wash steps. Many FRET-based peptide assays have been developed to monitor renin activity and for the identification of renin inhibitors. The FRET peptide-substrate sequence was derived from the native renin cleavage site in the N-terminal region of angiotensinogen, Ile-His-Pro-Phe-His-Leu-Val-Ile-His-Thr. The octapeptide, labeled with either Mca-Ala/Dnp or EDANS/DABSYL, was used as FRET peptide. Another continuous assay of renin activity was developed using a FRET peptide with a fluorophore 5-FAM and quencher, QXL™ 520 (R-E(EDANS)-IHPFHLVIHT-K(DABCYL)-R, SensoLyte® 520 Renin Assay Kit, AnaSpec). Upon cleavage into two separate fragments by renin, the fluorescence of 5-FAM is recovered and can be monitored at Ex/Em of 490/520 nm. This assay is about 50-fold more sensitive than an EDANS/DABSYL-based assay.

BACE-1 β-Amyloid precursor protein cleaving enzyme-1 (BACE-1, or β-secretase) is a transmembrane protein containing pepsin-like Asp protease activity [76]. BACE-1 is considered the initial and rate-limiting enzyme responsible for the proteolytic cleavage of amyloid precursor protein (APP) to a membrane-bound APP C-terminal fragment (APPCTFβ or C99), which is further processed by presenilins (γ-secretase) to generate amyloidogenic peptides Aβ40 and Aβ42. The latter peptides constitute the major components of amyloid plaques accumulating in the brain of Alzheimer's disease (AD) patients. BACE-1 is expressed as a preproenzyme, and its maturation involves removal of signal peptide, followed by the cleavage of the prodomain and sorting of BACE-1 by the secretory system to plasma membrane and clusters within lipid rafts. Various inhibitors to BACE-1 have been designed based on the X-ray crystallographic structure, which shows a high degree of homology to that of other human Asp proteases like pepsin and renin. The cleavage site specificity sequences of BACE-1 resides mainly at the S1′ subsite, which requires small side chains such as alanine (Ala), Ser, and Asp. Most of the assays for BACE-1 use short peptides based on an extensively characterized Swedish mutation of the 695 amino acid of the APP isoform (KM to NL), a variant that is a 100-fold better substrate for BACE-1 cleavage compared with the wild-type sequence. The FRET substrate specific for BACE-1, Rh-EV**NL**DAEFKQuencher [77] or EVNLDAEF peptide labeled with EDANS/DABSYL pair, has been used to determine the potency of synthetic BACE inhibitors. After cleavage by BACE, the product (peptide–EDANS) is brightly fluorescent and can be easily analyzed using a fluorescence plate reader or a fluorometer with Ex wavelengths of 335–345 nm and Em wavelengths of 485–510 nm. Highly hydrophobic BACE-1 FRET peptide has low solubility, resulting in their use at much below Km value and low assay sensitivities and use of high enzyme concentrations. Alternate protease substrates like casein–FITC and BPNA–FITC have also been reported for use as BACE-1 substrates for *in vitro* activity assays [78]. Pietrak et al. [79] reported an optimized peptide cleavage sequence, NFEV, which was cleaved by very low picomolar levels of BACE-1 about 10 times more efficiently than was the substrate containing the Swedish (NLDA) sequence [80]. An HTRF-based assay was reported for assaying BACE-1 activity using an 18-amino acid peptide substrate, EuK-APPsw (EuK-KTEEISEVNLDAEFRHDKC-biotin), that was labeled with Eu cryptate at the N-terminus and was biotin labeled at the C-terminus [81]. The streptavidin-coupled cross-linked allophycocyanin (SA-XL665) protein was used as an energy acceptor. In these assays, the EuK-peptide was incubated with recombinant BACE-1 and SA-XL665 protein. The EuK fluorophore is excited at 337 nm, and after a 50 μs delay, the 620 and 665 nm Em were measured. When the peptide is intact, resonance energy transfer occurs between EuK and XL665, resulting in quenched 620 nm fluorescence and increased 665 nm and a high 665/620 nm ratios. FRET is disrupted by cleavage at the scissile bond by BACE-1. Several cell culture models for assaying secretase activity were described in a recent report by Volbracht et al. [76]. In one such model, the BACE-1 activity in HEK293 cells was determined in cells transfected with constructs expressing secreted alkaline phosphatase (SEAP)–APP in which the extracellular domain of APP was substituted with SEAP. Secreted alkaline phosphatase was quantitated in the cell culture supernatants after cleavage of SEAP–APP by BACE-1. The SEAP hydrolyzes *para*-nitrophenyl-phosphate (*p*NPP) to yield phosphate and *p*-nitrophenolate, a yellow chromophore that absorbs at 405 nm. An ELISA-based assay is available for assaying BACE-1 activity in brain cell extracts or cell lysates.

This solid-phase sandwich ELISA uses two kinds of high specific antibodies against BACE-1. The plates coated with one of the anti-BACE-1 antibody (N42) is incubated with cell extracts and, after several washes, is incubated with HRP-labeled BACE-1 antibody (C) (Immuno-Biological Laboratories Co., Ltd). After washes, the bound HRP activity is measured using tetra methyl benzidine (TMB), and the strength of color development is in proportion to the amount of BACE1.

B. Metalloproteases

Matrix metalloproteinases are a family of zinc-dependent endoproteinases secreted by cells and are responsible for much of the turnover, remodeling of components of the ECM and basement membrane. MMPs cleave a peptide bond before a hydrophobic amino acid residue (Leu, Ile, Met, Phe, Tyr) [82]. MMPs are modular proteins which are secreted as inactive proenzymes and contain a prodomain, a catalytic domain, and hemopexin domains. The prodomain harbors a Cys switch motif PRCGXPD, in which the Cys sulfhydryl group interacts with the zinc ion and blocks active site. Removal of the 10 kDa propeptide by proteolysis leads to activation of the enzyme [83]. The catalytic domain contains catalytic zinc which is coordinated by three histidine (His) residues in the active site motif HEXXHXXGXXH. The carbonyl group of the scissile peptide bond coordinates with the zinc atom, displacing water. A glutamate adjacent to the first His in the active site motif serves as a general base to draw a proton from the displaced water molecule on the carbonyl carbon. The S1′ pocket which accommodates the side chain of the substrate residue varies in size among the MMPs and determine the substrate specificity. All MMPs, except MMP-7 and MMP-26, have regulatory α-propeller hemopexin domains, which mediate protein–protein interactions and contribute to substrate specificity. MMP activities are measured using full-length collagens/gelatins, short synthetic linear peptides, or synthetic mini-collagens as substrates [84]. Zymography, based on collagens/gelatin substrates copolymerized in polyacrylamide gels, is the most commonly employed technique for in-gel detection of enzyme, proenzyme forms of MMP activity [85]. Gelatins are used for detection of MMP-2, MMP-9, and to a lower extent MMP-8. Collagen-embedded gels are used for MMP-1 and MMP-13, while casein–SDS-PAGE are the preferred substrate gels for MMP-11 and MMP-1, MMP-3, MMP-7, and MMP-12. Addition of heparin to the sample either at time of gel loading or post-electrophoresis improves detection of very low levels (low ngs to pgs) of MMP-7 (matrilysin) and, for collagenases, MMP-1 and MMP-13, respectively. A number of short peptides derived from MMP sites on collagen were initially designed for fluorescence-based assays: Dnp-Pro-Leu-Gly-Ile-Ala-Gly for MMP-8 and Dnp-Pro-Gln-Gly-Ile-Ala for MMP-9, containing the MMP cleavage site, the Gly-Ile bond. A large number of assays based on the use of FRET substrates [86] have been developed with various donor and quencher combinations (Fig. 8). Most of the linear peptides are labeled at the N-terminus or in the P3 position with Dnp as the quenching group. The donor groups are varied: Mca ((7-methoxycoumarin-4-yl)acetyl) and Nma (*N*-methylanthranilic acid). The peptide substrate, now marketed by Enzo, is extensively used for assaying almost all MMPs of commercial interest. The solubility

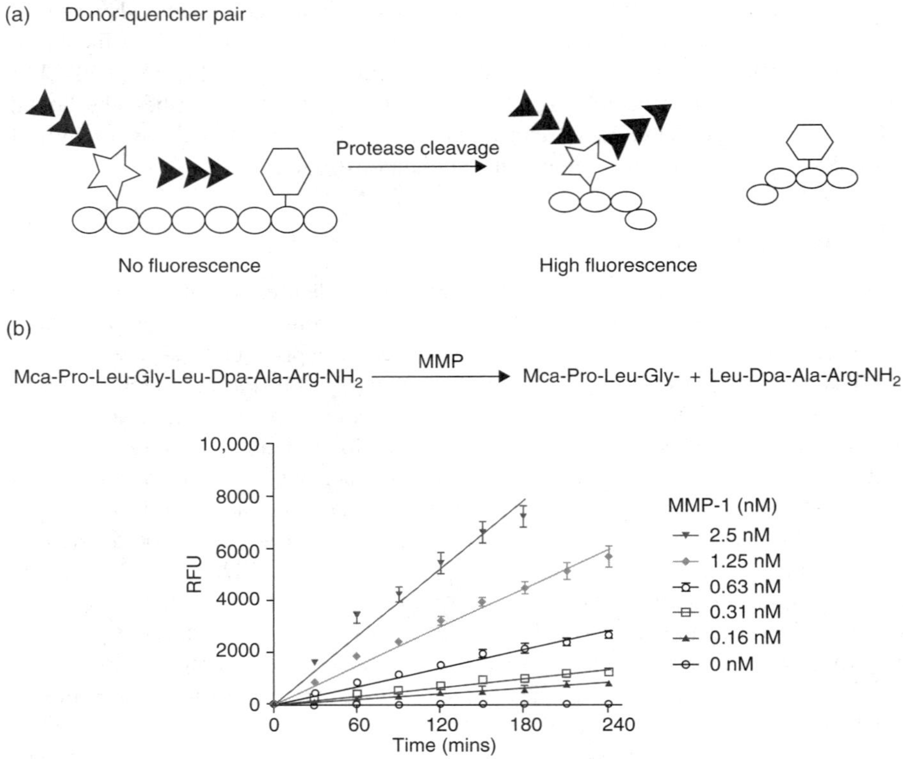

FIGURE 8 Matrix metalloproteinase assay. (a) Donor–quencher pair. Most of the MMP assays use FRET peptides in which the fluorescence from the donor is quenched by the quencher. (b) Detection of MMP-1 activity using an MCA/ Dpa FRET peptide as substrate. The graph shows kinetics of increase in fluorescence of hydrolyzed peptide which is monitored at Ex/Em = 340/490 nm. The increase in luminescence after hydrolysis is shown for different concentrations of MMP-1. (*See insert for color representation of part b.*)

of this peptide was improved by the addition of an N-terminal lysine (Lys). The cleavage by MMP-1, MMP-2, MMP-7, MMP-8, MMP-9, MMP-10, MMP-12, MMP-13, or MMP-14 at the Gly-Leu bond separates the Dpa, resulting in an increase in fluorescence (Fig. 8b). FRET peptides labeled with EDANS/DABSYL as donor–quencher pair are also used as MMP-assay substrates (Dabcyl-Gaba-Pro-Gln-Gly-Leu-Cys-Ala-Lys-NH2) with EDANS conjugated to the side chain of Glu or at the C-terminus of the peptides [87]. A derivative of the nonspecific metalloproteinase inhibitor, actinonin, was used as a universal probe to develop a generic selectivity profiling platform for HTS of MMP inhibitors. Actinonin, which binds to most of the MMP active sites, was used to design a probe for use in FP-based competition assay to obtain a selectivity profiling of inhibitors against a panel of MMP enzymes [88]. A colorimetric assay was developed by Weingarten et al. [23,24] by introducing a thioester bond at the MMP cleavage site. MMP-mediated hydrolysis of the thiopeptide produced a sulfhydryl group, which reacts with DTNB, Ellman's reagent, to form a colored compound, 2-nitro-5-thiobenzoic acid, detectable at 412 nm. An Enzo kit based on the concept of the thiopeptide chromogenic substrate (Ac-PLG-[2-mercapto-4-methyl-pentanoyl]-LG-OC2H5) allows activity determination on majority of MMP-3, MMP-12, MMP-1, MMP-7, MMP-2, MMP-9, MMP-13, MMP-8, MMP-14, MMP-10, and MMP-20 enzymes. AlphaScreen™-based MMP assay has been developed for monitoring degradation of the substrate aggrecan, an important structural component of joint cartilage [89]. At least four antibodies specific to the carbohydrate side chains of aggrecan were used to create a macromolecular complex whereby aggrecan could form a cross-link between donor and acceptor AlphaScreen detector beads. Digested aggrecan, which failed to form a cross-link, generated no signal, so that inhibitors of the digestion could be detected as a restoration of signal. The assay was rapid and well suited for HTS and gave signal-to-background values around 20:1 and an overall Z′ factor of 0.8.

C. Serine Proteases

Around one-third of the known proteases are Ser proteases, which are predominantly endoproteases. All Ser proteases share common structural features of their active sites: a catalytic triad, a selectivity S1 pocket, a proximal hydrophobic S2 pocket, and a distal hydrophobic S3 pocket. All Ser proteases can be grouped into three types: trypsin-like, which cleave proteins after basic amino acids; chymotrypsin-like, which cleave after hydrophobic amino acids; and elastase-like, which cleave after small hydrophobic amino acids. The hydroxyl group of Ser acts as a nucleophile that attacks the carbonyl moiety of the substrate peptide bond to form an acyl–enzyme intermediate. In addition to Ser, the catalytic machinery requires a proton donor (His) and a third residue, like Asp or another His. The catalytic triad is a hallmark of Ser proteases though many Ser proteases employ a simpler dyad mechanism where Lys (proton donor) or His is paired with the catalytic Ser. Antithrombotic agents like anticoagulants and antiplatelet agents targeting Ser proteases, the coagulation factors thrombin and FXa, are very prominent drugs for cardiovascular disease. Factor Xa catalyzes the hydrolysis of the Arg-Thr and then Arg-Ile bonds in prothrombin to yield active thrombin, with specificity for the recognition sequence Ile-Glu (or Asp)-Gly-Arg-X [90,91]. Several short peptide substrates are available for FXa assays: *N*-benzoyl-Ile-Glu-Gly-Arg-*p*-Na, Boc-Ile-Glu-Gly-Arg-7-amido-4-methylcoumarin (Sigma), methoxycarbonyl-cyclohexylglycyl-Gly-Arg-NHPhNO2 (Spectrozyme Xa), methanesulfonyl-D-Leu-Gly- Arg-NHPhNO2 (CBS 31.39), and Bz-Ile-Glu-(piperidine amide)-Gly-Arg-NHPhNO2 (S2337). Thrombin is the last protease in the coagulation cascade and is activated by a cleavage of its prodomain by FXa. Thrombin triggers clot formation by cleaving N-termini of A and B chains of fibrinogen and releases two corresponding oligopeptides, fibrinopeptides A and B. It also activates factor XIII which cross-links fibrinogen monomers and stabilizes the clot. Activity of thrombin is assayed colorimetrically using chromogenic peptide substrates like D-Phe-Pip-Arg-NHPhNO2, S-2238 (Chromogenix), Tos-Gly-Pro-Arg-NHPhNO2, Chromozym-TH (Boehringer Mannheim), or AFC thrombin substrate (AnaSpec). Argatroban is an arginine (Arg)-based peptidomimetic small-molecule inhibitor of thrombin.

D. Threonine Proteases: Proteasome Degradation Enzymes

Ubiquitination Threonine proteases are very similar to Ser proteases since the catalytic mechanism is identical, but a Thr is used in place of Ser. Threonine proteases are components of the proteasome, a multi-subunit complex which plays a central role in ubiquitin (Ub)-mediated degradation of Ub-tagged cytoplasmic proteins. Proteasome activity regulates cell cycle progression, apoptosis, and inflammation and has a basic housekeeping function of removing damaged or misfolded proteins. Proteasome subunits are upregulated in many cancers, and proteasome inhibitors are known to block cell cycle proliferation. The proteasome has 7α and 7β subunits (proteolytic chamber), and the core is made up of six proteolytic sites with three distinct substrate specificities [92,93]. The catalytic activity in β5-subunits of the proteasome has chymotrypsin-like activity and can hydrolyze bonds on the carboxyl side of Tyr or Phe at the peptide carbonyl or P1 position. The β2-subunits harbor trypsin-like activity with preference for basic amino acids Arg or Lys at the P1 site. The β1-subunits have caspase-like activity with preference for glutamate or acidic (peptidyl-glutamyl peptide bond hydrolyzing) amino acid residue [94,95]. The proteasome uses the N-terminal Thr of its catalytic β-subunit as a nucleophile for peptide bond hydrolysis. The cleavage of the propeptide upon activation exposed Thr as the N-terminal residue. Misfolded or other target proteins targeted for degradation are first acted upon by ubiquitination enzymes which add polyubiquitin chains to internal Lys. The polyubiquitin chains interact with the degradation signals such as misfolded domains, small motifs, or posttranslational signals. The N-end rule and ubiquitin fusion degradation (UFD) signals [96,97] are modular domains that were inserted in GFP [98] for developing fluorescence-based *in vitro* assays for monitoring proteasome activity (Fig. 9). The presence of one or several degradation (UFD) signals fused to GFP reporter, along with an N-terminal Ub moiety as the acceptor for polyubiquitin chains, shortens the half-life since the proteasome degrades the GFP, resulting in loss of fluorescence. The treatment of UFD–GFP-expressing cells with proteasome inhibitors led to stabilization of the GFP reporter with a resultant up to a 1000-fold increase in fluorescent intensity [35]. This increase correlated with cell cycle arrest and apoptosis, which is an established result of proteasome activity arrest. The fact that these fluorescent proteins follow the same pathway as endogenous substrates allows functional analysis of the Ub–proteasome system in cells. The UFD–GFP system was also used for *in vivo* analysis of the Ub–proteasome system in mouse strains transgenic for (GFP) reporter fused to constitutively active degradation signal [99]. *In vitro* assays for proteasome activity have been based on utilization of fluorescently AMC-tagged tri-/tetrapeptides (Table 2). Substrates like LLVY-AMC are also acted upon by other nonspecific protease activities in whole cell

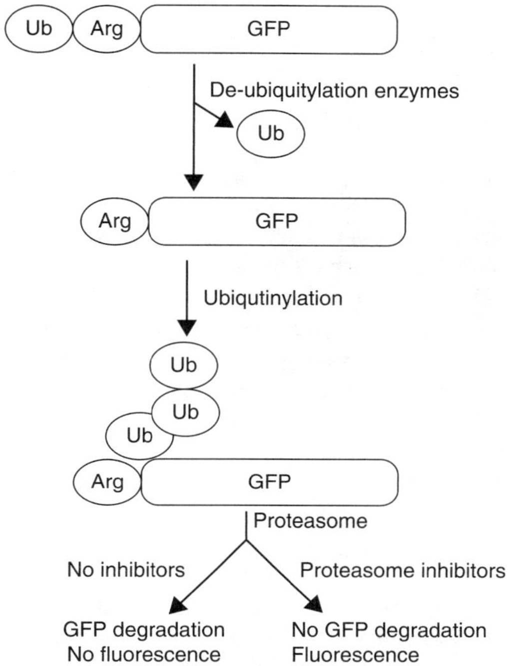

FIGURE 9 Proteasome assay using GFP. The Ub at the amino-terminus of the fusion protein is removed by deubiquitinating enzymes, thus exposing the N-terminal Arg residue. The Arg residue promotes ubiquitination and proteasome-mediated degradation of GFP.

lysates [100] but are specific for use in assaying circulating proteasome activity in the blood of chronic lymphocytic leukemia patients [101]. FRET peptides have been used to assay proteasome activity *in vitro*. The FRET peptides contain donor fluorescein and acceptor tetramethylrhodamine (TMR), which have spectral overlaps and are separated by six amino acids. In the undegraded peptide, Ex of fluorescein results in energy transfer to TMR and measurable acceptor fluorescence is obtained. After proteasome-mediated degradation, the donor and acceptor pair separates accompanied with an increased Em from the donor and a concomitant loss of acceptor fluorescence. The fluorescent peptides that contain a donor and a quencher have also been used for assaying proteasome activity. Cleavage of such peptides by proteasome separates the donor and quencher with a resultant increase in fluorescence of the donor over very low backgrounds. The small peptides can also be injected into cells and proteolysis followed by imaging techniques. Since the fluorescent assays are less sensitive than bioluminescent assays, many luminogenic substrates have been developed by Promega Corp. for testing proteasome activity in homogeneous bioluminescent assays. The Suc-LLVY-aminoluciferin, Z-LRR-aminoluciferin, and Z-norleucineLPnLD-aminoluciferin have been used to monitor chymotrypsin-like, trypsin-like, and caspase-like activities of proteasome. The rate of proteasome cleavage of the substrate was found to be equal to the rate of luciferase utilization of the released aminoluciferin [92].

E. Cystine Proteases

Cysteine proteases use a catalytic dyad comprising of a thiolate ion in active site Cys as a nucleophile which is activated by imidazole group of a His (proton donor). The Cys proteases stabilize an oxyanion of tetrahedral transition state via hydrogen bond interactions with active sites Cys and glycine (Gly). Cys proteases are grouped into six clans based on their structure of the active site and include calpain families as well as Ub-processing peptidases, the cytosolic cysteinyl aspartate-specific proteases, or caspases. Caspases are Cys proteases responsible for proteolysis associated with programmed cell death or apoptosis, whereas lysosomal cathepsins are involved in protein degradation.

Caspases Caspases are a family of cytosolic Cys proteases involved in the initiation and execution of apoptosis by targeting a discrete set of proteins for proteolysis which occurs after an Asp residue [102]. The caspase family in mammals has at least 15 members, which are either inflammatory (caspase-1, caspase-4, caspase-5, caspase-12, caspase-13, and caspase-14) or apoptotic caspases. The apoptotic caspases are further characterized as either initiator caspases (caspase-2, caspase-8, caspase-9, and caspase-10) or effector caspases (caspase-3, caspase-6, and caspase-7). Caspases are present as inactive proenzymes, most of which are activated by proteolytic cleavage or via autoproteolysis at the Asp-X bonds. X-ray crystallography studies and sequence analysis have shown that the caspases are modular with a 17–21 kDa catalytic domain (p20), a 10–13 kDa small catalytic domain (p10), and a 3–24 kDa amino-terminal prodomain (death domain). Both the large (p20) and the small (p10) catalytic subunits contribute residues to form substrate binding sites (S4, S3, S2, S1, S1′). All caspases show an absolute requirement for the aspartate residue in position P1 for catalysis [103]. However, the substrate specificity of human caspases is determined by their preferences for residues at the P4 site. The group I pro-inflammatory caspases (caspase-1, caspase-4, caspase-5, and caspase-13) prefer large aromatic/hydrophobic amino acids in P4, group II initiator caspases (caspase-2, caspase-8, caspase-9, and caspase-10) require Asp at P4, and the group III apoptotic caspases (caspase-3, caspase-6, and caspase-7) prefer branched-chain aliphatic amino acids at P4 [103]. The general cleavage site for all caspases is X-Glu-X-Asp: caspase-1, caspase-4, and caspase-5 prefer WEHD; caspase-2, caspase-3, and caspase-7 cleave DEXD; and caspase-6, caspase-8, and caspase-9 cleave specifically at (L/V)EXD [103]. Caspase activity measurement is widely used in apoptosis, and numerous *in vitro* and cell-based assay methods are available. In cell line apoptotic models, the relative abundance and activity of caspases in a cell actually dictates which activity is being measured primarily in the cell lysates [104]. It has been reported that caspase-3 is most active for many nonspecific substrates and is also the most abundant caspase in cell lysates [103]. The activity of initiator caspases is minimal in dilute cell extracts due to dimer dissociation. As a result, most of the activity measured in cell lysates is due to one or more caspases and does not allow specifically associating a cleavage event with any one caspase. The short tetrapeptide substrates are more relevant in the context of purified recombinant caspases and for determination of enzyme kinetic parameters. Recombinant caspase-2, caspase-3, caspase-6, and caspase-7 exist as active dimers in solution, whereas the recombinant caspase-8 and caspase-9 exist as mixtures of inactive monomers and active dimers in solution [105]. Various kosmotropic salts like sodium or ammonium citrate or dithiothreitol are added to the buffers to stabilize these protein mixtures as dimers or change protein conformation to increase substrate binding and decrease the Km values. Caspase activity is measured with homogeneous platforms using labeled peptides and in cell FRET assays [104,106]. Although most caspases have overlapping substrate specificities, a number of assay kits are available from various vendors that use different peptides for different members of caspase family. The various tetrapeptide sequences have been defined as optimal for different caspases: DEVD is used for assays with caspase-3 and caspase-7, YVAD for caspase-1, VDVAD for caspase-2, LEVD for caspase-4, WEHD for caspase-5, VEID for caspase-6, IETD for caspase-8, LEHD for caspase-9, AEVD for caspase-10, and ATAD for caspase-10. The peptides are labeled with chromogenic group like *p*-NA for colorimetric assays; fluorofluors like AFC, AMC, EDANS, and MCA for fluorescence assays; or aminoluciferin for bioluminescent assays (Fig. 10). Luminescence-based assays developed by Promega Corp. utilize the same tetrapeptide sequences for conjugation with aminoluciferin. The peptide is solubilized in a buffer containing luciferase, $MgSO_4$, and ATP. The purified recombinant caspases or caspase mixtures from lysates of cells undergoing apoptosis are added to the mix and luminescence is measured.

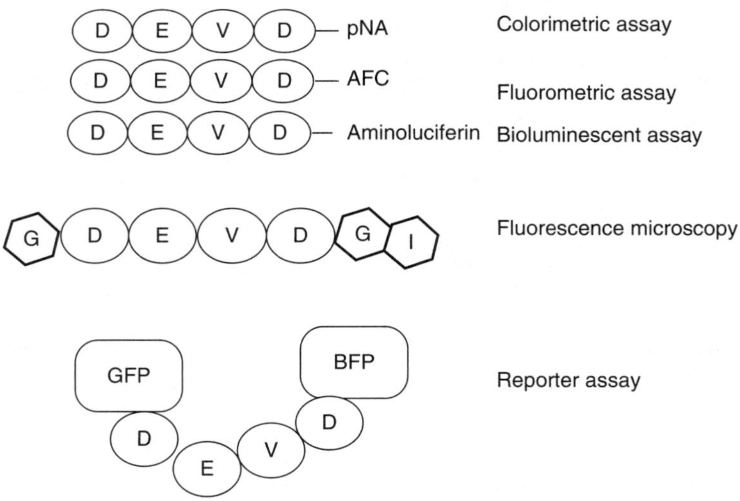

FIGURE 10 Substrates for caspase assay. Various conjugations with chromogenic, fluorescent, and luminescence tags to the same peptide generate substrates for all major assay formats.

Cathepsins Many cathepsins are lysosomal proteases that play an important role in antigen presentation, tissue remodeling, and enzyme activation/inactivation [107]. Upregulation of one or more cathepsins has been implicated in cancer, AD, and rheumatoid arthritis. Structurally, cathepsins A and G are members of the Ser peptidase family, cathepsins D and E are Asp proteases, and cathepsins B, C (DPPI), F, L, S, H, O, V, X, W, and K belong to the Cys protease family [108–110]. All cathepsins belong to papain family and are synthesized as inactive zymogens. Lysosomal Cys proteinases are active in weakly acidic environment, and the substrate binds in an extended conformation with specificity defined by S2 and S1′ in the active sites. Cathepsin B, which has been studied most extensively, has both endo- and exopeptidase activities and has less activity for substrates with Pro or Arg at P1′ [110]. Figure 11 lists minimal peptide substrates for assaying some of the cathepsins. The commonly used formats include fluorescence intensity measurements using short peptides labeled with AFC or AMC. In addition, many Mca/Dnp FRET peptides are available commercially (AnaSpec) for measurement of cathepsin activity. Cathepsin B is assayed using substrates like Z-Arg-Arg-AFC or AMC, as cathepsin B specifically cleaves the Arg-fluorophore bond in the fluorometric assay. Cathepsins H, L, and S show very little activity against these peptides. Cathepsin B also preferentially cleaves Z-Phe-Arg-AFC at pH 7.5–8. The substrate specificity for cathepsin H has not been clearly defined. Cathepsin C, an aminodipeptidase which requires halide ions for optimal activity, cleaves dipeptides from N-terminal of peptides and is inactive for substrates with Arg, Lys, or Pro at N-terminal positions. Cathepsin L has broad substrate tolerance and prefers substrates with hydrophobic amino acids in P2 and P3 residues [110]. Cathepsin L is active at pH 3–6.5 in the presence of thiol compounds. A generic benzyloxycarbonyl-protected dipeptide Leu-Arg-AMC (Peptide Institute) is used as a fluorogenic substrate for monitoring activity of cathepsins B, S, K, and V using Ex/Em wavelengths of 355/465 nm. Longer peptides like Mca-Gly-Lys-Pro-Ile-Leu-Phe-Phe-Arg-Leu-Lys(Dnp)-D-Arg-NH2 are substrates for cathepsin D/E. Some of the cathepsins have been targeted for drug development. Cathepsin C or dipeptidyl protease I (DPPI) activity is critical for differentiation of precursor promyelocytes into mature neutrophils as DPPI mediates production of neutrophil elastase, proteinase 3, and cathepsin G, which are potential drug targets for the treatment of chronic obstructive pulmonary disease (COPD). In most cases, cathepsin

Cathepsin B + RR-AFC ⟶ RR + AFC Ex 400 nm/Em 505 nm
Cathepsin K + LR-AFC ⟶ LR + AFC Ex 400 nm/Em 505 nm
Cathepsin H + R-AFC ⟶ R + AFC Ex 400 nm/Em 505 nm
Cathepsin L + FR-AFC ⟶ FR + AFC Ex 400 nm/Em 505 nm
Cathepsin S + VVR-AFC ⟶ VVR + AFC Ex 400 nm/Em 505 nm

FIGURE 11 Substrates for cathepsin assay. The preferred dipeptide sequence for cathepsin assays is linked to the fluorescent group to generate substrates for fluorescence intensity assays.

C activity has been measured using fluorescent substrates like Gly-Phe-AFC. Cathepsin K, a key enzyme in bone resorption, is a new therapeutic target for the treatment of osteoporosis. Odanacatib (Merck) is a powerful, reversible nonpeptidic biaryl inhibitor of cathepsin K which was highly selective over cathepsins B, L, and S in enzyme assays. Based on structure–activity data, a series of epoxysuccinate derivatives have been reported as Cys cathepsin-specific inhibitors [111].

IV. COMPUTATIONAL APPROACHES FOR PROTEASE IDENTIFICATION AND CHARACTERIZATION

Just as the analytical technologies required in the identification and characterization of proteases and their function are diverse, a substantial variety of different computational tools provide complementary insight. In silico methods can be applied to *de novo* prediction of biochemical structure and function, but their greatest power is found when they are used in concert with laboratory analyses. Key contributions include helping to interpret or deconvolute complex data, to extrapolate observations from a limited set of systems toward a more general species manifold, and to assimilate observations made from multiple different analytical platforms. Such analysis helps to address numerous specific questions about proteases, including:

- Which of the myriad of characterized and postulated proteins exhibit proteolytic activity?
- What type of catalytic function is any given protease likely to support?
- What substrate (or range of substrates) can a given protease process?
- What is the proteolytic mechanism of action?
- What types of chemicals are likely to serve as competitive inhibitors?
- Are there foreseeable mechanisms for noncompetitive inhibition or agonist effects?
- If so, which chemicals are candidates for accomplishing such modulation?

The importance of having computational tools to address toward such questions is well established. For example, fewer than 4% of putative enzymes currently have experimentally characterized catalytic sites or binding data [112], but for those that have experimentally characterized crystallographic or NMR structure, it is possible to computationally perceive specific cavities that have a size and shape potentially suitable for ligand binding [113]. Furthermore, a much larger number of proteins have been sequenced and annotated within large systematic sequence databases, and a reasonable fraction have been at least qualitatively characterized in terms of general protein family and general biochemical function. Cross-references between sequence and function have enabled bioinformatics discovery of plausible sequence motifs (including either phylogenetically conserved sequence patterns or rigorously conserved sequence substrings) that can be used to identify with reasonable accuracy which proteins are likely to support several of the known enzymatic functions [114].

Consequently today there are many examples of specific sequence substrings that are indicative (with varying degrees of certainty) of a given enzyme family. The most comprehensive collection of such motifs is probably housed within the PROSITE database (PROSITE: Database of protein domains, families and functions; http://www.expasy.ch/prosite/). Querying a sequence in PROSITE gives one a reasonable first-pass prediction of whether a given protein has a significant chance of belonging to a known enzyme family and, if so, which class would it belong to. Such analysis also identifies the specific amino acid sequence strings on which the protein family class assignment was made. A specific motif may either be labeled as an *active site*, in which case the constituent amino acids are expected to play a direct role in enzyme catalytic function (information that can be useful to elucidating the mechanism of action), or as a *signature*, in which case the amino acids may still play an important role in function but they may be somewhat spatially peripheral to the active site or have indirect implications toward the reactive mechanism (e.g., perhaps facilitating substrate admission to the active site or fixing it in place for reaction or expedite product evacuation). In general one can assume that enzymes within a certain class will accept substrate species that all share some similar chemical functionality. In other words, proteases are expected to admit amino acid chains for processing, whereas phosphatases should be selective toward phosphate-containing substrates, kinases act on species that are capable of receiving a phosphate group, hydrolases process ligands that can be made susceptible to aqueous hydrolysis, etc. Identification of a key sequence motif can thus be used to project not only that a given protein is likely to be an enzyme but also that it is likely to process some members of given chemotype, for example, esters, phosphates, lipids, and specific amino acids. In general, experimental characterization or computational prediction of the precise range of substrates that a given enzyme will process (e.g., alkyl vs. aryl esters, ATP vs. GTP) and the exact nature of the enzyme functionality/mechanism can be significantly more challenging. Some members of the protease family have unusually consistent function (e.g., trypsin, which is capable of digesting an immense variety of different peptides as long as they have at least one nonterminal Lys or Arg residue), while others are much more limited in their scope of action. The fact that there can be a tremendous amount of variation in the extent of observed substrate specificity from one enzyme family to another, and even within some specific families, makes it difficult to confidently state whether or not a given substrate-like molecule will or will not be processed by a given enzyme unless that molecule has been subjected to direct assay in the presence of that enzyme. This challenge has not prevented the computational community from attempting to address the issue however. Since a key component of enzyme specificity is determined by the three-dimensional structure of the active site cavity, it is very helpful to have a plausible spatial receptor model that can guide a computational assessment of whether a given chemical has a size and shape that is amenable to occupying the active site. Chromotrypsin, for example, has a sizeable hydrophobic pocket located strategically close to the catalytically active Ser residue. This location strongly favors the binding of lipophilic amino acids such as tryptophan, Phe, and Tyr, thus leading to reasonably specific cleavage points adjacent to such residues. This specificity contrasts with elastase, which preferentially cleaves on the carboxy side of much smaller neutral residues such as Gly, Ala, and valine. Such size- and shape-dependent specificity is not always sufficient evidence to proclaim that a given molecule will definitely be an enzyme–substrate, but the converse can be stated more confidently: a molecule that is too large, or has the wrong shape, to occupy an active site will very likely not be a substrate. Applying these criteria toward an understanding of substrate limits of specific proteases has straightforward prerequisites: a reasonable representation of the three-dimensional receptor structure (i.e., Cartesian spatial coordinates and elemental types for all enzyme atoms) and a computational algorithm for sampling and energetically assessing conformations of prospective substrate species interacting with the receptor. The best enzyme receptor spatial models will obviously be those that have been experimentally resolved via crystallographic or NMR studies. Such species constitute a fairly minor fraction of the total population of known or putative enzymes and are often those about which substrate activity and specificity are already well characterized. One computational tool, comparative protein modeling, is particularly helpful in extending knowledge on this limited subset base toward a broader range of analogous species. The guidelines and protocols for use of comparative modeling in producing spatial models of putative enzymes are surveyed elsewhere [115], but it should be mentioned that in order to generate a model capable of supporting high-confidence structure–activity assessments, it is generally assumed that you need an experimentally resolved template structure whose amino acid sequence is at least 70% identical to the putative enzyme of interest, with minimal sequence gaps between target and template. Computational studies on species that do not have a template with that high degree of homology may still be possible (and are frequently pursued) but must proceed with the caveat of lower confidence in their resulting analytical and predictive capacity. Comparative models based on templates with less than 30% sequence identity to the target may theoretically yield plausible structures, but the chances that they will offer meaningful structure–activity guidance are considered to be thin.

There are numerous molecular docking software programs capable of computationally evaluating the capacity of small-molecule or peptidic ligands (including prospective substrate species) to access the enzyme receptor model. A comprehensive review of these programs is beyond the scope of this chapter, but it is reasonable to assert that most of them are theoretically capable of addressing the relatively simple-minded negative test we specified earlier (i.e., whether one can show that the putative substrate does not have a suitable shape and size for accessing the receptor model site). In order to approach such a test rigorously, however, one may wish to modify the docking approach from the default settings which are typically geared toward effecting the converse (i.e., positive) test for evaluating the probability that a molecule *will* occupy a receptor. If calculations based on the default settings do lead to conformers docked within the enzyme active site cavity, then the test is over in that you have failed to rule out the possibility that your molecule is a substrate and have successfully suggested that it might be. If the default settings do not report a suitable docked conformer, however, then a rigorous negative test will require more stringent conditions. Such a test can be achieved in a number of ways, with the simplest being to give the molecule more opportunities to find a cavity-centered pose, that is, by simply requesting a larger number (≥100) pose trials. Many docking programs also provide some mechanism for biasing the pose toward a specific receptor location, for example, by specifying a specific H-bonding or hydrophobic interaction within the receptor that must be satisfied in order for a pose to be considered or by permitting one to pre-position the ligand directly within the cavity and test to see if a subsequent pose evolution retains the molecule in this cavity instead of preferentially diffusing outward.

If molecular docking can be used for a negative test to rule out species as possible substrates, it is natural to ask whether one can use similar methods to accurately quantify the likelihood that the species actually is a substrate. The simple answer is no. Molecular docking can provide some insight into whether a given ligand might be a viable enzyme–substrate, but the task of predicting reactive viability or quantitative catalytic kinetics requires treatment of multiple underlying factors, for which docking alone cannot fully account. In the broadest enzymological sense, key factors include:

1. The relative amount of attraction between potentially relevant enzyme peripheral sites and specific substrate candidates (best quantified as a free energy of interaction that accounts for the fact that many enzymes electrostatically or lipophilically attract substrate molecules to a region close to their active site, from which then can recruit specific molecules into their active site for catalytic processing).
2. The suitability of substrate candidates for progressing from a peripheral location into the active site (kinetic transport rate constant; note that some enzymes rely on passive substrate diffusion, while others engage in proactive dynamics that help to guide substrate into position for catalysis).
3. The capacity of a substrate molecule within the active site to orient itself in a position suitable for catalysis (measured by free energy of interaction).
4. The propensity for covalent reaction between enzyme and substrate, leading to the desired catalytic products (determined by the activation barrier computed as an enthalpy difference between the reactant species and transition states progressing toward and relevant intermediate states and the final product state).
5. The capacity of the products to evacuate the enzyme, leaving it free to process additional substrate species (again, quantified as a transport rate constant). In truth, the dominant factor underlying the mechanism by which some proteolytic substrates access their intended catalytic site may be passive diffusion, in which case some of these considerations would be superfluous. However, for the sake of more sophisticated proteolytic systems with tangible intra-enzyme coupling effects, it is worth examining more generalizable analytical techniques.

Molecular docking techniques are very well designed for addressing factor 1. In terms of a compromise between reasonable prediction accuracy and reasonable computational efficiency, they are very likely the best currently available tools for this step, offering a major savings in computational expense relative to the potentially more quantitatively accurate option of molecular dynamics simulations. Most molecular docking methods are not suited for the probing transport kinetics underlying factors 2 and 5, because they rely on algorithmic shortcuts (e.g., random positional sampling and fragment-based searches to identify a binding site). Programs such as AutoDock [116] that treat ligands as whole molecules whose translational, rotational, and torsional coordinates are subjected to pseudodirectional variation as dictated by energy-weighted stochastic sampling techniques (e.g., genetic algorithms, Monte Carlo) may be able to shed some insight into the relative propensity of different ligands to effect such transport in that the programs can be trained to yield a trajectory that semi-realistically traces the progress of a ligand through a macromolecular environment [117], but we are unaware of any systematic validation of whether this approach can quantitatively reproduce transport-related factors in protease kinetics. Molecular dynamics simulations represent a more obvious rigorous avenue for transport kinetics modeling, although one encounters the practical problem of how to bridge the time scale gap between practical simulations (fully atomistic MD simulations of proteins are currently limited by computational expense to time scales on the order $<1.0\ \mu S$) and real applications (depending on the specific enzyme, transport processes may take orders of magnitude more time). One reasonable option for semiquantitative estimates of transport rate constants is statistical extrapolation of MD trajectories. As an illustration of this extrapolation, we performed a molecular dynamics simulation [118] of the acetylcholine substrate interacting with the active site gorge of *Torpedo californica* acetylcholinesterase, a Ser hydrolase enzyme closely functionally related to Ser proteases, modeled from crystal structure [119] using the AMBER suite (AMBER charges and force field [120], fully explicit TIP3P solvent model [121] periodicity modeled via Ewald sums, 20 ps volumetric equilibration followed by 20 ps thermal equilibration and 1.0 nS analysis) to probe the time scale within which the reactive ligand ester carbonyl C would approach within a covalent-bonding distance ($R_{CO} < 1.5$ Å) of the catalytic Ser200 side chain oxygen atom. Figure 12 reports a statistical sampling of the R_{CO} distance recorded every 100 fs over the course of the 1.0 nS analysis run. Visual inspection of the plot suggests a biphasic distribution, for which each phase corresponds fairly nicely to a quadratic curve fit. Taking into account the logarithmic scaling of the abscissa, this curve suggests that the simulation yielded two main conformational families, both of which displayed approximately normal R_{CO} distributions. The more populous conformational family further to the right in this figure is a reasonable representation of the positional distribution experienced by the ligand as it strives to transit from a peripherally bound location near the active site before ultimately penetrating into the catalytic receptor region (smaller leftmost conformational family). The quadratic curve fit to the rightmost distribution yields a logarithmic frequency expression that can be expressed as $\ln[F(R_{CO})] = -4.06R_{CO}^2 + 41.62R_{CO} - 20.78$. Assuming that the barrier between this conformational family and the one represented by shorter R_{CO} distances occurs approximately at $R_{CO} = 4.0$ Å, this expression yields a value of $\ln[F(R_{CO})] = 0.35$, which translates to a frequency of 1.41/nS, that is, the molecule would be expected to pass this barrier en route to accessing the catalytic site approximately 1.4 times per nS, which is qualitatively reasonable since we observed such a transit to occur only once during the underlying 1.0 nS simulation. The primary

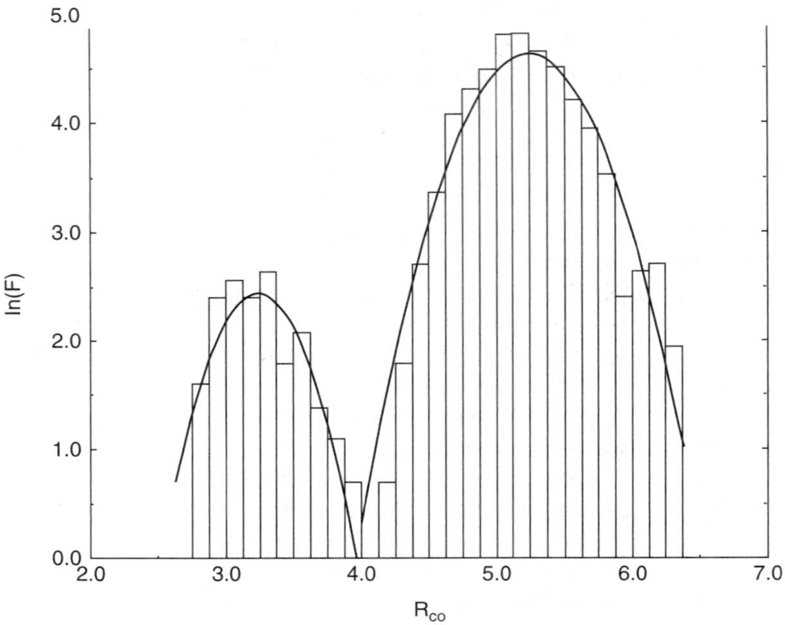

FIGURE 12 Logarithmic frequency of the distance between acetylcholine reactive C to acetylcholinesterase catalytic Ser 200 O during a 1 nS MD simulation.

drawback for such calculations is that even a relatively modest time scale such as 1 nS requires a significant investment in computer time (several days of CPU time on a reasonably efficient processor circa 2009) and thus is not amenable to high-throughput processing of a diverse range of prospective substrate candidates. Given a reasonable level of quantitative reliability in carefully constructed molecular dynamics calculations, however, it may be possible to perform simulations on a moderate number of prospective substrate systems and use the results as the basis for statistically training a quantitative structure–property relationship (QSPR) model to reproduce MD-derived transport rate constants. Some suitable descriptors for such QSPR models might entail volumetric and surfacial properties such as VolSurf metrics [122] (broadly used for transport and permeability properties), as well as terms describing intra-chain interactions that can impact peptide molecular flexibility.

The MD simulation described earlier also lends itself very well to redress of factor 3 (assessment of the capacity of a substrate molecule to orient itself within the active site in a conformation suitable for covalent reaction) in that the leftmost curve in Figure 12 tracks the conformationally dictated distribution of R_{CO} distances fairly close to the covalent limit (~1.5 Å), essentially representing substrate binding modes that are possible precursors to the Michaelis complex. This latter curve yields a logarithmic frequency quantified as $\ln(R_{CO}) = -3.45R_{CO}^2 - 0.33R_{CO} + 3.96$. By this relationship, the incidence rate for $R_{CO} < 1.5$ Å separation is predicted to be ~0.1/nS (i.e., once every 10 nS), which can be assumed to provide a rough estimate for the rate at which substrate will approach the reactive Ser200 oxygen closely enough to form a covalent bond. This rate is obviously a coarse approximation since the classical MD force field represents the atoms as fairly firm spheres governed by repulsive van der Waals potentials, while more realistic close contacts of reactive species should exhibit polarized electron densities that are often more permissive toward further approach. However, within a family of prospective substrate molecules with similar core electronic structure, the errors should be systematic; thus, it is reasonable to expect that the trends dictated strictly by conformational access to the Michaelis complex should correlate fairly well with observed substrate reactivity. It should also be noted that MD simulations such as these can yield quite accurate interaction free-energy prediction via rigorous analysis of variations in the computed classical electrostatic and van der Waals enthalpy terms between the ligand and receptor (e.g., [123]).

Although MD simulations are clearly superior to molecular docking for the purpose of addressing transport-related factors 2 and 5, it is not clear whether this advantageous performance is also observed for prediction of molecular propensity for forming the Michaelis complex. Many conventional docking programs provide a suitable medium for probing the capacity of a putative substrate molecule to orient itself within the active site cavity in a position susceptible to covalent interaction with the receptor and for gauging the relative stability of covalently bound enzyme–substrate intermediates. Essentially all small-molecule docking programs support the generation of precovalent enzyme–ligand Michaelis interaction complexes, and a selection of docking programs (e.g., AutoDock [116], FlexX [124], GOLD [125], and MacDOCK [126]) enables the projection of covalently bound complexes based on specification of one ligand atom that must bind to one particular receptor atom. While such calculations do not take into account the covalent bond-formation energy, they otherwise enable qualitative assessment of whether a given putative substrate has the right shape and electrostatic profile for effecting a covalently bound complex and enable comparative evaluation of enthalpy trends across a series of putative substrates. Such docking experiments cannot yield quantitative kinetic rate constants in the manner of MD simulations; however, their score models have been trained (via a QSPR-like process across diverse sets of ligands and receptors) to reproduce binding free-energy trends. One possible drawback with these models is that they are primarily geared toward treating protein inhibitor or protein agonist interactions, and at present we are unaware of rigorous validation of such score models for reproducing trends among enzyme–substrate complexes.

Computationally quantifying the covalent activated enzyme–substrate complex kinetics underlying factor 4 is most rigorously achieved via quantum chemical transition state calculations [127]. While these methods can be quite accurate and are a tempting tool for the pursuit of insight into specific protease–substrate pairings, an important caveat for such approaches is that they are very computationally demanding and low throughput. As can be surmised from the earlier discussion, computationally derived insight into the specific factors underlying the process of protease function is challenging to achieve. Furthermore, since many of the specific factors are also difficult to experimentally quantify, one frequently lacks quality analytical points of validation with which to evaluate specific models. By sacrificing fundamental detail, however, it is nonetheless possible to obtain fairly reliably predictive computational models of enzyme kinetics by focusing directly on readily measurable rate constants such as the Michaelis constant, K_M, that reflect the overall substrate processing capacity of the enzyme. Such observables can serve as the basis for the successful development of general, computationally efficient quantitative structure–activity relationship (QSAR) models (e.g., [128–130]). Constructing a reasonable model that accounts for substrate transport, Michaelis complex formation, and covalent reactivity requires a flexible basis of QSAR descriptors. Highly generalized descriptors such as the VolSurf parameters are very useful for such a requirement, although many instances with significant variation in the electronic structures of the reactive core may also require quantification of substrate atomic charge, bond strength, and polarizability terms via quantum chemical calculations. In cases where protease–substrate specificity appears to hinge on subtle aspects of the ligand–receptor interaction, the comparative binding energy (COMBINE) method [131,132] could provide important additional structure–activity insight. QSAR is a potentially very powerful tool for efficient and reliable prediction and characterization of enzyme–substrate processes, but nonetheless has fundamental limitations in practice. The predictive value of QSAR models can vary greatly as a function of many technical aspects relating to the manner in which they were trained. The general level of predictivity of QSAR models can be gauged through assessment of the statistical significance of the model and by performing cross-validation and blind-test studies using the model; however, an important additional caveat arises in assessing the scope of the model, that is, whether a model that has been trained based on one set of molecules is suitable for predicting the behavior of a distinct set of molecules. The general rule is that one is not advised to attempt a QSAR-based prediction on a molecule for which there are no reasonably similar analogs within the model training set, but at present we are unaware of studies that attempt to rigorously quantify how many analogs should be present in a training set and what is a reasonable similarity threshold by which to consider a given training set molecule to be an analog.

Most of the methods discussed this far with potential relevance to understanding protease catalytic function, kinetics, and substrate selectivity are equally applicable to computational assessment or design of ligand-based protease modulation and inhibition. Modeling studies on the latter are actually far more common than those applied toward assessment and prediction of substrate kinetics, perhaps in part due to the substantially greater availability of quantitative inhibition and modulation kinetics data. We can thus reprise the general list of key factors underlying enzyme kinetics discussed earlier, except this time in the context of non-substrate interactions:

1. Peripheral site interactions quantified as ligand–receptor interaction free energies are likely to be as important for quantifying non-substrate interactions.
2. Ligand transport from peripheral sites into the active site (for inhibitors) or adjunct subsites (modulators or noncompetitive inhibitors) is likely to be as qualitatively important.
3. The interaction free energy for a substrate molecule binding to the active site or adjunct subsite remains a key component.

4. The propensity for ligand–receptor covalent binding is only relevant in those cases (generally in the minority) when the inhibitor or modulator acts according to a covalent binding mode.
5. Ligand evacuation from the receptor can sometimes (although need not always) be important in inhibitor and modulator design or characterization.

The aforementioned considerations are largely sufficient for understanding the fundamental basis for competitive inhibition of proteases. Sometimes competitive inhibitors act by complexing with a peripheral or intermediate site in such a way as to bar access by substrate species to points further along (in particular the catalytic active site). In such cases the inhibitor mechanism of action is dictated largely by factor 1 and perhaps also by factor 2 (in scenarios where the inhibitor transits part of the way toward the active site). More often, however, the inhibitor possesses physical attributes adequately similar to those of substrate molecules so as to permit it to occupy the active site, but not so similar as to allow it to be catalytically processed and expelled in the same manner as a substrate. In this case the key actions by the inhibitor are determined by factors 1–3 and possibly also (in the case of a covalent inhibitor) by factor 4. The only other significant modeling considerations that may need to be entertained arise from the question of how the inhibitor is delivered into the general vicinity of the enzyme. In the case of *in vivo* delivery, it may thus be important to take into account the full range of absorption, distribution, and metabolic properties of the prospective ligand. Such properties are generally modeled quite well for standard drug-like and natural product-like compounds via established QSAR/QSPR models that have been assembled for drug design. The aforementioned VolSurf suite [122], for example, provides well-validated models for ligand solubility, albumin binding, blood–brain barrier permeation, volume distribution, cell permeation, etc.

In silico characterization and prediction of modulator and noncompetitive inhibitor kinetics pose a somewhat greater challenge in that the mechanism by which they affect enzyme function can be more complex than just physically blocking access by the substrate to its corresponding active site. Some cases of noncompetitive inhibition entail cases where inhibitors occupy enzyme sites intended for enzymatic cofactors, such as is being considered for targeting the NS2B–NS3 dengue virus protease [133]. In this case, kinetic characterization is really no more complicated than the competitive inhibition scenario since it can be adequately quantified by assessing the capacity of the inhibitor to approach, enter, and persist in the cofactor subsite. In other cases, however, the mechanism typically entails induction by the inhibitor/modulator of an enzymatic conformation change that modifies the capacity (reduced in the case of inhibitors, amplified in the case of activators) of substrate molecules to access the catalytic site and be covalently processed. QSAR-based studies of enzyme modulation are commonly carried out, with reasonable levels of predictive performance [134,135]; however, detailed understanding of the underlying biochemical basis for such effects is really only possible either from analytical characterization (i.e., crystallography or NMR of the enzyme in a modulated state) or via molecular dynamics simulations that may extrapolate a plausible modulated state from a ligand–enzyme complex assembled and initiated from an apo state structure, provided the molecular conformational transformation occurs within a computationally feasible time frame.

Finally, any comprehensive treatment of computational enzymology should include at least some mention of mutagenesis. Along the amino acid chain from which an enzyme is composed, many residue positions have little bearing on the ultimate biomolecular structure and function since amino acid substitution (and sometimes also deletion or insertion) at these points has little effect on quantitative observations of substrate catalytic, kinetics, or inhibition/modulation profiles; however, some sites yield significant (sometimes critical) sensitivity. Determination of the structural effects underlying such mutational sensitivity can shed very important light into the basic mechanism of protease function or modulation, and computational tools can play important roles in unraveling this insight. The results of molecular docking simulations can sometimes demonstrate sensitivity commensurate with experimental observations [136] especially in cases where the mutation is in close proximity to a ligand binding site and directly impinges on the capacity of a substrate, inhibitor, or modulator to complex with the site. Molecular dynamics simulations are even better suited to such mutations in that they can translate such structural modifications into revised projections for ligand transport kinetics and binding site complexation and can capture small- to medium-sized protein-wide conformational shifts that might arise from relaxation around the mutation site [137], provided the shift occurs in a computationally feasible time frame ($<1.0\ \mu S$). One of the most promising computational resources for guiding the conception and deconvoluting the results of mutagenetic studies is the aforementioned COMBINE technique whose analytical strength, in addition to providing potentially valuable information about the underlying ligand structure–activity relationship, is the capacity for framing this SAR in the context of specific ligand–residue interactions. Thus, a COMBINE study on activity trends observed for a wild-type protease can identify specific residues that appear to play a critical role in determining the SAR and thus provide the basis for subsequent mutagenetic profiling. Furthermore, the COMBINE model can be retrained to account for trends observed in subsequent screening results on the various resulting mutational variants, thus leading to a comprehensive model that should yield reasonably predictive insight into the target system both as a function of variations in receptor and ligand structure. Some examples of COMBINE models applied to specific proteases include work done on HIV-1 protease [138,139] and trypsin [140] and our own work on proteasome 20S where we extended the basic COMBINE technique to account for covalent inhibitors [141].

V. CONCLUSIONS

Proteases perform a wide range of functions essential for cell survival and homeostasis. Since the peptide bond hydrolysis is an irreversible process, all protease activities are tightly regulated. Dysregulation of protease activity has been reported in cancer, inflammatory, and neurodegenerative diseases and in cardiovascular anomalies. A number of drugs targeting proteases are available for indications where the protease activity is unequivocally associated with the disease. The development of safe and efficacious drugs targeting therapeutically relevant proteases requires detailed characterization of physiological protease substrates as well as protease expression and activation in normal and diseased states. A variety of biochemical, cell-based, and genomic approaches as well as *in vivo* animal models to study gain or loss of function have been useful in complete characterization of proteases. The mechanistic studies on proteolytic catalysis on different substrates require assay development and optimization. Majority of commonly used activity assays for proteases are based on short peptide substrates which may be generic or specific for a protease or for the protease family. Short peptide substrates designed for colorimetric-, fluorometric-, and luminescence-based assays give significant information on protease activity and its modulation by peptidomimetics or small molecules. For proteases with well-defined active site chemistry, ABPs have been utilized to detect active proteases in the cell after treatment with inhibitors or during disease progression or for determination of *in vivo* selectivity. Another modification of the assay platform uses quenched fluorescent peptide substrates immobilized on AuNP, in which fluorescence is observed post proteolysis. While the *in vitro* assays are useful for understanding mechanisms of proteolysis and their modulators, structural studies on protease–substrate interactions have helped refine specificity requirements in potential inhibitor scaffolds. In addition, computational tools clearly have a significant role to play in the characterization of normal and modulated protease function. While we must still take care to consider their current practical limitations, it should also be noted that active algorithmic development and the constant increase in available computational power are rendering in silico techniques ever more accurate and effective and are increasing the scope of applications to which they may be addressed. The day when computational analysis will no longer require complementary experimental analysis as a source of parameterization and validation is not yet conceivable; however, in silico methods are distinguishing themselves as ever more equal partners with experiment toward the complex goals of unraveling protease structure–function relationships and designing novel homologs with specially tailored functional attributes. An all-encompassing approach ranging from biochemical aspects of substrate specificity and assay development to optimization in conjunction with information from other genomics, proteomics, and degradomics will prove valuable in understanding proteases, designing assays, and ultimately identifying specific or pan-inhibitors which can be developed into safe and efficacious drugs in the future.

Disclaimer: Any reference to vendor products is merely for information purposes and is not a statement of endorsement.

REFERENCES

1. AJ Barrett, ND Rawlings, JF Woessner. *Handbook of Proteolytic Enzymes*. Amsterdam: Academic Press, 1998.
2. AJ Barrett, ND Rawlings. Types and families of endopeptidases. Biochem Soc Trans 19:707–715, 1991.

3. NM Hooper. Proteases: a primer. Essays Biochem 38:1–8, 2002.
4. C Lopez-Otin, JS Bond. Proteases: multifunctional enzymes in life and disease. J Biol Chem 283:30433–30437, 2008.
5. XS Puente, LM Sanchez, CM Overall, C Lopez-Otin. Human and mouse proteases: a comparative genomic approach. Nat Rev Genet 4:544–558, 2003.
6. V Quesada, GR Ordonez, LM Sanchez, XS Puente, C Lopez-Otin. The degradome database: mammalian proteases and diseases of proteolysis. Nucleic Acids Res 37:D239–D243, 2009.
7. ND Rawlings, AJ Barrett, A Bateman. MEROPS: the peptidase database. Nucleic Acids Res 38:D227–D233, 2010.
8. AJ Barrett. Classification of peptidases. Meth Enzymol 244:1–15, 1994.
9. I Schechter. Mapping of the active site of proteases in the 1960s and rational design of inhibitors/drugs in the 1990s. Curr Protein Pept Sci 6:501–512, 2005.
10. I Schechter, A Berger. On the size of the active site in proteases. I. Papain. Biochem Biophys Res Commun 27:157–162, 1967.
11. I Schechter, A Berger. On the active site of proteases. 3. Mapping the active site of papain; specific peptide inhibitors of papain. Biochem Biophys Res Commun 32:898–902, 1968.
12. A Berger, I Schechter. Mapping the active site of papain with the aid of peptide substrates and inhibitors. Philos Trans R Soc Lond B Biol Sci 257:249–264, 1970.
13. B Turk. Targeting proteases: Successes, failures and future prospects. Nat Rev Drug Discov 5:785–799, 2006.
14. I Chowdhury, B Tharakan, GK Bhat. Caspases – an update. Comp Biochem Physiol B Biochem Mol Biol 151:10–27, 2008.
15. CS Craik, MJ Page, EL Madison. Proteases as therapeutics. Biochem J 435:1–16, 2011.
16. M Drag, GS Salvesen. Emerging principles in protease-based drug discovery. Nat Rev Drug Discov 9:690–701, 2010.
17. SS Cotrin, L Puzer, WA de Souza Judice, L Juliano, AK Carmona, MA Juliano. Positional-scanning combinatorial libraries of fluorescence resonance energy transfer peptides to define substrate specificity of carboxydipeptidases: assays with human cathepsin b. Anal Biochem 335:244–252, 2004.
18. WZ Mari Enoksson, GS Salvesen, ed. *Identification of Protease Substrates by Mass Spectrometry Approaches-1*. New York: Springer Science, 2008, pp. 67–81.
19. CM Overall, CP Blobel. In search of partners: linking extracellular proteases to substrates. Nat Rev Mol Cell Biol 8:245–257, 2007.
20. O Schilling, CM Overall. Proteome-derived, database-searchable peptide libraries for identifying protease cleavage sites. Nat Biotechnol 26:685–694, 2008.
21. M Niinobe, Y Hitomi, S Fujii. A sensitive colorimetric assay for various proteases using naphthyl ester derivatives as substrates. J Biochem 87:779–783, 1980.
22. AK Hazra, SP Chock, RW Albers. Protein determination with trinitrobenzene sulfonate: a method relatively independent of amino acid composition. Anal Biochem 137:437–443, 1984.
23. H Weingarten, J Feder. Spectrophotometric assay for vertebrate collagenase. Anal Biochem 147:437–440, 1985.
24. H Weingarten, R Martin, J Feder. Synthetic substrates of vertebrate collagenase. Biochemistry 24:6730–6734, 1985.
25. DM Jameson, JC Croney, PD Moens. Fluorescence: basic concepts, practical aspects, and some anecdotes. Meth Enzymol 360:1–43, 2003.

26. RP Haugland. *Hand Book of Fluorescent Probes and Chemical Research.* Eugene: Molecular Probes, Inc., 1996.
27. AJ Pope, UM Haupts, KJ Moore. Homogeneous fluorescence readouts for miniaturized high-throughput screening: theory and practice. Drug Discov Today 4:350–362, 1999.
28. DM Jameson, JC Croney. Fluorescence polarization: past, present and future. Comb Chem High Throughput Screen 6:167–173, 2003.
29. LM Levine, ML Michener, MV Toth, BC Holwerda. Measurement of specific protease activity utilizing fluorescence polarization. Anal Biochem 247:83–88, 1997.
30. SK Grant, JG Sklar, RT Cummings. Development of novel assays for proteolytic enzymes using rhodamine-based fluorogenic substrates. J Biomol Screen 7:531–540, 2002.
31. J Karvinen, P Hurskainen, S Gopalakrishnan, D Burns, U Warrior, I Hemmila. Homogeneous time-resolved fluorescence quenching assay (LANCE) for caspase-3. J Biomol Screen 7:223–231, 2002.
32. S Mizukami, K Tonai, M Kaneko, K Kikuchi. Lanthanide-based protease activity sensors for time-resolved fluorescence measurements. J Am Chem Soc 130:14376–14377, 2008.
33. AB Cubitt, R Heim, SR Adams, AE Boyd, LA Gross, RY Tsien. Understanding, improving and using green fluorescent proteins. Trends Biochem Sci 20:448–455, 1995.
34. BA Pollok, R Heim. Using GFP in FRET-based applications. Trends Cell Biol 9:57–60, 1999.
35. J Neefjes, NP Dantuma. Fluorescent probes for proteolysis: tools for drug discovery. Nat Rev Drug Discov 3:58–69, 2004.
36. R Onuki, A Nagasaki, H Kawasaki, T Baba, TQ Uyeda, K Taira. Confirmation by FRET in individual living cells of the absence of significant amyloid beta -mediated caspase 8 activation. Proc Natl Acad Sci USA 99:14716–14721, 2002.
37. K Moin, M Samen, C Jedeszko, Q Li, MB Olive, RR Mattingly, BF Sloane. Images of cleavage: tumor proteases in action.In: D Edwards, G Høyer-Hansen, F Blasi, BF Sloane, eds. *The Cancer Degradome: Proteases and Cancer Biology.* New York: Springer Science, 2008, pp. 137–154.
38. DD Nolting, JC Gore, W Pham. Near-infrared dyes: probe development and applications in optical molecular imaging. Curr Org Synth 8:521–534, 2011.
39. D Maxwell, Q Chang, X Zhang, EM Barnett, D Piwnica-Worms. An improved cell-penetrating, caspase-activatable, near-infrared fluorescent peptide for apoptosis imaging. Bioconjug Chem 20:702–709, 2009.
40. M Funovics, R Weissleder, CH Tung. Protease sensors for bioimaging. Anal Bioanal Chem 377:956–963, 2003.
41. M Fonovic, M Bogyo. Activity based probes for proteases: applications to biomarker discovery, molecular imaging and drug screening. Curr Pharm Des 13:253–261, 2007.
42. WP Heal, TH Dang, EW Tate. Activity-based probes: discovering new biology and new drug targets. Chem Soc Rev 40:246–257, 2011.
43. D Kato, KM Boatright, AB Berger, T Nazif, G Blum, C Ryan, KA Chehade, GS Salvesen, M Bogyo. Activity-based probes that target diverse cysteine protease families. Nat Chem Biol 1:33–38, 2005.
44. D Greenbaum, KF Medzihradszky, A Burlingame, M Bogyo. Epoxide electrophiles as activity-dependent cysteine protease profiling and discovery tools. Chem Biol 7:569–581, 2000.
45. NA Thornberry, HG Bull, JR Calaycay, KT Chapman, AD Howard, MJ Kostura, DK Miller, SM Molineaux, JR Weidner, J Aunins, et al. A novel heterodimeric cysteine

protease is required for interleukin-1 beta processing in monocytes. Nature 356:768–774, 1992.

46. N Jessani, M Humphrey, WH McDonald, S Niessen, K Masuda, B Gangadharan, JR Yates, 3rd, BM Mueller, BF Cravatt. Carcinoma and stromal enzyme activity profiles associated with breast tumor growth in vivo. Proc Natl Acad Sci USA 101:13756–13761, 2004.

47. A Saghatelian, N Jessani, A Joseph, M Humphrey, BF Cravatt. Activity-based probes for the proteomic profiling of metalloproteases. Proc Natl Acad Sci USA 101:10000–10005, 2004.

48. V Gurtu, SR Kain, G Zhang. Fluorometric and colorimetric detection of caspase activity associated with apoptosis. Anal Biochem 251:98–102, 1997.

49. A Prinz, G Reither, M Diskar, C Schultz. Fluorescence and bioluminescence procedures for functional proteomics. Proteomics 8:1179–1196, 2008.

50. MA O'Brien, RA Moravec, TL Riss, RF Bulleit. Homogeneous, bioluminescent proteasome assays. Methods Mol Biol 414:163–181, 2008.

51. MA O'Brien, WJ Daily, PE Hesselberth, RA Moravec, MA Scurria, DH Klaubert, RF Bulleit, KV Wood. Homogeneous, bioluminescent protease assays: caspase-3 as a model. J Biomol Screen 10:137–148, 2005.

52. WJD Martha, A O'Brien, P Eric Hesselberth, RA Moravec, MA Scurria, DH Klaubert, RF Bulleit, KV Wood. Homogeneous, bioluminescent protease assays: caspase-3 as a model. J Biomol Screen 10:137–148, 1995.

53. BR Branchini, JC Rosenberg, DM Ablamsky, KP Taylor, TL Southworth, SJ Linder. Sequential bioluminescence resonance energy transfer–fluorescence resonance energy transfer-based ratiometric protease assays with fusion proteins of firefly luciferase and red fluorescent protein. Anal Biochem 414:239–245, 2011.

54. O Salata. Applications of nanoparticles in biology and medicine. J Nanobiotechnology 2:3, 2004.

55. C Guarise, L Pasquato, V De Filippis, P Scrimin. Gold nanoparticles-based protease assay. Proc Natl Acad Sci USA 103:3978–3982, 2006.

56. C Guarise, L Pasquato, P Scrimin, V De Filippis. Reversible aggregation/deaggregation of gold nanoparticles induced by a cleavable dithiol linker gold nanoparticles-based protease assay. Langmuir 21:5537–5541, 2005.

57. A Laromaine, L Koh, M Murugesan, RV Ulijn, MM Stevens. Protease-triggered dispersion of nanoparticle assemblies. J Am Chem Soc 129:4156–4157, 2007.

58. E Dulkeith, AC Morteani, T Niedereichholz, TA Klar, J Feldmann, SA Levi, FC van Veggel, DN Reinhoudt, M Moller, DI Gittins. Fluorescence quenching of dye molecules near gold nanoparticles: radiative and nonradiative effects. Phys Rev Lett 89:203002, 2002.

59. CH Tung, Y Lin, WK Moon, R Weissleder. A receptor-targeted near-infrared fluorescence probe for in vivo tumor imaging. Chembiochem 3:784–786, 2002.

60. BW Lee, MR Olin, GL Johnson, RJ Griffin. In vitro and in vivo apoptosis detection using membrane permeant fluorescent-labeled inhibitors of caspases. Meth Mol Biol 414:109–135, 2008.

61. J Xie, F Zhang, M Aronova, L Zhu, X Lin, Q Quan, G Liu, G Zhang, KY Choi, K Kim, et al. Manipulating the power of an additional phase: a flower-like AU-FE3O4 optical nanosensor for imaging protease expressions in vivo. ACS Nano 5:3043–3051, 2011.

62. L Josephson, MF Kircher, U Mahmood, Y Tang, R Weissleder. Near-infrared fluorescent nanoparticles as combined MR/optical imaging probes. Bioconjug Chem 13:554–560, 2002.

63. M Zhao, L Josephson, Y Tang, R Weissleder. Magnetic sensors for protease assays. Angew Chem Int Ed Engl 42:1375–1378, 2003.

64. T Asefa, CT Duncan, KK Sharma. Recent advances in nanostructured chemosensors and biosensors. Analyst 134:1980–1990, 2009.

65. K Welser, J Grilj, E Vauthey, JW Aylott, WC Chan. Protease responsive nanoprobes with tethered fluorogenic peptidyl 3-arylcoumarin substrates. Chem Commun (Camb): 6: 671–673, 2009.

66. J Anderson, C Schiffer, SK Lee, R Swanstrom. Viral protease inhibitors. Handb Exp Pharmacol 189:85–110, 2009.

67. UH Jörg Eder, F Cumin, B Martoglio, B Gerhartz. Aspartic proteases in drug discovery. Curr Pharma Des 13:271–285, 2007.

68. R Lapatto, T Blundell, A Hemmings, J Overington, A Wilderspin, S Wood, JR Merson, PJ Whittle, DE Danley, KF Geoghegan, et al. X-ray analysis of HIV-1 proteinase at 2.7 a resolution confirms structural homology among retroviral enzymes. Nature 342:299–302, 1989.

69. NE Kohl, EA Emini, WA Schleif, LJ Davis, JC Heimbach, RA Dixon, EM Scolnick, IS Sigal. Active human immunodeficiency virus protease is required for viral infectivity. Proc Natl Acad Sci USA 85:4686–4690, 1988.

70. AP Lea, D Faulds. Ritonavir. Drugs 52:541–546; discussion 547–548, 1996.

71. ED Matayoshi, GT Wang, GA Krafft, J Erickson. Novel fluorogenic substrates for assaying retroviral proteases by resonance energy transfer. Science 247:954–958, 1990.

72. AV Broadhurst, NA Roberts, AJ Ritchie, BK Handa, C Kay. Assay of HIV-1 proteinase: a colorimetric method using small peptide substrates. Anal Biochem 193:280–286, 1991.

73. CA Sanoski. Aliskiren: an oral direct renin inhibitor for the treatment of hypertension. Pharmacotherapy 29:193–212, 2009.

74. JR Stockigt, RD Collins, EG Biglieri. Determination of plasma renin concentration by angiotensin I immunoassay. Diagnostic import of precise measurement of subnormal renin in hyperaldosteronism. Circ Res 28(Suppl. 2):175–191, 1971.

75. E Cauchon, S Liu, MD Percival, SE Rowland, D Xu, C Binkert, P Strickner, J-P Falgueyret. Development of a homogeneous immunoassay for the detection of angiotensin I in plasma using alphalisa acceptor beads technology. Anal Biochem 388:134–139, 2009.

76. C Volbracht, S Penzkofer, D Mansson, KV Christensen, K Fog, S Schildknecht, M Leist, J Nielsen. Measurement of cellular [beta]-site of APP Cleaving enzyme 1 activity and its modulation in neuronal assay systems. Anal Biochem 387:208–220, 2009.

77. C Garino, T Tomita, N Pietrancosta, Y Laras, R Rosas, G Herbette, B Maigret, G Quéléver, T Iwatsubo, J-L Kraus. Naphthyl and coumarinyl biarylpiperazine derivatives as highly potent human β-secretase inhibitors. Design, synthesis, and enzymatic BACE-1 and cell assays. J Med Chem 49:4275–4285, 2006.

78. F Mancini, M Naldi, V Cavrini, V Andrisano. Multiwell fluorometric and colorimetric microassays for the evaluation of beta-secretase (BACE-1) inhibitors. Anal Bioanal Chem 388:1175–1183, 2007.

79. BL Pietrak, M-C Crouthamel, K Tugusheva, JE Lineberger, M Xu, JM DiMuzio, T Steele, AS Espeseth, SJ Stachel, CA Coburn, et al. Biochemical and cell-based assays for characterization of BACE-1 inhibitors. Ana Biochem 342:144–151, 2005.

80. F Mancini, A De Simone, V Andrisano. Beta-secretase as a target for Alzheimer's disease drug discovery: an overview of in vitro methods for characterization of inhibitors. Anal Bioanal Chem 400:1979–1996, 2011.

81. ME Kennedy, W Wang, L Song, J Lee, L Zhang, G Wong, L Wang, E Parker. Measuring human [beta]-secretase (BACE1) activity using homogeneous time-resolved fluorescence. Anal Biochem 319:49–55, 2003.
82. H Nagase, JF Woessner, Jr. Matrix metalloproteinases. J Biol Chem 274:21491–21494, 1999.
83. TL Blundell. Metalloproteinase superfamilies and drug design. Nat Struct Biol 1:73–75, 1994.
84. C Lombard, J Saulnier, J Wallach. Assays of matrix metalloproteinases (mmps) activities: a review. Biochimie 87:265–272, 2005.
85. K Kupai, G Szucs, S Cseh, I Hajdu, C Csonka, T Csont, P Ferdinandy. Matrix metalloproteinase activity assays: importance of zymography. J Pharmacol Toxicol Methods 61:205–209, 2010.
86. S Netzel-Arnett, SK Mallya, H Nagase, H Birkedal-Hansen, HE Van Wart. Continuously recording fluorescent assays optimized for five human matrix metalloproteinases. Anal Biochem 195:86–92, 1991.
87. B Beekman, JW Drijfhout, HK Ronday, JM TeKoppele. Fluorogenic mmp activity assay for plasma including mmps complexed to α2-macroglobulin. Ann NY Acad Sci 878:150–158, 1999.
88. C Antczak, C Radu, H Djaballah. A profiling platform for the identification of selective metalloprotease inhibitors. J Biomol Screen 13:285–294, 2008.
89. J Peppard, F Glickman, Y He, SI Hu, J Doughty, R Goldberg. Development of a high-throughput screening assay for inhibitors of aggrecan cleavage using luminescent oxygen channeling (alphascreen). J Biomol Screen 8:149–156, 2003.
90. RJ Butkowski, J Elion, MR Downing, KG Mann. Primary structure of human prethrombin 2 and alpha-thrombin. J Biol Chem 252:4942–4957, 1977.
91. DA Walz, D Hewett-Emmett, WH Seegers. Amino acid sequence of human prothrombin fragments 1 and 2. Proc Natl Acad Sci USA 74:1969–1972, 1977.
92. RA Moravec, MA O'Brien, WJ Daily, MA Scurria, L Bernad, TL Riss. Cell-based bioluminescent assays for all three proteasome activities in a homogeneous format. Anal Biochem 387:294–302, 2009.
93. S Murata, H Yashiroda, K Tanaka. Molecular mechanisms of proteasome assembly. Nat Rev Mol Cell Biol 10:104–115, 2009.
94. A Liggett, LJ Crawford, B Walker, TCM Morris, AE Irvine. Methods for measuring proteasome activity: current limitations and future developments. Leuk Res 34:1403–1409, 2010.
95. M Unno, T Mizushima, Y Morimoto, Y Tomisugi, K Tanaka, N Yasuoka, T Tsukihara. The structure of the mammalian 20S proteasome at 2.75 Å resolution. Structure 10:609–618, 2002.
96. ES Johnson, PC Ma, IM Ota, A Varshavsky. A proteolytic pathway that recognizes ubiquitin as a degradation signal. J Biol Chem 270:17442–17456, 1995.
97. A Varshavsky. The n-end rule pathway of protein degradation. Genes Cells 2:13–28, 1997.
98. NP Dantuma, K Lindsten, R Glas, M Jellne, MG Masucci. Short-lived green fluorescent proteins for quantifying ubiquitin/proteasome-dependent proteolysis in living cells. Nat Biotechnol 18:538–543, 2000.
99. K Lindsten, V Menendez-Benito, MG Masucci, NP Dantuma. A transgenic mouse model of the ubiquitin/proteasome system. Nat Biotechnol 21:897–902, 2003.

100. J Reidlinger, AM Pike, PJ Savory, RZ Murray, AJ Rivett. Catalytic properties of 26 s and 20 s proteasomes and radiolabeling of mb1, lmp7, and c7 subunits associated with trypsin-like and chymotrypsin-like activities. J Biol Chem 272:24899–24905, 1997.

101. W Ma, H Kantarjian, S O'Brien, I Jilani, X Zhang, Z Estrov, A Ferrajoli, M Keating, F Giles, M Albitar. Enzymatic activity of circulating proteasomes correlates with clinical behavior in patients with chronic lymphocytic leukemia. Cancer 112:1306–1312, 2008.

102. B Howley, HO Fearnhead. Caspases as therapeutic targets. J Cell Mol Med 12:1502–1516, 2008.

103. NA Thornberry, TA Rano, EP Peterson, DM Rasper, T Timkey, M Garcia-Calvo, VM Houtzager, PA Nordstrom, S Roy, JP Vaillancourt, et al. A combinatorial approach defines specificities of members of the caspase family and granzyme B. Functional relationships established for key mediators of apoptosis. J Biol Chem 272:17907–17911, 1997.

104. C Pop, GS Salvesen, FL Scott. Caspase assays: identifying caspase activity and substrates in vitro and in vivo. Methods Enzymol 446:351–367, 2008.

105. JB Denault, GS Salvesen. Apoptotic caspase activation and activity. Methods Mol Biol 414:191–220, 2008.

106. AL Niles, RA Moravec, TL Riss. Caspase activity assays. Methods Mol Biol 414:137–150, 2008.

107. V Turk, B Turk, D Turk. Lysosomal cysteine proteases: facts and opportunities. Embo J 20:4629–4633, 2001.

108. S Conus, HU Simon. Cathepsins and their involvement in immune responses. Swiss Med Wkly 140:w13042, 2010.

109. MM Mohamed, BF Sloane. Cysteine cathepsins: multifunctional enzymes in cancer. Nat Rev Cancer 6:764–775, 2006.

110. HH Otto, T Schirmeister. Cysteine proteases and their inhibitors. Chem Rev 97:133–172, 1997.

111. N Katunuma. Structure-based development of specific inhibitors for individual cathepsins and their medical applications. Proc Jpn Acad Ser B Phys Biol Sci 87:29–39, 2011.

112. P Liewlom, T Rakthanmanon, K Waiyamai. Prediction of enzyme class by using reactive motifs generated from binding and catalytic sites. Lect Notes Artif Int 4632:442–453, 2007.

113. GP Brady, Jr., PF Stouten. Fast prediction and visualization of protein binding pockets with pass. J Comput Aided Mol Des 14:383–401, 2000.

114. S Jones, JM Thornton. Searching for functional sites in protein structures. Curr Opin Chem Biol 8:3–7, 2004.

115. GH Lushington. Comparative modeling of proteins. Methods Mol Biol 443:199–212, 2008.

116. GM Morris, DS Goodsell, RS Halliday, R Huey, WE Hart, RK Belew, AJ Olson. Automated docking using a lamarckian genetic algorithm and an empirical binding free energy function. J Comp Chem 19:1639–1662, 1998.

117. P Cunningham, I Afzal-Ahmed, RJ Naftalin. Docking studies show that D-glucose and quercetin slide through the transporter GLUT1. J Biol Chem 281:5797–5803, 2006.

118. JX Guo, Lushington, G.H. Unpublished results. 2004.

119. JL Sussman, M Harel, F Frolow, C Oefner, A Goldman, L Toker, I Silman. Atomic structure of acetylcholinesterase from Torpedo californica: a prototypic acetylcholine-binding protein. Science 253:872–879, 1991.

120. DA Case, JW Caldwell, TE Cheatham, III, WS Ross, CL Simmerling, TA Darden, KM Merz, RV Stanton, AL Cheng, JJ Vincent, et al. (1999) Amber 6: University of California, San Francisco.

121. WL Jorgensen, C Jenson. Temperature dependence of TIP3P, SPC, AND TIP4P water from NPT Monte Carlo simulations: seeking temperatures of maximum density. J Comp Chem 19:1179–1186, 1998.

122. G Cruciani, M Pastor, W Guba. Volsurf: a new tool for the pharmacokinetic optimization of lead compounds. Eur J Pharm Sci 11(Suppl. 2):S29–S39, 2000.

123. GD Szklarz, MD Paulsen. Molecular modeling of cytochrome P450 1A1: enzyme-substrate interactions and substrate binding affinities. J Biomol Struct Dyn 20:155–162, 2002.

124. M Rarey, B Kramer, T Lengauer, G Klebe. A fast flexible docking method using an incremental construction algorithm. J Mol Biol 261:470–489, 1996.

125. GW Jones, RC Glen, AR Leach, R Taylor. Development and validation of a genetic algorithm for flexible docking. J Mol Biol 267:727–748, 1997.

126. X Fradera, J Kaur, J Mestres. Unsupervised guided docking of covalently bound ligands. J Comput Aided Mol Des 18:635–650, 2004.

127. L Ridder, BA Palfey, J Vervoort, IM Rietjens. Modelling flavin and substrate substituent effects on the activation barrier and rate of oxygen transfer by p-hydroxybenzoate hydroxylase. FEBS Lett 478:197–201, 2000.

128. Z Anusevicius, AEMF Soffers, N Cenas, J Sarlauskas, M Martinez-Julvez, IMCM Rietjens. Quantitative structure activity relationships for the electron transfer reactions of Anabaena PCC 7119 ferredoxin-NADP+ oxidoreductase with nitrobenzene and nitrobenzimidazolone derivatives: mechanistic implications. FEBS Lett 450:44–48, 1999.

129. A Costescu, CD Moldovan, G Katona, M Diudea. QSAR modeling of human catechol O-methyltransferase enzyme kinetics. J Math Chem 45:287–294, 2009.

130. J Sipila, AM Hood, MW Coughtrie, J Taskinen. CoMFA modeling of enzyme kinetics: Km values for sulfation of diverse phenolic substrates by human catecholamine sulfotransferase SULT1A3. J Chem Inf Comput Sci 43:1563–1569, 2003.

131. GH Lushington, JX Guo, JL Wang. Whither COMBINE? New opportunities for receptor-based QSAR. Curr Med Chem 14:1863–1877, 2007.

132. AR Ortiz, MT Pisabarro, F Gago, RC Wade. Prediction of drug binding affinities by comparative binding energy analysis. J Med Chem 38:2681–2691, 1995.

133. G Katzenmeier. Inhibition of the ns2b-ns3 protease – towards a causative therapy for dengue virus diseases. Dengue Bull 28:58–67, 2004.

134. M Araaujo de Brito, CR Rodrigues, JJV Cirino, R Bicca de Alencastro, HC Castro, MG Albuquerque. 3D-QSAR COMFA of a series of DABO derivatives as HIV-1 reverse transcriptase non-nucleoside inhibitors. J Chem Inf Model 48:1706–1715, 2008.

135. J Singh, B Shaik, S Singh, S Sikhima, VK Agrawal, PV Khadikar, CT Supuran. QSAR studies on the activation of the human carbonic anhydrase cytosolic isoforms I and II and secretory isozyme VI with amino acids and amines. Bioorg Med Chem 15:6501–6509, 2007.

136. S Repo, J Jyrkkärinne, JT Pulkkinen, R Laatikainen, P Honkakoski, MS Johnson. Ligand specificity of constitutive androstane receptor as probed by induced-fit docking and mutagenesis. J Med Chem 51:7119–7131, 2008.

137. P Kollman. Computer simulation methods as applied to site-specific mutations. Protein Des Dev New Ther Vaccines: 229–236, 1990.

138. S Nakamura, I Nakanishi, K Kitaura. Binding affinity prediction of non-peptide inhibitors of HIV-1 protease using combine model introduced from peptide inhibitors. Bioorg Med Chem Lett 16:6334–6337, 2006.

139. C Pérez, M Pastor, AR Ortiz, F Gago. Comparative binding energy analysis of HIV-1 protease inhibitors: incorporation of solvent effects and validation as a powerful tool in receptor-based drug design. J Med Chem 41:836–852, 1998.

140. MBM Peters, M Kenneth. Semiempirical comparative binding energy analysis (SE-COMBINE) of a series of trypsin inhibitors. J Chem Theor Comput 2:383–399, 2006.

141. JL Wang, A Datta, GH Lushington. Structure-based quantitative structure activity relationship analysis of omuralide analogs in the 20S proteasome: a covalent inhibitor COMBINE study. Lett Drug Des Discov 4:417–421, 2007.

2

DISCOVERY AND DEVELOPMENT OF ISOZYME-SELECTIVE INHIBITORS INVOLVED IN LIPID METABOLISM

TAICHI OHSHIRO

Department of Microbial Chemistry, Graduate School of Pharmaceutical Sciences, Kitasato University, Tokyo, Japan; Section on Lipid Sciences, Department of Pathology, Wake Forest University School of Medicine, Winston-Salem, North Carolina

HIROSHI TOMODA

Department of Microbial Chemistry, Graduate School of Pharmaceutical Sciences, Kitasato University, Tokyo, Japan

I. INTRODUCTION

Recent advances in gene technology and molecular biology have revolutionized strategies of drug discovery in the last decade. There has been a vast increase both in the number of researchers who use these technologies and, accordingly, in the number of important achievements made. Recently, several enzymes, including those for lipid metabolism, have been identified as isozymes, which are enzymes that differ in amino acid sequences but catalyze the same enzymatic reaction. Techniques of molecular biology and biochemistry have been used to study the differences of isozymes in their distribution, function, and regulation in cells and individuals. Furthermore, gene knockout studies and transgenic mice have also helped in identifying isozymes involved in some diseases. This progress implies that isozyme-selective inhibitors can be used as potential therapeutic agents. Representative isozyme-selective inhibitors are summarized in Table 1 and Figure 1.

Enzyme Technologies: Pluripotent Players in Discovering Therapeutic Agents, First Edition.
Edited by Hsiu-Chiung Yang, Wu-Kuang Yeh, and James R. McCarthy.

TABLE 1 Isozyme-Selective or Isozyme-Nonselective Inhibitors

Inhibitor	Target enzyme	Adaptation	References
COX inhibitor			
Aspirin (**1**)	Nonselective	Anti-inflammatory	9
Indomethacin (**2**)	Nonselective	Anti-inflammatory	9
Celecoxib (**3**)	COX-2	Anti-inflammatory	9,10
		Rheumatoid arthritis	9,10
PLA_2 inhibitor			
Varespladib (**4**)	$sPLA_2$	Acute coronary syndrome (ACS)	19,20
Darapladib (**5**)	$Lp\text{-}PLA_2$	Atherosclerosis	21
PDE inhibitor			
IBMX (**6**)	Nonselective		34,35
Vinpocetine (**7**)	PDE1	Hypertension	31
EHNA (**8**)	PDE2		31
Milrinone (**9**)	PDE3	Asthma	38
Rolipram (**10**)	PDE4	Depression	31
Sildenafil (**11**)	PDE5	Erectile dysfunction (ED) and	36,37
		Pulmonary arterial hypertension	36,37
DPP inhibitor			
Sitagliptin (**12**)	DPP-4	Type 2 diabetes	43
Vildagliptin (**13**)	DPP-4	Type 2 diabetes	46
Saxagliptin (**14**)	DPP-4	Type 2 diabetes	47
Alogliptin (**15**)	DPP-4	Type 2 diabetes	44
Linagliptin (**16**)	DPP-4	Type 2 diabetes	45

Cyclooxygenase (COX, EC 1.14.99.1), one of the most well-known targets for an isozyme-selective inhibitor, is responsible for the formation of important physiological and pathological mediators called prostanoids (including prostaglandins, prostacyclin, and thromboxane) [1]. Inhibition of prostaglandins by nonsteroidal anti-inflammatory drugs (NSAIDs) was first demonstrated in 1971 [2]. The mechanism of action (MOA) of the NSAIDs is based on the inhibition of COX. After this initial report, interest in COX grew steadily. The first purified COX enzyme was reported in 1976 [3]. More than a decade later, the COX gene was cloned in 1988 [4–6]. In the 1990s, an inducible isozyme now called COX-2 was discovered [7]. This discovery led to the hypothesis that anti-inflammatory prostaglandins were produced through constitutive expression of the COX-1 isozyme, whereas proinflammatory prostaglandins were produced via induction of the COX-2 isozyme. From extensive studies involving biochemical experiments and animal models, it became clear that the COX-2 isozyme is associated with inflammation, whereas COX-1 does not appear to be involved with the inflammatory process. Furthermore, the gastrointestinal side effects associated with traditional NSAIDs are due to the inhibition of gastroprotective prostaglandins synthesized via the COX-1 pathway [8]. Thus, scientists from the academic community and pharmaceutical companies focused their efforts on the design of COX-2-selective inhibitors in order to develop superior anti-inflammatory and analgesic agents with reduced adverse effects compared to

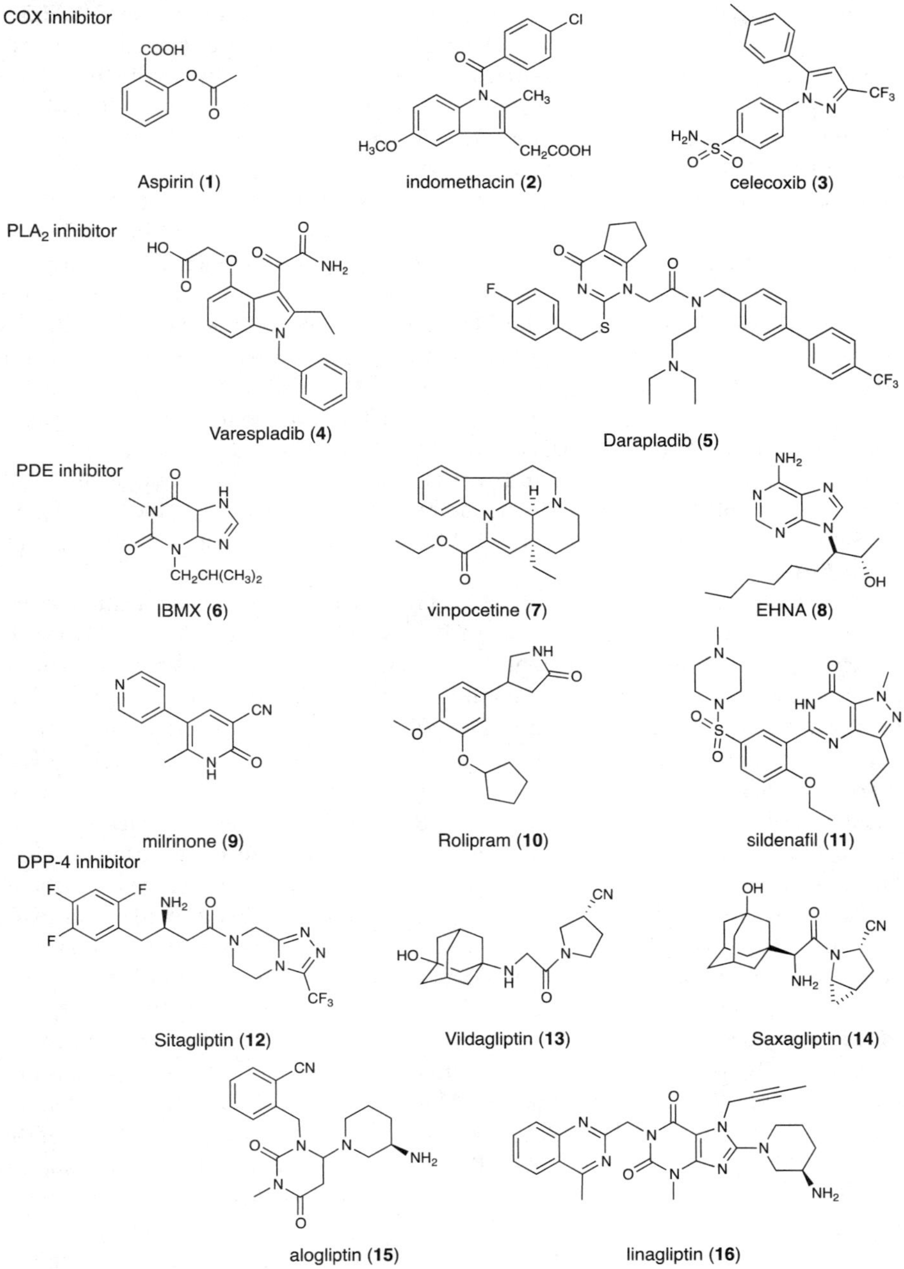

FIGURE 1 Representative isozyme-selective inhibitors.

traditional NSAIDs. Aspirin (**1**) and indomethacin (**2**), the most popular NSAIDs, inhibit both COX-1 and COX-2 isozymes; G.D. Searle and Pfizer launched the first COX-2-selective inhibitor, celecoxib (**3**), in 1999 (Fig. 1 and Table 1) [9,10].

Phospholipase A_2 (PLA_2, EC 3.1.1.4) is an enzyme that catalyzes the hydrolysis of phospholipids in the *sn*-2 position to generate arachidonic acid and lysophospholipid [11,12]. (Yang et al. have a sophisticated review entitled "Design and Development of a Selective Assay System for the Phospholipase A_2 Superfamily" in this series of books [11].) PLA_2 inhibitors are shown in Table 1 and Figure 1. The PLA_2 family, comprising more than 30 isozymes, plays important roles in phospholipid metabolism under various conditions such as host defense, inflammation, and atherosclerosis [13–18]. Varespladib (**4**) inhibits secretory PLA_2 ($sPLA_2$) ($sPLA_2$-IIA, $sPLA_2$-V, and $sPLA_2$-X isozymes) and reduces atherosclerosis and aneurysm formation in atherogenic animal models [19,20]. Darapladib (**5**), which selectively inhibits lipoprotein-associated PLA_2 (Lp-PLA_2) (no inhibitory activity against $sPLA_2$-IIA, $sPLA_2$-V, or $sPLA_2$-X isozymes) [21], was reported to retard the progression of coronary atherosclerosis in hypercholesterolemic and diabetic swine [21]. Recently, varespladib, a potent $sPLA_2$ inhibitor, and darapladib, a potent Lp-PLA_2 inhibitor, were evaluated in some clinical trials for the treatment of cardiovascular (CV) disease [22–26].

Phosphodiesterase (PDE, EC 3.1.4.1) is an enzyme involved in degradation of the intracellular second messengers cyclic AMP (cAMP) and cyclic GMP (cGMP). The increase of cAMP and cGMP triggers signaling pathways responsible for various physiological functions [27,28]. Accordingly, PDE is considered to be a promising therapeutic target of various diseases, including heart failure, hypertension, depression, asthma, inflammation, and erectile dysfunction (ED) [29–31]. To date, 11 PDE isozymes (PDE1 to PDE11) have been characterized. Furthermore, the data from genetic, molecular biological, and biochemical studies indicate that each isozyme has distinct expression with relevant roles and different substrate affinities [32]. Furthermore, each isozyme has a C-terminal catalytic domain conserved throughout all PDE isozymes, while each has a unique N-terminal regulatory domain. In PDE isozymes, PDE1 to PDE5 have been proven to be pharmacologically important targets [31,33]. PDE isozyme-selective inhibitors are also summarized in Figure 1 and Table 1 [31]. For example, IBMX (**6**) broadly inhibits PDE1 to PDE5, PDE7, and PDE11 isozymes with IC_{50} values of 10–50 μM [34,35]. On the other hand, sildenafil (**10**), clinically used for the treatment of ED and pulmonary arterial hypertension, inhibits the PDE5 isozyme with an IC_{50} value of 1.6 nM. Sildenafil shows highly selective inhibition toward the PDE5 isozyme with a selective index (SI) of 150–10 000-fold [36,37]. Milrinone (**8**), a PDE3-selective inhibitor, has been found to improve cardiac contractility and may be used for the treatment of congestive heart failure [38].

Dipeptidyl peptidases (DPPs), which cleave a peptide bond from the N-terminal dipeptide, belong to the serine protease family. This family has DPP-1, DPP-2 (known as quiescent cell proline dipeptidase (QPP) or DPP-7), DPP-3, DPP-4, fibroblast activation protein (FAP), DPP-8, and DPP-9 isozymes [39,40]. However, all these isozymes remain poorly understood except the DPP-4 isozyme. DPP-4

(EC 3.4.14.5) inactivates incretin hormones, glucagon-like peptide-1 (GLP-1), and glucose-dependent insulinotropic polypeptide (GIP). GLP-1 and GIP stimulate insulin release by the pancreas, reduce glucagon secretion, and improve glucose homeostasis. Therefore, selective inhibition of DPP-4 leads to improved fasting and postprandial hyperglycemia in type 2 diabetes patients. DPP-4-selective inhibitors are shown in Figure 1 and Table 1 [41,42]. Sitagliptin (**12**) was launched as the first DPP-4-selective inhibitor, followed by vildagliptin (**13**). Interestingly, sitagliptin (**12**), alogliptin (**15**), and linagliptin (**16**) are non-covalently bound to the active site of DPP-4 [43–45], while vildagliptin (**13**) and saxagliptin (**14**) are covalently bound to the site [46,47].

In lipid metabolism, there is an elegant balance in the levels of the end-product lipids, as enzymes and genes involved in their biosynthesis and metabolism maintain homeostasis. When the balance is lost, obesity or hyperlipidemia will develop, leading to a variety of serious diseases including atherosclerosis, diabetes, hypertension, and functional depression of certain organs. Therefore, the control of lipid metabolism by drugs could lead to the prevention or treatment of these diseases [48,49]. Statins, inhibitors of 3-hydroxy-3-methylglutaryl coenzyme A (HMG-CoA) reductase [50], have proved clinically useful for the treatment of hypercholesterolemia and atherosclerosis for 20 years. However, even in patients treated with statins, the risk of complications and death from CV events is reduced by only 30% [51]. Consequently, the quest for novel pharmacological agents that target selective steps of lipid metabolism has intensified significantly in recent years.

Our recent studies have focused on the biosynthesis of neutral lipids, triacylglycerol (TG) and cholesteryl ester (CE) [52–56]. These are the final storage forms of cholesterol and long-chain fatty acids in mammals. Excessive accumulation of neutral lipids due to a fat-rich diet or sedentary lifestyle causes obesity, nonalcoholic fatty liver disease (NAFLD), and atherosclerosis. Specifically, TG accumulates in the adipocytes and in the liver of obese and NAFLD patients, and CE accumulates in the arteries (macrophages and smooth muscle cells) in atherosclerosis patients. These conditions contribute to lifestyle-related diseases and metabolic syndrome, to which much attention has been paid recently because of their current importance in health care. The final enzymes involved in TG and CE biosyntheses are diacylglycerol acyltransferase (DGAT) and acyl-CoA:cholesterol acyltransferase (ACAT), respectively. They are membrane proteins of the endoplasmic reticulum (ER) and are difficult to purify to homogeneity. In spite of the lack of detailed biochemical and structural information about these enzymes, the ACAT and DGAT genes have been identified, and knowledge about the functions of DGAT and ACAT in mammals has accumulated. Recent molecular biological studies have revealed that there are two types of DGAT (DGAT1 and DGAT2 isozymes, see following text, [57]) and ACAT (ACAT1 and ACAT2 isozymes, see following text, [58]). It has been suggested that each isozyme of both enzymes has a different function. As a matter of course, isozyme-selective inhibitors are candidates as anti-obesity and anti-atherosclerotic agents. However, the selectivity of the inhibitors, even synthetic ones, toward the isozymes has not been fully studied.

In this chapter, we summarize the methods of discovery of isozyme-selective inhibitors of DGAT and ACAT and also describe their future prospects.

II. DIACYLGLYCEROL ACYLTRANSFERASE (DGAT)

TG synthesis is important in many processes, including lactation, energy storage in adipose tissue and muscle, fat absorption in the intestine, and the assembly of lipoprotein particles from the liver and intestines [59]. Excessive accumulation of TG due to a fat-rich diet or sedentary lifestyle is associated with diseases such as obesity, NAFLD, diabetes, and hypertriglyceridemia [60]. Therefore, inhibitors of TG biosynthesis may protect patients from obesity and other disorders. Diacylglycerol acyltransferase (acyl-CoA:1,2-diacyl-*sn*-glycerol *O*-acyltransferase, DGAT, EC 2.3.1.20) is a membrane-bound enzyme that catalyzes the reaction of acyl residue transfer from acyl-CoA to diacylglycerol (DG) to form TG [61]. This reaction is the final step in *de novo* TG biosynthesis. However, the DGAT enzyme has not been purified to date because it is a hydrophobic and transmembrane protein in the ER.

A. DGAT1 and DGAT2 Isozymes

Recently, two DGAT isozymes were identified in mammals: DGAT1 and DGAT2 [57]. Farese's groups identified the first gene encoding a mammalian DGAT (DGAT1 isozyme) from its homology to ACAT genes [62,63]. A full-length cDNA encodes a 498-amino acid protein that is about 20% identical to mouse ACAT1, with the most highly conserved region in the C-terminus. From the predicted protein sequence, DGAT1 has a potential N-linked glycosylation site, a putative tyrosine phosphorylation site, and an active site serine residue also found in ACAT. The protein has multiple hydrophobic domains and two or three possible transmembrane domains [64,65]. Its mRNA is detected in every mammalian tissue, and the highest expression levels are found in the adipose tissue and small intestine. These findings are consistent with a proposed role for DGAT in intestinal fat absorption. However, mRNA expression is relatively low in the livers of humans despite high DGAT activity, suggesting the existence of a second DGAT in the liver. Accordingly, the second mammalian DGAT (DGAT2) gene was identified by mammalian homologs [66] and also by a fungal DGAT enzyme [67]. DGAT2 showed no homology with the DGAT1 isozyme, but showed homology with monoacylglycerol acyltransferase (MGAT). The protein with 387 amino acids in humans has one or two possible transmembrane domain(s) [68]. Its mRNA is also detected in every mammalian tissue, with high expression levels in adipose tissue and liver. Interestingly, DGAT2 activity is completely inhibited by a high concentration of $MgCl_2$ (100 mM) [66].

Since the discovery of the DGAT isozymes, data from extensive genetic studies, including knockout mice, have afforded us new information about DGAT. Newborn DGAT2 knockout ($Dgat2^{-/-}$) mice die soon after birth [69]. Carcass TG content is severely reduced by about 90%, and the mice lack substrates for oxidative metabolism. They also lack essential fatty acids, resulting in abnormalities in skin lipid and epidermal barrier function. As a result, newborn $Dgat2^{-/-}$ mice rapidly become dehydrated, which hastens their demise. On the other hand, DGAT1 knockout ($Dgat1^{-/-}$) mice are viable and have more modest reduction in tissue TG [70]. Adult $Dgat1^{-/-}$ mice have about 50% less adipose mass and smaller adipocytes than wild-type mice on a chow diet [70,71]. Although tissue TG levels are decreased in the mice, the levels of DG and free fatty acid (FFA) are not significantly elevated, and serum TG levels are normal. Thus, DGAT2 plays an essential role in TG metabolism, while DGAT1 plays a limited role. Interestingly, it was found that *Dgat1*$^{-/-}$ mice are resistant to diet-induced obesity and that insulin sensitivity is increased in *Dgat1*$^{-/-}$ mice [72].

Studies on DGAT-deficient mice indicate that pharmacological intervention of DGAT1 may be a useful strategy for the treatment of obesity and type 2 diabetes, whereas inhibition of DGAT2 as an anti-obesity therapy may face the obstacle of a narrow therapeutic window [73].

B. Assay for DGAT Isozyme Activity

Before the discovery of DGAT isozymes, several groups, including our group, reported on DGAT inhibitors. Unfortunately, the early DGAT inhibitors were evaluated by enzyme or cell-based assays in which DGAT1 and DGAT2 isozymes appeared mixed. However, we now know that it is important to determine the selectivity of inhibitors toward the two isozymes for the development of anti-obesity, anti-NAFLD, antidiabetes, and anti-hypertriglyceridemia agents. Only a few groups have reported assay systems to measure the activity of the two DGAT isozymes.

Enzyme Assay Inokoshi et al. reported an assay system for the selectivity of DGAT isozymes using human DGAT1- or DGAT2-expressing yeasts [74]. Human DGAT1 and DAGT2 cDNAs were cloned from HeLa cells and a cDNA library of human liver, respectively, using PCR, and were introduced into the yeast expression vector pYES-DES52 by *in vitro* recombination. The constructed vectors were transformed into DGAT-deficient *Saccharomyces cerevisiae* to establish DGAT1- and DGAT2-expressing *S. cerevisiae*. It was confirmed that DGAT-deficient *S. cerevisiae* completely lack DGAT activity, while TG synthetic activity was fully recovered in the microsomes prepared from DGAT1- and DGAT2-expressing *S. cerevisiae*. Thus, it is possible to measure human DGAT1 and DGAT2 activities by utilizing these transformed yeasts. The activity of both DGATs was more potent than that of rat liver microsomes. Other groups have also established similar cell lines using insect cells (e.g., Sf9 or Hi5) in which DGAT1 and DGAT2 isozymes (e.g., human, rat, or mouse) were transfected [75,76].

Cell-Based Assay Cao et al. reported a cell-based assay for DGAT1 activity using human DGAT1-expressing HEK293A cells [77]. However, they showed no data about DGAT2 activity or the selectivity of inhibitors toward the isozymes in the cell-based assay. HEK293A has low endogenous DGAT activity. Human DGAT1 cDNA with an in-frame N-terminal FLAG tag was introduced into the mammalian expression vector pcDNA3,1(-)/Hygro by *in vitro* recombination. The constructed vector was transformed into HEK293A cells to establish DGAT1-expressing HEK293A cells. For the cell-based assay, DGAT1-expressing HEK293A cells were pretreated with an inhibitor for 1 h, followed by the addition of [^{14}C]oleic acid. After a 4-h incubation, [^{14}C]TG was mainly produced by the reaction of the DGAT1 isozyme. Similarly, another group also reported the establishment of human DGAT1-expressing Chinese hamster ovary (CHO) K1 cells and measured the radioactivity of [^{14}C]TG synthesized from [^{14}C]palmitic acid [78,79]. These two groups focused only on the inhibitory activity of compounds against DGAT1 without accounting for isozyme selectivity in the cell-based assay.

C. DGAT Inhibitors

As described earlier, DGAT inhibitors had been studied before there was understanding of the distinct functions of DGAT1 and DGAT2 isozymes. However, it is important to measure the selectivity of such inhibitors toward the isozymes. Accordingly, we reevaluated their inhibitory activities against the isozymes in the enzyme assay using DGAT1- and DGAT2-expressing *S. cerevisiae*. The structures and isozyme selectivity of DGAT inhibitors are summarized in Figure 2 and Table 2, respectively.

Amidepsines (**17–28**), isolated from the culture broth of fungal strains *Humicola* spp. FO-2942 [**80–82**] and FO-5969 [83], have a common skeleton of tridepside gyrophoric acid attached to an amino acid (alanine or valine) or a sugar moiety (D-glucose or 2-*O*-acetyl-D-glucose) [84]. Non-glycosylated amidepsines A to D (**17–20**) and J (26) inhibit both DGAT1 and DGAT2 activities with selectivity index (SI, log [IC_{50} for DGAT1]/[IC_{50} for DGAT2]) values of −0.20 to 0.00, whereas glycosylated amidepsines F to I (**22–25**) show very weak inhibitory activity against both isozymes [80]. Roselipins (**29–32**), isolated from the culture broth of fungus *Gliocladium roseum* KF-1040 [85–87], are glycolipids composed of a highly methylated fatty acid, a hexose, and alditol moieties. Interestingly, roselipins have been found to selectively inhibit the DGAT2 isozyme [74]. No other compounds tested to date have shown such DGAT2-selective inhibition (Table 2). Although a lot of herbal compounds have been isolated as DGAT inhibitors, their selectivity toward DGAT isozymes has rarely been reported. Xanthohumol (**33**), isolated from the methanol extract of hops *Humulus lupulus* (L.) [88], was reevaluated in the enzyme assay, revealing that the compound inhibits both DGAT1 and DGAT2 isozymes with similar potency [74].

Recently, a number of synthetic DGAT inhibitors have been reported to show DGAT1 selectivity. T863 (**34**) [77], a 2-phenyl-5-trifluoromethyloxazole-4-carboxamide derivative (**35**) [78], and a 2-phenyloxazole derivative (**36**) [79] selectively inhibit human DGAT1 isozyme in the enzyme assay with SI values of −2.74, −3.24, and −2.72, respectively. T863 (**34**) and a 2-phenyl-5-trifluoromethyloxazole-4-carboxamide derivative (**35**) inhibit mouse and rat DGAT1 isozymes, respectively [77,78]. Furthermore, these synthetic inhibitors (34–36) show inhibitory activity for TG synthesis in a cell-based assay using human DGAT1-overexpressing HEK293A or CHO K1 cells [77–79]. Interestingly, 2-phenyl-5-trifluoromethyloxazole-4-carboxamide derivative (**34**) treatment shows not only inhibition of weight gain in diet-induced obese (DIO) rats but also improved glucose tolerance in rats [78]. DGAT1-selective inhibitors may be effective against hypertriglyceridemia, obesity, NAFLD, and diabetes in humans [55,73,89].

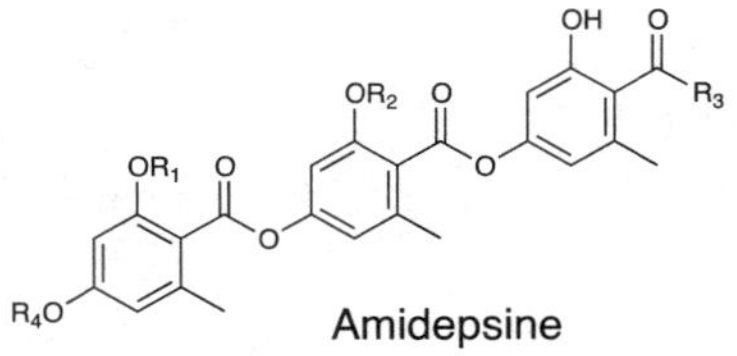

Amidepsine

Amidepsine	R_1	R_2	R_3	R_4
A (**17**)	-CH_3	-H	Ala	-CH_3
B (**18**)	-H	-H	Ala	-CH_3
C (**19**)	-H	-H	Val	-CH_3
D (**20**)	-CH_3	-H	-OH	-CH_3
E (**21**)	-CH_3	-CH_3	Ala	-CH_3
F (**22**)	-CH_3	Glc	Ala	-CH_3
G (**23**)	-H	Glc	Ala	-CH_3
H (**24**)	-CH_3	Acglc	Ala	-CH_3
I (**25**)	-H	Acglc	Ala	-CH_3
J (**26**)	-CH_3	-H	Val	-CH_3
K (**27**)	-H	-H	Ala	-CH_3
Gyr (**28**)	-H	-H	Ala	-H

Glc: D-glucose; Acglc: 2-O-acetyl-D-glucose
Ala: L-alanine; Val: L-valine; Gyr: gyrophoric acid

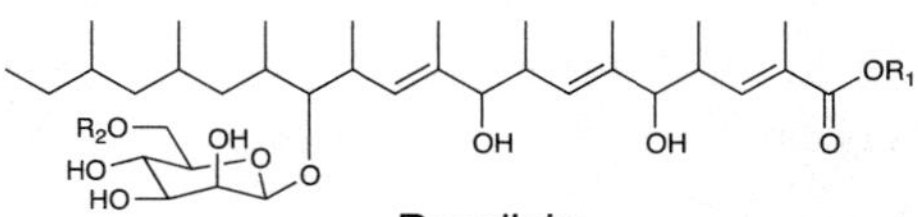

Roselipin

Roselipin	R_1	R_2
1A (**29**)	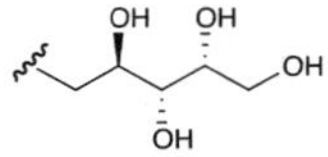	H
1B (**30**)	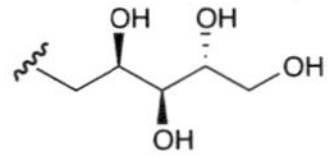	H
2A (**31**)	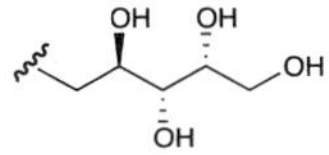	$COCH_3$
2B (**32**)	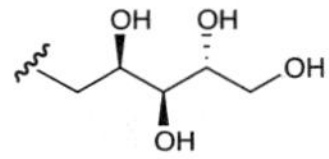	$COCH_3$

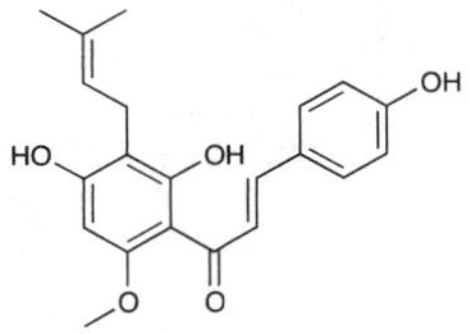

Xanthohumol (**33**)

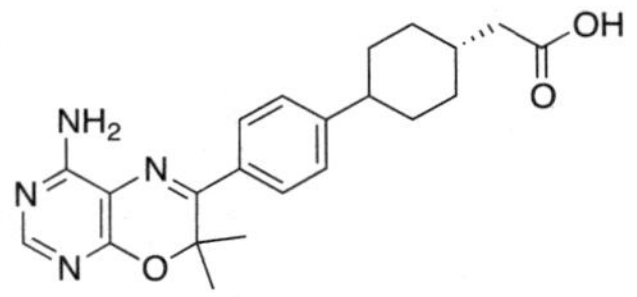

T863 (**34**)

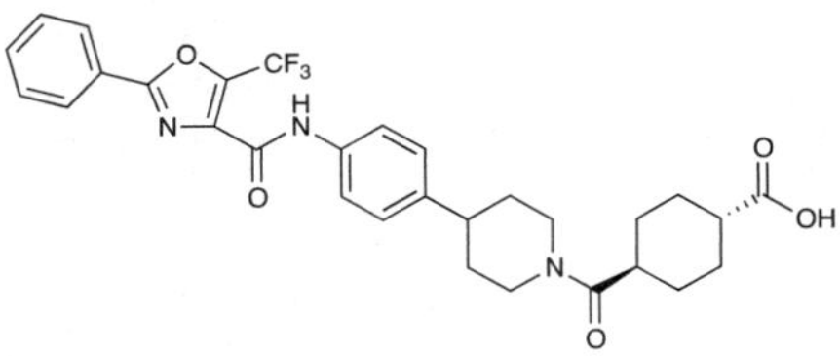

35

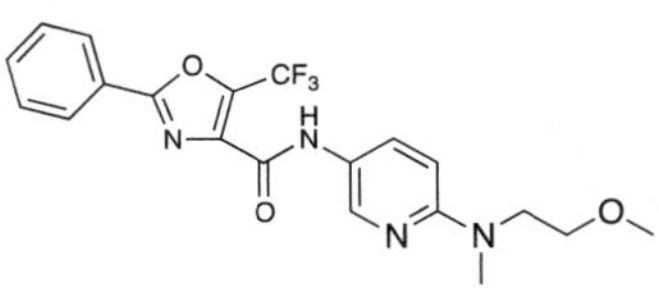

36

FIGURE 2 Representative DGAT inhibitors.

III. ACYL-COA: CHOLESTEROL ACYLTRANSFERASE (ACAT)

The enzyme ACAT (EC 2.3.1.26), which catalyzes the synthesis of CE from free cholesterol and long-chain fatty acyl-CoA, is considered to be a promising therapeutic target [90–95]. ACAT plays important roles in mammals. For example, it aids in the absorption of dietary cholesterol from the intestines, in the production of very low-density lipoprotein (VLDL) in the liver, and in the formation of foam cells from

TABLE 2 Selectivity of DGAT Inhibitors Toward DGAT1 and DGAT2 Isozymes

Inhibitor	Assay system[a]	Species[b]	IC_{50} (μM) DGAT1	IC_{50} (μM) DGAT2	SI[c]	References[d]
Microbial origin						
Amidepsine A (**17**)	E	Hu	40	70	−0.24	74, 82
Amidepsine B (**18**)	E	Hu	30	60	−0.30	74, 82
Amidepsine C (**19**)	E	Hu	170	170	+0.00	74, 82
Amidepsine D (**20**)	E	Hu	20	30	−0.18	74, 82
Amidepsine E (**21**)	E	Hu	—	—	—	74, 82
Amidepsine F (**22**)	E	Hu	>200	>200	—	82
Amidepsine G (**23**)	E	Hu	>200	>200	—	82
Amidepsine H (**24**)	E	Hu	>200	>200	—	82
Amidepsine I (**25**)	E	Hu	>200	>200	—	82
Amidepsine J (**26**)	E	Hu	40	40	+0.00	82
Amidepsine K (**27**)	E	Hu	>200	>200	—	82
Gyrophoric acid (**28**)	E	Hu	>100	50	> +0.30	82
Roselipin 1A (**29**)	E	Hu	>200	30	> +0.82	74
Roselipin 1B (**30**)	E	Hu	>200	40	> +0.70	74
Roselipin 2A (**31**)	E	Hu	>200	45	> +0.65	74
Roselipin 2B (**32**)	E	Hu	>200	50	> +0.60	74
Herb origin						
Xanthohumol (**33**)	E	Hu	40	40	+0.00	74
Synthetic origin						
T863 (**34**)	E	Hu	0.018	>10	−2.74	77
	E	Mo	0.009	ND	—	77
	C	Hu	0.122	ND	—	77
35	E	Hu	0.057	>100	−3.24	78
	C	Hu	0.493	ND	—	78
36	E	Hu	0.038	>20	−2.72	79
	E	Rt	0.120	>20	−2.22	79
	C	Hu	0.660	ND	—	79

ND: no data.
[a]E, enzyme assay; C, cell-based assay.
[b]Hu, human; Rt, rat; Mo, mouse.
[c]Selectivity index (SI): log [IC_{50} for DGAT1/IC_{50} for DGAT2].
[d]IC_{50} values are cited from references.

macrophages in the arterial wall [58]. Therefore, ACAT inhibition is expected not only to lower plasma cholesterol levels but also to have a direct effect at the arterial wall. Thus, many pharmaceutical companies have great interest in ACAT as a promising target enzyme for the development of novel anti-atherosclerotic agents. Accordingly, a large number of synthetic ACAT inhibitors, containing ureas, imidazoles, and amide moieties, have been reported [95]. Unfortunately, no inhibitors have been clinically used so far.

A. ACAT1 and ACAT2 Isozymes

Two ACAT isozymes, ACAT1 and ACAT2, have been identified in various mammals. Chang's group succeeded in cloning a cDNA for the first ACAT gene (ACAT1) from human monocytes [96]. The second ACAT gene (ACAT2) was then cloned from the liver and intestine by three different groups [63,97–98]. The human ACAT1 gene encodes a 550-amino acid protein, while the human ACAT2 gene encodes proteins of 522 amino acids. Human ACAT2 is about 40% identical to human ACAT1. Both ACAT isozymes are membrane proteins localized in the ER and are assumed to possess several hydrophobic domains that can penetrate the ER membrane.

Since the discovery of the ACAT1 and ACAT2 isozymes, data from extensive genetic studies, including knockout mice, have indicated that the two isozymes have distinct functions. ACAT1 is ubiquitous and is expressed most strongly in the sebaceous glands, steroidogenic tissues, and macrophages, while ACAT2 is expressed in the liver and intestine. Controversy has existed over which isozyme is predominant in the human liver (hepatocytes and Kupffer cells). However, a recent study using human liver biopsy samples and an ACAT2-selective inhibitor (pyripyropene A (PPPA), as described in the following text) showed that ACAT2 is expressed as the dominant ACAT isozyme in hepatocytes, whereas ACAT1 is expressed only in Kupffer cells [99]. To further understand the roles of the ACAT1 isozyme, ACAT1 knockout mice ($Acat1^{-/-}$ mice) were created by two different groups [100–102]. ACAT1-deficient ($Acat1^{-/-}$) mice are viable and fertile, and ACAT activity is present in the liver and intestines, but completely absent in the adrenal cortex and peritoneal macrophages. However, in addition to having dry eye syndrome [100], $Acat1^{-/-}$ mice lacking apolipoprotein E or the low-density lipoprotein receptor, as atherogenic mouse models, are susceptible to atherosclerosis and have extensive cutaneous xanthomatosis with loss of hair [101,102]. Free cholesterol content is significantly increased in their skin. In contrast, ACAT activity is completely absent in the liver and intestines of ACAT2-deficient ($Acat2^{-/-}$) mice, and these mice are completely resistant to diet-induced hypercholesterolemia and cholesterol gallstone formation [103,104].

Studies on these knockout mice have indicated that the ACAT1 isozyme contributes greatly to macrophage-derived foam cell formation in atherosclerosis and that the ACAT2 isozyme is responsible for the absorption of cholesterol from the intestines and lipoprotein production by the liver. However, after the failure of clinical trials of the nonselective ACAT inhibitors avasimibe (**46**) [105] and pactimibe (**47**) [106,107], Fazio and Linton observed [108] that inhibition of ACAT1 activity leads to cytotoxicity toward macrophages and vascular cells because of excessive accumulation of free cholesterol in these cells [101,102,109]. On the other hand, ACAT2-selective inhibition may be a useful strategy for the treatment of atherosclerosis and hypercholesterolemia [109–112].

B. Assay for ACAT Isozyme Activity

Many ACAT inhibitors have been discovered without a complete understanding of the ACAT1 and ACAT2 isozymes as described earlier. Therefore, the selectivity of ACAT inhibitors toward the two isozymes is important for their development as novel anti-atherosclerotic and anti-hypercholesterolemic agents. However, only a few studies have focused on this selectivity so far [113–115].

Enzyme Assay Generally, most groups have established similar animal-derived cell lines (e.g., CHO, Sf9, or Hi5) in which mammalian ACAT1 and ACAT2 isozymes have been transfected, and have evaluated the selectivity of inhibitors toward ACAT1 and ACAT2 isozymes in enzyme assays using the microsome fractions prepared from ACAT1- and ACAT2-expressing cells [115–117]. For example, Lada et al. established ACAT1- or ACAT2-expressing CHO cells (designated ACAT1-CHO and ACAT2-CHO). ACAT1 and ACAT2 genes from human or African green monkey (AGM) were transfected into the ACAT-deficient CHO cell line, AC29 [113]. Briefly, each ACAT cDNA was introduced into the mammalian expression vector pOPSRV1 by *in vitro* recombination, and the constructed vectors were transformed into AC29 cells to establish ACAT1-CHO and ACAT2-CHO cells. African green monkey ACAT1 and ACAT2 isozymes have a high degree of homology with the human enzymes (98.4% for ACAT1 and 96% for ACAT2). Since AGM ACAT isozymes show higher stability than human ACAT isozymes [99,113], we have mainly used AGM ACAT1- and ACAT2-CHO cells for our experiments. In an enzyme assay, the microsome fractions were prepared from ACAT1- and ACAT2-CHO cells as enzyme sources and used to measure each ACAT activity by quantifying [^{14}C]CE from the substrate [^{14}C]oleoyl-CoA [114].

Cell-Based Assay We established a cell-based assay using ACAT1- and ACAT2-CHO cells and the substrate [^{14}C]oleic acid [114]. In this cell-based assay, [^{14}C]CE was produced by the reaction catalyzed by ACAT1 or ACAT2 isozymes, because the parental AC29 cells cannot synthesize CE.

C. ACAT Inhibitors

In the 1980s, researchers had great hopes for ACAT as an ideal therapeutic target for atherosclerosis and hypercholesteremia, and ACAT inhibitors were screened from natural sources or synthetic compounds in enzyme or cell-based assays [53,54, 93–95]. As a result, a large number of ACAT inhibitors were discovered and reported, but development of most of these inhibitors for clinical use was terminated in the 1990s before our current understanding of the presence of ACAT isozymes with distinct functions in mammals. Therefore, we reevaluated the selectivity of ACAT available inhibitors (Fig. 3) toward the two isozymes in the enzyme and cell-based assays. The IC_{50} values of inhibitors against ACAT1 and ACAT2 are summarized in Table 3. The data obtained from the cell-based assays using monkey and human ACAT1- and ACAT2-CHO cells appeared essentially comparable. Furthermore, the IC_{50} values of most inhibitors, except beauveriolides, obtained from the enzyme assay using microsomes were similar to those from the cell-based assay.

As shown in Table 3, most ACAT inhibitors (purpactin A (**38**) [118–120], terpendole C (**39**) [121–123], glisoprenin A (**40**) [124–126], beauvericin (**41**) [127], spylidone (**44**) [128], CL-283,546 (**45**) [129], avasimibe (**46**) [130], pactimibe (**47**) [116], and esculeogenin A (**50**) [117]) were nonselective, inhibiting both ACAT1 and ACAT2 isozymes with similar potency ($-1.00 \leq$ SI (log [IC_{50} for ACAT1]/[IC_{50} for ACAT2]) $\leq +1.00$) in the enzyme and cell-based assays. In contrast, Wu-V-23 (**48**), K-604 (**49**), and fungal beauveriolides I (**42**) and III (**43**) [131–133] appeared to be ACAT1-selective inhibitors with SI values of −2.18, −2.36, −1.52, and < −1.35, respectively [113–115]. Interestingly, the enzyme assay using microsomes indicated

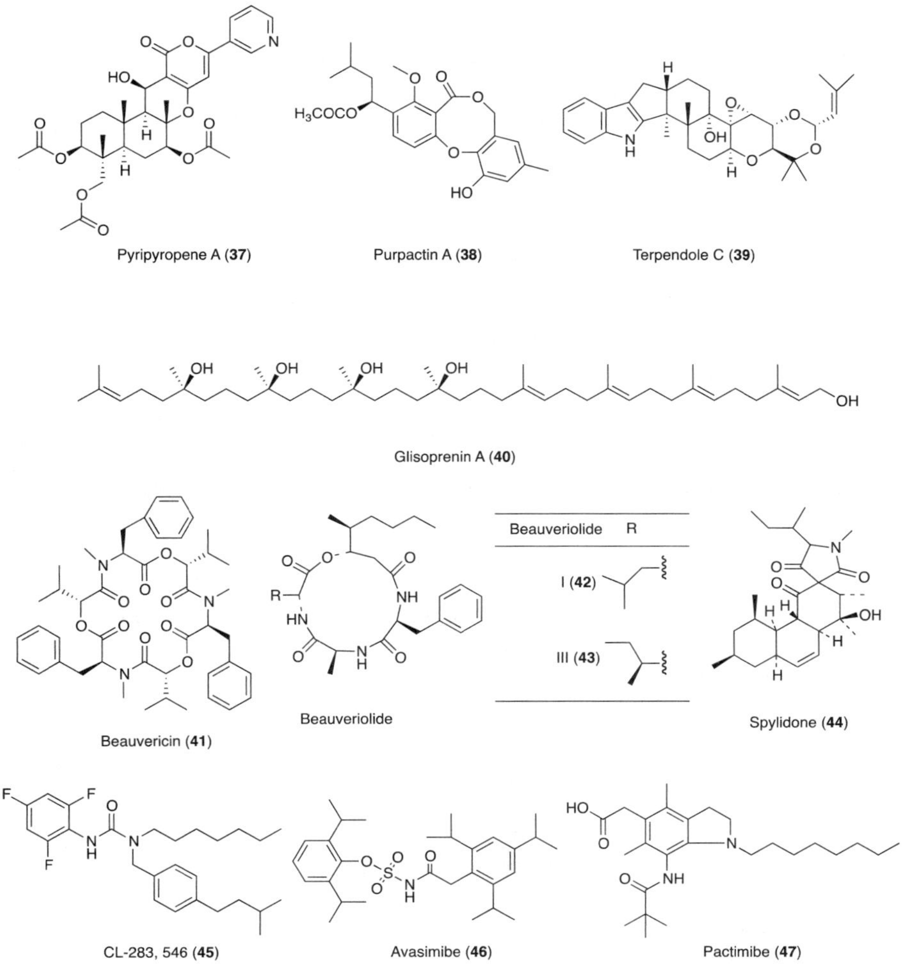

FIGURE 3 Representative ACAT inhibitors.

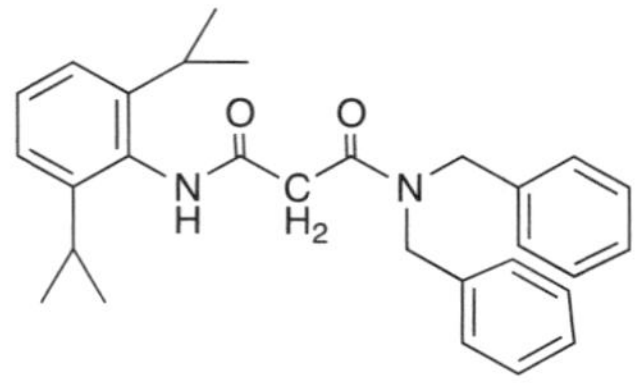

Wu-V-23 (**48**)

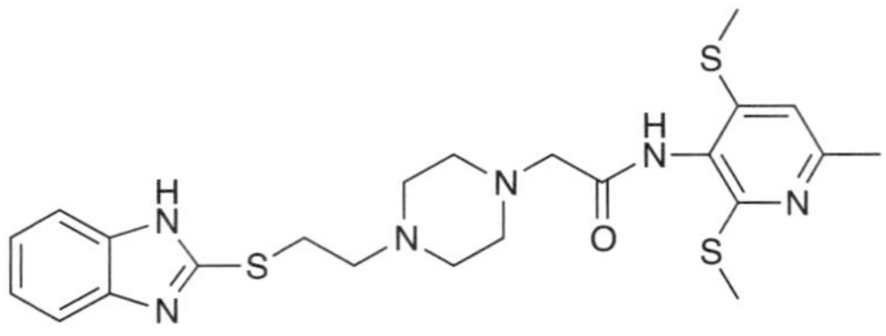

K-604 (**49**)

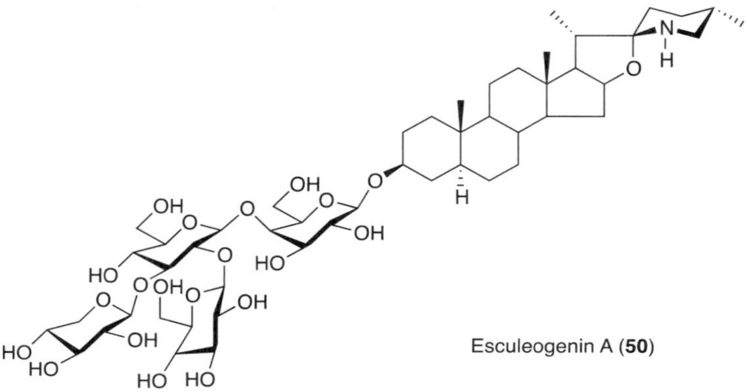

Esculeogenin A (**50**)

FIGURE 3 (*Continued*)

that beauveriolides inhibited both ACAT1 and ACAT2 isozymes almost equally (SI, 0.06 and 0.00), in accordance with an enzyme assay using mouse liver (expressing ACAT2) microsomes and mouse macrophage (expressing ACAT1) microsomes as previously reported [134]. ACAT isozymes are assumed to possess several hydrophobic domains that can penetrate the ER membrane. However, no consensus exists on the structural model of the proteins because of the difficulty of their purification and lack of X-ray crystallographic analyses, as well as the lack of structural information for similar proteins. Several groups have described the key active site residues (consensus motif) of each isozyme [135–138]. It is commonly agreed that the highly conserved FYXDWWN motif is essential for ACAT activity. Das et al. reported that the FYXDWWN motif of the ACAT1 isozyme is located in the cytoplasm, while that of ACAT2 isozyme is located in the transmembrane [138]. On the other hand, Chang et al. reported that those of both isozymes are located in the cytoplasm [136,137]. The different inhibition of beauveriolides toward ACAT isozymes between the cell-based and enzyme assays might be due to the different distribution of the active site in the ER between the two isozymes. Further study is necessary to clarify the structure and active site topology of the isozymes.

Among the ACAT inhibitors tested, fungal PPPA (37) strongly and selectively inhibited the ACAT2 isozyme (IC_{50}, 70 nM for AGM ACAT2 and 59 nM for human ACAT2 isozyme; SI value, > +3.06 and +3.13, respectively) [114]. Notably, no

TABLE 3 Selectivity of ACAT Inhibitors Toward ACAT1 and ACAT2 Isozymes

Inhibitor	Assay system[a]	Species[b]	IC_{50} (μM) ACAT1	ACAT2	SI[c]	References[d]
Microbial origin						
Pyripyropene A (**37**)	E	AGM	>30	0.060	> +2.63	114
	C	AGM	>80	0.070	> +3.06	114
	C	Hu	>80	0.059	> +3.13	114
Purpactin A (**38**)	E	AGM	0.9	1.8	−0.30	114
	C	AGM	2.5	1.5	0.23	114
Terpendole C (**39**)	E	AGM	8.0	0.5	1.20	114
	C	AGM	10	10	0.00	114
Glisoprenin A (**40**)	E	AGM	0.4	1.3	−0.51	114
	C	AGM	4.3	10	−0.37	114
Beauvericin (**41**)	E	AGM	0.35	0.35	0.00	114
	C	AGM	2.0	0.70	0.46	114
Beauveriolide I (**42**)	E	AGM	2.2	1.9	0.06	114
	C	AGM	0.60	20	−1.52	114
	C	Hu	0.61	>20	< −1.52	114
Beauveriolide III (**43**)	E	AGM	3.0	3.0	0.00	114
	C	AGM	0.90	>20	<−1.35	114
	C	Hu	0.51	>20	<−1.59	114
Spylidone (**44**)	E	AGM	7.0	3.0	0.37	114
	C	AGM	25	5.0	0.70	114
Synthetic origin						
CL-283,546 (**45**)	E	AGM	0.12	0.060	+0.30	114
	C	AGM	0.10	0.090	+0.05	114
Avasimibe (**46**)	E	Hu	24	9.2	+0.41	115
Pactimibe (**47**)	E	Hu	4.9	3.0	+0.21	116
Wu-V-23 (**48**)	E	AGM	0.010	1.5	−2.18	113
K-604 (**49**)	E	Hu	0.45	102	−2.36	115
Herb origin						
Esculeogenin A (**50**)	E	Hu	About 50% at 100 μM	About 50% at 100 μM	–	117

[a]E, enzyme assay; C, cell-based assay.
[b]AGM, African green monkey; Hu, human.
[c]Selectivity index (SI): log [IC_{50} for ACAT1/IC_{50} for ACAT2].
[d]IC_{50} values are cited from references.

ACAT2-selective inhibitor other than PPPA has been reported to date. Recently, PPPA proved orally active in atherogenic mouse models, reducing total cholesterol in plasma and atherosclerosis lesions without causing side effects [139]. Therefore, PPPA is expected to be a novel lead compound for the treatment of atherosclerosis and hypercholesterolemia, including homozygous familial hypercholesterolemia (FH) [56].

IV. CONCLUSIONS AND FUTURE PERSPECTIVES

Recent advances in gene technology and molecular biology have provided researchers with new strategies and methodologies for drug discovery. To date, many isozymes have been identified. Mammalian cells or microorganisms are easily manipulated to overexpress target isozymes. Such cellular models with overexpressing of the enzyme of interest or recombinant enzymes are easily utilized in automatic robotic or high-throughput screening (HTS) systems for drug discovery.

In the field of lipid metabolism, statins are the standard of care for the treatment of hypercholesterolemia and atherosclerosis [50]. However, they have several problems, in addition to only reducing the risk of complications and death from CV events by 30% [51]. Recently, the US Food and Drug Administration (FDA) reported an increased risk of muscle injury (myopathy and rhabdomyolysis) in patients treated with high-dose statins [140,141]. Furthermore, homozygous FH patients have reduced efficacy on statin therapy compared with heterozygous FH patients and other hyperlipidemia patients [143–145]. Therefore, the search for novel pharmacological agents with a specific target different from HMG-CoA reductase is of interest [48,145]. Thus, the ACAT2 isozyme is a potential target for the treatment of atherosclerosis and hypercholesterolemia, including homozygous FH, since PPPA shows *in vivo* efficacy using atherogenic mice models without causing side effects [139], in addition to studies of knockout mice and liver-specific knockdown of ACAT2 using an antisense oligonucleotide approach [103,104,146].

Pyripyropenes, including natural products and semisynthetic derivatives, are the first and only small molecules that show high ACAT2-selective inhibition [114,147]. Among them, PPPA exhibits the most selective ACAT2 inhibition (Fig. 3 and Table 3). Therefore, we have revisited the synthesis of PPPA derivatives to identify more potent and highly ACAT2-selective inhibitors. Recently, certain PPPA derivatives with more potent and selective inhibition of the ACAT2 isozyme than PPPA have been developed, and an *in vivo* study is underway [56]. Hopefully, some of these agents will proceed to clinical trials to provide new types of anti-atherosclerosis and anti-hypercholesterolemia therapies. Additionally, PPPA is highly useful in basic research for the study of ACAT isozymes. Das et al. reported that PPPA is non-covalently bound to a specific site (Gln492-Val493-Ser494) of ACAT2 isozymes [148].

In conclusion, it is important to determine the selectivity of inhibitors toward different isozymes, which can lead to new drug discoveries or useful probes to study isozyme function.

REFERENCES

1. D Georgiou, BH Brundage. Regression of left ventricular mass in systemic hypertension. Clin Cardiol 15:5–16, 1992.
2. JR Vane. Inhibition of prostaglandin synthesis as a mechanism of action for aspirin-like drugs. Nat New Biol 231:232–235, 1971.

3. T Miyamoto, N Ogino, S Yamamoto, O Hayaishi. Purification of prostaglandin endoperoxide synthetase from bovine vesicular gland microsomes. J Biol Chem 251:2629–2636, 1976.

4. DL DeWitt, WL Smith. Primary structure of prostaglandin G/H synthase from sheep vesicular gland determined from the complementary DNA sequence. Proc Natl Acad Sci USA 85:1412–1416, 1988.

5. JP Merlie, D Fagan, J Mudd, P Needleman. Isolation and characterization of the complementary DNA for sheep seminal vesicle prostaglandin endoperoxide synthase (cyclooxygenase). J Biol Chem 263:3550–3553, 1988.

6. C Yokoyama, T Takai, T Tanabe. Primary structure of sheep prostaglandin endoperoxide synthase deduced from cDNA sequence. FEBS Lett 231:347–351, 1988.

7. JL Masferrer, BS Zweifel, K Seibert, P Needleman. Selective regulation of cellular cyclooxygenase by dexamethasone and endotoxin in mice. J Clin Invest 86:1375–1379, 1990.

8. CE Eberhart, RN Dubois. Eicosanoids and the gastrointestinal tract. Gastroenterology 109:285–301, 1995.

9. DL Simmons, RM Botting, T Hla. Cyclooxygenase isozymes: the biology of prostaglandin synthesis and inhibition. Pharmacol Rev 56:387–437, 2004.

10. TD Penning, JJ Talley, SR Bertenshaw, JS Carter, PW Collins, S Docter, MJ Graneto, LF Lee, JW Malecha, JM Miyashiro, et al. Synthesis and biological evaluation of the 1,5-diarylpyrazole class of cyclooxygenase-2 inhibitors: identification of 4-[5-(4-methylphenyl)-3-(trifluoromethyl)-1H-pyrazol-1-yl]benze nesulfonamide (SC-58635, celecoxib). J Med Chem 40:1347–1365, 1997.

11. HC Yang, M Mosior, E Dennis. Design and development of a selective assay system for the phospholipase A2 superfamily. In: HA Kirst, WK Yeh, MJ Zmijewski, Jr., eds. *Enzyme Technology for Pharmacological and Biotechnological Application*. New York: Marcel Dekker Inc., 2001, pp. 379–396.

12. JE Burke, EA Dennis. Phospholipase A2 structure/function, mechanism, and signaling. J Lipid Res 50(Suppl.):S237–S242, 2009.

13. VJ Laine, DS Grass, TJ Nevalainen. Protection by group II phospholipase A2 against Staphylococcus aureus. J Immunol 162:7402–7408, 1999.

14. B Balestrieri, A Maekawa, W Xing, MH Gelb, HR Katz, JP Arm. Group V secretory phospholipase A2 modulates phagosome maturation and regulates the innate immune response against Candida albicans. J Immunol 182:4891–4898, 2009.

15. NM Munoz, AY Meliton, JP Arm, JV Bonventre, W Cho, AR Leff. Deletion of secretory group V phospholipase A2 attenuates cell migration and airway hyperresponsiveness in immunosensitized mice. J Immunol 179:4800–4807, 2007.

16. WR Henderson, Jr., EY Chi, JG Bollinger, YT Tien, X Ye, L Castelli, YP Rubtsov, AG Singer, GK Chiang, T Nevalainen, et al. Importance of group X-secreted phospholipase A2 in allergen-induced airway inflammation and remodeling in a mouse asthma model. J Exp Med 204:865–877, 2007.

17. MA Bostrom, BB Boyanovsky, CT Jordan, MP Wadsworth, DJ Taatjes, FC de Beer, NR Webb. Group v secretory phospholipase A2 promotes atherosclerosis: evidence from genetically altered mice. Arterioscler Thromb Vasc Biol 27:600–606, 2007.

18. M Murakami, Y Taketomi, H Sato, K Yamamoto. Secreted phospholipase A2 revisited. J Biochem 150:233–255, 2011.

19. H Fraser, C Hislop, RM Christie, HL Rick, CA Reidy, ML Chouinard, PI Eacho, KE Gould, J Trias. Varespladib (A-002), a secretory phospholipase A2 inhibitor, reduces atherosclerosis and aneurysm formation in ApoE-/- mice. J Cardiovasc Pharmacol 53:60–65, 2009.
20. JO Leite, U Vaishnav, M Puglisi, H Fraser, J Trias, ML Fernandez. A-002 (Varespladib), a phospholipase A2 inhibitor, reduces atherosclerosis in guinea pigs. BMC Cardiovasc Disord 9:7, 2009.
21. RL Wilensky, Y Shi, ER Mohler, 3rd, D Hamamdzic, ME Burgert, J Li, A Postle, RS Fenning, JG Bollinger, BE Hoffman, DJ Pelchovitz, J Yang, RC Mirabile, CL Webb, L Zhang, P Zhang, MH Gelb, MC Walker, A Zalewski, CH Macphee. Inhibition of lipoprotein-associated phospholipase A2 reduces complex coronary atherosclerotic plaque development. Nat Med 14:1059–1066, 2008.
22. RS Rosenson, C Hislop, D McConnell, M Elliott, Y Stasiv, N Wang, DD Waters. Effects of 1-H-indole-3-glyoxamide (A-002) on concentration of secretory phospholipase A2 (PLASMA study): a phase II double-blind, randomised, placebo-controlled trial. Lancet 373:649–658, 2009.
23. RS Rosenson, C Hislop, M Elliott, Y Stasiv, M Goulder, D Waters. Effects of varespladib methyl on biomarkers and major cardiovascular events in acute coronary syndrome patients. J Am Coll Cardiol 56:1079–1088, 2010.
24. RS Rosenson, M Elliott, Y Stasiv, C Hislop. Randomized trial of an inhibitor of secretory phospholipase A2 on atherogenic lipoprotein subclasses in statin-treated patients with coronary heart disease. Eur Heart J 32:999–1005, 2011.
25. PW Serruys, HM Garcia-Garcia, P Buszman, P Erne, S Verheye, M Aschermann, H Duckers, O Bleie, D Dudek, HE Botker, et al. Effects of the direct lipoprotein-associated phospholipase A(2) inhibitor darapladib on human coronary atherosclerotic plaque. Circulation 118:1172–1182, 2008.
26. ER Mohler, 3rd, CM Ballantyne, MH Davidson, M Hanefeld, LM Ruilope, JL Johnson, A Zalewski. The effect of darapladib on plasma lipoprotein-associated phospholipase A2 activity and cardiovascular biomarkers in patients with stable coronary heart disease or coronary heart disease risk equivalent: the results of a multicenter, randomized, double-blind, placebo-controlled study. J Am Coll Cardiol 51:1632–1641, 2008.
27. JA Beavo. Cyclic nucleotide phosphodiesterases: functional implications of multiple isoforms. Physiol Rev 75:725–748, 1995.
28. SH Soderling, JA Beavo. Regulation of cAMP and cGMP signaling: new phosphodiesterases and new functions. Curr Opin Cell Biol 12:174–179, 2000.
29. M Conti, G Nemoz, C Sette, E Vicini. Recent progress in understanding the hormonal regulation of phosphodiesterases. Endocr Rev 16:370–389, 1995.
30. C Mehats, CB Andersen, M Filopanti, SL Jin, M Conti. Cyclic nucleotide phosphodiesterases and their role in endocrine cell signaling. Trends Endocrinol Metab 13:29–35, 2002.
31. S Uckert, M Oelke. Phosphodiesterase (PDE) inhibitors in the treatment of lower urinary tract dysfunction. Br J Clin Pharmacol 72:197–204, 2011.
32. TJ Torphy. Phosphodiesterase isozymes: molecular targets for novel antiasthma agents. Am J Respir Crit Care Med 157:351–370, 1998.
33. R Gupta, G Kumar, RS Kumar. An update on cyclic nucleotide phosphodiesterase (PDE) inhibitors: phosphodiesterases and drug selectivity. Methods Find Exp Clin Pharmacol 27:101–118, 2005.

34. L Fawcett, R Baxendale, P Stacey, C McGrouther, I Harrow, S Soderling, J Hetman, JA Beavo, SC Phillips. Molecular cloning and characterization of a distinct human phosphodiesterase gene family: PDE11A. Proc Natl Acad Sci USA 97:3702–3707, 2000.

35. H Wang, Y Liu, Y Chen, H Robinson, H Ke. Multiple elements jointly determine inhibitor selectivity of cyclic nucleotide phosphodiesterases 4 and 7. J Biol Chem 280:30949–30955, 2005.

36. JD Corbin, SH Francis. Cyclic GMP phosphodiesterase-5: target of sildenafil. J Biol Chem 274:13729–13732, 1999.

37. D Kouvelas, A Goulas, G Papazisis, C Sardeli, C Pourzitaki. PDE5 inhibitors: in vitro and in vivo pharmacological profile. Curr Pharm Des 15:3464–3475, 2009.

38. BE Jaski, MA Fifer, RF Wright, E Braunwald, WS Colucci. Positive inotropic and vasodilator actions of milrinone in patients with severe congestive heart failure. Dose–response relationships and comparison to nitroprusside. J Clin Invest 75:643–649, 1985.

39. MD Gorrell. Dipeptidyl peptidase IV and related enzymes in cell biology and liver disorders. Clin Sci (Lond) 108:277–292, 2005.

40. SH Havale, M Pal. Medicinal chemistry approaches to the inhibition of dipeptidyl peptidase-4 for the treatment of type 2 diabetes. Bioorg Med Chem 17:1783–1802, 2009.

41. JJ Neumiller. Differential chemistry (structure), mechanism of action, and pharmacology of GLP-1 receptor agonists and DPP-4 inhibitors. J Am Pharm Assoc (2003) 49(Suppl. 1):S16–S29, 2009.

42. CF Deacon. Dipeptidyl peptidase-4 inhibitors in the treatment of type 2 diabetes: a comparative review. Diabetes Obes Metab 13:7–18, 2011.

43. D Kim, L Wang, M Beconi, GJ Eiermann, MH Fisher, H He, GJ Hickey, JE Kowalchick, B Leiting, K Lyons, F Marsilio, ME McCann, RA Patel, A Petrov, G Scapin, SB Patel, RS Roy, JK Wu, MJ Wyvratt, BB Zhang, L Zhu, NA Thornberry, AE Weber. (2R)-4-oxo-4-[3-(trifluoromethyl)-5,6-dihydro[1,2,4]triazolo[4,3-a]pyrazin-7(8H)-yl]-1-(2,4,5-trifluorophenyl) butan-2-amine: a potent, orally active dipeptidyl peptidase IV inhibitor for the treatment of type 2 diabetes. J Med Chem 48:141–151, 2005.

44. J Feng, Z Zhang, MB Wallace, JA Stafford, SW Kaldor, DB Kassel, M Navre, L Shi, RJ Skene, T Asakawa, K Takeuchi, R Xu, DR Webb, SL Gwaltney, 2nd. Discovery of alogliptin: a potent, selective, bioavailable, and efficacious inhibitor of dipeptidyl peptidase IV. J Med Chem 50:2297–2300, 2007.

45. M Eckhardt, E Langkopf, M Mark, M Tadayyon, L Thomas, H Nar, W Pfrengle, B Guth, R Lotz, P Sieger, H Fuchs, F Himmelsbach. 8-(3-(R)-aminopiperidin-1-yl)-7-but-2-ynyl-3-methyl-1-(4-methyl-quinazolin –2-ylmethyl)-3,7-dihydropurine-2,6-dione (BI 1356), a highly potent, selective, long-acting, and orally bioavailable DPP-4 inhibitor for the treatment of type 2 diabetes. J Med Chem 50:6450–6453, 2007.

46. I Brandt, J Joossens, X Chen, MB Maes, S Scharpe, I De Meester, AM Lambeir. Inhibition of dipeptidyl-peptidase IV catalyzed peptide truncation by Vildagliptin ((2S)-{[(3-hydroxyadamantan-1-yl)amino]acetyl}-pyrrolidine-2-carbonitrile). Biochem Pharmacol 70:134–143, 2005.

47. YB Kim, LM Kopcho, MS Kirby, LG Hamann, CA Weigelt, WJ Metzler, J Marcinkeviciene. Mechanism of Gly-Pro-pNA cleavage catalyzed by dipeptidyl peptidase-IV and its inhibition by saxagliptin (BMS-477118). Arch Biochem Biophys 445:9–18, 2006.

48. P Libby, PM Ridker, GK Hansson. Progress and challenges in translating the biology of atherosclerosis. Nature 473:317–325, 2011.

49. JP Despres, I Lemieux. Abdominal obesity and metabolic syndrome. Nature 444:881–887, 2006.

50. A Endo. A historical perspective on the discovery of statins. Proc Jpn Acad Ser B Phys Biol Sci 86:484–493, 2010.

51. P Libby. The forgotten majority: unfinished business in cardiovascular risk reduction. J Am Coll Cardiol 46:1225–1228, 2005.

52. H Tomoda, S Omura. Screening for inhibitors of lipid metabolism. In: HA Kirst, WK Yeh, MJ Zmijewski, Jr., eds. *Enzyme Technology for Pharmacological and Biotechnological Application*. New York: Marcel Dekker Inc., 2001, pp. 343–378.

53. H Tomoda, I Namatame, S Omura. Microbial metabolites with inhibitory activity against lipid metabolism. Proc Jpn Acad Ser B 78:217–240, 2002.

54. H Tomoda, S Omura. Potential therapeutics for obesity and atherosclerosis: inhibitors of neutral lipid metabolism from microorganisms. Pharmacol Ther 115:375–389, 2007.

55. D Matsuda, H Tomoda. DGAT inhibitors for obesity. Curr Opin Investig Drugs 8:836–841, 2007.

56. T Ohshiro, H Tomoda. Isoform-specific inhibitors of ACATs: recent advances and promising developments. Future Med Chem 3:2039–2061, 2011.

57. CL Yen, SJ Stone, S Koliwad, C Harris, RV Farese, Jr. Thematic review series: glycerolipids. DGAT enzymes and triacylglycerol biosynthesis. J Lipid Res 49:2283–2301, 2008.

58. LL Rudel, RG Lee, TL Cockman. Acyl coenzyme A: cholesterol acyltransferase types 1 and 2: structure and function in atherosclerosis. Curr Opin Lipidol 12:121–127, 2001.

59. R Lehner, A Kuksis. Biosynthesis of triacylglycerols. Prog Lipid Res 35:169–201, 1996.

60. SH Choi, HN Ginsberg. Increased very low density lipoprotein (VLDL) secretion, hepatic steatosis, and insulin resistance. Trends Endocrinol Metab 22:353–363, 2011.

61. RM Bell, RA Coleman. Enzymes of triacylglycerol formation in mammals. Enzymes 16:87–111, 1983.

62. S Cases, SJ Smith, YW Zheng, HM Myers, SR Lear, E Sande, S Novak, C Collins, CB Welch, AJ Lusis, et al. Identification of a gene encoding an acyl CoA:diacylglycerol acyltransferase, a key enzyme in triacylglycerol synthesis. Proc Natl Acad Sci USA 95:13018–13023, 1998.

63. P Oelkers, A Behari, D Cromley, JT Billheimer, SL Sturley. Characterization of two human genes encoding acyl coenzyme A:cholesterol acyltransferase-related enzymes. J Biol Chem 273:26765–26771, 1998.

64. PJ McFie, SL Stone, SL Banman, SJ Stone. Topological orientation of acyl-CoA:diacylglycerol acyltransferase-1 (DGAT1) and identification of a putative active site histidine and the role of the n terminus in dimer/tetramer formation. J Biol Chem 285:37377–37387, 2010.

65. HR Wurie, L Buckett, VA Zammit. Evidence that diacylglycerol acyltransferase 1 (DGAT1) has dual membrane topology in the endoplasmic reticulum of HepG2 cells. J Biol Chem 286:36238–36247, 2011.

66. S Cases, SJ Stone, P Zhou, E Yen, B Tow, KD Lardizabal, T Voelker, RV Farese, Jr. Cloning of DGAT2, a second mammalian diacylglycerol acyltransferase, and related family members. J Biol Chem 276:38870–38876, 2001.

67. KD Lardizabal, JT Mai, NW Wagner, A Wyrick, T Voelker, DJ Hawkins. DGAT2 is a new diacylglycerol acyltransferase gene family: purification, cloning, and expression in insect cells of two polypeptides from Mortierella ramanniana with diacylglycerol acyltransferase activity. J Biol Chem 276:38862–38869, 2001.
68. SJ Stone, MC Levin, RV Farese, Jr. Membrane topology and identification of key functional amino acid residues of murine acyl-CoA:diacylglycerol acyltransferase-2. J Biol Chem 281:40273–40282, 2006.
69. SJ Stone, HM Myers, SM Watkins, BE Brown, KR Feingold, PM Elias, RV Farese, Jr. Lipopenia and skin barrier abnormalities in DGAT2-deficient mice. J Biol Chem 279:11767–11776, 2004.
70. SJ Smith, S Cases, DR Jensen, HC Chen, E Sande, B Tow, DA Sanan, J Raber, RH Eckel, RV Farese, Jr. Obesity resistance and multiple mechanisms of triglyceride synthesis in mice lacking Dgat. Nat Genet 25:87–90, 2000.
71. HC Chen, SJ Smith, B Tow, PM Elias, RV Farese, Jr. Leptin modulates the effects of acyl CoA:diacylglycerol acyltransferase deficiency on murine fur and sebaceous glands. J Clin Invest 109:175–181, 2002.
72. HC Chen, SJ Smith, Z Ladha, DR Jensen, LD Ferreira, LK Pulawa, JG McGuire, RE Pitas, RH Eckel, RV Farese, Jr. Increased insulin and leptin sensitivity in mice lacking acyl CoA:diacylglycerol acyltransferase 1. J Clin Invest 109:1049–1055, 2002.
73. HC Chen, RV Farese, Jr. Inhibition of triglyceride synthesis as a treatment strategy for obesity: lessons from DGAT1-deficient mice. Arterioscler Thromb Vasc Biol 25:482–486, 2005.
74. J Inokoshi, K Kawamoto, Y Takagi, M Matsuhama, S Omura, H Tomoda. Expression of two human acyl-CoA:diacylglycerol acyltransferase isozymes in yeast and selectivity of microbial inhibitors toward the isozymes. J Antibiot 62:51–54, 2009.
75. AM Birch, S Birtles, LK Buckett, PD Kemmitt, GJ Smith, TJ Smith, AV Turnbull, SJ Wang. Discovery of a potent, selective, and orally efficacious pyrimidinooxazinyl bicyclooctaneacetic acid diacylglycerol acyltransferase-1 inhibitor. J Med Chem 52:1558–1568, 2009.
76. Y Nakada, M Ogino, K Asano, K Aoki, H Miki, T Yamamoto, K Kato, M Masago, N Tamura, M Shimada. Novel acyl coenzyme A: diacylglycerol acyltransferase 1 inhibitors-synthesis and biological activities of N-(substituted heteroaryl)-4-(substituted phenyl)-4-oxobutanamides. Chem Pharm Bull (Tokyo) 58:673–679, 2010.
77. J Cao, Y Zhou, H Peng, X Huang, S Stahler, V Suri, A Qadri, T Gareski, J Jones, S Hahm, et al. Targeting Acyl-CoA:diacylglycerol acyltransferase 1 (DGAT1) with small molecule inhibitors for the treatment of metabolic diseases. J Biol Chem 286:41838–41851, 2011.
78. Y Qian, SJ Wertheimer, M Ahmad, AW Cheung, F Firooznia, MM Hamilton, S Hayden, S Li, N Marcopulos, L McDermott, et al. Discovery of orally active carboxylic acid derivatives of 2-phenyl-5-trifluoromethyloxazole-4-carboxamide as potent diacylglycerol acyltransferase-1 inhibitors for the potential treatment of obesity and diabetes. J Med Chem 54:2433–2446, 2011.
79. W Yun, M Ahmad, Y Chen, P Gillespie, K Conde-Knape, S Kazmer, S Li, Y Qian, R Taub, SJ Wertheimer, et al. Discovery and optimization of 2-phenyloxazole derivatives as diacylglycerol acyltransferase-1 inhibitors. Bioorg Med Chem Lett 21:7205–7209, 2011.
80. H Tomoda, M Ito, N Tabata, R Masuma, Y Yamaguchi, S Omura. Amidepsines, inhibitors of diacylglycerol acyltransferase produced by Humicola sp. FO-2942. I. Production, isolation and biological properties. J Antibiot 48:937–941, 1995.

81. H Tomoda, N Tabata, M Ito, S Omura. Amidepsines, inhibitors of diacylglycerol acyltransferase produced by Humicola sp. FO-2942. II. Structure elucidation of amidepsines A, B and C. J Antibiot 48:942–947, 1995.
82. J Inokoshi, Y Takagi, R Uchida, R Masuma, S Omura, H Tomoda. Production of a new type of amidepsine with a sugar moiety by static fermentation of Humicola sp. FO-2942. J Antibiot 63:9–16, 2010.
83. H Tomoda, Y Yamaguchi, N Tabata, T Kobayashi, R Masuma, H Tanaka, S Omura. Amidepsine E, an inhibitor of diacylglycerol acyltransferase produced by Humicola sp. FO-5969. J Antibiot 49:929–931, 1996.
84. T Nagamitsu, K Marumoto, A Nagayasu, T Fukuda, S Arima, R Uchida, T Ohshiro, Y Harigaya, H Tomoda, S Omura. Total synthesis of amidepsine B and revision of its absolute configuration. J Antibiot 62:69–74, 2009.
85. S Omura, H Tomoda, N Tabata, Y Ohyama, T Abe, M Namikoshi. Roselipins, novel fungal metabolites having a highly methylated fatty acid modified with a mannose and an arabinitol. J Antibiot 52:586–589, 1999.
86. H Tomoda, Y Ohyama, T Abe, N Tabata, M Namikoshi, Y Yamaguchi, R Masuma, S Omura. Roselipins, inhibitors of diacylglycerol acyltransferase, produced by Gliocladium roseum KF-1040. J Antibiot 52:689–694, 1999.
87. N Tabata, Y Ohyama, H Tomoda, T Abe, M Namikoshi, S Omura. Structure elucidation of roselipins, inhibitors of diacylglycerol acyltransferase produced by Gliocladium roseum KF-1040. J Antibiot 52:815–826, 1999.
88. N Tabata, M Ito, H Tomoda, S Omura. Xanthohumols, diacylglycerol acyltransferase inhibitors, from Humulus lupulus. Phytochemistry 46:683–687, 1997.
89. AM Birch, LK Buckett, AV Turnbull. DGAT1 inhibitors as anti-obesity and anti-diabetic agents. Curr Opin Drug Discov Devel 13:489–496, 2010.
90. T Ohshiro, H Tomoda. Isoform-specific inhibitors of ACATs: recent advances and promising developments. Future Med Chem 3:2021–2038, 2011.
91. TY Chang, BL Li, CC Chang, Y Urano. Acyl-coenzyme A:cholesterol acyltransferases. Am J Physiol Endocrinol Metab 297:E1–E9, 2009.
92. MC Meuwese, R Franssen, ES Stroes, JJ Kastelein. And then there were acyl coenzyme A:cholesterol acyl transferase inhibitors. Curr Opin Lipidol 17:426–430, 2006.
93. A Miyazaki, T Kanome, T Watanabe. Inhibitors of acyl-coenzyme a: cholesterol acyltransferase. Curr Drug Targets Cardiovasc Haematol Disord 5:463–469, 2005.
94. M Alegret, G Llaverias, JS Silvestre. Acyl coenzyme A:cholesterol acyltransferase inhibitors as hypolipidemic and antiatherosclerotic drugs. Methods Find Exp Clin Pharmacol 26:563–586, 2004.
95. DR Sliskovic, JA Picard, BR Krause. ACAT inhibitors: the search for a novel and effective treatment of hypercholesterolemia and atherosclerosis. Prog Med Chem 39:121–171, 2002.
96. CC Chang, HY Huh, KM Cadigan, TY Chang. Molecular cloning and functional expression of human acyl-coenzyme A:cholesterol acyltransferase cDNA in mutant Chinese hamster ovary cells. J Biol Chem 268:20747–20755, 1993.
97. RA Anderson, C Joyce, M Davis, JW Reagan, M Clark, GS Shelness, LL Rudel. Identification of a form of acyl-CoA:cholesterol acyltransferase specific to liver and intestine in nonhuman primates. J Biol Chem 273:26747–26754, 1998.

98. S Cases, S Novak, YW Zheng, HM Myers, SR Lear, E Sande, CB Welch, AJ Lusis, TA Spencer, BR Krause, et al. ACAT-2, a second mammalian acyl-CoA:cholesterol acyltransferase. Its cloning, expression, and characterization. J Biol Chem 273:26755–26764, 1998.
99. P Parini, M Davis, AT Lada, SK Erickson, TL Wright, U Gustafsson, S Sahlin, C Einarsson, M Eriksson, B Angelin, et al. ACAT2 is localized to hepatocytes and is the major cholesterol-esterifying enzyme in human liver. Circulation 110:2017–2023, 2004.
100. H Yagyu, T Kitamine, J Osuga, R Tozawa, Z Chen, Y Kaji, T Oka, S Perrey, Y Tamura, K Ohashi, et al. Absence of ACAT-1 attenuates atherosclerosis but causes dry eye and cutaneous xanthomatosis in mice with congenital hyperlipidemia. J Biol Chem 275:21324–21330, 2000.
101. M Accad, SJ Smith, DL Newland, DA Sanan, LE King, Jr., MF Linton, S Fazio, RV Farese, Jr. Massive xanthomatosis and altered composition of atherosclerotic lesions in hyperlipidemic mice lacking acyl CoA:cholesterol acyltransferase 1. J Clin Invest 105:711–719, 2000.
102. S Fazio, AS Major, LL Swift, LA Gleaves, M Accad, MF Linton, RV Farese, Jr. Increased atherosclerosis in LDL receptor-null mice lacking ACAT1 in macrophages. J Clin Invest 107:163–171, 2001.
103. KK Buhman, M Accad, S Novak, RS Choi, JS Wong, RL Hamilton, S Turley, RV Farese, Jr. Resistance to diet-induced hypercholesterolemia and gallstone formation in ACAT2-deficient mice. Nat Med 6:1341–1347, 2000.
104. EL Willner, B Tow, KK Buhman, M Wilson, DA Sanan, LL Rudel, RV Farese, Jr. Deficiency of acyl CoA:cholesterol acyltransferase 2 prevents atherosclerosis in apolipoprotein E-deficient mice. Proc Natl Acad Sci U S A 100:1262–1267, 2003.
105. JC Tardif, J Gregoire, PL L'Allier, TJ Anderson, O Bertrand, F Reeves, LM Title, F Alfonso, E Schampaert, A Hassan, et al. Effects of the acyl coenzyme A:cholesterol acyltransferase inhibitor avasimibe on human atherosclerotic lesions. Circulation 110:3372–3377, 2004.
106. SE Nissen, EM Tuzcu, HB Brewer, I Sipahi, SJ Nicholls, P Ganz, P Schoenhagen, DD Waters, CJ Pepine, TD Crowe, et al. Effect of ACAT inhibition on the progression of coronary atherosclerosis. N Engl J Med 354:1253–1263, 2006.
107. MC Meuwese, E de Groot, R Duivenvoorden, MD Trip, L Ose, FJ Maritz, DC Basart, JJ Kastelein, R Habib, MH Davidson, et al. ACAT inhibition and progression of carotid atherosclerosis in patients with familial hypercholesterolemia: the CAPTIVATE randomized trial. JAMA 301:1131–1139, 2009.
108. S Fazio, M Linton. Failure of ACAT inhibition to retard atherosclerosis. N Engl J Med 354:1307–1309, 2006.
109. DE Dove, YR Su, W Zhang, WG Jerome, LL Swift, MF Linton, S Fazio. ACAT1 deficiency disrupts cholesterol efflux and alters cellular morphology in macrophages. Arterioscler Thromb Vasc Biol 25:128–134, 2005.
110. LL Rudel, RV Farese, Jr. ACAT inhibition and the progression of coronary atherosclerosis. N Engl J Med 354:2616–2617; author reply 2616–2617, 2006.
111. RV Farese, Jr. The nine lives of ACAT inhibitors. Arterioscler Thromb Vasc Biol 26:1684–1686, 2006.
112. P Parini, M Eriksson, LL Rudel. Carotid atherosclerosis progression and ACAT inhibition. JAMA 302:255; author reply 256–257, 2009.

113. AT Lada, M Davis, C Kent, J Chapman, H Tomoda, S Omura, LL Rudel. Identification of ACAT1- and ACAT2-specific inhibitors using a novel, cell-based fluorescence assay: individual ACAT uniqueness. J Lipid Res 45:378–386, 2004.
114. T Ohshiro, LL Rudel, S Omura, H Tomoda. Selectivity of microbial acyl-CoA: cholesterol acyltransferase inhibitors toward isozymes. J Antibiot 60:43–51, 2007.
115. M Ikenoya, Y Yoshinaka, H Kobayashi, K Kawamine, K Shibuya, F Sato, K Sawanobori, T Watanabe, A Miyazaki. A selective ACAT-1 inhibitor, K-604, suppresses fatty streak lesions in fat-fed hamsters without affecting plasma cholesterol levels. Atherosclerosis 191:290–297, 2007.
116. N Terasaka, A Miyazaki, N Kasanuki, K Ito, N Ubukata, T Koieyama, K Kitayama, T Tanimoto, N Maeda, T Inaba. ACAT inhibitor pactimibe sulfate (CS-505) reduces and stabilizes atherosclerotic lesions by cholesterol-lowering and direct effects in apolipoprotein E-deficient mice. Atherosclerosis 190:239–247, 2007.
117. Y Fujiwara, N Kiyota, M Hori, S Matsushita, Y Iijima, K Aoki, D Shibata, M Takeya, T Ikeda, T Nohara, et al. Esculeogenin A, a new tomato sapogenol, ameliorates hyperlipidemia and atherosclerosis in ApoE-deficient mice by inhibiting ACAT. Arterioscler Thromb Vasc Biol 27:2400–2406, 2007.
118. H Tomoda, H Nishida, R Masuma, J Cao, S Okuda, S Omura. Purpactins, new inhibitors of acyl-CoA:cholesterol acyltransferase produced by Penicillium purpurogenum. I. Production, isolation and physico-chemical and biological properties. J Antibiot 44:136–143, 1991.
119. H Nishida, H Tomoda, J Cao, S Okuda, S Omura. Purpactins, new inhibitors of acyl-CoA:cholesterol acyltransferase produced by Penicillium purpurogenum. II. Structure elucidation of purpactins A, B and C. J Antibiot 44:144–151, 1991.
120. H Nishida, H Tomoda, J Cao, S Araki, S Okuda, S Omura. Purpactins, new inhibitors of acyl-CoA:cholesterol acyltransferase produced by Penicillium purpurogenum. III. Chemical modification of purpactin A. J Antibiot 44:152–159, 1991.
121. XH Huang, H Tomoda, H Nishida, R Masuma, S Omura. Terpendoles, novel ACAT inhibitors produced by Albophoma yamanashiensis. I. Production, isolation and biological properties. J Antibiot 48:1–4, 1995.
122. XH Huang, H Nishida, H Tomoda, N Tabata, K Shiomi, DJ Yang, H Takayanagi, S Omura. Terpendoles, novel ACAT inhibitors produced by Albophoma yamanashiensis. II. Structure elucidation of terpendoles A, B, C and D. J Antibiot 48:5–11, 1995.
123. H Tomoda, N Tabata, DJ Yang, H Takayanagi, S Omura. Terpendoles, novel ACAT inhibitors produced by Albophoma yamanashiensis. III. Production, isolation and structure elucidation of new components. J Antibiot 48:793–804, 1995.
124. H Tomoda, XH Huang, H Nishida, R Masuma, YK Kim, S Omura. Glisoprenins, new inhibitors of acyl-CoA: cholesterol acyltransferase produced by Gliocladium sp. FO-1513. I. Production, isolation and physico-chemical and biological properties. J Antibiot 45:1202–1206, 1992.
125. H Nishida, XH Huang, H Tomoda, S Omura. Glisoprenins, new inhibitors of acyl-CoA: cholesterol acyltransferase produced by Gliocladium sp. FO-1513. II. Structure elucidation of glisoprenins A and B. J Antibiot 45:1669–1676, 1992.
126. I Ghosh, Y Kishi, H Tomoda, S Omura. Use of a chiral praseodymium shift reagent in predicting the complete stereostructure of glisoprenin A. Org Lett 6:4719–4722, 2004.

127. H Tomoda, XH Huang, J Cao, H Nishida, R Nagao, S Okuda, H Tanaka, S Omura, H Arai, K Inoue. Inhibition of acyl-CoA: cholesterol acyltransferase activity by cyclodepsipeptide antibiotics. J Antibiot 45:1626–1632, 1992.
128. N Koyama, T Nagahiro, Y Yamaguchi, T Ohshiro, R Masuma, H Tomoda, S Omura. Spylidone, a novel inhibitor of lipid droplet accumulation in mouse macrophages produced by Phoma sp. FKI-1840. J Antibiot 58:338–345, 2005.
129. VG DeVries, SA Schaffer, EE Largis, MD Dutia, CH Wang, JD Bloom, AS Katocs, Jr. Potential antiatherosclerotic agents. 5. An acyl-CoA:cholesterol O-acyltransferase inhibitor with hypocholesterolemic activity. J Med Chem 29:1131–1133, 1986.
130. HT Lee, DR Sliskovic, JA Picard, BD Roth, W Wierenga, JL Hicks, RF Bousley, KL Hamelehle, R Homan, C Speyer, et al. Inhibitors of acyl-CoA: cholesterol O-acyl transferase (ACAT) as hypocholesterolemic agents. CI-1011:an acyl sulfamate with unique cholesterol-lowering activity in animals fed noncholesterol-supplemented diets. J Med Chem 39:5031–5034, 1996.
131. I Namatame, H Tomoda, S Si, Y Yamaguchi, R Masuma, S Omura. Beauveriolides, specific inhibitors of lipid droplet formation in mouse macrophages, produced by Beauveria sp. FO-6979. J Antibiot 52:1–6, 1999.
132. I Namatame, H Tomoda, N Tabata, S Si, S Omura. Structure elucidation of fungal beauveriolide III, a novel inhibitor of lipid droplet formation in mouse macrophages. J Antibiot 52:7–12, 1999.
133. D Matsuda, I Namatame, H Tomoda, S Kobayashi, R Zocher, H Kleinkauf, S Omura. New beauveriolides produced by amino acid-supplemented fermentation of Beauveria sp. FO-6979. J Antibiot 57:1–9, 2004.
134. I Namatame, H Tomoda, S Ishibashi, S Omura. Antiatherogenic activity of fungal beauveriolides, inhibitors of lipid droplet accumulation in macrophages. Proc Natl Acad Sci U S A 101:737–742, 2004.
135. CW Joyce, GS Shelness, MA Davis, RG Lee, K Skinner, RA Anderson, LL Rudel. ACAT1 and ACAT2 membrane topology segregates a serine residue essential for activity to opposite sides of the endoplasmic reticulum membrane. Mol Biol Cell 11:3675–3687, 2000.
136. S Lin, X Lu, CC Chang, TY Chang. Human acyl-coenzyme A:cholesterol acyltransferase expressed in chinese hamster ovary cells: membrane topology and active site location. Mol Biol Cell 14:2447–2460, 2003.
137. ZY Guo, S Lin, JA Heinen, CC Chang, TY Chang. The active site His-460 of human acyl-coenzyme A:cholesterol acyltransferase 1 resides in a hitherto undisclosed transmembrane domain. J Biol Chem 280:37814–37826, 2005.
138. A Das, MA Davis, LL Rudel. Identification of putative active site residues of ACAT enzymes. J Lipid Res 49:1770–1781, 2008.
139. T Ohshiro, D Matsuda, K Sakai, C Degirolamo, H Yagyu, LL Rudel, S Omura, S Ishibashi, H Tomoda. Pyripyropene A, an acyl-coenzyme A:cholesterol acyltransferase 2-selective inhibitor, attenuates hypercholesterolemia and atherosclerosis in murine models of hyperlipidemia. Arterioscler Thromb Vasc Biol 31:1108–1115, 2011.
140. J Armitage, L Bowman, K Wallendszus, R Bulbulia, K Rahimi, R Haynes, S Parish, R Peto, R Collins. Intensive lowering of LDL cholesterol with 80 mg versus 20 mg simvastatin daily in 12,064 survivors of myocardial infarction: a double-blind randomised trial. Lancet 376:1658–1669, 2010.

141. A Egan, E Colman. Weighing the benefits of high-dose simvastatin against the risk of myopathy. N Engl J Med 365:285–287, 2011.

142. HH Hobbs, DW Russell, MS Brown, JL Goldstein. The LDL receptor locus in familial hypercholesterolemia: mutational analysis of a membrane protein. Annu Rev Genet 24:133–170, 1990.

143. HH Hobbs, MS Brown, JL Goldstein. Molecular genetics of the LDL receptor gene in familial hypercholesterolemia. Hum Mutat 1:445–466, 1992.

144. FJ Raal, GJ Pilcher, DR Illingworth, AS Pappu, EA Stein, P Laskarzewski, YB Mitchel, MR Melino. Expanded-dose simvastatin is effective in homozygous familial hypercholesterolaemia. Atherosclerosis 135:249–256, 1997.

145. KJ Moore, I Tabas. Macrophages in the pathogenesis of atherosclerosis. Cell 145:341–355, 2011.

146. TA Bell, 3rd, JM Brown, MJ Graham, KM Lemonidis, RM Crooke, LL Rudel. Liver-specific inhibition of acyl-coenzyme a:cholesterol acyltransferase 2 with antisense oligonucleotides limits atherosclerosis development in apolipoprotein B100-only low-density lipoprotein receptor−/− mice. Arterioscler Thromb Vasc Biol 26:1814–1820, 2006.

147. T Ohshiro, S Ohte, D Matsuda, M Ohtawa, T Nagamitsu, T Sunazuka, Y Harigaya, LL Rudel, S Omura, H Tomoda. Selectivity of pyripyropene derivatives in inhibition toward acyl-CoA:cholesterol acyltransferase 2 isozyme. J Antibiot 61:503–508, 2008.

148. A Das, MA Davis, H Tomoda, S Omura, LL Rudel. Identification of the interaction site within acyl-CoA:cholesterol acyltransferase 2 for the isoform-specific inhibitor pyripyropene A. J Biol Chem 283:10453–10460, 2008.

3

COVALENT ENZYME INHIBITION IN DRUG DISCOVERY AND DEVELOPMENT

SHUJAATH MEHDI

Immunoinflammation Therapeutic Strategy Unit, Sanofi Pharmaceuticals, Bridgewater, New Jersey

I. INTRODUCTION

Enzyme inhibition is best viewed as a continuum (Fig. 1). At one extreme are inhibitors for which the binding equilibrium is achieved rapidly (within seconds) and on the other are inhibitors whose complexes with the enzyme dissociate either not at all or too slowly to be easily measured. In addition to drugs that fall at either extreme, there are many that fall in-between, for which the dissociation half-life is of the order of several minutes to hours. (Such slowly reversible inhibitors are sometimes referred to as being "quasi-irreversible.") For example, many ACE inhibitors, statins, and COX inhibitors exhibit slow-binding behavior (referring to the rate at which binding equilibrium is achieved, generally also having slow dissociation rates), to form noncovalent complexes that dissociate slowly; these are not discussed in the review. Irreversible inhibitors have been referred to as "covalent drugs"; however, covalent binding is neither necessary nor sufficient for irreversibility – there are examples of molecules that are noncovalently but irreversibly bound and also many examples of inhibitors that make covalent bonds that are more or less reversible. The key to reversibility is the type of covalent bond and its chemical stability, and its stability is a biological context (i.e., on the enzyme) – this is different for different types

Enzyme Technologies: Pluripotent Players in Discovering Therapeutic Agents, First Edition.
Edited by Hsiu-Chiung Yang, Wu-Kuang Yeh, and James R. McCarthy.

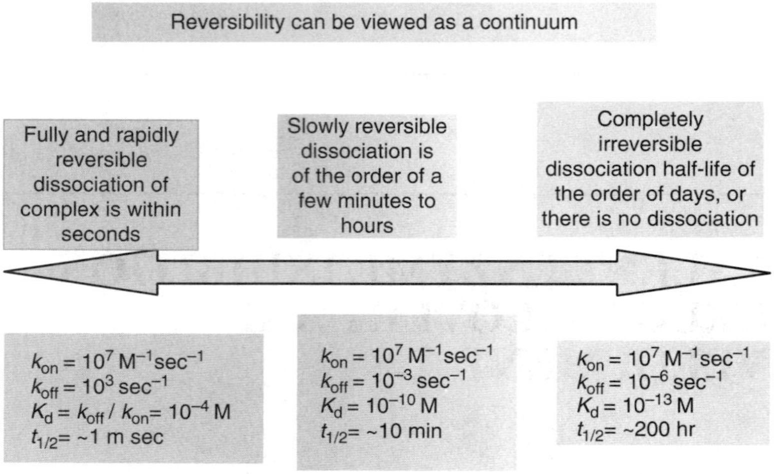

FIGURE 1 The "continuum" of enzyme inhibition, representing the time scale for reactivation. In these illustrative examples, k_{on} and k_{off} are taken to be microscopic rate constants. In the general case, the rates of inactivation and regain of enzyme activity may be described by rate constants which may be composites of rate constants for individual steps, including those for bond-breaking or bond-making.

Addition of an enzyme nucleophile to an electrophilic center such as aldehyde, nitrile, electrophilic ketone, a boron center
Chemically and biologically reversible - fast or slow reversibility
Displacement at a carbon, such as at an epoxide (alkylation)
Chemically and biologically irreversible
Addition or conjugate addition to a C–C double bond (e.g. Michael addition)
Chemically reversible or irreversible; biologically fully irreversible
Carbamoylation
Chemically stable (at neutral pH), biologically (i.e. on the enzyme) Š range of reversibility (e.g. acetlycholinesterase inhibitors)
Acylation
Chemically and biologically somewhat labile
Mixed disuflide
Chemically and biologically stable but can be reversed by athiol by disulfide exchange
Phosphorylation or phosphonylation
Chemically and biologically stable

FIGURE 2 Types of covalent bonds between inhibitors and enzyme nucleophiles. Biological stability refers to stability on the enzyme.

of linkages (Fig. 2). Reversibility of drug action at the target has an impact on duration of action (a longer duration may or may not be desirable in a particular context) and, relatedly, on frequency of dosing (with a reduced frequency being desirable from a patient compliance and marketing perspective) and safety. In this chapter, a perspective is given on covalent inhibitors, which in general fall in the middle and to the right in the continuum – that is, that are slowly reversible or fully irreversible.

Inhibitors that make covalent bonds with the target enzyme not only constitute important drugs but also provide interesting insights into chemical mechanisms of enzymatic reactions. Irreversible inhibitors have also been very useful as tools for target or pathway validation, for target discovery, or for establishing the selectivity of reversible inhibitors, and their use as such will be described.

References are given where appropriate to reviews where aspects of the subject matter are treated more thoroughly. Most examples given are drugs, although other inhibitors are used as examples for illustrative purposes. Different types of inhibitors from the point of view of inhibition concept or inhibition principle are first discussed. Conceptual and practical aspects of the kinetics of slow binding and of irreversible inhibitors are discussed, followed by specific examples of inhibitors classed by type of covalent attachment. The review ends with the discussion of special topics relating to irreversible inhibition and with guidelines that may help in decision-making when encountering irreversible inhibition in a project.

II. MECHANISM OF INHIBITION: MECHANISTIC TYPES BY PRINCIPLE OF INHIBITION

Different types of mechanisms – both chemical and kinetic – characterize enzyme–inhibitor (E–I) interactions: it is important to understand these mechanisms if one is to rationally design or modify inhibitors for a particular system, but functionally (from a pharmacological viewpoint) the most pertinent factor is not necessarily the actual chemical mechanism but the k_{off} (the off-rate constant characterizing the dissociation of inhibitor) or, more generally, the reactivation rate (the recovery of enzyme activity when free inhibitor is removed). *In vitro*, these may be simply described by a single or a combination of rate constants, but *in vivo*, the rate of return of enzyme activity also depends on the rate of synthesis of new protein. This parameter describes the duration of action on the target (sometimes referred to as the "residence time"), with its implications as stated earlier. Two reviews discuss the relevance of residence time in the pharmacology of slowly dissociating inhibitors [1,2].

A. Affinity-Labeling Agents

Aside from natural products, the history of irreversible inhibitors begins with synthetic affinity-labeling agents. Many early examples of these types of molecules are described in the monograph by Baker [3]. These types of molecules are inherently reactive and often react indiscriminately with nucleophiles. Some of these types of molecules are sulfonyl fluorides, phosphates or phosphonates with a good leaving group on phosphorus, and α-halomethylketones. Typically, these types of reagents were first used with serine-dependent hydrolases (esterases, lipases, amidases, proteases, peptidases) which, owing to the placement and enhanced nucleophilic reactivity of the hydroxyl of the active-site catalytic serine residue, are ready targets for such types of molecules. Examples of such molecules used to inhibit serine proteases are tosyl-L-lysine chloromethyl ketone (TLCK), tosyl-L-phenylalanine chloromethyl ketone (TPCK), diisopropyl fluorophosphate (DIFP), and phenylmethylsulfonyl fluoride (PMSF) (Fig. 3). In a study of chymotrypsin inhibition by Cbz-Ala-Phe-CH_2Cl [4], it was established that the mechanism involves the initial attack of the active-site serine –OH to the carbonyl to form a hemiketal. In a subsequent step, attack of the oxyanion of the hemiketal at the alpha carbon to displace chloride gives an epoxide, which opens by attack of a histidine to give the stable alkylated enzyme product. Supporting this mechanism is the evidence that there is overall retention of configuration at the alpha carbon, indicating that two displacements have occurred at this center. Thus, direct displacement by histidine, the simpler and more obvious mechanism, is not operative.

Affinity-labeling agents related to certain nerve gases were also developed against acetylcholinesterase (AChE) and are still in use today as pesticides, targeting the enzyme in agricultural pests. In these types of molecules, displacement at a phosphorous center occurs to give a stable phosphorylated or phosphonylated enzyme.

Attempts to modulate the reactivity of molecules with leaving groups at the alpha position to a ketone by changing the leaving group ability are recounted by Powers et al. [5]. This review also gives a general and comprehensive account of irreversible inhibitors of serine, cysteine, and threonine proteases.

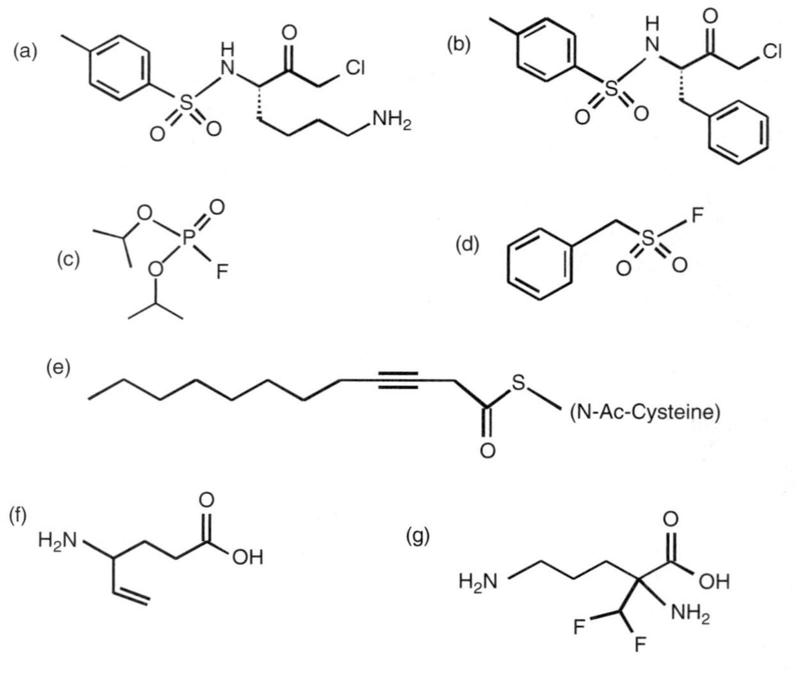

FIGURE 3 (a) TLCK; (b) TPCK; (c) DIFP; (d) PMSF; (e) the first mechanism-based inhibitor to be characterized as such (ref Bloch); (f) vinyl GABA; (g) DFMO.

B. Mechanism-Based Inhibitors (Suicide Inhibitors, Suicide Substrates)

In contrast to affinity-labeling agents, which have inherent reactivity, mechanism-based inhibitors are unreactive until acted upon by the target enzyme to produce a reactive group in the active site, usually an electrophile, which then reacts with a nucleophile of the enzyme and inactivates it. The first synthetic molecules that were characterized as mechanism-based inhibitors were β,γ-acetylenic thioester inhibitors of β-hydroxydecanoyl thioester dehydrase [6,7]. The first two rationally designed drugs that are mechanism-based inhibitors are γ-vinyl GABA (GABA = γ-aminobutyric acid) and α-(difluoromethyl)ornithine (DFMO; Fig. 3), both designed as inhibitors of pyridoxal 5′-phosphate (PLP)-dependent enzymes [8,9]. Vinyl GABA was designed to inhibit GABA aminotransferase. As with the substrate GABA, vinyl GABA forms an imine with the aldehyde of the cofactor PLP. The resulting adduct is set up for conjugate addition, with electron flow into the pyridinium ring. This drug has safety issues, but has earned a niche use in certain types of epilepsy. DFMO inhibits ornithine decarboxylase, another PLP enzyme. As DFMO is decarboxylated, a negative charge develops on the alpha carbon (and is able to delocalize into the conjugated system), resulting in the elimination of one mole of fluoride to give a vinyl fluoride. The vinyl fluoride is intercepted by conjugate addition of an enzyme nucleophile followed by the expulsion of a second mole of fluoride. The molecule was designed to inhibit the polyamine pathway, which is important in cellular proliferation. It was not successful as a broad antitumor drug as hoped, but found a niche as an antiparasitic drug. Merrell Dow (one of the predecessor companies of Sanofi Pharmaceuticals), where both vinyl GABA and DFMO (Fig. 3) were invented, had an agreement with the World Health Organization for the manufacture and distribution of DFMO for treatment of trypanosomiasis. The monograph by Silverman [10] and two reviews [11,12] give an account of early developments in the field of suicide inhibitors.

Mechanism-based inhibition (MBI) is a particular issue for the drug-metabolizing cytochrome-P450-dependent enzymes (abbreviated here as CYPs, though there are other cytochrome-P450-dependent enzymes that are not involved in xenobiotic metabolism) [13]. Most enzymes that are targets for drugs have limited substrate specificity. Mechanism-based inhibitors for such enzymes therefore must resemble the substrate (like vinyl GABA and DFMO), not just to be recognized for binding but also for activation by catalytic processing. The scope for chemical optimization of such molecules by structure modification and variation to resolve potential metabolism, safety, solubility, or bioavailability issues is therefore limited. This may be one reason that mechanism-based inhibitors are not more widespread as drugs. Another factor is that the reactions catalyzed by many enzymes are not such that an electrophilic center can be generated in a potential suicide substrate to allow the design of an appropriate inhibitor. CYPs, however, have a broad substrate specificity (to deal with a large variety of xenobiotic structures) and also frequently catalyze reactions in which an electrophile such as a quinone (by hydroxylation of an appropriately substituted aromatic nucleus) or an epoxide is generated. MBI of CYPs is therefore more common than of other enzymes that may be therapeutic targets. MBI of CYPs is discussed in a later section.

While the term MBI has become well defined and understood as describing enzyme-activated irreversible inhibition (inactivation) (and is used as such in this chapter), many inhibitors rely in one or another manner on the normal catalytic mechanism of the enzyme to achieve their effect (or as a factor in their design) and so are, in this broader sense, "mechanism based." The reactivity of an inhibitor, the nucleophilicity of a residue, or the stability of an adduct may all be enhanced by the same factors such as acid–base catalysis, protein conformational shifts, proximity effects, and induced bond strain and bond polarization that are used in the normal catalytic mechanism to enhance the reaction rates of substrates.

Transition-state analogs, discussed later, are also in a sense mechanism based in that they resemble the transition state for the normal catalytic mechanism. The type of transition-state analogs in which a hemi(thio)ketal/acetal or (thio)ketal/acetal is formed relies on the reactivity and position of the catalytic serine or cysteine and also on the stabilization of the tetrahedral adduct by the enzyme, in much the same way that the tetrahedral species along the normal catalytic pathway may be stabilized.

C. Transition-State Analogs

By consideration of over-the-barrier catalysis only and not through-the-barrier quantum tunneling, enzymes achieve enormous rate enhancements by lowering the energy of – stabilizing – the transition state in relation to substrate or product. The transition state is a theoretical construct with only fleeting physical existence, in which bonds are partially made or broken. However, one can try to establish the structure of the transition state by the effects of substituents on the reaction rates, by measurement of isotope effects, from the mechanism of the reaction (the transition state must bear resemblance to some degree to substrates, products, or reaction intermediates to which it is close in energy, according to Hammond's postulate), or by computation. Transition-state analogs are stable molecules that attempt to capture some feature or features of the (hypothetical) transition state such that the enzyme will bind them tightly [14,15]. In general, it may be more accurate to describe such types of inhibitors as resembling some species, of whatever duration of existence or of whatever placement along the reaction coordinate, to which the enzyme binds tightly.

Though many transition-state analogs bind potently, many of them do not react covalently with the enzyme. One class of covalent transition-state inhibitors constitutes molecules in which a peptide or peptidomimetic incorporates an electrophilic moiety in place of the amide carbonyl of the cleaved bond. The catalytic serine (or cysteine) adds to this electrophilic moiety, producing a tetrahedral adduct that may be said to resemble the transition state at least to some degree [16,17]. The transition-state analog is formed in the active site; in solution many of these types of molecules exist as hydrates, but the species that binds is the carbonyl form that is present in solution in equilibrium with the hydrate [16]. There are many examples of both natural products and synthetic inhibitors that follow this mechanism [5]. Of the former, leupeptin and chymostatin are examples. Both are aldehydes, and the mechanism of inhibition involves the addition of an active-site thiol in the case of cysteine proteases or a serine –OH to the aldehyde group, to give reversible or slowly reversible inhibition. Based on this principle, several investigators designed inhibitors such as fluoroketones, α-ketoamides, α-ketoesters, 1,2-diketones, and ketones bearing an electron-withdrawing heterocycle at the alpha carbon [5]. A broader application of this principle, where an electrophilic center, not necessarily an activated carbonyl, is present at the center where the amide carbonyl in a peptide substrate would normally be, has yielded two new classes of drugs: bortezomib (bearing the electron-deficient boron) and the nitrile inhibitors of dipeptidyl peptidase IV (DPP-IV).

There are other examples of covalent transition-state analogs. The crystal structure of a class A TEM-1 β-lactamase covalently bound to a phosphonate via the active-site (catalytic) Ser700 has been determined [18]. The complex is believed to represent an analog of the transition state for the breakdown of the tetrahedral species formed during catalysis of β-lactam hydrolysis via an acyl–enzyme intermediate. A crystal structure has also been obtained of a complex of vanadate with the *Yersinia pestis* protein tyrosine phosphatase and shows the vanadate covalently bound with pentavalent geometry, mimicking the transition state for substrate hydrolysis via a thiophosphoryl enzyme intermediate [19].

D. Slow Substrate (Alternative Substrates)

This type of inhibition is best exemplified by the clinically relevant carbamate inhibitors of AChE and by inhibitors of the enzyme FAAH (fatty acid amide hydrolase), both of which are serine-dependent hydrolases [20,21]. The molecules are substrates for the enzyme, and the active-site serine is carbamoylated. The carbamoyl enzyme is hydrolyzed more or less slowly, and thus the enzyme appears inhibited when its activity is queried by another more efficient or a physiological substrate.

Although the mechanism of regulation of proteinase activity by serpins is complex and involves intricate conformational changes, one component of their action is acylation of the target enzyme to form quasi-stable covalent complexes with it [22]. Thus, they are slow or alternate substrates.

E. Slow-Binding Inhibitors

The term "slow binding" refers to a kinetic rather than a chemical mechanism of inhibition. All binding reactions have a pre-equilibrium phase, before the binding equilibrium is established. In cases in which the binding equilibrium is established rapidly, the pre-equilibrium phase is not observed on the time scale (minutes) of a typical assay and falls in the realm of rapid kinetics. In some cases however the binding equilibrium is slow to establish and is observed in the minutes time scale. These types of inhibitors are said to be "slow binding." This term does not imply that the on-rate constant is necessarily small; in the case of a single-step slow binding, for example, the (slow) rate of approach to equilibrium, given by k_{obs}, is actually a function of both k_{on} (the second-order rate constant for association of inhibitor with enzyme) and k_{off} (the first-order rate constant for dissociation of the E–I complex) and also on the inhibitor concentration. This may seem counterintuitive, but can be thought of in this way: at the start of the binding reaction, there is contribution only from the k_{on} term; as the binding proceeds to equilibrium, the k_{off} term starts to become significant and contributes to the rate at which equilibrium is approached. When equilibrium is reached, the forward (on) rate and the back (off) *rates* (not rate constants) become equal. In the simplest case of slow binding (one-step binding, with an inhibitory potency in the nM range or smaller), the slow binding is typically due to a small k_{off} and the low inhibitor concentration in the assay (necessitated by the high potency). In some cases however slow binding is a result of a slow conformational change of the complex initially formed to a more stable structure or due to the displacement of a tightly bound water molecule. In such cases, the k_{off} is no longer the simple microscopic rate constant, but a composite of individual rate constants [1,2]. In general, the key characteristic of slow-binding inhibitors is that the dissociation of the E–I complex is slow, and the inhibition is only slowly reversible, at rates that vary over the spectrum shown in Figure 1.

F. Natural Products

Several natural products are known that react irreversibly with enzymes, and several others that are quasi-irreversible. All inhibition mechanisms described earlier are represented by natural products. A recent review highlights some of the former types of molecules [23]. Some of these are actually marketed drugs in their natural forms (aspirin, natural β-lactams), and others are marketed in a somewhat modified form (tetrahydrolipstatin, compactin, rivastigmine). Still others such as E-64 and other protease inhibitors have been extremely useful as tools [5]. The β-lactams are acylating agents by virtue of a reactive carbonyl group and are discussed in a later section. Both aspirin, which acylates and inactivates COX enzymes, and β-lactamases are affinity-labeling agents. Whereas β-lactams may also be considered to be alternate/slow substrates of β-lactamases, this is not the case with aspirin: the acetylated serine is not a catalytic residue, and the normal function of COX is not hydrolysis.

III. KINETICS: CONCEPTUAL AND EXPERIMENTAL CONSIDERATIONS

The kinetics of slow-binding inhibitors and irreversible inhibitors is discussed. It will be seen that the two types of inhibitors have some similarities from a kinetics point of view, but differ conceptually. For the purposes of the discussion, it will be assumed that the inhibitors are binding in the active site and are competitive with substrate. Also the simplifying assumption is made that [I] >> [E], that is, the "tight-binding" condition is not pertinent. In spite of the term, the tight-binding condition, where [E] ≃ [I], can arise even when the K_d is not small: for example, an insensitive assay or an enzyme of low specific activity, or both, can create a situation in which a large amount of enzyme is

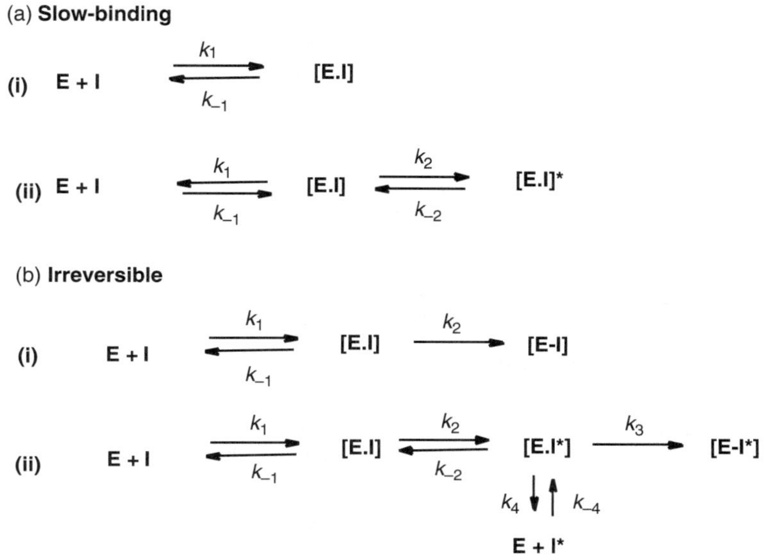

FIGURE 4 Kinetic mechanisms for (a) slow-binding and (b) irreversible inhibition (inactivation).

needed in the assay and whose concentration can approach the order of the concentration of the inhibitor in the assay.

The monograph by Copeland [24] has an excellent treatment of inhibition kinetics and may be consulted for more details on the mechanisms described in this review and for the rate equations that describe these mechanisms.

A. Slow Binding

Two main mechanisms of slow binding can be considered (Fig. 4).

Single-Step Binding In this mechanism (Fig. 4a (i)), binding occurs in a single step, with no formation of an encounter complex (defined as a relatively rapidly formed initial complex prior to the slower formation of the final complex). The rate constant describing the forward (binding) step is the microscopic second-order rate constant k_{on}, and the forward rate is given by $[I].k_{on}$. The progress curve at any inhibitor concentration is nonlinear at first, going to a linear final rate when binding equilibrium has been achieved. Figure 5 shows the appearance of such a progress curve for the inhibition of the enzyme DPP-IV by an aldehyde inhibitor A000366444R for one concentration of inhibitor (raw data and fitted curve) [25]. The first-order exponential rate at which equilibrium is achieved, k_{obs}, can be measured for a range of inhibitor concentrations, and the values of k_{obs} plotted versus [I]. In the case of the one-step mechanism, this replot is a straight line, with the y-intercept$=k_{off}$, and the slope$=k_{on\ (app)}$ (Fig. 6) [25]. After correction for the substrate concentration relative to its K_m, a true k_{on} is obtained.

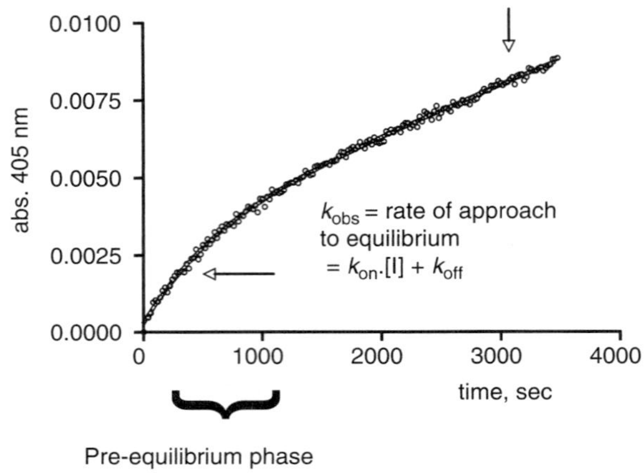

A pre-equilibrium phase exists for all binding reactions; in rapid binding situations this pre-equilibrium phase may occur in the << second time frame.

FIGURE 5 A progress curve for a slow-binding inhibitor. Data shown are for an aldehyde DPP-IV inhibitor [25]. The pre-equilibrium phase is nonlinear; the rate after binding equilibrium has been established is linear. The rate of approach to binding equilibrium depends on both k_{on} and k_{off} as well as the inhibitor concentration. Binding equilibrium has been reached: forward (off) *rate* = reverse (on) *rate.*

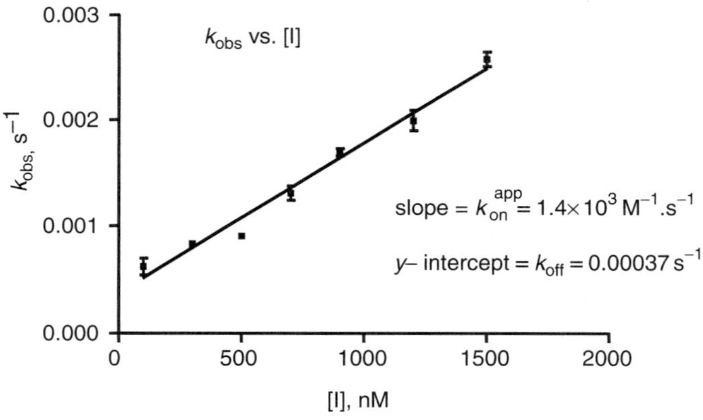

$$k_{obs} = [I].\ k_{on}^{app} + k_{off} \qquad k_{on}^{app} = 1.4\times10^3\ M^{-1}.s^{-1}$$

$$k_{on} = k_{on}^{app}.\ (1 + [S]/K_m) \qquad k_{on} = 7.6\times10^3\ M^{-1}.s^{-1}$$

FIGURE 6 For the DPP-IV inhibitor, a replot of k_{obs} (obtained from a series of progress curves such as the one in Fig. 4) versus inhibitor concentration is linear, and k_{on} and k_{off} can be obtained from this plot [26]. The K_i expressed as the ratio of k_{off} and k_{on} ($K_i = k_{off}/k_{on} = 49$ nM) is in good agreement with the K_i calculated from the final rates of progress curves. A replot of k_{obs} vs inhibitor concentration gives k_{off} and k_{on}.

The final slope in the progress curve (i.e., the rate after binding equilibrium is reached) can be used for K_i or IC_{50} calculation.

Two-Step Binding In this mechanism (Fig. 4a (ii)), the enzyme and inhibitor form a rapid-equilibrium complex, which then reorganizes to a tighter complex. The overall k_{on} and k_{off} are no longer single, microscopic rate constants but composites of individual rate constants. For this type of mechanism, a plot of k_{obs} versus inhibitor concentration is a hyperbola. In this plot, the intersection of the hyperbola with the y-axis is the value of k_{-2}, and as the inhibitor concentration tends to infinity, the value of k_{obs} tends to the sum $k_2 + k_{-2}$. The leveling off of k_{obs} at high inhibitor concentration signifies saturation, which is a characteristic of two-step binding. The one-step binding mechanism may be considered a limiting case of the two-step mechanism in which an encounter complex may exist but is too weak to be kinetically significant at the inhibitor concentrations tested. The symbol K^*_i is sometimes used to designate the strength of the binding in the "final" complex, that is, after equilibrium has been fully achieved.

When the dissociation of the inhibitor is very small, k_{-2} is very small, and the intercept of the hyperbola with the y-axis is close to 0. Such a small k_{-2}, although it may be nonzero, may be too small to measure experimentally on an ordinary time scale. Unlike the case for the one-step binding, in which there is no initial complex, the initial rates of the progress curves for a series of inhibitor concentrations for the two-step binding mechanism are different, as the inhibitor is competing with substrate for binding to the initial complex.

The rate of reactivation of the enzyme (rate of recovery of enzyme activity), as discussed in the introduction, is a key parameter, regardless of the chemical or kinetic mechanism of inhibition. Conceptually, first forming the complex, then removing excess inhibitor by some means, and observing the rate of return of activity measures the reactivation rate or the k_{off}. This is in accord with Le Chatelier's principle: the equilibrium is being perturbed by reducing the concentration of one of the species in the equilibrium (the inhibitor), and the equilibrium shifts accordingly. Because (free) inhibitor is no longer present, and the complex is diluted to the extent that dissociation does not lead to the buildup of significant free inhibitor concentration, the only process being measured is the dissociation and thus the reactivation of the enzyme activity. Experimentally, there are many ways to reduce free inhibitor concentration, the simplest being dilution of the E–I complex severalfold, then monitoring return of enzyme activity. If a continuous assay is available, the initially formed complex can be directly diluted into the assay mixture – the presence of substrate additionally prevents released inhibitor from going back on (with effectiveness of competition depending on the factor $[S]/K_m$). Another method that can be used to achieve dilution is rapid gel filtration. Rapid microcentrifugation followed by a rapid wash through a filtration membrane can also be used. In SPR (Biacore) measurements, the "dilution" is achieved by a wash once the formation of the complex on the chip has been achieved. Dialysis does not have the temporal resolution to measure reactivation rates that are typically less than a few hours; it is best suited for demonstrating complete irreversibility (no return of activity following extensive dialysis over an extended time period). In cases where the E–I complex has a bond such as a carbamate, the reactivation rate may include or be predominated by the (chemical) rate of hydrolysis of such a bond. Given the pharmacological importance of the reactivation time course, this parameter should be measured (or a limit – "faster than…" or "slower than…" – set on it) for any drug candidates, for representatives of lead chemical series, or for making a selection from a series. Its relative ease of measurement puts it in easy reach of most labs.

B. Fully Irreversible Inhibition (Inactivation)

It makes sense intuitively that for an irreversible inhibitor of this type, the progress curve will be curved (the inhibition will be time dependent), with the final rate = 0. The following formalism makes this clear. The rate of the reaction at any time point t (which equals the slope of the tangent to the curve at that time t) is proportional to the concentration of free enzyme. As the irreversible inactivation progresses, less and less free enzyme is present, and the slope of the tangent progressively diminishes, until there is no free enzyme left and the slope is 0. Time dependence is thus a reflection of irreversibility and may be used as an initial screen for such type of inhibition. However, the nonlinearity may be missed if the inactivation rate is slow or the inhibitor is present at low concentration, so lack of time dependence should not be used to rule out irreversible inhibition. Instead, one of the methods described in the previous section should be used for assessing reversibility (using a sufficiently high concentration of inhibitor and a sufficiently long incubation to fully form the inhibited complex).

Two mechanisms for irreversible inhibition may be considered (Fig. 4b). These describe inhibitors that are either "quiescent" affinity labels or mechanism-based inhibitors. A third mechanism, analogous to the first slow-binding mechanism discussed earlier in which there is no encounter complex, is not relevant to the kind of enzyme inhibitors described in this review. Inhibitors that follow this type of one-step kinetic scheme tend to be nonspecific inactivators, such as the reagents iodoacetamide or *N*-ethylmaleimide that react nonspecifically with protein thiols, and such enzyme inhibitors are generally not useful as drugs. Inhibitors that follow this mechanism do not show saturation – a plot of k_{obs} (the first-order rate of inactivation at a particular inhibitor concentration) versus the inhibitor concentration will give a straight line, starting at the origin ($y = 0$) because the inactivation is fully irreversible.

Two parameters describe all irreversible inhibitors: k_{inact} and K_I. The definition of k_{inact} is that it is the maximal rate of inactivation (k_{obs} will tend to this value as inhibitor concentration tends to infinity), and K_I is defined as the inhibitor concentration giving half the maximal rate. These are analogous to the substrate parameters K_m and V_{max}, and, also in parallel with the substrate parameters, the best comparison of irreversible inhibitor efficiency is a comparison of k_{inact}/K_I values (analogous to comparison of V/K for substrates). This ratio has the units of a second-order rate constant and may be thought of as taking into account both potency and rate. Like the analogous constants for a substrate, k_{inact} and K_I depending on the mechanism can be made up of individual rate constants; k_{inact} may not reflect the rate of a single step, and likewise, K_I may not reflect a true affinity. It is best to think of k_{inact} and K_I as operational and functional constants, without assigning further meaning to them (such as affinity, or the rate of a single step) in the absence of knowledge of the kinetic and chemical mechanism.

K_I and k_{inact} can be determined from plots of k_{obs} (observed rate of inactivation) versus [I]. For both mechanisms described later, this plot is a hyperbola (indicating saturation at high [I]), and these constants can be determined by fitting of this curve to the equation for a hyperbola. In the case of inhibitors following the one-step mechanism, which do not show saturation in the range of the inhibitor concentrations, the slope of the linear plot of k_{obs} versus [I] directly gives k_{inact}/K_I.

Two-Step Inactivation In this type of inhibition (Fig. 4b (i)), an initial complex is formed, generally rapidly, between enzyme and inhibitor before the chemical inactivation step. "Quiescent" affinity labeling follows this mechanism – they have significant-enough binding interaction in the active site as to enable the formation of an initial, noncovalent complex, before the inactivation step. As noted earlier, a plot of k_{obs} versus inhibitor concentrations yields the parameters k_{inact} and K_I. Alternately, under conditions such that $[I] << K_I$, the initial slope of the hyperbola, which is k_{inact}/K_I, can be used as a basis for comparison of inhibitor efficiency of a series of compounds.

For any first-order or pseudo-first-order process (association, dissociation, inactivation), the first-order rate constant (either microscopic or a composite) describing it can be converted to the more easily conceptualized parameter $t_{1/2}$ (the time of half completion, or half-life, of the process) by $t_{1/2} = \ln 2/k_{obs} = 0.693/k_{obs}$.

Mechanism-Based Inhibition Mechanism-based inhibitors are typically unreactive molecules that become reactive when they are acted on by the enzyme. Thus, they are substrates for the enzyme, hence the term "suicide substrates" that is sometimes used for them. Usually the reactive species that is generated by the enzyme is an electrophile, though free radical species may be involved in some cases. The reactivity may be reversed in a few cases where a nucleophile derived from the inhibitor reacts with an electrophilic site on a cofactor. The kinetic scheme describing these (Fig. 4b (ii)) includes an extra step, that for the catalytic step that converts I to I*, the reactive moiety. An additional feature of MBI is that the enzyme-bound reactive species can have two fates: it can dissociate from the enzyme, or it can react with an enzyme nucleophile and inactivate the enzyme. If the former were exclusively occurring, the molecule would be a true substrate and not an inhibitor. If the latter were exclusively occurring, the molecule would be a mechanism-based inhibitor with 100% efficiency. Most inhibitors are somewhere in-between and are characterized by the "partition ratio," defined as the ratio of these two limiting events.

As with inhibitors following mechanism a (ii) in Figure 4, the plot of k_{obs} versus inhibitor concentration gives a hyperbola, and k_{inact} and K_I can be derived from this plot. Kinetically, one cannot distinguish mechanism-based inhibitors (Fig. 4b (ii)) from inherently reactive affinity-labeling agents (such as those following mechanism b (i)). Several experimental criteria can be used to distinguish these, and these are given in Silverman [10] and Copeland [24]. A chemist's knowledge of the reactivity of the molecule, coupled with the knowledge of the reaction catalyzed by the enzyme, can sometimes give a clue as to which type of inhibition to expect, as the examples of vinyl GABA and DFMO illustrate. A chemist will recognize that neither molecule is especially activated toward nucleophilic attack but will also recognize that in the case of vinyl GABA, Schiff-base formation with the PLP cofactor sets it up for conjugate addition. In the case of DFMO, the chemist will recognize that Schiff-base formation with the PLP cofactor in the active site will labilize the molecule for decarboxylation, resulting in fluoride elimination and conjugate addition–elimination. Such prediction, however, is not always straightforward, particularly in the case of CYPs and their substrates and inhibitors. Although the reactions catalyzed by CYPs are well known (e.g., hydroxylation, hydroxylation-O-dealkylation, epoxidation), whether a given molecule will be a substrate or at which site it may be modified can be cryptic and not necessarily easy to predict. Even if one identifies the major site of reaction on the molecule, MBI may be the result of modification at a minor site. The problem is compounded *in vivo*, where multiple CYP isoforms exist and where detection of a minor metabolite responsible for MBI may be difficult.

Molecules that show slow binding with very small off rates and fully irreversible inhibitors (inactivators) cannot be easily distinguished on the ordinary laboratory time scale, but there is nonetheless a conceptual difference between them. The term "slow binding" refers to a slow approach to a binding equilibrium – the key here is that the final state is an equilibrium, no matter how far to one side it lies. Given it is an equilibrium, it can be perturbed (such as by reducing free inhibitor concentration) and a new equilibrium state will be achieved, no matter how slowly. For irreversible inhibitors (inactivators), the final state is not an equilibrium; it can be unequivocally stated in most cases that the dissociation rate is 0; this cannot be categorically stated for a slow-binding inhibitor – the k_{off} may approach 0 and the complex may dissociate over days or months when free inhibitor is removed, but it is nonetheless eventually going to dissociate. A variety of physical methods, including the use of labels, may be used to demonstrate that a stable covalent bond is indeed formed – two of these are X-ray crystallography and LC–MS. In the latter method, a mass shift of the order of a few hundred Da can be easily observed. Since tight, noncovalent binders may be observable by LC–MS, first digestion of the complex to peptides before MS analysis will ensure that noncovalently bound molecules will not be observed. This method also has the advantage that the amino acid residue or residues labeled by the inhibitor can be identified.

For mechanism-based inhibitors and affinity-labeling agents, the classical way to determine k_{obs} for a given inhibitor concentration is to incubate inhibitor with enzyme, take out a portion at different time points, and assay for enzyme activity remaining. An exponential decay of activity with time is seen, and k_{obs} derived from this (a semilog plot of %activity of control vs. time is linear). A series of such experiments at varying inhibitor concentrations furnishes data for a plot of k_{obs} versus [I], which then yields k_{inact}, K_I, or directly the ratio k_{inact}/K_I. (The monograph by Silverman describes special cases in which the rate of loss of activity is not exponential but either slows down or speeds up with time with respect to a single exponential [10].) This method of determining inactivation parameters was first described by Kitz and Wilson, and the method and associated plot bear their names [27]. For slow-binding inhibitors with a very slow k_{off}, the same method can be applied for determining k_{obs}, especially where a continuous assay is not available. Leytus et al. [28] and Liu and Tsou [29] first described the determination of the inactivation parameters for irreversible inhibitors directly from progress curves (continuous monitoring of enzyme activity with substrate in the presence of inhibitor). The relevant equations are described in the monograph by Marangoni [30]. This is a convenient way of deriving these constants without resorting to the cumbersome Kitz and Wilson method. As with any type of progress curve analysis and fitting for determining inhibition parameters, one needs to be sure that the curvature is due only to interaction of the enzyme with the inhibitor being studied and not due to enzyme instability, substrate depletion, or product inhibition. When a continuous assay is available, progress curve analysis is also the preferred way of determining k_{obs} for a series of inhibitor concentrations for slow-binding inhibitors. Primary data for the progress curve of either type of inhibitor (a data array of, e.g., fluorescence or absorbance readings at each time point collected) can be imported into a program such as GraphPad Prism® and then fitted to the appropriate equation. It may be necessary to input these equations using the capability of the software that lets the user define the equations to be used for the curve fitting.

A third parameter, the partition ratio, is useful in studying mechanism-based inhibitors [10]. It is a measure of the efficiency of inhibition – it is the ratio of turnover events (with product release) to inactivation events. Mechanism-based inhibitors, as noted earlier, are turned over by the enzyme as though they are substrates. For 100% efficient inhibitors, every catalytic cycle of the inhibitor results in inactivation of the enzyme. For less efficient inhibitors, the product escapes the active site and the enzyme will be inactivated only occasionally. When the product released is reactive, as is sometimes the case of MBI of CYPs, the partition ratio has significance for the chemical modification by such reactive species in the general biological milieu. For CYPs the partition ratio represents a lose–lose situation: a high efficiency results in potent inhibition of the CYP, with the potential for drug–drug interactions; a low efficiency results in less potent CYP inhibition, but reactive molecules (generally electrophiles) are released from the active site and can react with surrounding proteins, causing loss of protein function or an immune reaction. In the case of acetaminophen, the formation of an iminoquinone with poor efficiency of inhibition in the active site results in release of this molecule. In some instances (high doses, in susceptible individuals), the result is fulminant liver failure, likely due to a triggering of an immune response by the modified proteins [31] (see Section XI on haptenation).

IV. SPECIFIC EXAMPLES OF COVALENT INHIBITORS

Inhibitors that make covalent bonds with their targets are discussed according to the classification in Figure 2. As mentioned earlier, these can be of either the quasi-irreversible or fully irreversible type. Covalent irreversible inhibitors were reviewed by Johnson et al. [32].

A. Addition to an Electrophilic Carbonyl or Aldehyde, Nitrile, or Boron Center

These types of inhibitors constitute examples of inhibitors that make a covalent link with the enzyme, and yet are reversible, with varying degrees of dissociation half-life. All molecules of this type may be considered to be transition-state analogs.

Bortezomib (Velcade®) Boron, belonging to Group 13 in the periodic table, is electron deficient in the +3 oxidation state. It acts as a Lewis acid and can stably bond with nucleophiles, which contribute a pair of electrons into a vacant orbital and complete its octet. In bortezomib (Fig. 7), the C-terminal carboxyl group of a dipeptide is replaced by a boronate function [33]. This molecule, when bound to the catalytic subunit of the 20S proteasomal hydrolase, can add the –OH group of the active-site threonine (in this case the N-terminal residue) to the boron center and form a stable tetrahedral complex. The molecule is a slow-binding inhibitor with a dissociation half-life of ~110 min [34]. The drug is approved for the second-line treatment of malignant melanoma and mantle cell lymphoma. A next generation of proteasome 20S catalytic complex inhibitors is in trials, exemplified by carfilzomib [35], which, like the natural product epoxomicin, is an epoxyketone (see Section IVB). Bortezomib resistance in the clinic is an issue, and the availability of other inhibitors, particularly those with a different inhibition mechanism and structurally distinct, could be clinically useful.

DPP-IV Inhibitors A number of DPP-IV inhibitors that are nitriles have reached the market (vildagliptin, saxagliptin; denagliptin in late development) (Fig. 7). These are peptidomimetics bearing the electrophilic nitrile function (the

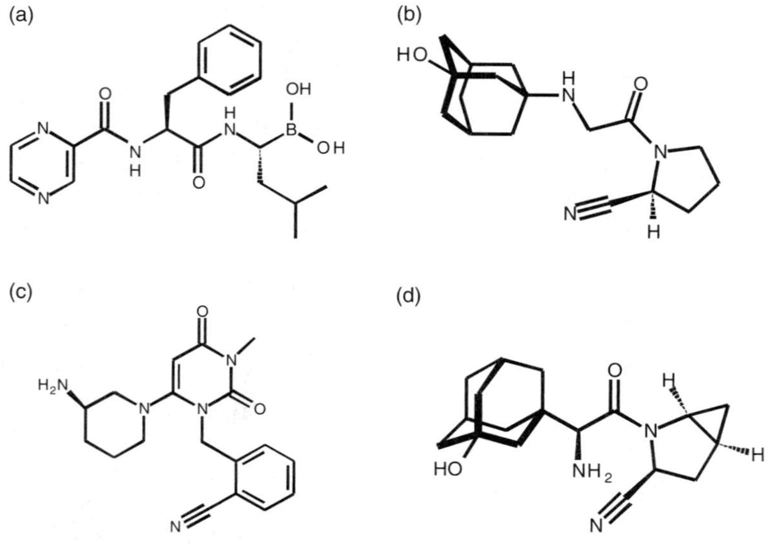

FIGURE 7 (a) Bortezomib; (b) vildagliptin; (c) alogliptin; (d) saxagliptin.

electron-withdrawing nitrogen makes the carbon electrophilic) to which the active-site serine (serine 630) of DPP-IV can add, more or less reversibly. The first DPP-IV inhibitor to reach the market, sitagliptin (Januvia®), is not a nitrile and is a rapidly reversible inhibitor that does not show slow binding [36]. DPP-IV is an N-terminal dipeptidase that cleaves (and inactivates), among other substrates, the incretin hormone glucagon-like peptide-1 (GLP-1). Inhibition of DPP-IV increases the half-life of this hormone, which is produced after a meal, and improves control of blood glucose in diabetics [37]. Because it is an aminopeptidase, optimal substrates and inhibitors have a protonatable amino group (corresponding to the N-terminal amino group of the substrate) that binds to glutamates 206 and 205 in the active site. This requirement for an amino group to be present, and the presence of an electrophile the right distance apart, means that DPP-IV inhibitors are apt to undergo entropically favorable intramolecular cyclization in solution, if geometrically and sterically possible. This was shown for the Pro-Pro analog Pro-Pro-B-$(OH)_2$ inhibitor [38,39] and was also observed for the pentafluoroethylketone analog [25], as evidenced by the gradual loss of inhibition in the assay. Vildagliptin is slow binding, with $K^*_i = 11$ nM (human), $k_{on} = 1.3 \times 10^{-5}$ M^{-1}s^{-1}, and $k_{off} = 1.3 \times 10^{-3}$ s^{-1} [40]. *In vivo*, it has a relatively short pharmacokinetic half-life but a longer duration of action [41]. The crystal structure of saxagliptin with human DPP-IV shows the covalent attachment of the serine –OH to the nitrile carbon [42]. The molecule with the same scaffold but without the nitrile is ~20-fold less potent. It is surprising that there is no evidence suggesting substantive hydrolysis of nitrile inhibitors by DPP-IV by enzyme-catalyzed attack of water on the imidate formed with the serine and the nitrile.

Hepatitis C Protease Inhibitors The first report of a synthetic peptidic α-ketoamide protease inhibitor was by Hu and Abeles [43]. Subsequently, many groups reported such types of inhibitors [5,17]. α-Ketoesters, α-ketoamides, 1,2-diketones, fluoroketones, and ketobenzoxazoles, among others, are extensions of the concept of the electrophilic aldehydic peptides, in which the carbonyl is rendered electrophilic such that an active-site serine or cysteine can add to it. This principle was first elucidated using naturally occurring aldehydes such as leupeptin [5]. Two natural products (cyclotheonamide and postatin) bearing the α-ketoamide group were discovered around the time of the first report of synthetic α-ketoamides. Recently, the crystal structures of cyclotheonamide A complexed with human α-thrombin and bovine β-trypsin were reported [44]. In both structures the addition of the active-site serine to the ketone of the ketoamide group is apparent.

Both marketed drugs for hepatitis C viral infection, inhibiting the HCV (NS3) serine protease, are α-ketoamides. Boceprevir (Fig. 8), developed by Schering-Plough (now part of Merck), is reported to be a slow-binding inhibitor [45]. Telaprevir (Fig. 8), marketed by Vertex, is also a ketoamide and reported to be a slow-binding inhibitor of the NS3 protease [46]. The $t_{1/2}$ for dissociation of the inhibitor from the NS3 protease genotype 1a is 58 min ($K^*_i \sim 10$ nM). It also inhibits proteases of genotypes 2a and 3a, albeit with reduced potency and a shorter dissociation half-life. Narlaprevir, a second-generation inhibitor in development, also from

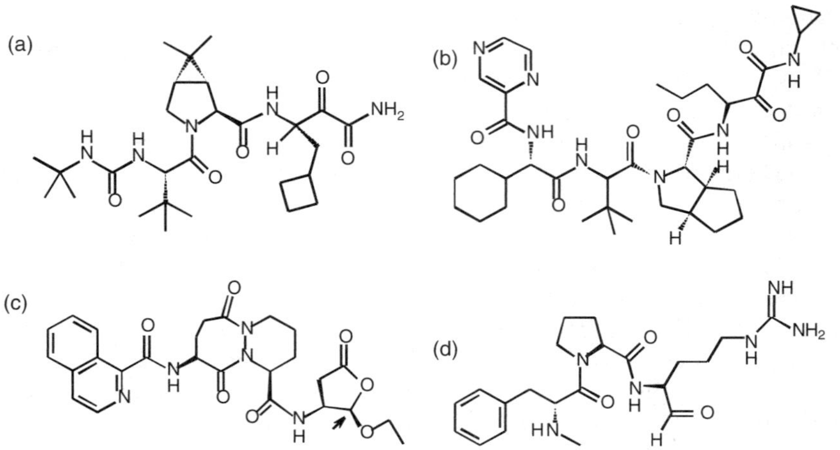

FIGURE 8 (a) Boceprevir; (b) telaprevir; (c) pralnacasan; the arrow shows the aldehydic carbon; (d) efegatran.

Schering-Plough, is a slow-binding inhibitor with K^*_i ~7 nM, with a lower limit for the dissociation half-life of the E–I complex of 1–2 h for genotype 1b NS3 protease [47]. The crystal structure of narlaprevir with the NS3 protease shows addition of Ser139 [47].

Pralnacasan (VX-740) This aldehyde inhibitor (masked as a lactol-acetal, with the lactol being formed intramolecularly with an aspartate carboxylate) is in development from Vertex as an inhibitor of caspase-1 (interleukin-1β-converting enzyme, ICE) (Fig. 8). ICE is responsible for the cleavage of the IL-1β precursor to release mature IL-1β, a potent proinflammatory cytokine. Wannamaker et al. [48] reported that the molecule inhibits IL-1β (and IL-18) release from cytokines *in vitro* and is efficacious in several animal models of inflammation (skin inflammation, collagen-induced arthritis model, and inhibition of LPS-induced cytokine release).

The molecule appears to be in clinical trials not for classical inflammatory disease but for treatment-resistant epilepsy. In 2011 Vertex announced the results of a phase IIa study – based on safety, tolerability, and preliminary clinical efficacy, the company decided at that time to enter into further trials with this molecule [49].

Efegatran (LY-294468) Efegatran (Fig. 8) is a tripeptide aldehyde (N-methylated at the N-terminus) that is a thrombin inhibitor. Reports from clinical trials were published in 1999 and 2002 [50,51]. It appears to have been stopped for reasons of lack of superior efficacy over standard treatments and not because of target-unrelated safety issues.

Aldehydes have not been favored as drugs due to the perceived safety risks associated with reactivity with biological amines and other nucleophiles. However, such imine (or acetal or hemiacetal) formation is reversible and would reverse when the concentration of the drug is depleted by metabolism or excretion. There is an exception however in the case of sugars such as glucose bearing a hydroxyl group alpha to the aldehyde – initial formation of the imine is followed by a rearrangement (the Amadori rearrangement) that results in the formation of a stable aminoketone [52].

B. Alkylations (Direct Displacement at Carbon)

The most relevant molecules in this class are epoxides, acting by alkylation of a nucleophilic residue of the enzyme with ring opening.

Fumagillin Fumagillin (Fig. 9) is a natural product, isolated from *Aspergillus fumigatus*. Fumagillin covalently inhibits the enzyme methionine aminopeptidase type 2 (MetAP-2, a binuclear metalloenzyme), first shown by labeling with a biotinylated fumagillin analog [53]. This inhibition is responsible both for its anti-angiogenic activity and likely for its anti-microsporidial activity. MetAP-2 as the relevant target for the inhibition of endothelial cell proliferation by fumagillin was cast in doubt by a report in 2004 in which depletion of MetAP-2 by siRNA did not

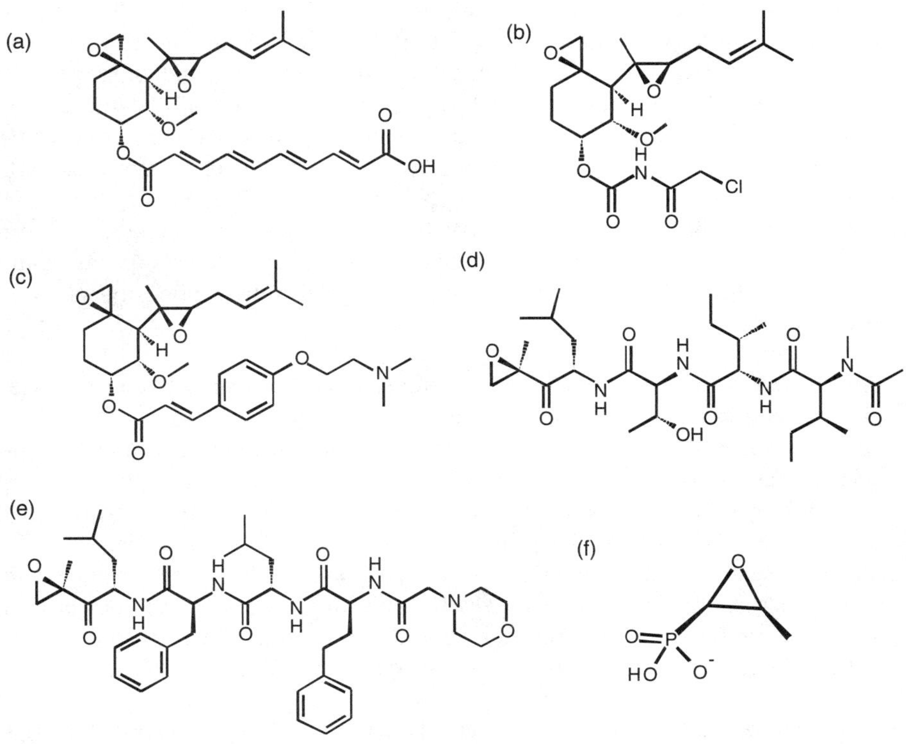

FIGURE 9 (a) Fumagillin; (b) TNP-470; (c) beloranib; (d) epoxomicin; (e) carfilzomib; (f) fosfomycin.

alter the response of endothelial cell growth or the effect of fumagillin [54]. In a more definitive subsequent report in 2006, however, using targeted and conditional knockout mice, it was established that MetAP-2 is indeed necessary for endothelial cell proliferation and vascular development and that it is the target for the fumagillin analog TNP-470 [55].

Fumagillin has been used in humans to treat intestinal microsporidiosis, which is a chronic diarrhea and wasting caused in immunocompromised individuals by the unicellular parasite *Enterocytozoon bieneusi* [56]. It has been sold as Fumidil B® for a microsporidial disease, nosema, in honeybees [57], though recently withdrawn in the EU. Nosema infection may have a role in the recent, commercially catastrophic colony collapse disorder (CCD) in honeybees [57]. Co-crystal structures have been reported with fumagillin covalently bound to the microsporidian MatAP-2 [58] and of fumagillin bound to human MetAP-2 [59]. The *spiro* epoxide is opened by attack of the thiol of Cys25. The review by Drahl et al. gives an account of fumagillin and its associated pharmacology and biology [23]. Like aspirin, the target residue of fumagillin is not a catalytic residue.

Fumagillin Analogs The first synthetic fumagillin analog to enter human clinical trials for oncology indications was TNP-470 (Fig. 9) [60]. A study using a combination of chemical and genetic means established that TNP-470 (and presumably also fumagillin) inhibition of MetAP-2 interferes with noncanonical Wnt signaling, which has a role in development, cell differentiation, and tumorigenesis [61]. The molecule reached phase III clinical trials in 2000. Following earlier reports on the effects of this molecule on obesity, it was recently shown that TNP-470 reduces caloric intake and weight gain in mice fed with a fatty diet [62]. Another fumagillin analog, ZGN-4033 (beloranib, from Zafgen, Fig. 9), is in clinical trials to treat obesity in humans, with entry into phase II predicted for 2012 [63]. It would appear that adipogenesis, like organogenesis, depends on angiogenesis, and thus MetAP-2 inhibition at least partly explains the antiobesity action of these molecules.

Epoxomicin Epoxomicin (Fig. 9) is a natural product that is an α',β'-epoxyketone. This unique juxtaposition of functionalities makes epoxomicin an irreversible inhibitor of the proteasome 20S catalytic activity with an unusual mechanism and resulting in an unusual structure of the bound inhibitor [64]. The co-crystal structure shows the formation of a morpholine ring, and the proposed mechanism involves the addition of the –OH of Thr-1 of the β-5 subunit to the keto function to give a hemiketal, with concomitant addition of the N-terminal amine of Thr-1 to the epoxide with ring opening. An epoxomicin analog, carfilzomib (Fig. 9), is in development as a proteasome inhibitor, in the expectation that it will be effective against tumor cells that have developed resistance to bortezomib (see Section IVA).

Phosphomycin (Fosfomycin) This relatively structurally simple epoxide (Fig. 9) is an antibiotic and is sold as Monural® (also Monurol®) in certain countries for the treatment of urinary tract infection and cystitis in women caused by the susceptible bacteria *Escherichia coli*, *Enterococcus faecalis*, and other susceptible bacteria [65]. The molecule inhibits the enzyme MurA [26], which catalyzes the first step in bacterial cell-wall synthesis. MurA catalyzes the unusual transfer of the enolpyruvyl group of phosphoenolpyruvate to a sugar–OH of UDP-*N*-acetylglucosamine to make the enol ether enolpyruvyl-UDP-*N*-acetylglucosamine. Cys115 is modified in this enzyme by fosfomycin [26].

C. Addition/Conjugation Addition to a C–C Double or Triple Bond, Including Michael Addition and Addition to Quinones, and Addition–Elimination (Including Possible SN2′)

Vinyl GABA As discussed earlier, vinyl GABA and DFMO were the first rationally designed mechanism-based inhibitors to be marketed. The target of vinyl GABA is GABA aminotransferase, a PLP-dependent enzyme. Inhibition of this enzyme prevents the degradation of GABA and increases its levels in the brain, thus making the enzyme an attractive target for inhibition for the treatment of epilepsy. The mechanism of inhibition most likely involves the formation of a Schiff base with the PLP, setting up the molecule for conjugate addition at the terminal alkene carbon [8]. A crystal structure of porcine liver enzyme inhibited by vinyl GABA shows the adduct of Lys329 at this carbon in the inhibitor–PLP complex [66]. Vinyl GABA (vigabatrin, Sabril®) has been marketed in many countries except the United States for a long time and was approved in 2009 for the first time in the United States for the treatment of infantile spasms and certain types of seizures in adults.

DFMO (Alpha-(difluoromethyl)ornithine) DFMO was demonstrated to be a mechanism-based inhibitor of ornithine decarboxylase [9]. This enzyme catalyzes the first step in the biosynthesis of polyamines, which are important for cellular proliferation. Although not efficacious as a broad anticancer agent, the molecule found a niche for the treatment of trypanosomiasis (sleeping sickness). This PLP-dependent enzyme makes a Schiff base with the inhibitor, making the alpha hydrogen labile for abstraction and causing the elimination of the first mole of fluoride. The resulting vinyl fluoride is set up for conjugate addition–elimination, and a nucleophile of the enzyme adds to this vinyl fluoride and results in elimination of a second fluoride and irreversible inhibition. The co-crystal structure of the *Trypanosoma brucei* enzyme with DFMO has been solved and shows that the nucleophile that adds is Cys360 [67].

Aromatase Inhibitors Aromatase is a cytochrome-P450-dependent enzyme whose role is the biosynthesis of female steroid hormones from androgens. In a three-step reaction, the A ring of the steroid nucleus is aromatized, and the C-19 angular methyl group is cleaved off [68]. Aromatase inhibitors are strong contenders as first-line treatment of postmenopausal estrogen-dependent breast cancer, set to replace the classical drug tamoxifen (estrogen receptor antagonist) in this role. The latest generation steroid-skeleton-based aromatase inhibitor is typified by exemestane (Aromasin®) (Fig. 10).

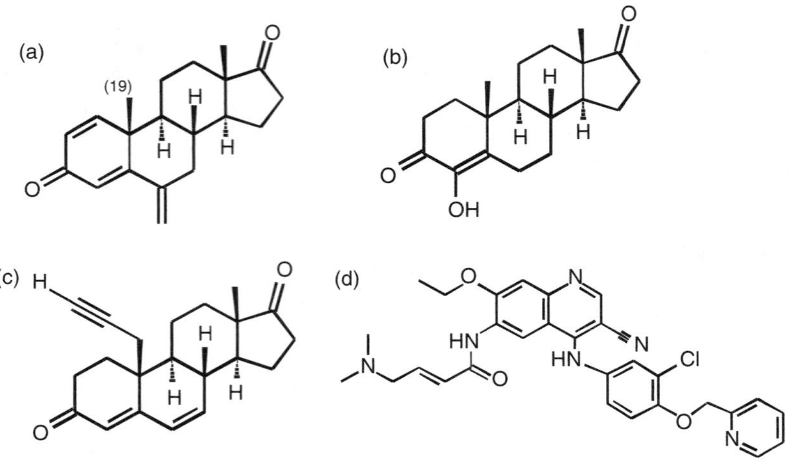

FIGURE 10 (a) Exemestane; (b) formestane; (c) plomestane; (d) neratinib (HKI-272).

This molecule is described as an irreversible inhibitor of aromatase [69,70]. Other aromatase mechanism-based inhibitors are formestane (4-hydroxyandrostenedione; Lentaron®) (Fig. 10) [70,71] and the first molecule characterized as an aromatase mechanism-based inhibitor, the acetylenic compound MDL 18 962 (plomestane) (Fig. 10) [72–74]. Formestane was first thought to be a reversible inhibitor, but later reports characterized it as a mechanism-based inhibitor. The mechanism of inhibition by these molecules is not clear; presumably conjugate addition or SN2′ displacement is involved. In fact the details of the multistep transformation catalyzed by aromatase have also not been fully delineated [68]. A commonly used criterion for deciding whether an inhibitor is mechanism based or not for aromatase (or other CYP450-dependent enzymes) is to test the molecule for inhibition in the presence and in the absence of NADPH (or NADPH regenerating system). Because NADPH as an electron donor (via the auxiliary flavoenzyme cytochrome P450 reductase) is required for catalytic turnover, inhibition only in its presence but not in its absence is taken as an indication of MBI. A recent paper discusses the structural basis of exemestane inhibition and the aromatase reaction [75]. Aromatase inhibitors are subject to abuse for off-label bodybuilding purposes, and an underground market for these appears to be thriving for those desiring to use these as chemical enhancers.

Kinase Inhibitors: EGFR Family and Other Kinases In recent years, a number of kinase inhibitors have been developed that are irreversible inhibitors. The farthest along in terms of clinical trials are inhibitors of EGFR (HER1) kinase and its family members (HER2 and HER4). The residue for covalent attachment is Cys797 (or Cys773, depending on the numbering system) in HER1 and Cys805 in HER2 [76]. HER3 does not have this conserved, noncatalytic, kinase domain residue. All of these inhibitors are Michael acceptors, and some of the compounds in clinical trials are neratinib (HKI-272; Fig. 10), dacomitinib, and afatinib. Activating mutations in the intracellular kinase domains of these receptors are associated with a number of aggressive human cancers. One of the driving forces for the development of irreversible inhibitors of these targets was the appearance in the clinic of mutations arising that rendered the targets or tumors resistant to the marketed reversible inhibitors such as Iressa® and Tarceva® [77]. Indeed, it was shown in early reports that irreversible inhibitors were effective against proteins with one of the common resistance-conferring mutations in EGFR, the T790M mutation [77,78]. The T790M mutation causes resistance by increasing the affinity for ATP [78], and it was recently shown that this same mutation appeared *in vitro* and in a corresponding clinical case and conferred resistance to an irreversible inhibitor [79,80]. These irreversible inhibitors also bind at the ATP site, and, at low inhibitor concentrations, those that are physiologically pertinent will be outcompeted just as effectively as a reversible inhibitor and the K_I will increase: although the k_{inact} remains the same, at low inhibitor concentration, the molecule will be a less effective inactivator.

In chemistry, Michael additions can be reversible or irreversible, depending on the conditions and the nature of the nucleophile and the acceptor. On enzymes, however, Michael adducts appear to be uniformly described as "irreversible" though hard data on this point (e.g., rate of reactivation) are lacking. Clearly a covalent bond has formed [78,81], and the Michael adduct may be stable by virtue of the optimal positioning of the cysteine with respect to the acceptor and the stabilization of the adduct by other, nonbonding interactions with the enzyme. In contrast, where there is no such stabilization in an active site or an optimally positioned nucleophile, the Michael addition might be expected to be reversible; indeed, in one report, the covalent reaction of neratinib with human serum albumin was shown to be reversible (however, the rate of reactivation was not determined) [82]. A recent paper by Serafimova et al. addresses the question of the reversibility of Michael addition in enzyme inhibitors [83]. This paper reinforces that it is possible to have covalent addition and still see rapidly reversible inhibition in the case of Michael addition. This type of rapid reversibility may also be observed with the types of inhibitors described in Section IVA.

Other kinases have been targeted with this type of approach as well, including Btk [84], p38-α [85], and JNK [81]. The company Avila (now part of Celgene) has been an advocate of the concept of covalent inhibition in the context of their Btk inhibitor [86,87]. The same company has also reported on Michael acceptor-type inhibitors of the HCV protease [88].

DNA Hypomethylation Agents: 5-Azacytidine, 5-aza-2′-Doexycytidine, and Zebularine Upon incorporation into DNA (directly or after conversion to the deoxyribo form), these molecules (Fig. 11) form a covalent bond with DNA methyltransferases. The normal methylation mechanism involves the addition of an enzyme cysteine at the C6 position of the cytosine base, followed by nucleophilic attack of the carbanion generated at the C5 position on the methyl group of the cofactor S-adenosylmethionine [89]. Abstraction of the proton at C5 followed by expulsion of the added enzyme nucleophile gives the C5-methylated cytosine product. In the case of azaC, the enzyme thiol

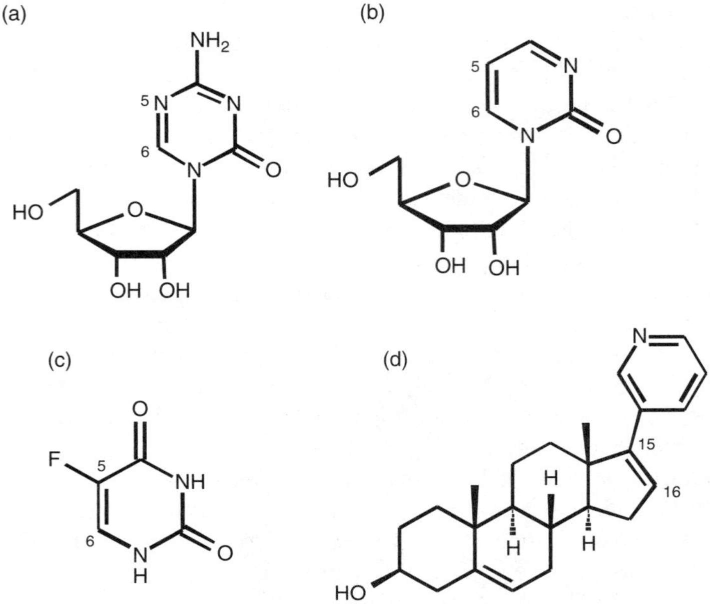

FIGURE 11 (a) 5-Azacytidine; (b) zebularine; (c) 5-FU; (d) abiraterone.

adds at C6, accompanied by methylation at C5 [89–92]. The adduct remains stable. In the case of zebularine, the enzyme nucleophile adds at C6 of this pyrimidone base, followed by protonation at C5, giving a stable adduct [89]. The crystal structure of zebularine complexed with the *M. Hha* I enzyme shows the adduct and the absence of methylation at C5 [89]. Early work by Santi established that in the case of azaC, the adduct with the bacterial methyltransferase from *Hpa* II was stable for at least 3 days and survived denaturation by sodium dodecyl sulfate [90]. Both the 5-aza compounds are in use as marketed anticancer drugs as azacitidine (Vidaza®) and decitabine (Dacogen®); these appear to be the only marketed drugs thus far that target the methylome.

5-Fluorouracil 5-Fluorouracil (5-FU) (Fig. 11), an anticancer agent (Adrucil®), after metabolic conversion to 5-fluoro-2′-deoxyuridylate, inhibits the enzyme thymidylate synthase. The net reaction catalyzed by this enzyme, the conversion of 2′-deoxyuridylate to thymidylate using the cofactor methylenetetrahydrofolate, is the last step in the biosynthesis of thymidylate [68]. The inactivation of this enzyme by 5-fluoro-2′-deoxyuridylate involves the addition of the thiol of a cysteine at C6 accompanied by attack of the incipient carbanion at C5 on the methylene carbon of the cofactor N^5,N^{10}-methylenetetrahydrofolate to give a stable adduct. *In vivo*, this adduct has a half-life of decomposition of 14 h (Reference [12] and references therein).

These drugs based on modified nucleobases have multiple modes of action as antitumor agents, with inhibition of the enzymes as discussed earlier representing one of their actions. For example, in the case of 5-azacytidine, the major metabolic pathway leads to incorporation into RNA, with DNA incorporation being a minor pathway [92].

Abiraterone (Zytiga®) Abiraterone (Fig. 11) is an inhibitor of 17-α-hydroxylase/17, 20 lyase (CYP17, CYP17A1), an enzyme that catalyzes the conversion of progesterone to androstenedione and pregnenolone to dehydro*epi*androsterone [93]. The enzyme first catalyzes the hydroxylation at the 17 position (which bears a methyl ketone group). The subsequent reaction cleaves the C–C bond between the steroid and the ketone, and the product is the 17-keto molecule. Phase III trials of abiraterone were halted prematurely due to an interim successful outcome, and it was approved in 2011 for the treatment of castration-resistant prostate cancer.

Abiraterone is described as a mechanism-based irreversible inhibitor [94], but its mechanism of inhibition is not clear. It is included in this section due to the possibility that the pyridine is N-oxidized by the enzyme, setting it up for a Michael addition to the pyridine N-oxide ring. Another possibility is an epoxidation at the 16,17 double bond, as it was shown that this double bond is essential for time-dependent inhibition [94]. However, an epoxide analog at this position was synthesized and was only a moderately potent inhibitor with $IC_{50} = 250$ nM with no time dependence [94]. Addition of an enzyme nucleophile at the 16 position with electrons flowing to the positive nitrogen of the putative pyridine N-oxide is not possible because the point of attachment of the pyridine is *meta*. The requirement of the double bond may be for the correct geometric positioning or binding of the pyridine ring.

D. Carbamates/Ureas: Slow Substrates

FAAH (Fatty Acid Amide Hydrolase) Inhibitors Among the substrates of this serine-dependent hydrolase/amidase is *N*-arachidonylethanolamine (AEA, anandamide), one of two well-characterized endocannabinoids. In recent years, a number of groups gave designed inhibitors of this enzyme for the treatment of pain, anxiety, and depression, with the therapeutic benefit resulting from an increased central concentration of the endocannabinoid by inhibition of its breakdown. FAAH inhibitors were reviewed by Ahn et al. in 2009 [21]. Compounds have progressed into the clinic from a number of companies, for example, the Pfizer compound PF-04457845 (Fig. 12) [95], but their current status is uncertain, and none is close to approval. Carbamates and ureas carbamoylate the active-site serine FAAH, which is then only slowly decarbamoylated by hydrolysis. This type of inhibitor represents the slow substrate/alternate substrate mechanism. Attack of the catalytic serine at the carbamate (urea) carbonyl followed by breaking of the C–N bond (ureas) or C–O bond (carbamates) results in transfer of the carbamoyl group to the enzyme. The stability toward hydrolysis by the enzyme of the carbamoyl enzyme is variable. The mechanism of inhibition by carbamates was established by Alexander and Cravatt [96]. MALDI-MS mapping of a tryptic digest of FAAH inhibited by carbamates showed carbamoylation (and not oxyacylation by cleavage of the C–N bond) of Ser241.

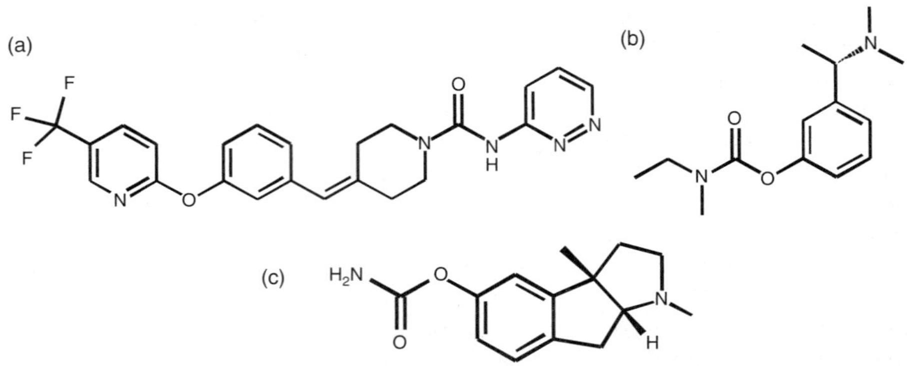

FIGURE 12 (a) PF-04457845; (b) rivastigmine; (c) physostigmine.

Acetylcholinesterase Inhibitors AChE, a serine-dependent hydrolase, is a target for inhibition for the improvement of cognitive function in Alzheimer's disease, to improve muscle tone in myasthenia gravis, and for glaucoma. Several drugs in use for it are carbamates; these are distigmine, neostigmine, physostigmine (Fig. 12), pyridostigmine, and rivastigmine (Fig. 12). They have variable selectivity between acetyl- and butyrylcholinesterases. Physostigmine is a natural product; pyridostigmine, distigmine, and neostigmine have positive charges and so are non-CNS penetrant – they are useful only as peripheral AChE inhibitors. Like FAAH inhibitors, carbamate inhibitors of AChE are covalent inhibitors by carbamoylation of the active-site serine, with variable rates of reactivation.

Studies on the inhibition of human AChE by rivastigmine were reported by Bar-On et al. [97]. They report the rate of inhibition of the recombinant enzyme as a bimolecular rate constant of 3300 M^{-1} min^{-1}. The inactivation rate reported for human butyrylcholinesterase was 30-fold faster. For AChE, no reactivation could be observed by dialysis, and a 25% return of activity was detected after 48 h after gel filtration. For human butyrylcholinesterase, <10% activity returned after 24 h. These authors also determined the crystal structure of the torpedo eel enzyme inhibited by rivastigmine. Not only is the carbamoyl link with the enzyme observed, as expected, but the leaving group part is also (noncovalently) bound in the active site in the structure. The leaving group moiety is not found in the active site of FAAH inhibited by the carbamate inhibitor PF-750 [21].

There are carbamate and organophosphate insecticides in use that inhibit AChE. These are of commercial importance for agriculture, but they also directly and adversely impact human health because of the danger of accidental exposure. Organophosphates are discussed in a later section.

E. Acylation

Aspirin (Acetylsalicylic Acid) Aspirin (Fig. 13) was first recognized by Smith and Lands in 1971 as a time-dependent inhibitor of prostaglandin synthesis by a dioxygenase in sheep vesicular gland preparation using an oxygen electrode to measure

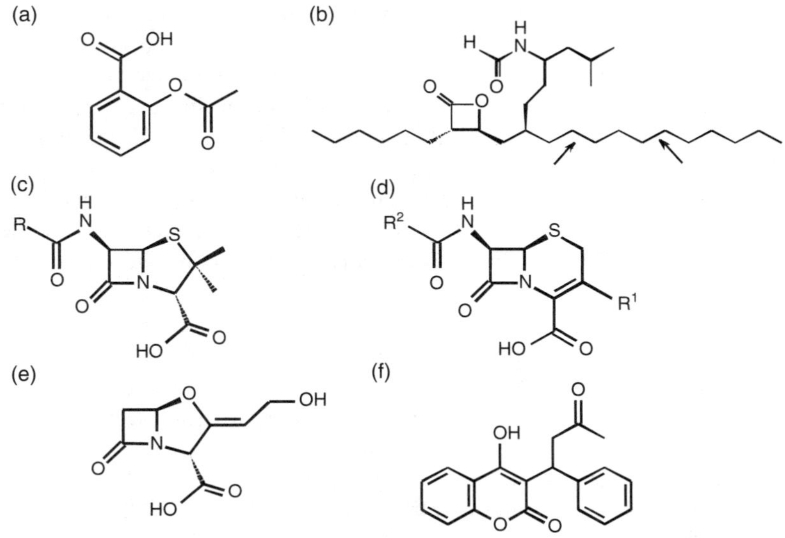

FIGURE 13 (a) Aspirin; (b) tetrahydrolipstatin; the arrows show the positions of the double bonds in the parent natural product lipstatin; (c) penicillin core structure; (d) a cephalosporin; (e) clavulanic acid; (f) warfarin.

oxygen consumption [98]. Acetylation of cyclooxygenase (COX) by aspirin (with a $t_{1/2}$ of inactivation of ~15 min at 100 μM) and its protection by competition by arachidonic acid were first demonstrated by Roth et al. using radiolabeled aspirin [99]. The molecule acetylates Ser530 of COX-1 and Ser516 of COX-2 [100,101], preventing the binding of the substrate arachidonic acid to the enzymes. Inhibition of COX-1 and COX-2 and the consequent inhibition of the synthesis of prostaglandins are responsible in large part for both the beneficial and adverse effects of aspirin. Aspirin has selectivity for COX-1 over COX-2, whereas newer generation of COX inhibitors aims for the opposite selectivity to minimize GI side effects. The history of the elucidation of aspirin's mode of action is nicely recounted in a brief review [102]. Given the abundance of nucleophiles (water, amines, exposed protein hydroxyls and amines), it is remarkable that aspirin is (relatively) selective for the particular serine in COX, especially as that serine is not catalytic and therefore does not have enhanced nucleophilicity. An elaborate hydrogen bonding and ionic interaction network positions the aspirin molecule exactly and enhances its reactivity toward Ser530 (COX-1) [23] – a local "perfect storm" for the reaction to occur. The interaction of the carboxylate of aspirin with an arginine not only serves to position the molecule but would also make the phenolate a better leaving group.

Tetrahydrolipstatin (Orlistat, Xenical®, Alli®) Lipstatin is a naturally occurring β-lactone, first isolated from the actinobacterium *Streptomyces toxytricini*. Tetrahydrolipstatin (orlistat; Fig. 13) is a hydrogenated derivative of lipstatin and is commercially sold as a prescription weight-loss drug under the name Xenical® and over the counter as a smaller dosage form as Alli®. Xenical® is one of only a few prescription drugs currently marketed for weight loss. Orlistat is minimally absorbed systemically after oral dosing and acts to inhibit pancreatic and gastric lipases in the GI tract. This lack of absorption is probably responsible for the lack of systemic side effects that might arise from nondiscriminate inhibition of lipases [103]. The hydrolysis of dietary triglycerides to free fatty acids is inhibited in the GI tract to the extent of ~30% by orlistat, thus preventing their full utilization [104]. As the acyl–enzyme is slowly hydrolyzed and the enzyme eventually deacylated, and given that the natural function of the enzyme is ester bond hydrolysis, orlistat is an alternate substrate/slow substrate or type of inhibitor, while it is also an affinity-labeling agent [105,106].

β-Lactam Antibiotics and β-Lactamase Inhibitors The β-lactam ring in this class of antibiotics containing this moiety is strained; in addition, the nitrogen lone pair is incapable of optimally participating in amide bond resonance due to the nonplanarity of the ring system. As a consequence, the carbonyl moiety in such systems is activated toward acylation. The core structures of penicillins and of a cephalosporin are shown in Figure 13. Woodward has suggested a measure of reactivity – the degree of nonplanarity as defined by the height of a trigonal pyramid, in which the nitrogen is the apex and the three connecting carbon atoms are the base of the pyramid. The greater the nonplanarity, the less the degree of amide bond resonance and the greater the degree of acylation potential [107]. The target of β-lactam antibiotics is a group of bacterial proteins, the penicillin-binding proteins (PBPs) that were originally identified by covalent labeling with radiolabeled penicillin. At least some of these proteins (bacteria have a variable number of them – *E. coli* has 12) are transpeptidases, enzymes that catalyze the cross-linking and thus strengthening of the cell-wall peptidoglycan. Transpeptidases catalyze transacylation, and an active-site serine residue forms an acyl–enzyme intermediate during catalysis [108]. It is this serine that is acylated by β-lactam antibiotics, inactivating the transpeptidase and preventing cross-linking. For example, the PBP2x of *E. coli* is rapidly acylated, and the rate of deacylation (hydrolysis) is such that the half-life of the acyl PBP is ~24 h [108].

Several mechanisms are linked to resistance to β-lactam antibiotics. One of these is the expression of the enzymes β-lactamases, which can efficiently hydrolyze and inactivate β-lactams. A strategy has evolved to coadminister a β-lactam antibiotic together with a β-lactamase inhibitor (itself a β-lactam, albeit one that efficiently inhibits the enzyme). One example of such a combination is Augmentin®, which is a combination of amoxicillin and clavulanic acid. Clavulanic acid (Fig. 13) acylates the β-lactamase, and this adduct can further react with another serine residue. Thus, it has two points of covalent attachment to the β-lactamase in the inactivated enzyme [109].

Allergy to β-lactam antibiotics is a significant health problem. Allergic reactions run the gamut of hypersensitivity types (I–IV), but one of the important allergenic mechanisms is mediated by antibodies developed against the molecule [110]. Such antibodies result from haptenation, which will be discussed in a later section. Haptenation is the result of covalent attachment of a small molecule to a protein. It is a concern with small molecules that are protein reactive and can be the cause of adverse immune reactions.

Wafarin This coumarin (Fig. 13) has long been used as an anticoagulant. It inhibits vitamin K epoxide reductase (encoded by the gene *VKORC1*), an enzyme that recycles oxidized vitamin K to its reduced form. Vitamin K is a cofactor for the gamma-glutamyl carboxylation of several blood coagulation factors such as prothrombin and factor VII and two anticoagulant proteins [111]. These proteins are hypoactive in their functions until they are posttranslationally modified. Silverman has shown that warfarin is an irreversible inhibitor of vitamin K reductase, involving a protonation by the enzyme followed by acylation of an enzyme nucleophile [112,113]. After dosing, the effect of warfarin takes ~1 day to evolve as the active clotting factors that are already present decay according to their respective half-lives [114]. There is a hypercoagulability phase, because in addition to inactivation of the procoagulant factors by warfarin, the anticoagulant protein S and protein C are also inactivated and the already present active proteins S and C decay at different half-lives than the procoagulant proteins. An analogy is that a river has been blocked (irreversible inhibition), and the downstream rivulets dry up at different rates.

Warfarin has a narrow therapeutic window, and an incorrect dose can have disastrous consequences. For this reason, warfarin was the first widely used drug to receive a pharmacogenomic label (in 2007) warning of polymorphisms in *VKORC1* and *CYP2C9* (CYP2C9 being one of the enzymes that metabolize warfarin) that could affect the dose needed for efficacy without toxicity [115]. (Other drugs such as some anticancer agents had received such a label before, but none of these was as widely used as warfarin.) Several noncoding polymorphisms are found in *VKROC1* that likely affect the transcription of the gene and are associated with a reduced required dose [116], but at least one coding polymorphism exists (Asp36Tyr) which is associated with resistance [117]. The effect of this mutation on the catalytic activity of the enzyme or the effect on its inhibition by warfarin does not appear to have been studied. Warfarin is used as a rodenticide, and it is interesting to note that resistance arises in some strains of rats, one mechanism for which is a change in mechanism of inhibition from irreversible to (at least partly) reversible [118,119].

Long the mainstay (together with heparin and low-molecular-weight heparin) for anticoagulation therapy, warfarin may be destined to be replaced by the newly approved direct, reversible inhibitors of the proteolytic activity of thrombin and other clotting factors due to its small therapeutic window and adverse safety profile.

F. Mixed Disulfide

Omeprazole (Prilosec®) Prilosec (omeprazole; Fig. 14) is an example of an enzyme inhibitor prodrug whose active metabolite makes a disulfide with the target enzyme [120]. (Another important class of drugs, of which clopidogrel is an example, uses this mechanism – the target is the ADP receptor on platelets. Compared to enzyme inhibitors in use as drugs, covalent inhibition of receptors and ion channels is rare, although noncovalent ligands are known for both that dissociate slowly.) Omeprazole is converted in an acidic environment to a sulphenamide form (Fig. 14) that can react with cysteine sulfhydryl groups – Cys813 in particular – on the proton pump ($H+/K+$-ATPase) to form a mixed disulfide and inhibit the activity of the proton

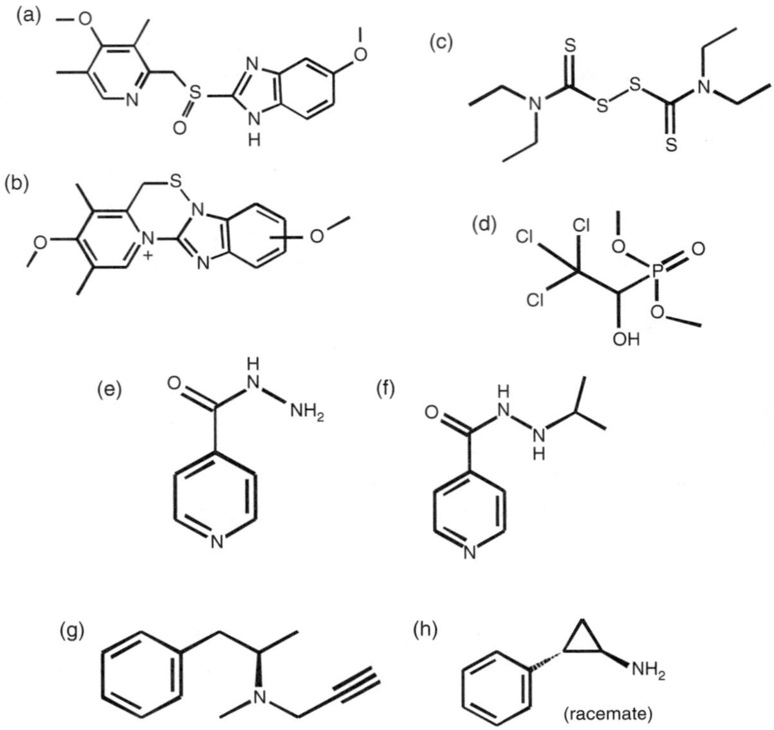

FIGURE 14 (a) Prilosec; (b) active metabolite of Prilosec; (c) disulfiram; (d) metrifonate; (e) isoniazid; (f) iproniazid; (g) selegiline; (h) tranylcypromine.

pump [120]. Esomeprazole (Nexium®) is the *S*-enantiomer of omeprazole (sulfur stereocenter) with better bioavailability.

Disulfiram (Antabuse®) Disulfiram (Fig. 14) is an older example of a drug that is an irreversible inhibitor by virtue of formation of mixed disulfide with an enzyme [121]. This molecule (or its metabolite) irreversibly inhibits aldehyde dehydrogenase [121] by disulfide exchange with a cysteine and is used for the treatment of alcohol dependence. Inhibition of this enzyme prevents acetaldehyde (formed from ethanol by alcohol dehydrogenase) from being further metabolized to the acid. Accumulating acetaldehyde results in a highly uncomfortable physiological reaction, thus creating an aversion to ingested alcohol.

G. Phosphorylation/Phosphonylation

Synthetic peptidic inhibitors have been designed and made that phosphonylate the active-site serine of serine-dependent hydrolases [122]. Phosphonylation of a β-lactamase was mentioned in Section IIC. In all these types of molecules, there is an O–P bond with the enzyme. If the other bonds to the phosphorus are all P–O bonds, then the reaction is a phosphorylation; if one of the other bonds is a P–C bond, then it is a phosphonylation.

Nerve gases (e.g., sarin) and a number of organophosphate pesticides that have been used commercially (e.g., malathion) are also in this category. Generally, both types of agents target AChE. These have a very significant impact on human health – through accidental overexposure of pesticides or potentially due to deliberately caused exposure in chemical warfare or a terrorist act [123,124].

One interesting organophosphorous molecule is metrifonate (Fig. 14), which has been used in animals and in humans to control parasitic infection. It has been in trials in humans for Alzheimer's disease [125], but longer-term administration, in contrast to the short-term administration sufficient for treating parasitic infection, led to unacceptable mechanism-related side effects. Metrifonate is a prodrug, being nonenzymatically converted to 2,2-dichlorovinyl dimethyl phosphate.

Antidotes exist that can be administered that reverse the inhibition of AChE by nucleophilic attack of a group of the antidote at the phosphorus of the inhibited enzyme and undergoing transphosphorylation and liberating free enzyme. One such antidote is pralidoxime, an oxime, in which the oxime hydroxyl attacks phosphorus; this molecule is only useful peripherally [124].

H. Radical Mechanisms

Radical mechanisms of inhibition have not been as easy to elucidate as some of the mechanisms involving heterolytic cleavage, and there are still some unanswered questions around these mechanisms.

Monoamine Oxidase (MAO) Inhibitors There are three basic types of irreversible inhibitors of monoamine oxidases (MAOs) A and B: acetylenes, cyclopropylamines, and hydrazines. The modern antidepressant era began after World War II (WW2), when the hydrazine-based drugs isoniazid and iproniazid (Fig. 14) were tested as antituberculosis agents [126]. (The story – possibly apocryphal – goes that chemists wanted to develop a hydrazine-derived drug, the manufacture of which would use some of the surplus hydrazine rocket fuel that was left over after WW2 [126].) They were efficacious as such, but it was also noted that the mood of some of the patients was elevated when they took the drug [126]. Subsequently, both agents were pressed into use as antidepressants. It was later discovered that the drugs are inhibitors of MAO (at that time the two isoforms A and B had not been distinguished). MAOs metabolize biogenic amines such as serotonin and dopamine (in addition to metabolizing certain dietary amines such as tyramine) by oxidative deamination [127]. Inhibition of MAO results in an increase of these amines at the synapse and restores a chemical imbalance. Isoniazid is no longer in use as an antidepressant but is still used as a first-line treatment in tuberculosis. Its mechanism as an antituberculosis agent is discussed in a later section. The acetylene type of inhibitor is typified by selegiline (L-deprenyl; Fig. 14) [128], which is used both as an antidepressant and in Parkinson's disease. It is an irreversible inhibitor selective for the B enzyme and thus avoids the infamous “cheese effect” [128] believed to be due to irreversible inhibition of MAO A (inhibiting this enzyme prevents the breakdown of tyramine, which is present in certain foods, and this precipitates a hypertensive crisis). Reversible inhibitors of MAOs are available as antidepressants, but reportedly these are not as efficacious for this use as the irreversible inhibitors. The cyclopropylamine type of irreversible inhibitor is exemplified by tranylcypromine (Fig. 14). MAOs likely act via a radical mechanism, involving a single-electron transfer to the flavin cofactor from the substrate [68]. Both selegiline and tranylcypromine are rendered reactive through a radical mechanism and attach covalently to the flavin cofactor (through the N5 nitrogen and the C4a carbon, respectively) and do not appear to make a stable covalent link directly with the enzyme [129,130]. It should be noted that in the MAOs, the flavin is covalently bound, so there is no possibility of reactivation of the enzyme by exchange with free flavin.

Gemcitabine (2′,2′-Difluorodeoxycytidine) Gemcitabine (Fig. 15) is converted sequentially to the monophosphate, diphosphate, and triphosphate by kinases. At the level of the diphosphate, it is a mechanism-based irreversible inhibitor of ribonucleotide reductase (RNR) [131], the enzyme that catalyzes the conversion of nucleotides to deoxynucleotides (at the diphosphate level). Gemcitabine is used in the treatment of certain types of cancer. As with modified nucleobases, modified sugars have multiple actions, and one of the other actions of the molecule is incorporation into DNA, causing stalling of DNA synthesis, resulting in chain termination. The mechanism of

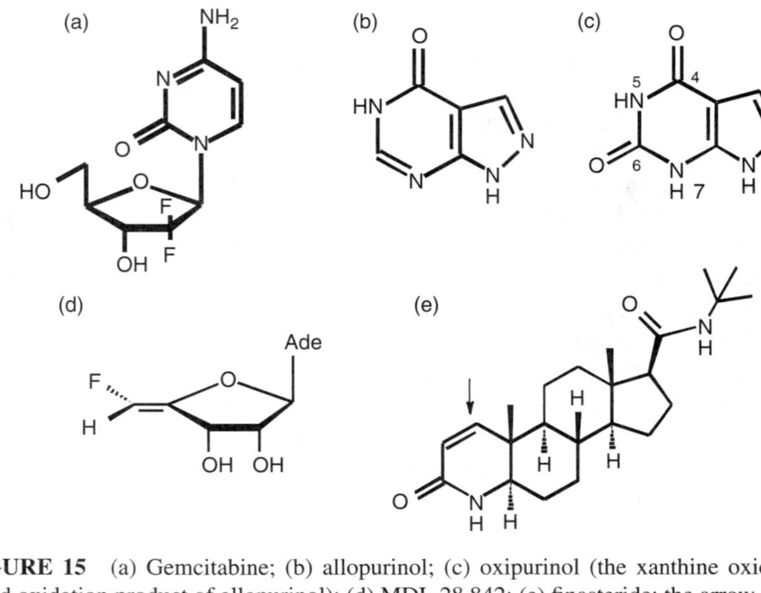

FIGURE 15 (a) Gemcitabine; (b) allopurinol; (c) oxipurinol (the xanthine oxidase-catalyzed oxidation product of allopurinol); (d) MDL 28,842; (e) finasteride; the arrow shows the point of attachment to the 4 position of the cofactor (Ade = adenine).

inhibition of RNRs by gemcitabine has been elucidated by Stubbe and coworkers and involves in the first step the generation of a radical at the 4′ position [131].

V. COVALENT ADDUCT WITH COFACTOR ALONE

In some of the instances given earlier, the inactivator may be bound covalently to the enzyme (via a protein nucleophile) while also remaining covalently bound to a cofactor (DFMO, vinyl GABA, 5-FU). In the case of MAOs, the inhibitor is bound to the flavin cofactor alone, but the cofactor is covalently attached to the enzyme (this is not the case with all flavoenzymes). There are however two interesting examples of molecules that bind covalently to a cofactor that is not covalently bound to the enzyme; the covalent complex of inhibitor and cofactor then tightly and noncovalently binds to the target enzyme for the pharmacological effect.

A. Finasteride

Finasteride (Fig. 15) is an inhibitor of the NADPH-dependent enzyme 5-α-reductase and sold as Proscar® for benign prostatic hyperplasia. The enzyme catalyzes the conversion of testosterone to dihydrotestosterone by reduction of the 4,5 double bond. Finasteride was originally believed to be a direct slow-binding inhibitor [132], but studies by a group from Merck established that it as a mechanism-based inhibitor that reacted with NADP in the active site of the enzyme [133]. After hydride transfer to finasteride, the enolate formed adds to the 4 position of the pyridinium ring of the cofactor. The resulting bisubstrate analog is tightly but noncovalently bound to the enzyme (dissociation half-life of 1 month at 37 °C). It is a very efficient inhibitor, with a partition ratio of 1.07, that is, nearly every turnover results in inactivation [133]. A study was subsequently published on the inhibition of the related enzyme 5-β-reductase [134]. In this case the inhibitor does not form a covalent bond with the NADP cofactor. The crystal structure of the enzyme bound to NADP and finasteride shows that because they are not close to one another, NADPH cannot reduce finasteride to dihydrofinasteride.

B. Isoniazid

Isoniazid, whose antituberculosis property was discovered in the early 1950s, remains a first-line therapy for treating tuberculosis. The discovery of its antidepressant properties occurred at around the same time and is described in the section on MAO inhibitors. Isoniazid inhibits the enzyme InhA, an enoyl reductase from *Mycobacterium tuberculosis*. It is not however a direct inhibitor. Isoniazid is converted to a reactive species, believed to be the isonicotinyl acyl radical, which in solution reacts with NAD at the 4 position of the pyridine ring to form an adduct [135]. This adduct is then a potent slow-binding inhibitor of InhA, with a $K^*_i = 0.75$ nM and a dissociation rate constant for the complex of $0.016\,\text{min}^{-1}$. Because the reactive species is formed by a different enzyme and reacts with the cofactor in solution before binding to InhA, it is not truly a mechanism-based inhibitor in the sense that finasteride is.

VI. FULLY IRREVERSIBLE NONCOVALENT INHIBITION

A. SAH Hydrolase (*S*-Adenosylhomocysteine Hydrolase, SAHase) Inhibition

SAH hydrolase catalyzes the conversion of *S*-adenosylhomocysteine (the product derived from *S*-adenosylmethionine after methyl transfer) to homocysteine and adenosine. This reaction which on the face of it looks like a displacement by water at the 5′ position to break the thioether bond is actually an elimination facilitated by oxidation of the 3′ hydroxyl to a ketone, followed by abstraction of the 4′ proton and elimination of homocysteine, Michael addition of water at the double bond, and finally reduction back to the –OH at the 3′ position [136]. The cofactor NAD is used in a catalytic fashion, undergoing cyclical reduction and then reoxidation in one catalytic cycle while remaining enzyme bound. The enzyme adopts a closed conformation during the catalytic cycle, presumably to keep the intermediates from diffusing out of the active site during the catalytic cycle [137]. Certain inhibitors of SAHase that can be oxidized at the 3′ position remain bound irreversibly and noncovalently in the active site. One such inhibitor is the vinyl fluoride MDL 28 842 [138] (Fig. 15), an efficient inhibitor with a partition ratio of ~0 [139]. Another is the molecule neplanocin, a natural product [137].

B. Allopurinol

Inhibition of xanthine oxidase by allopurinol (Fig. 15) is included here as another example because it was shown in early studies to be converted to oxipurinol (Fig. 15), which remained noncovalently bound to the reduced enzyme in a rearranged state from which dissociation is not possible [10]. More recently, a co-crystal structure with bovine enzyme confirmed the formation of oxipurinol bound in the active site, with strong electron density between the active-site molybdenum and the N2 nitrogen of oxipurinol with a distance of 2.3 Å between them [140]. In addition, E802 was seen to be H-bonded to the carbonyl at the 4 position, and R880 H-bonded to the carbonyl at the 6 position. In an earlier study [141] using human enzyme, when the corresponding residues E803 and E881 (corresponding to residues E802 and E880 of the bovine enzyme) were mutated, a low catalytic activity remained to convert allopurinol to oxipurinol, but the product was no longer an inhibitor. This suggests that the strength of the N2–Mo bond is not sufficient to sustain inhibition without the additional interactions and raises the question of whether this is a true example of covalent inhibition. ACE inhibitors, for example, in which there is a coordination interaction between an inhibitor thiol or carboxylate with the active-site zinc, would not be called covalent inhibitors. The oxipurinol–xanthine oxidase complex reactivates with a $t_{1/2}$ of 300 min at 25 °C due to spontaneous reoxidation of the molybdenum center [140]. Although it can be debated whether this is an example of covalent inhibition, allopurinol is clearly a mechanism-based inhibitor.

Certain noncovalent inhibitors that lie to the far side of the continuum (Fig. 1) without actually having, strictly speaking, a zero dissociation rate would fall under this category. One example is the finasteride–NADPH complex discussed earlier.

The complex is noncovalently bound to the 5-α-reductase, but the binding is so tight ($K_d = 1 \times 10^{-13}$ M) and dissociation so slow ($t_{1/2} \sim 1$ month) that on an ordinary time scale and for practical purposes, this would be called an irreversible inhibitor, although it would not qualify to be conceptually classed as an inactivator. *In vivo*, if the half-life for protein resynthesis is significantly smaller than the half-life for reactivation by dissociation, then the molecule is functionally fully irreversible.

VII. REVERSIBILITY AND DRUG RESISTANCE

In any rapidly proliferating system (tumor cells, bacteria, virus), mutations arising in an essential protein that is the target of inhibitors can be selected for if the mutation confers resistance to the inhibitor. It has been suggested that irreversible inhibitors (inactivators) will be less prone to giving rise to resistance in the clinic. While such inhibitors may indeed be effective against mutations that arise against other reversible inhibitors just by virtue of being different chemotypes, there does not however seem to be a sound basis for the suggestion that irreversible inhibitors will be less likely to produce resistant mutations by virtue of their inhibition mechanism. The PBP of bacteria susceptible to penicillins can mutate and give rise to resistance despite the irreversible nature of penicillin binding. Another example is the tyrosine kinase activity associated with EGFR – resistance has been found in the clinic to an irreversible inhibitor, as discussed in Section IVC. What probably can be stated with reasonable certainty is that the nucleophilic residue that is modified by an irreversible inhibitor is not likely to mutate, provided that it is critical for catalytic activity and provided that the enzymatic activity is critical for proliferation or replication. (Another way to state this is that mutations in such residues are likely to result in nonviable cells or virus particles and so will not be observed.) Consistent with this thinking is the recent demonstration that the transition state for substrate hydrolysis, as measured by isotope effects, appears to be unchanged in inhibitor-resistant HIV protease [142]. Aside from mutations in the target proteins, resistance can arise by other means (such as by efflux) that are independent of the E–I interaction mechanism. The development of resistance is complex however, with both primary and secondary mechanisms occurring, and it is always a good thing to have a number of drugs with different chemotype and mechanism type against the same target at the disposal of the prescribing clinician to use in a cocktail in the face of potential development of resistance.

VIII. PSEUDO-IRREVERSIBLE INHIBITION

One particularly common instance of loss of enzyme activity mediated by a small molecule that may mimic irreversible inhibition by active-site binding inhibitors is redox cycling [143]. Commonly seen with enzymes such as cysteine proteases or protein tyrosine phosphatases that rely on a catalytic cysteine, redox cyclers are redox-active molecules with the right redox potential to shuttle electrons between oxygen and DTT (present typically in assays of these types of enzymes to maintain the active-site thiol in a reduced and catalytically active state), resulting in oxidation of DTT and concomitant reduction of oxygen to hydrogen peroxide. Excess hydrogen peroxide then oxidizes the critical thiol of the enzyme. If the oxidation goes all the way to a sulfonic acid, the inhibition will be irreversible in addition to being time dependent. Examples are found for the phosphatases CD45 and protein tyrosine phosphatase 1B [144] and for caspase-3 [145]. These types of redox cyclers are typically present in large compound collections and will turn up as positives in screening. Sometimes they can be recognized from their chemical structure as being redox active (flavins, quinones) but not in all cases. One counter screen that can detect and weed out these types of pseudo-inhibitors is the analytical measurement of the (catalytic) depletion of a stoichiometric excess of DTT in oxygenated enzyme-free buffer by the compound [146]. These types of compounds also may produce a spectral change in the UV or visible region when mixed with DTT [146]. This type of inhibitor is covalent in the sense that covalent modification of a residue of the enzyme results in inhibition of activity, but not in the sense that the inhibitor molecule is actually binding in the active site to have its effect.

IX. IRREVERSIBLE INHIBITORS AS TOOLS: ACTIVITY-BASED PROTEOMICS (ABP)

Many enzymes, particularly proteinases, are regulated by inhibition by endogenous inhibitors or by virtue of being present as inactive (zymogen) precursors until activation. For enzymes that are regulated in this fashion, the study of proteomics does not distinguish between the inactive and active forms. For example, in order to understand a particular disease mechanism, it is important to be able to distinguish these forms and to know which enzymes are present in their active forms. One goal of activity-based proteomics (ABP) is to map enzymes that are catalytically active. It is accurately assumed in ABP that nonactive enzymes have blocked access to the active site, through, for example, being bound to an endogenous inhibitor. A class-specific active-site reactive probe will label only those enzymes whose active sites are accessible and therefore are catalytically active. Thus, ABP sits at the top of the -omics hierarchy (genomics → transcriptomics → proteomics → ABP). Several reviews and the references within highlight the many facets of ABP and its utility [147–152].

ABP probes were at first based on irreversible inhibitors (containing active-site reactive functionality) linked to a tag such as biotin or a fluorescent tag. In a biological milieu such as a crude cell or tissue homogenate, an ABP probe will ideally selectively and uniformly label all active enzymes belonging to a particular class. Detection of these by gel electrophoresis allows one to identify which enzymes are active. Dissection of bands followed by elution, digestion, and mass spectrometry allows assignment of bands. (Probes for enzymes whose active sites do not have a uniquely reactive nucleophile contain a photoactivatable functionality as part of an active-site-directed molecule. Irradiation generates a reactive carbene or nitrene, which can

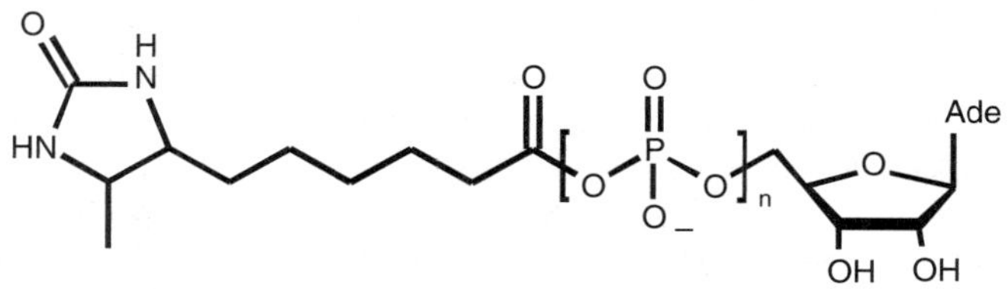

FIGURE 16 Acyl-phosphate ABP probe for kinases from ActivX Biosciences (Ade = adenine).

insert into a C–H bond and label the enzyme.) Aside from the use of ABP for understanding disease and the function of a protein or identifying new targets, one interesting use for this type of probe is to establish the selectivity of reversible inhibitors for an entire class of enzymes at once. Older examples of this exist for hydrolytic enzymes; one recent paper reports the application of this method to determine the selectivity of kinase inhibitors using an irreversible probe that is an acyl nucleotide (Fig. 16) [153]. This is a mixed acid anhydride and, as such, reactive with a nucleophile in the active site. A surprising result was obtained; the inhibition profile of certain kinase inhibitors was different using this ABP probe in a cellular context with endogenous enzymes than what was expected from the inhibitory profile determined using purified, recombinant enzymes.

There are numerous examples of the use of irreversible inactivators for validation of specific enzymes as targets for drug design or discovery. Being chemical agents, they mimic a pharmacological intervention perhaps better than a genetic knockout. It may be said that a genetic knockout heterozygote mimics reversible inhibition and a homozygote mimics irreversible inhibition (provided that in the heterozygote, the protein is present at approximately half the level of the homozygote; this appears to generally be the case). One of the caveats of using genetic knockout studies to validate a target for enzyme inhibition is the possibility of developmental compensation. Another is the tissue- or organ-selective knockout of the enzyme, whereas a pharmacological agent, by design or by chance, may favor distribution in a specific location or locations. (Both of these can in principle be addressed by using a conditional knockout and by using a tissue-specific promoter for tissue- or organ-specific knockout.) However, the reliability of the data gathered using an irreversible inhibitor depends on the selectivity of inhibition of the molecule, something that is difficult to establish in an absolute sense.

X. MBI OF CYPs

MBI of the CYP enzymes is generally undesirable, due to the potential for resulting drug–drug interaction – the MBI of a CYP can significantly affect the metabolism of another drug that is metabolized by that CYP [154]. As a result, the second drug may reach concentrations that pose a safety problem, especially if it has a narrow therapeutic window. One striking example of this case is the drug terfenadine (Seldane®) [154]. At high concentrations, this drug can cause an irregular heartbeat, and it was found that other drugs that inhibit its metabolism (e.g., erythromycin, a mechanism-based CYP inhibitor), when taken together, cause this problem, and even some deaths were reported. Seldane® has an active metabolite that does not have the cardiac issue, and terfenadine was later replaced by its active metabolite fexofenadine (Allegra®). The drug–drug interaction potential is not restricted to mechanism-based inhibitors; reversible inhibitors also have the potential to cause this problem. In the case of MBI, however, the problem is more likely to arise because of the irreversible nature of the inhibition – prolonged and unable to be completely prevented by high substrate concentration. In a contemporary drug development program, reversible CYP inhibition may be tolerated, depending on which CYPs are inhibited and also on whether a good separation (a good safety window) can be achieved between the CYP inhibition potency and the potency at the target. MBI, however, generally spells death for a drug candidate, hence the importance of getting inhibition data early in the project.

There is at least one example of the deliberate introduction of MBI, the purpose of which is to reduce the metabolism of a coadministered drug and increase its exposure. The drug in question is ritonavir, an HIV protease inhibitor and also a mechanism-based inhibitor of CYP3A4, which is used at a low dose in a cocktail with other HIV protease inhibitors to increase their efficacy. Two examples are ritonavir in combination with tipranavir and in combination with saquinavir [155]. When combined at a low dose with saquinavir, ritonavir increases the exposure of its partner drug by 20-fold.

XI. THE PROBLEM OF HAPTENATION

Proteins (and nucleic acid) contain nucleophiles. In general, free, reactive electrophiles in biological systems (compared with the electrophiles that are generated in enzyme active sites with controlled and directed reactivity and which are in a protected environment) can have various problematic consequences, by reacting with these nucleophiles [156–160]. One function of glutathione and the glutathione-conjugating system is to remove such reactive electrophiles, including those formed by the CYPs, but this is not a fail-safe system. As mentioned in the section on β-lactams, one of the problems that free electrophiles can create is haptenation, in which the normally immune-silent small molecule or endogenous protein becomes immunogenic by irreversible binding. In general, it cannot be predicted which molecule is likely to cause such a problem or which individuals may be more affected than others. Sources of reactive electrophiles include molecules such as affinity-labeling agents or other molecules that are inherently reactive and molecules that are inherently unreactive but can be metabolically converted into reactive species. Both categories could include drugs, other man-made chemicals (pesticides, food additives), and natural products. Problems with electrophiles other than haptenation can occur (e.g., the carcinogenicity of polyaromatics by epoxidation and reaction of the epoxide with nucleic acid or the loss of protein function by nonenzymatic nonspecific glycosylation of proteins as a result of high blood glucose). Haptenation can result in different immune responses, from chemical sensitivities (e.g., poison ivy) to the more serious specific or disseminated autoimmune response, which can lead to organ failure. Haptenation should be recognized as a potential liability of irreversible protein binders, and the consequences may occur with unpredictable frequency or severity. It should also be recognized, however, that such reactions are not inevitable and problems do not occur in all cases, given the number of covalent drugs in use and the relative paucity of such events.

XII. CONCLUSION

A general decision tree (Fig. 17) is proposed in dealing with irreversible inhibitors in a program. Clearly the decision whether or not to bring these types of inhibitors forward is context specific (in the context of the target, the pathway, and the disease), and each case must be considered individually. However, it is often the case that all the required information for making decisions is not available, and the decision to develop or not to develop an irreversible inhibitor is made by empirical testing for efficacy and safety. In the ideal world, one would have inhibitors available that have different modes of action – reversible (both noncompetitive and competitive – allosteric, orthosteric), slowly

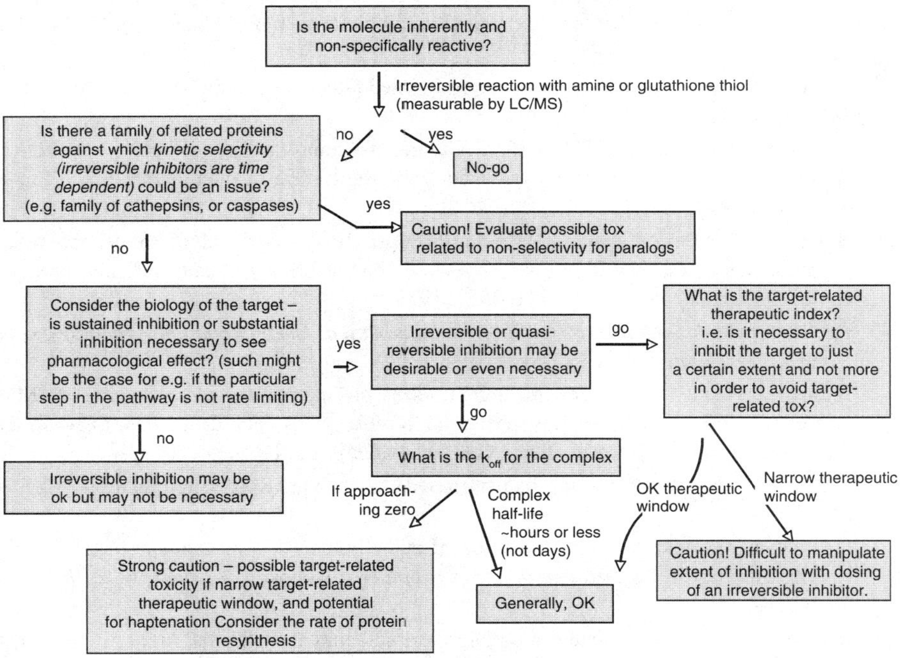

FIGURE 17 Proposed decision tree for irreversible inhibitors in a drug discovery program.

reversible, and irreversible – and these would all be tested for finding the optimal drug candidate [161].

The fear of toxicity of irreversible inhibitors may have been justified for early types of nonspecifically reactive inhibitors, but inhibitors that have more nuanced reactivity that are irreversible or slowly reversible have been successful drugs. Such drugs offer unique advantages such as long duration of action and will continue to be designed or discovered and developed.

REFERENCES

1. RA Copeland, DL Pompliano, TD Meek. Drug-target residence time and its implications for lead optimization. Nat Rev Drug Discov 5:730–739, 2006.
2. PJ Tummino, RA Copeland. Residence time of receptor-ligand complexes and its effect on biological function. Biochemistry 47:5481–5492, 2008.
3. BR Baker. *Design of Active-Site-Directed Irreversible Enzyme Inhibitors*. New York: John Wiley and Sons, 1967.
4. M Prorok, A Albeck, BM Foxman, RH Abeles. Chloroketone hydrolysis by chymotrypsin and *N*-methylhistidyl-57-chymotrypsin: implications for the mechanism of chymotrypsin inactivation by chloroketones. Biochemistry 33:9784–9790. 1994.
5. JC Powers, JL Asgian, OD Ekici, KE James. Irreversible inhibitors of serine, cysteine, and threonine proteases. Chem Rev 102:4639–4750, 2002.
6. GM Helmkamp, Jr., RR Rando, DJH Brock, K Bloch. β-Hydroxydecanoyl thioester dehydrase. Specificity of substrates and acetylenic inhibitors. J Biol Chem 243:3229–3231, 1968.
7. K Bloch. Enzymatic synthesis of monounsaturated fatty acids. Acc Chem Res 7:193–202, 1969.
8. B Lippert, BW Metcalf, MJ Jung, P Casara. 4-Amino-hex-5-enoic acid, a selective catalytic inhibitor of 4-aminobutyric-acid aminotransferase in mammalian brain. Eur J Biochem 74:441–445, 1977.
9. BW Metcalf, P Bey, C Danzin, MJ Jung, P Casara, JP Vevert. Catalytic irreversible inhibition of mammalian ornithine decarboxylase (E.C. 4.1.1.17) by substrate and product analogs. J Am Chem Soc 100:2551–2553, 1978.
10. RB Silverman. *Mechanism-Based Enzyme Inactivation: Chemistry and Enzymology, vols 1 and 2*. Boca Raton, FL: CRC Press, 1988.
11. P Bey. Mechanism-based enzyme inhibitors as an approach to drug design. In: MG Palfreyman, PP McCann, W Lovenberg, JG Temple, Jr., A Sjoerdsma, eds. *Enzymes as Targets for Drug Design*. San Diego, CA: Academic Press, 1989, pp. 59–83.
12. C Walsh. Suicide substrates: mechanism-based enzyme inactivators. Tetrahedron 38:871–909, 1982.
13. D Rock, LC Wienkers, Characterization of cytochrome P450 mechanism-based inhibition. In: AF Nassar, ed. *Drug Metabolism Handbook*. Hoboken, NJ: John Wiley & Sons, 2009.
14. R Wolfenden. Transition state analog inhibitors and enzyme catalysis. Annu Rev Biophys Bioeng 5:271–306, 1976.

15. VL Schramm. Enzymatic transition-state analogs, dynamics, thermodynamics and lifetimes. Annu Rev Biochem 80:703–732, 2011.
16. K Brady, RH Abeles. Inhibition of chymotrypsin by peptidyl trifluoromethyl ketones: determinants of slow-binding kinetics. Biochemistry 29:7608–7617, 1990.
17. S Mehdi. Synthetic and naturally occurring protease inhibitors containing an electrophilic carbonyl group. Bioorg Chem 21:249–259, 1991.
18. L Maveyraud, RF Pratt, J-P Samama. Crystal structure of an acylation transition-state analog of the TEM-1 β-lactamase. Mechanistic implications for class A β-lactamases. Biochemistry 37:2622–2628, 1998.
19. JM Denu, DL Lohse, J Vijayalakshmi, MA Saper, JE Dixon. Visualization of intermediate and transition-state structures in protein-tyrosine phosphatase catalysis. Proc Natl Acad Sci USA 93:2493–2498, 1996.
20. VP Nair, JM Hunter. Anticholinesterases and anticholinergic drugs. Continuing education in anaesthesia. Crit Care Pain 4:164–168, 2004.
21. K Ahn, DS Johnson, BF Cravatt. Fatty-acid amide hydrolase as a potential therapeutic target for the treatment of pain and CNS disorders. Expert Opin Drug Discov 4:763–784, 2009.
22. PGW Gettins. Serpin structure, mechanism, and function. Chem Rev 102:4751–4803, 2002.
23. C Drahl, BF Cravatt, EJ Sorensen. Protein-reactive natural products. Angew Chem Int Ed 44:5788–5809, 2005.
24. RA Copeland. *Evaluation of Enzyme inhibitors in Drug Discovery*. Hoboken, NJ: John Wiley and Sons, 2005.
25. GA Flynn, S Mehdi. Unpublished data .
26. S Eschenburg, M Priestman, E Schonbrunn. Evidence that the fosfomycin target Cys115 in UDP-N-acetylglucosamine enolpyruyl transferase (MurA) is essential for product release. J Biol Chem 280:3757–3763, 2005.
27. R Kitz, IB Wilson. Esters of methanesulfonic acid as irreversible inhibitors of acetylcholinesterase. J Biol Chem 237:3245–3249, 1962.
28. SP Leytus, DL Toledo, WF Mangel. Theory and experimental method for determining individual constants for fast-acting, irreversible proteinase inhibitors. Biochim Biophys Acta 788:74–86, 1984.
29. W Liu, C Tsou. Determination of rate constants for the irreversible inhibition of acetylcholine esterase by continuously monitoring the substrate reaction in the presence of the inhibitor. Biochim Biophys Acta 870:185–190, 1986.
30. AG Marangoni. *Enzyme Kinetics*. Hoboken, NJ: John Wiley and Sons, 2003.
31. CD Davis, UM Hanumegowda. The role of drug metabolism in toxicity. In: AF Nassar, ed. *Drug Metabolism Handbook*. Hoboken, NJ: John Wiley & Sons, 2009, pp. 561–628.
32. DS Johnson, E Weerapana, BF Cravatt. Strategies for discovering and derisking covalent, irreversible enzyme inhibitors. Future Med Chem 2:949–964, 2010.
33. J Adams, M Behnke, S Chen, AA Cruickshank, LR Dick, L Grenier, JM Klunder, Y-T Ma, L Plamondon, RL Stein. Potent and selective inhibitors of the proteasome: dipeptidyl boronic acids. Bioorg Med Chem Lett 8:333–338, 1998.
34. E Kupperman, EC Lee, Y Cao, B Bannerman, M Fitzgerald, A Berger, J Yu, Y. Yang, P Hales, F Bruzzese, et al. Evaluation of the proteasome inhibitor MLN9708 in preclinical models of human cancer. Cancer Res 70:1970–1980, 2010.

35. DJ Kuhn, Q Chen, PM Voorhees, JS Strader, KD Shenk, CM Sun, SD Demo, MK Bennett, FWB van Leeuwen, et al. Potent activity of carfilzomib, a novel, irreversible inhibitor of the ubiquitin-proteasome pathway, against preclinical models of multiple myeloma. Blood 110:3281–3290, 2007.
36. JA Davis, S Singh, S Sethu, S Roy, S Mitra, G Rayasam, V Bansal, J Sattigeri, A Ray. Nature of action of sitagliptin, the dipeptidyl peptidase-IV inhibitor in diabetic animals. Indian J Pharmacol 42:229–233, 2010.
37. CF Deacon, TE Hughes, JJ Holst. Dipeptidyl peptidase IV inhibition potentiates the insulinotropic effect of glucagon-like peptide 1 in the anesthetized pig. Diabetes 47:764–769, 1998.
38. WG Gutheil, WW Bachovchin. Separation of L-Pro-DL-boroPro into its component diastereomers and kinetic analysis of their inhibition of dipeptidyl peptidase IV. A new method for the analysis of slow, tight-binding inhibition. Biochemistry 32:8723–8731, 1993.
39. JL Sudmeir, UL Gunther, WG Gutheil, SJ Coutts, RJ Snow, RW Barton, WW Bachovchin. Solution structures of active and inactive forms of the DP IV (CD26) inhibitor Pro-boroPro determined by NMR spectroscopy. Biochemistry 33:12427–12438, 1994.
40. TE Hughes, MD Mone, ME Russell, SC Weldon, EB Villhauer. NVP-DPP728 (1-[[[2-[(5-Cyanopyridin-2-yl)amino]ethyl]amino]acetyl]-2-cyano-(S)-pyrrolidine), a slow-binding inhibitor of dipeptidyl peptidase IV. Biochemistry 38:11597–11603, 1999.
41. P Hu, Q Yin, F Deckert, J Jiang, D Liu, L Kjems, WP Dole, Y-L He. Pharmacokinetics and pharmacodynamics of vildagliptin in healthy Chinese volunteers. J Clin Pharm 49:39–49, 2009.
42. WJ Metzler, J Yanchunas, C Weigelt, K Kish, HE Klei, D Xie, Y Zhang, M Corbett, JK Tamura, B He, et al. Involvement of DPP-IV catalytic residues in enzyme-saxagliptin complex formation. Protein Sci 17:240–250, 2008.
43. LY Hu, RH Abeles. Inhibition of cathepsin B and papain by peptidyl alpha-keto esters, alpha-keto amides, alpha-diketones, and alpha-keto acids. Arch Biochem Biophys 281:271–274, 1990.
44. V Ganesh, AY Lee, J Clardy, A Tulinsky. Comparison of the structures of the cyclotheonamide A complexes of human α-thrombin and bovine β-trypsin. Protein Sci 5:825–835, 1996.
45. FG Njoroge, KX Chen, N-Y Shih, JJ Piwinski. Challenges in modern drug discovery: a case study of boceprevir, an HCV protease inhibitor for the treatment of hepatitis C virus infection. Acc Chem Res 41:50–59, 2008.
46. RB Perni, SJ Almquist, RA Byrn, G Chandorkar, PR Chaturvedi, LF Courtney, CJ Decker, K Dinehart, CA Gates, SL Harbeson, et al. Preclinical profile of VX-950, a potent, selective, and orally bioavailable inhibitor of hepatitis virus NS3-4A serine protease. Antimicrob Agents Chemother 50:899–909, 2006.
47. X Tong, A Arasappan, F Bennett, R Chase, B Feld, Z Guo, A Hart, V Madison, B Malcom, J Pichardo, et al. Preclinical characterization of the antiviral activity of SCH 900518 (narlaprevir), a novel mechanism-based inhibitor of hepatitis C virus NS3 protease. Antimicrob Agents Chemother 54:2365–2370, 2010.
48. W Wannamaker, R Davies, M Namchuk, J Pollard, P Ford, G Ku, C Decker, P Charifson, P Weber, UA Germann, et al. (S)-1-((S)-2-{[1-(4-amino-3-chloro-phenyl)-methanoyl]-amino}-3,3-dimethyl-butanoyl)-pyrrolidine-2-carboxylic acid ((2R,3S)-2-ethoxy-5-oxo-tehtrahydro-furan-3-yl)-amide (VX-765), an orally available selective interleukin (IL)-converting enzyme/caspase-1 inhibitor, exhibits potent anti-inflammatory activities by inhibiting the release of IL-1β and IL-18. J Pharmacol Exp Ther 321:509–516, 2007.

49. Vertex Pharmaceuticals website: http://www.vrtx.com (accessed July 3, 2013).
50. P Klootwijk, T Lenderink, S Meij, H Boersma, R Melkert, VAW Umans, J Stibbe, EJ Muller, KJ Poortermans, JW Deckers, et al. Anticoagulant properties, clinical efficacy, and safety of efegatran, a direct thrombin inhibitor, in patients with unstable angina. Eur Heart J 20:1101–1111, 1999.
51. The PRIME Investigators. Multicenter, dose-ranging study of efegatran sulfate versus heparin with thrombolysis for acute myocardial infarction: the promotion of reperfusion in myocardial evolution (PRIME) trial. Am Heart J 143:95–105, 2002.
52. P Nacharaju, AS Acharaya. Amadori rearrangement potential of hemoglobin at its glycation sites is dependent on the three-dimensional structure of protein. Biochemistry 31:12673–12679, 1992.
53. N Sin, L Meng, MQW Wang, JJ Wen, WG Bornmann, CM Crews. The anti-angiogenic agent fumagillin covalently binds and inhibits the methionine aminopeptidase, MetAP-2. Proc Natl Acad Sci USA, 94:6099–6103, 1997.
54. S Kim, K LaMontagne, M Sabio, S Sharma, RW Versace, N Yusuff, PE Phillips. Depletion of methionine aminopeptidase 2 does not alter cell response to fumagillin or bengamides. Cancer Res 64:2984–2987, 2004.
55. J-RJ Yeh, R Ju, KM Brdlik, W Zhang, Y Zhang, ME Matyskiela, JD Shotwell, CM Crews. Targeted gene disruption of methionine aminopeptidase 2 results in an embryonic gastrulation defect and endothelial cell growth arrest. Proc Natl Acad Sci USA 103:10379–10384, 2006.
56. J-M Molina, M Tourneur, C Sarfati, S Chevret, A de Gouvello, J-G Gobert, S Balkan, F Derouin. Fumagillin treatment of intestinal microsporidiosis. N Eng J Med 346:1963–1969, 2002.
57. M Higes, R Martin-Hernandez, E Garrido-Bailon, AV Gonzalez-Porto, P Garcia-Palencia, A Meana, MJ Del Nozal, R Mayo, JL Bernal. Honeybee colony collapse due to *Nosema ceranae* in professional apiaries. Environ Microbiol Rep 1:110–113, 2009.
58. JJ Alvarado, A Nemkal, JM Sauder, M Russell, DE Akiyoshi, W Shi, SC Almo, LM Weiss. Structure of a microsporidian methionine aminopeptidase type 2 complexed with fumagillin and TNP-470. Mol Biochem Parasitol 168:158–167, 2009.
59. S Liu, J Widom, CW Kemp, CM Crews, J Clardy. Structure of human methionine aminopeptidase-2 complexed with fumagillin. Science 282:1324–1327, 1998.
60. BJ Dezube, JH Von Roenn, J Holden-Wiltse, TW Cheung, SC Remick, TP Cooley, J Moore, JP Sommadossi, SL Shriver, CW Suckow, et al. Fumagillin analog in the treatment of Kaposi's sarcoma: a phase I AIDS clinical trial group study. AIDS clinical trial group No. 215 team. J Clin Oncol 16:1444–1449, 1998.
61. YI Zhang, JR Yeh, A Mara, R Ju, JF Hines, P Cirone, HL Griesbach, I Schneider, DC Slusarski, SA Holley, et al. A chemical and genetic approach to the mode of action of fumagillin. Chem Biol 13:1001–1009, 2006.
62. HM White, AL Acton, RV Considine. The angiogenic inhibitor TNP-470 decreases caloric intake and weight gain in high-fat fed mice. Obesity (Silver Spring) 20:2003–2009, 2012.
63. http://clinicaltrials.gov/ct2/show/NCT1507077 (accessed July 3, 2013).
64. M Groll, KB Kim, N Kairies, R Huber, CM Crews. Crystal structure of epoxomicin:20S proteasome reveals a molecular basis for selectivity of α',β'-epoxyketone proteasome inhibitors. J Am Chem Soc 122:1237–1238, 2000.
65. NK Shrestha, JW Tomford. Fosfomycin: a review. Infect Dis Clin Pract 10:255–260, 2001.

66. P Storici, D De Biase, F Bossa, S Bruno, A Mozerelli, C Peneff, RB Silverman, T Schirmer. Structures of γ-aminobutyric acid (GABA) aminotransferase, a pyridoxal 5′-phosphate, and [2Fe-2S] cluster-containing enzyme, complexed with γ-ethynyl-GABA and with the antiepilepsy drug vigabatrin. J Biol Chem 279:363–373, 2004.

67. NV Grishin, AL Osterman, HB Brooks, MA Phillips, EJ Goldsmith. X-ray structure of ornithine decarboxylase from *Trypanosoma brucei*: the native structure and the structure in complex with α-difluoromethylornithine. Biochemistry 38:15174–15184, 1999.

68. RB Silverman. *The Organic Chemistry of Enzyme-Catalyzed Reactions*. San Diego, CA: Academic Press, 2002.

69. D Guidici, G Ornati, G Briatico, F Buzzetti, P Lombardi, E Di Salle. 6-Methyleneandrosta-1,4-diene-3,17-dione (FCE 24304): a new irreversible aromatase inhibitor. J Steroid Biochem 30:391–394, 1988.

70. AMH Brodie, WM Garrett, JR Hendrickson, C-W Tsai-Morris, PA Marcotte, CH Robinson. Inactivation of aromatase *in vitro* by 4-hydroxy-4-androstene-3,17-dione and 4-acetoxy-4-androstene-3,17-dione and sustained effects *in vivo*. Steroids 38:693–702, 1981.

71. RW Brueggemeir. Aromatase inhibitors – mechanisms of steroidal inhibitors. Breast Cancer Res Treat 30:31–42, 1994.

72. DF Covey, WF Hood, VD Parikh. 10β-propynyl-substituted steroids. Mechanism-based enzyme activated irreversible inhibitors of estrogen biosynthesis. J Biol Chem 256:1076–1079, 1981.

73. BW Metcalf, CL Wright, JP Burkhart, JO Johnston. Substrate-induced inactivation of aromatase by allenic and acetylenic steroids. J Am Chem Soc 103:3221–3222, 1981

74. PA Marcotte, CH Robinson. Synthesis and evaluation of 10β-substituted estr-4-ene 3,17-diones as inhibitors of human placenta microsomal aromatase. Steroids 39:325–344, 1982.

75. Y Hong, B Yu, M Sherman, Y-C Yuan, D Zhou, S Chen. Molecular basis for the aromatization reaction and exemestane-mediated irreversible inhibition of human aromatase. Mol Endocrinol 21:401–414, 2007.

76. A Wissner, TS Mansour. The development of HKI-272 and related compounds for the treatment of cancer. Arch Pharm Chem Life Sci 341:465–477, 2008.

77. EL Kwak, R Sordella, DW Bell, N Godin-Heymann, RA Okimoto, BW Brannigan, PL Harris, DW Driscoll, P Fidias, TJ Lynch, et al. Irreversible inhibitors of the EGF receptor may circumvent acquired resistance to gefitinib. Proc Natl Acad Sci U S A 102:7665–7670, 2005.

78. C-H Yun, KE Mengwasser, AV Toms, MS Woo, H Greulich, K-K Wong, M Meyerson, MJ Eck. The T790M mutation in EGFR kinase causes drug resistance by increasing the affinity for ATP. Proc Natl Acad Sci U S A 105:2070–2075, 2008.

79. Y Kim, J Ko, ZY Cui, A Abolhoda, JS Ahn, S-H Ou, M-J Ahn, K Park. The *EGFR* T790M mutation in acquired resistance to an irreversible second-generation EGFR inhibitor. Mol Cancer Ther 11:784–791, 2012.

80. N Godin-Heymann, L Ulkus, BW Brannigan, U McDermott, J Lamb, S Maheshwaran, J Settleman, DA Haber. The T790M "gatekeeper" mutation in *EGFR* mediates resistance to low concentrations of an irreversible EGFR inhibitor. Mol Cancer Ther 7:874–879, 2008.

81. T Zhang, F Inesta-Vaquera, M Niepel, J Zhang, SB Ficarro, T Machleidt. T Xie, JA Marto, N Kim, T Sim, et al. Discovery of potent and selective covalent inhibitors of JNK. Chem Biol 19:140–154, 2012.

82. J Wang, XX Li-Chan, J Atherton, L Deng, R Espina, L Yu, P Horwatt, S Ross, S Lockhead, S Ahmad, et al. Characterization of HK-272 binding to human serum albumin. Drug Metab Dispos 38:1083–1093, 2010.

83. IM Serafimova, MA Pufall, S Krishnan, K Duda, MS Cohen, RL Maglathlin, JM McFarland, RM Miller, M Frodin, J Taunton. Reversible targeting of noncatalytic cysteines with chemically tuned electrophiles. Nat Chem Biol 8:471–476, 2012, doi: 10.1038/NCHEMBIO.952.

84. Z Pan, H Scheerens, SJ Li, BE Schultz, PA Sprengler, LC Burrill, RV Mendonca, MD Sweeney, KC Scott, PG Grothaus, et al. Discovery of selective irreversible inhibitors for Bruton's tyrosine kinase. ChemMedChem 2:58–61, 2007.

85. J Li, TS Kaoud, C Laroche, KN Dalby, SM Kerwin. Synthesis and biological evaluation of P38α kinase-targeting dialkynylimidazoles. Bioorg Med Chem Lett 19:6293–6297, 2009.

86. J Singh, RS Petter, TA Baillie, A Whitty. The resurgence of covalent drugs. Nat Rev Drug Discov 10:307–317, 2011.

87. J Singh, RS Petter, AF Kluge. Targeted covalent drugs of the kinase family. Curr Opin Chem Biol 14:475–480, 2010.

88. M Hagel, D Niu, T St Martin, MP Sheets, L Qiao, H Bernard, RM Karp, Z Zhu, MT Labenski, P Chaturvedi, et al. Selective irreversible inhibition of a protease by targeting a noncatalytic cysteine. Nat Chem Biol 7:22–24, 2011.

89. L Zhou, X Cheng, BA Connolly, MJ Dickman, PJ Hurd, DP Hornby. Zebularine: a novel DNA methylation inhibitor that forms a covalent complex with DNA methytransferases. J Mol Biol 321:591–599, 2002.

90. DV Santi, A Norment, CE Garrett. Covalent bond formation between a DNA-cytosine methyltransferase and DNA containing 5-azacytidine. Proc Natl Acad Sci U S A 81:6993–6997, 1984.

91. AS Brank, R Eritja, RG Garcia, VE Marquez, JK Christman. Inhibition of *Hha*I DNA (cytosine-C5) methyltransferase by oligodeoxyribonucleotides containing 5-aza-2′-deoxycytidine: examination of the intertwined roles of co-factor, target, transition state structure and enzyme conformation. J Mol Biol 323:53–67, 2002.

92. JK Christman. 5-Azacytidine and 5-aza-2′-deoxycytidine as inhibitors of DNA methylation: mechanistic studies and their implication for cancer therapy. Oncogene 21:5483–5495, 2002.

93. T Yamazaki, T Ohno, T Sakaki, M Akiyoshi-Shibata, Y Yabusaki, T Imai, S Kominami. Kinetic analysis of successive reactions catalyzed by bovine cytochrome P450(17alpha, lyase). Biochemistry 37:2800–2806, 1998.

94. M Jarman, SE Barrie, JM Llera. The 16,17-double bond is needed for irreversible inhibition of human cytochrome $P450_{17\alpha}$by abiraterone (17-(3-pyridyl)androsta-5,16-dien-3β-ol) and related steroidal inhibitors. J Med Chem 41:5375–5381, 1998.

95. DS Johnson, C Stiff, SE Lazerwith, SR Kesten, LF Fay, M Morris, D Beidler, MB Limatta, SE Smith, DT Dudley, et al. Discovery of PF-04457845: a highly potent, orally bioavailable, and selective urea FAAH inhibitor. ACS Med Chem Lett 2:91–96, 2011.

96. JP Alexander, BF Cravatt. Mechanism of carbamate inactivation of FAAH: implications for the design of covalent inhibitors and in vivo functional probes for enzymes. Chem Biol 12:1179–1187, 2005.

97. P Bar-On, CB Millard, M Harel, H Dvir, A Enz, JL Sussman, I Silman. Kinetic and structural studies on the interaction of cholinesterases with the anti-Alzheimer drug rivastigmine. Biochemistry 41:3555–3564, 2002.

98. WL Smith, WEM Lands. Stimulation and blockade of prostaglandin biosynthesis. J Biol Chem 246:6700–6704, 1971.

99. GJ Roth, N Stanford, PW Majerus. Acetylation of prostaglandin synthase by aspirin. Proc Natl Acad Sci U S A 72:3073–3076, 1975.

100. T Shimokawa, WL Smith. Prostaglandin endoperoxide synthase. The aspirin acetylation region. J Biol Chem 267:12387–12392, 1992.

101. JA Mancini, GP O'Neill, C Bayly, PJ Vickers. Mutation of serine-516 in human prostaglandin G/H synthase-2 to methionine or aspirin acetylation of this residue stimulates 15-*R*-HETE synthesis. FEBS Lett 342:33–37, 1994.

102. N Kresge, RD Simoni, RL Hill. How aspirin interferes with cyclooxygenase activity: the work of William M. Smith. J Biol Chem 283:e12–e14, 2008.

103. K Al-Suwailem, AS Al-Tamimi, MA Al-Omar, MS Al-Suhibani. Safety and mechanism of action of orlistat (tetrahydrolipstatin) as the first local antiobesity drug. J Appl Sci Res 2:205–208, 2006.

104. R Guerciolini. Mode of action of orlistat. Int J Obes Relat Metab Disord 3(Suppl.): S12–S23, 1997.

105. B Borgstrom. Mode of action of tetrahydrolipstatin: a derivative of the naturally occurring lipase inhibitor lipstatin. Biochim Biophys Acta 962:308–316, 1988.

106. A Tiss, H Lengsfeld, F Carriere, R Verger. Inhibition of human pancreatic lipase by tetrahydrolipstatin: further kinetic studies showing its reversibility. J Mol Catal B Enzym 58:41–47, 2009.

107. RB Woodward. Penems and related substances. Phil Trans R Soc Lond 289:239–250, 1980.

108. AF Kluge, RC Petter. Acylating drugs: redesigning natural covalent inhibitors. Current Opin Chem Biol 14:421–427, 2010.

109. H Frase, M Toth, MW Champion, NT Antunes, SB Vakulenko. Importance of position 170 in the inhibition of GES-type β-lactamases by clavulanic acid. Antimicrob Agents Chemother 55:1556–1562, 2011.

110. HU Weltzien, E Padovan. Molecular features of penicillin allergy. J Invest Dermatol 110:203–206, 1998.

111. B Furie, BA Bouchard, BC Furie. Vitamin K-dependent biosynthesis of γ-carboxyglutamic acid. Blood 93:1798–1808, 1999.

112. RB Silverman. A model for a molecular mechanism of anticoagulant activity of 3-substituted 4-hydroxycoumarins. J Am Chem Soc 102:5421–5423, 1980.

113. RB Silverman. Model studies for a molecular mechanism of action of oral anticoagulants. J Am Chem Soc 103:3910–3915, 1981.

114. RE Regal, V Tsui. Optimal understanding of warfarin: beyond the nomogram. PT J 29:652–656, 2004.

115. F Kamali, H Wynne. Pharmacogenetics of warfarin. Annu Rev Med 61:63–75, 2010.

116. MJ Rieder, AP Reiner, BF Gage, DA Nickerson, CS Eby, HL McLeod, DK Blough, KE Thummel, DL Veenstra. Effect of *VKORC1* haplotypes on transcriptional regulation and warfarin dose. N Eng J Med 352:2285–2293, 2005.

117. R Lobstein, I Dvoskin, H Halkin, M Vecsler, A Lubetsky, G Rechavi, N Amariglio, Y Cohen, G Ken-Dror, S Almog, et al. A coding VKORC1 Asp36Tyr polymorphism predisposes to warfarin resistance. Blood 109:2477–2480, 2007.

118. TM Misenheimer, JW Suttie. Warfarin resistance in a Chicago strain of rats. Biochem Pharmacol 40:2079–2084, 1990.

119. HHW Thijssen. Warfarin-based rodenticides: mode of action and mechanism of resistance. Pestic Sci 43:73–78, 1995.

120. L Olbe, E Carlsson, P Lindberg. A proton-pump inhibitor expedition: the case histories of omeprazole and esomeprazole. Nat Rev Drug Discov 2:132–139, 2003.

121. ML Shen, JJ Lipsky, S Naylor. Role of disulfiram in the *in vitro* inhibition of rat liver mitochondrial aldehyde dehydrogenase. Biochem Pharmacol 60:947–953, 2000.

122. NS Sampson, PA Bartlett. Peptidic phosphonylating agents as irreversible inhibitors of serine proteases and models of the tetrahedral intermediates. Biochemistry 30:2255–2263, 1991.

123. LG Costa. Current issues in organophosphate toxicology. Clin Chim Acta 366:1–13, 2006.

124. J Geoghegan, JL Tong. Chemical warfare agents. Continuing education in anesthesia, Crit Care Pain 6:230–234, 2006.

125. JM Lopez-Arrieta, L Schneider. Metrifonate for Alzheimer's disease. Cochrane Database Syst Rev April 19(2):CD003155, 2006.

126. D Healy. *The Anti-Depressant Era*. Cambridge, MA: Harvard University Press, 1997.

127. W Weyler, Y-PP Hsu, XO Breakafield. Biochemistry and genetics of monoamine oxidase. Pharmacol Ther 47:391–417, 1990.

128. M Wimbiscus, O Kostenko, D Malone. MAO inhibitors: risks, benefits, and lore. Cleve Clin J Med 77:859–882, 2010.

129. C Binda, M Li, F Hubalek, N Restelli, DE Edmondson, AN Mattevi. Insights into the mode of inhibition of human mitochondrial monoamine oxidase B from high-resolution crystal structures. Proc Natl Acad Sci U S A 100:9750–9755, 2003.

130. RB Silverman. *The Organic Chemistry of Drug Design and Drug Action*, 2nd ed. Amsterdam: Elsevier Academic Press, 2004.

131. E Artin, J Wang, GJS Lohman, K Yokoyama, G Yu, RG Griffin, G Bar, J Stubbe. Insight into the mechanism of inactivation of ribonucleotide reductase by gemcitabine 5′-diphosphate in the presence and absence of reductant. Biochemistry 48:11622–11629, 2009.

132. ML Moss, P Kuzmic, JD Stuart, G Tian, AG Peranteau, SV Frye, SH Kadwell, TA Kost, LK Overton, IR Patel. Inhibition of human steroid 5α reductases type I and II by 6-aza-steroids: structural determinants of one-step vs two-step mechanism. Biochemistry 35:3457–3464, 1996.

133. HG Bull, M Garcia-Calvo, S Andersson, WF Baginsky, HK Chan, DE Ellsworth, RR Miller, RA Stearns, RK Bakshi, GH Rasmusson, et al.. Mechanism-based inhibition of human steroid 5α reductase by finasteride: enzyme-catalyzed formation of NADP-dihydrofinasteride, a potent bisubstrate analog inhibitor. J Am Chem Soc 118:2359–2365, 1996.

134. JE Drury, L Di Costanzo, TM Penning, DW Christianson. Inhibition of human steroid 5β reductase (AKR1D1) by finasteride and structure of the enzyme-inhibitor complex. J Biol Chem 284:19786–19790, 2009.

135. R Rawat, A Whitty, PJ Tonge. The isoniazid-NAD adduct is a slow, tight-binding inhibitor of InhA, the *Mycobacterium tuberculosis* enoyl reductase: adduct affinity and drug resistance. Proc Natl Acad Sci U S A 100:13881–13886, 2003.

136. JL Palmer, RH Abeles. The mechanism of action of S-adenosylhomocysteinase. J Biol Chem 754:1217–1226, 1979.

137. M Wang, JR Unruh, CK Johnson, K Kuczera, RL Schowen, RT Borchardt. Effects of ligand binding and oxidation on hinge-bending motions in S-adenosyl-L-homocysteine hydrolase. Biochemistry 45:7778–7786, 2006.
138. JR McCarthy, ET Jarvi, DP Matthews, ML Edwards, NJ Prakash, TL Bowlin, S Mehdi, PS Sunkara, P Bey. 4′,5′-Unsaturated 5′-fluoroadenosine nucleosides: potent mechanism-based inhibitors of S-adenosyl-L-homocysteine hydrolase. J Am Chem Soc 111:1127–1128, 1989.
139. S Mehdi, ET Jarvi, JR Koehl, JR McCarthy, P. Bey. The mechanism of inhibition of S-adenosyl-L-homocysteine hydrolase by fluorine-containing adenosine analogs. J Enzyme Inhib 4:1–13, 1990.
140. K Okamoto, BT Eger, T Nishino, EF Pai, T Nishino. Mechanism of inhibition of xanthine oxidoreductase by allopurinol: crystal structure of reduced bovine milk xanthine oxidoreductase bound with oxipurinol. Nucleosides Nucleotides Nucleic Acids 27:888–893, 2008.
141. Y Yamaguchi T Matsumura, K Ishida, K Okamoto, T Nishino. Human xanthine oxidase changes its substrate specificity to aldehyde oxidase type upon mutation of amino acid residues in the active site: roles of active site residues in binding and activation of purine substrate. J Biochem (Tokyo) 141:513–524, 2007.
142. DR Kipp, JS Hirschi, A Wakata, H Goldstein, VL Schramm. Transition states of native and drug-resistant HIV-protease are the same. Proc Natl Acad Sci U S A 109:6543–6548, 2012.
143. PA Johnston. Redox cycling compounds generate H_2O_2 in HTS buffers containing strong reducing agents – real hits or promiscuous artifacts? Curr Opin Chem Biol 15:174–182, 2011.
144. Q Wang, D Dube, RW Friesen, TG LeRiche, KP Bateman, L Trimble, J Sanghara, R Pollex, C Ramachandran, MJ Gresser, et al. Catalytic inactivation of protein tyrosine phosphatase CD45 and protein tyrosine phosphatase 1B by polyaromatic quinones. Biochemistry 43:4294–4303, 2004.
145. A Giovanni, F Wirtz-Brugger, Z Gao, S Mehdi. Caspase 3 activity: potent inhibition by redox cycling compounds. Soc for Neurosci Natl Mtg Abstracts 27(2):1740, 2001.
146. S Mehdi. Unpublished observations.
147. DA Jeffrey, M Bogyo. Chemical proteomics and its application to drug discovery. Curr Opin Biotechnol 14:87–95, 2003.
148. M Fonovic, M Bogyo, Activity based probes for proteases: applications to biomarker discovery, molecular imaging and drug screening. Curr Pharm Des 13:253–261, 2007.
149. J Krysiak, R Breinbauer. Activity-based protein profiling for natural product target discovery. Top Curr Chem 324:43–84, 2012.
150. BF Cravatt, AT Wright, JW Kozarich. Activity-based protein profiling: from enzyme chemistry to proteomic chemistry. Annu Rev Biochem 77:383–414, 2008.
151. AE Spears, BF Cravatt. Chemical strategies for activity-based proteomics. ChemBioChem 5:41–47, 2004.
152. DK Nomura, MM Dix, BJ Cravatt. Activity-based protein profiling for biochemical pathway discovery in cancer. Nat Rev Cancer 10:630–638, 2010.
153. MP Patricelli, TK Nomanbhoy, J Wu, H Brown, D Zhou, J Zhang, S Jaganathan, A Aban, E Okerberg, C Herring, et al. In situ kinase profiling reveals functionally relevant properties of native kinases. Chem Biol 18:699–710, 2011.

154. FP Guengerich. Role of cytochrome P450 enzymes in drug-drug interactions. Adv Pharmacol 43:7–35, 1997.

155. AI Veldkamp, RMW Hoetelmans, JH Beijnen, JW Mulder, PL Meenhorst. Ritonavir enables combined therapy with rifampin and saquinavir. Clin Infect Dis 29:1586, 1999.

156. JE Hopkins, DJ Naisbitt, NR Kitteringham, RJ Dearman, I Kimber, BK Park. Selective haptenation of cellular or extracellular protein by chemical allergens: Association with cytokine polarization. Chem Res Toxicol 18:375–381, 2005.

157. I Chipinda, JM Hettick, PD Siegel. Haptenation: chemical reactivity and protein binding. J Allergy 2011, article ID 839682.

158. J Uetrecht. Current trends in drug-induced autoimmunity. Autoimmun Rev 4:309–314, 2005.

159. JP Sanderson, DJ Naisbitt, BK Park. Role of bioactivation in drug-induced hypersensitivity reactions. AAPS J 8:E55–E64, 2006.

160. NW Palm, R Medzhitov. Immunostimulatory activity of haptenated proteins. Proc Natl Acad Sci U S A 106:4782–4787, 2009.

161. RM Rydzewski. *Real World Drug Discovery. A Chemists Guide to Biotech and Pharmaceutical Research*. Oxford: Elsevier Press, 2009.

4

PRECLINOMICS: ENZYME ASSAYS AND RODENT MODELS FOR METABOLIC DISEASES

WU-KUANG YEH AND RICHARD G. PETERSON
PreClinOmics, Inc., Indianapolis, Indiana

I. INTRODUCTION

The continuous improvement of human health and quality of life can be linked directly to discovery, development, manufacture, and applications of pharmaceutical agents. Though many drugs for some human diseases are available, there are still numerous unmet medical needs, providing plenty of career opportunities to researchers in the pharmaceutical and biotech industry. Drug discovery is a high-risk and potentially high-reward endeavor, costing approximately one billion US dollars in recent years for a new drug to reach the market place [1–3].

For a drug to reach patients, from a business perspective, it involves successive progression from discovery to development to commercialization. As a scientific endeavor, preclinical drug discovery programs focus on the identification of new chemical entities as drug candidates to enter clinical trials. Preclinical enzyme technologies are broadly defined in this chapter as those *in vitro*, *in vivo*, and *ex vivo* assays conducted prior to clinical trials. These enzyme technologies are applied in a variety of interdisciplinary fields including efficacy, pharmacokinetics and pharmacodynamics, toxicology, biomarkers, and computer-assisted structure–activity-based designs. For all of these preclinical enzyme technologies, many enzyme assays and

Enzyme Technologies: Pluripotent Players in Discovering Therapeutic Agents, First Edition.
Edited by Hsiu-Chiung Yang, Wu-Kuang Yeh, and James R. McCarthy.

animal models, especially small animals such as mice and rats, play critical roles in drug discovery and development.

After the introduction, the next section describes different types of enzyme assays to evaluate drug targets as well as to measure efficacy end points and biomarkers. Not as an extensive literature review, this section focuses on evolving enzyme assays from direct experience by the authors, particularly on applying enzyme assays to metabolic diseases. Section III involves development of a rat model for drug discovery related to diabetes, metabolic syndrome, and obesity. For this volume on **Enzyme Technologies**, as an extension of enzyme assays, the authors consider a rat or mouse model as a complex and integrative enzyme system commonly used for preclinical drug discovery and development. Such consideration is based on the biochemical understanding that enzymes are involved in metabolism, both biosynthesis and degradation, of all essential components in human and animals, including DNA, RNA, proteins, peptides/hormones, lipids, and carbohydrates. Section IV briefly describes testing of current antidiabetic and anti-obesity drugs. Using common enzyme assays and commercially available mouse and rat models, the fifth section provides one case study using control compounds in antidiabetic and anti-obesity drug discovery. Another case study in the sixth section describes similar enzyme and rodent model applications for a diabetic efficacy of a nutritional supplement in combination with an existing antidiabetic drug. The sixth section also describes a serendipitous case of testing the same supplement with a terminal cancer patient. This chapter should be informative to the readers in understanding the critical roles of the enzyme assays and rodent models used in drug discovery for metabolic diseases. Readers will recognize that many of the preclinical enzyme assays are also applicable for clinical trials and in fact for monitoring human health.

II. EVOLVING ENZYME ASSAYS

Enzyme assays have been evolving from academic to industrial applications, though overlapping in time and use. They play a significant role in advancing metabolic disciplines from metabolic evolution [4,5] to metabolic engineering [6] and combinatorial biosynthesis [7] and to metabolic diseases [8].

Enzymes are typically proteins that catalyze biological reactions. In addition to typical enzymes, there are also RNA biocatalysts and catalytic antibodies. In this chapter, the authors focus on protein biocatalysts for biotech and pharmaceutical applications. Enzyme assays have been evolving based on their experimental applications. One type of early enzyme assay involves measuring an enzymatic product by the optical density at wavelength corresponding to maximal absorbance. As a common example, the activity of lactic dehydrogenase was measured by the optical density at 340 nm which corresponds to the disappearance of NADH to NAD. This assay was developed over 60 years ago using a Beckman spectrophotometer [9]. Similarly, the activity of catechol 1,2-oxygenase was determined by the accumulation of its reaction product, *cis,cis*-muconate, quantitated by the optical density at 260 nm using a Gilford spectrophotometer over 40 years ago [10]. For measuring enzymatic product directly and for improving the sample throughput, several new formats of enzyme assay have been developed in the last several decades for drug discovery and development including HPLC, LC/MS, and fluorescence and luminescence coupled with micro-plate technology. Two HPLC-based enzyme assays are summarized in the following text to illustrate important metabolic engineering applications.

As one early antibiotic development application, the activity of macrocin *O*-methyltransferase was measured by formation of its reaction product, tylosin, by HPLC [11]. The HPLC assay was used to purify the enzyme and characterize its kinetic mechanism [12]. The insertion of one extra copy of the gene for this enzyme to the chromosome of the industrial strain of *Streptomyces fradiae* by "reverse engineering" resulted in a significant improvement in the yield of tylosin [13]. Similarly, in another early antibiotic development application, the activity of deacetoxycephalosporin synthase (expandase) was measured also by the formation of its reaction product, deacetoxycephalosporin C, by HPLC. Using the HPLC assay, the expandase was purified and characterized [14]. One copy of the gene for the enzyme was inserted back to the chromosome of *Cephalosporium acremonium*, and such extra gene copy insertion led to a significant yield improvement of cephalosporin C [15].

Different types of enzyme assays are described in the following section for preclinical drug discovery in metabolic diseases [8], broadly defined as those related to abnormal and pathological metabolism.

A. Enzymes as Drug Targets

Numerous enzymes are potentially direct targets for drug discovery. Two such enzymes and their assays are summarized in the following text.

β-Secretase One of the leading hypotheses for the pathogenesis of Alzheimer's disease is the accumulation and aggregation of β-amyloid peptides in the brain, especially in the hippocampal and cortical regions. Two proteases, β-secretase (also known as BACE1, memapsin 2) and γ-secretase [16], catalyze the sequential cleavages of β-amyloid precursor protein to generate β-amyloid peptide. These two proteases have been targeted in the last two decades for the treatment of Alzheimer's disease, a metabolic disorder. For β-secretase, at least one inhibitor has been tested in clinical trials [16]. To measure β-secretase activity, one common micro-plate assay monitors the cleaved peptide product by fluorescence resonance energy transfer (FRET, [17]). Application of FRET assay confirmed the same major cleavage site for mouse and human substrates by both mouse and human β-secretases, thus suggesting the usefulness of APP transgenic mouse models in the development of BACE inhibitors for the treatment of Alzheimer's disease [17].

11β-Hydroxysteroid Dehydrogenase Type 1 In the last decade, 11β-hydroxysteroid dehydrogenase type 1 (11β-HSD1), catalyzing interconversion of cortisone to cortisol, has become an important drug target for inhibitor discovery in diabetes, metabolic syndrome, obesity, and related complications such as retinopathy, nephropathy, and cognition impairment [18]. Cortisol, glucocorticoid dihydrocorticosterone, is one major stress hormone for human. Thus, lowering its level through inhibition of 11β-HSD1 has the potential to prevent metabolic diseases and their associated symptoms. Cortisol plays a regulatory role by increasing pyruvate enol phosphate carboxykinase and glucose

6-phosphatase, thus leading to more glucose production [19]. Multiple assay types, including radioactive and nonradioactive formats, have been used in basic research and drug discovery for this enzyme. To minimize the use of radioactive material and maintain high throughput, biotech and pharmaceutical researchers have used HPLC, LC/MS, and fluorescence [20–22] coupled with micro-plate technology for 11β-HSD1 assay applications. Several inhibitors for 11β-HSD1 have been advanced to clinical testing for several therapeutic indications [22,23].

B. Enzymes as Biocatalysts

For drug discovery and development in metabolic diseases, enzymes can be used as biocatalysts in the assays. Table 1 lists selective enzyme assays, with enzymes as biocatalysts for gluconeogenesis, lipid metabolism, liver toxicity, and urine metabolic biomarkers.

Enzyme and Immunological-Colorimetric Assays for Gluconeogenesis As the major disease biomarker for diabetes, glucose is measured by enzyme-catalyzed NADH formation (Scheme 1), typically using a clinical analyzer or a glucometer. Glucose is determined from blood, plasma, or urine. Hemoglobin A1C (HbA1c) is another common disease biomarker for diabetes. HbA1c has a longer half-life than glucose. The serum level of HbA1c is determined by the amount of HbA1c

TABLE 1 Selective Enzyme Assays: Enzymes as Biocatalysts

Enzyme assay	Enzymes[a] as biocatalysts	Assay kit sources
Gluconeogenesis		
Glucose	Hexokinase and G-6-P DH	Beckman
Lipid metabolism		
TG	GK, GPO, and catalase (Step 1)	Wako
	LPL, GK, GPO, and peroxidase (Step 2)	Beckman and Wako
Total CHOL	CE, CO, and peroxidase	Beckman and Wako
HDL CHOL	CE and peroxidase	Beckman
LDL CHOL	CE, CO, and peroxidase	Beckman
Free fatty acids	ACS, ACOD, and peroxidase	Wako
Glycerol	GK, GPO, and peroxidase	Zen-Bio
Ketone bodies	3-HBDH and diaphorase	Cayman
Liver toxicity		
AST	AST and MDH	Beckman
ALT	ALT and LDH	Beckman
Urine metabolic biomarkers		
BUN	Urease and GLDH	Beckman
Uric acid	Uricase and peroxidase	Beckman

HDL, high-density lipoprotein; LDL, low-density lipoprotein; ALT, alanine aminotransferase; AST, aspartate aminotransferase; BUN, blood urea nitrogen.

[a]See SCHEMES 1, 2 and 3 for the enzyme abbreviations. GLDH, glutamate dehydrogenase; 3-HBDH, 3-hydroxybutyrate dehydrogenase.

using an immunoassay divided by the amount of total hemoglobin using a colorimetric method with a clinical analyzer, for example, from Beckman Coulter.

Glucose + ATP $\xrightarrow{\text{Hexokinase}}$ G-6-P + ADP

G-6-P + NAD^+ $\xrightarrow{\text{G-6-PDH}}$ 6-PG + NADH + H^+

G-6-P, Glucose-6-Phosphate
G-6-PDH, Glucose-6-Phosphate Dehydrogenase
6-PG, 6-Phosphogluconate

SCHEME 1 Enzymatic assay for glucose.

Enzyme Assays for Lipid Metabolism Fat overload is one major hypothesis for insulin resistance and diabetes [24]. A few common lipids are used as indicators for rodent studies in drug discovery for diabetes, metabolic syndrome, and obesity. Triglycerides (TG), total cholesterol (CHOL), and free fatty acids are among the most frequently used lipid end points in rodent studies. Using enzyme-catalyzed reactions, TG (Scheme 2a, Step 2 for producing the red dye only) and total CHOL (Scheme 2b) from plasma are commonly analyzed by a clinical analyzer. Total CHOL from the liver and muscle can be determined using their homogenates with a clinical analyzer or a micro-plate reader, whereas TG can be determined from the homogenates using a micro-plate reader (Scheme 2a, Step 1 and Step 2 for producing the blue pigment) [25]. Similarly, free fatty acids from plasma, following three-step enzyme reactions (Scheme 2c), are measured with a micro-plate reader.

(a) **Triglycerides**

Step 1: Color A Reagent for Decomposing Free Glycerol in Sample

Free Glycerol + ATP $\xrightarrow{\text{GK}}$ Glycerol-3-Phosphate + ADP

Glycerol-3-Phosphate + O_2 $\xrightarrow{\text{GPO}}$ Dihydroacetone phosphate + H_2O_2

$2H_2O_2$ $\xrightarrow{\text{Catalase}}$ O_2 + H_2O

GK, Glycerol Kinase
GPO, Glycerol-3-Phosphate Oxidase

Step 2: Color B Reagent for Converting Triglyceride to Blue Pigment

Triglyceride + $3H_2O$ $\xrightarrow{\text{LPL}}$ Glycerol + 3Fatty Acids

Glycerol + ATP $\xrightarrow{\text{GK}}$ Glycerol-3-Phosphate + ADP

Glycerol-3-Phosphate + O_2 $\xrightarrow{\text{GPO}}$ Dihydroacetone phosphate + H_2O_2

$2H_2O_2$ + 4-Aminoantipyrine + HMMPS $\xrightarrow{\text{Peroxidase}}$ Blue Pigment

Or, **H_2O_2 + 4-Aminoantipyrine + DHBS $\xrightarrow{\text{Peroxidase}}$ Red Dye**

LPL, Lipoprotein Lipase
HMMPS, N-(3-sulfopropyl)-3-methoxy-5-methylaniline
DHBS, 3,5-dichloro-2-hydroxybenzenesulfonic acid

SCHEME 2 Enzymatic assays for (a) triglycerides, (b) total cholesterol, and (c) nonesterified free fatty acids.

(b) **Total Cholesterol**

Cholesterol Ester $\xrightarrow{\text{CE}}$ Cholesterol + Fatty Acid

Cholesterol + O_2 $\xrightarrow{\text{CO}}$ Cholesterol-3-one + H_2O_2

$2H_2O_2$ + 4-AAP + Phenol $\xrightarrow{\text{Peroxidase}}$ Quinoneimine + H_2O

CE, Cholesterol Esterase
CO, Cholesterol Oxidase
4-AAP, 4-Aminoantipyrine

(c) **Free Fatty Acids**

Fatty Acid + CoA $\xrightarrow{\text{ACS}}$ Acyl-CoA

Acyl-CoA + O_2 $\xrightarrow{\text{ACOD}}$ Oxidized Acyl-CoA + H_2O_2

H_2O_2 + 4-AAP + MEFA $\xrightarrow{\text{Peroxidase}}$ Purple Adduct

CoA, Coenzyme A
ACS, Acyl-CoA Synthetase
ACOD, Acyl-CoA Oxidase
4-AAP, 4-Aminoantipyrine
MEFA, 3-Methyl-N-Ethyl-N(β-Hydroxyethyl)-Aniline

SCHEME 2 (*Continued*).

Enzyme Assays for Liver Toxicity Biomarkers Alanine aminotransferase (ALT) and aspartate aminotransferase (AST) are enzymes that serve as common indicators of liver toxicity in rodent studies. Scheme 3 shows the two-step enzymatic reaction schemes of ALT and AST, the second step of each is converting NADH to NAD, resulting in a change in the optical density at 340 nm that is measured by a clinical analyzer.

(a) **AST**

L-Aspartate + α-Ketoglutarate $\xrightarrow{\textbf{AST}}$ Oxaloacetate + L-Glutamate

Oxaloacetate + NADH + H^+ $\xrightarrow{\textbf{MDH}}$ Malate + NAD^+

AST, Aspartate Aminotransferase
MDH, Malate Dehydrogenase

(b) **ALT**

L-Alanine + α-Ketoglutarate $\xrightarrow{\textbf{ALT}}$ Pyruvate + L-Glutamate

Pyruvate + NADH + H^+ $\xrightarrow{\textbf{LDH}}$ Malate + NAD^+

ALT, Alanine Aminotransferase
LDH, Lactate Dehydrogenase

SCHEME 3 Enzymatic assays for (a) AST and (b) ALT.

C. Enzymes as Reporters

The principle of a regular micro-plate ELISA involves binding of a peptide analyte to a polystyrene plate-coated specific antibody and reporting of the bound analyte by detecting antibody conjugated with an enzyme, typically horseradish peroxidase (HRP) or alkaline phosphatase (AP). In competitive micro-plate immunoassays, the small molecule or peptide/protein analyte competes with the tracer which is the same molecular entity conjugated with fluorescent dye or radioactive nuclei (as in EIA or RIA) for binding to a specific antibody pre-coated on the polystyrene micro-plate or beads. Table 2 lists selective analytes by both modes of micro-plate immunoassays for antidiabetic and anti-obesity drug discovery.

Enzyme Assays for Metabolic Biomarkers Insulin and glucose have been the most frequently used indicators for diabetic drug discovery *in vivo* and *in vitro*. Insulin is measured by micro-plate ELISA, using a specific antibody against insulin coated on a micro-plate and a reporting antibody conjugated with HRP that catalyzes the conversion of a substrate TMB to a yellow product measured by the optical density at 450 nm (Scheme 4). Since obesity contributes significantly to the development of diabetes [24], leptin, linked to obesity, is also a frequently used biomarker for diabetic and obesity drug discovery and can be similarly analyzed by micro-plate ELISA.

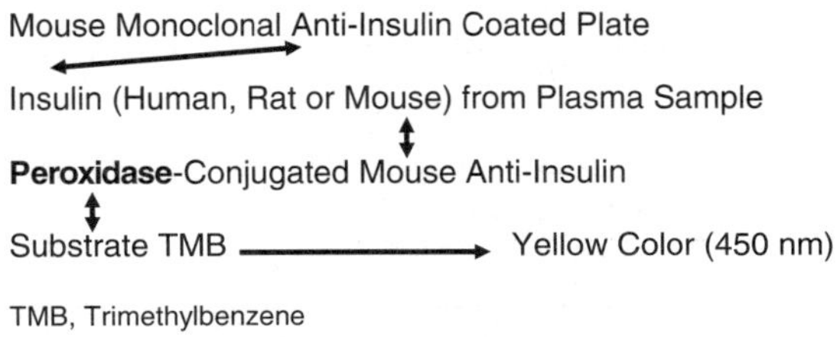

SCHEME 4 Insulin assay, enzyme (peroxidase) as reporter.

Enzyme Assays for Inflammation Biomarkers As an alternative hypothesis to fat overload, inflammation has also gained significance for insulin resistance and diabetes [24]. Many inflammation biomarkers have been used in early drug discovery for metabolic diseases; a few specific biomarkers such as IL-1β, IL-6, MCP-1, and TNFα are among the most frequently used indicators for rodent studies in diabetes, obesity, and cardiovascular diseases and can be analyzed by micro-plate immunoassays.

Multiplexing Enzyme Assays (Luminex) Micro-plate enzyme assay technology, typically 96-, 384-, or 1536-well plates, has significantly improved assay throughput for drug discovery. To further increase the assay throughput, in addition to micro-plate ELISA, Luminex is another micro-plate assay technology using coded beads allowing multiple assays for up to approximately 100 analytes to be run simultaneously from the same sample [26]. Luminex has become a popular choice for multiplexing assays

TABLE 2 Selective Enzyme Assays: Enzyme (Peroxidase) as Reporter Using Microplate/ELISA (Regular or Competitive Mode)

ELISA	Mode	Assay kit sources
Insulin	Regular	ALPCO
Leptin	Regular	ALPCO
GLP-1	Regular	ALPCO
GIP	Regular	Millipore
Adiponectin	Regular	ALPCO
PYY	Regular	ALPCO and Millipore
Ghrelin	Regular	ALPCO and Millipore
Micro-albumin	Competitive	ALPCO and Exocell
Epinephrine	Competitive	ALPCO
Norepinephrine	Competitive	ALPCO
Corticosterone	Competitive	Cayman
Angiotensin I and II	Regular	Phoenix
IL-1β	Regular	ALPCO
IL-6	Regular	ALPCO
MCP-1	Regular	ALPCO
C peptide	Regular	ALPCO and Millipore

GLP-1, glucagon-like protein 1; GIP, gastric inhibitory polypeptide; PYY, peptide YY; MCP-1, monocyte chemotactic protein-1.
Regular, signal is proportional to concentration of analyte in sample; competitive, signal is inversely proportional to concentration of analyte in sample.

TABLE 3 Selective Enzyme Assays: Enzyme (Peroxidase) as Reporter Using Luminex Multiplexing

2-Plex	5-Plex	7-Plex (A)	7-Plex (B)
Insulin	Insulin	Insulin	Insulin
GLP-1	Leptin	Leptin	Amylin
GIP	IL-6	IL-1β	Leptin
GLP-1	TNFα	IL-6	Ghrelin
MCP-1	Resistin	MCP-1	GIP
TNFα		PAI-1	PYY
		TNFα	PP

Luminex (Austin, Texas).
Commercial kits from Millipore (St. Louis, Missouri).
See Tables 2 and 4 for the analyte abbreviations.

in the last decade. Table 3 lists commonly used metabolic and inflammatory biomarkers analyzed by Luminex multiplexing.

Nonenzymatic Immunoassays To improve assay dynamic range and sensitivity, the nonenzymatic immunoassays from MSD have gained significant bio-analytical popularity in the last several years. MSD is based on the signal from electrochemiluminescence [27]. Its major advantage over Luminex is better sensitivity, wider dynamic range, and higher reproducibility most noticeably for GLP-1 (total and active) or insulin. However, the flexibility of multiplexing is not as great as compared to Luminex technology.

III. DEVELOPING NEW RAT MODEL FOR METABOLIC DISEASES

Metabolic diseases are broadly defined as pathological conditions caused by dysregulation of metabolic processes. They are commonly recognized as diabetes, metabolic syndrome, obesity, cardiovascular diseases, and diabetic complications. Together, they negatively affect the quality of life for about half of adult human population from the developed countries. It is imperative, for drug discovery, that developing and applying rodent models mimicking human conditions are required. We summarize here our development of the ZDSD rat, a new rat model for diabetes, metabolic syndrome, and obesity, without leptin and leptin receptor mutations. Currently, the ZDF rat, with a leptin receptor mutation, has been extensively used as a common preclinical animal model for antidiabetic drug discovery [28].

A. Breeding, Spontaneous Diabetic Development, and Characterization

The ZDSD rat was developed by crossing a homozygous lean ZDF rat, expressing β-cell failure with the fa/fa genotype, with a diet-induced obese (DIO) Sprague Dawley (SD) rat, exhibiting insulin resistance and polygenic obesity [29]. Rat selection from breeding was based on major characteristics for diabetes and obesity. Spontaneous hyperglycemia of male ZDSD rats occurs with an age range of 11–29 weeks (Fig. 1a). Insulin increases continuously until these rats become overtly diabetic and then drops gradually to the prediabetic level (Fig. 1b). The ZDSD rats, heavier than the ZDF rats, also have high TG and CHOL (Fig. 2). Food intake of the ZDSD rats is reduced in response to leptin, indicating a functional leptin pathway (Fig. 3). Such diabetic and obese characteristics as well as absence of a leptin or leptin receptor mutation render the ZDSD rats highly appropriate for preclinical rodent studies in diabetes, metabolic syndrome, and obesity. Evaluation of the ZDSD rats with current antidiabetic and anti-obesity drugs are described in Section IV.

B. Diet-Induced Diabetes and Obesity

The male ZDSD rats that have not become diabetic spontaneously within 15–21 weeks old can be induced to diabetic level by providing a high-fat diet for 2 weeks (Fig. 4, [29]). Such hyperglycemic synchronization allows the ZDSD rats to be used in more predictable and practical durations in antidiabetic drug discovery.

C. Determining Insulin Sensitivity

Development of diabetes may be preceded by insulin resistance, glucose intolerance, hyperlipidemia, and hypertension. Testing insulin sensitivity is an important aspect of preclinical antidiabetic drug discovery. In comparison to the Zucker fatty

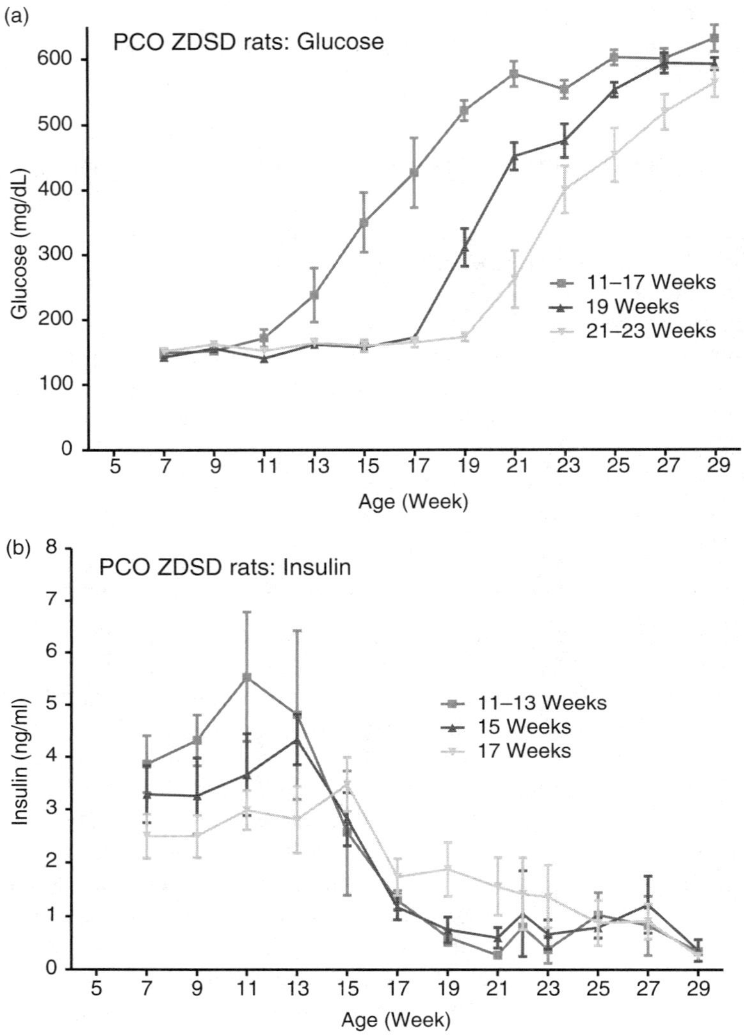

FIGURE 1 ZDSD rats: spontaneous diabetic development. Glucose (a) and insulin (b) were monitored for ZDSD male rats that became diabetic, using three different age ranges. (*See insert for color representation of the figure.*)

(ZF) rats, the rat standard for testing insulin sensitivity, the ZDSD rats show a similar insulin resistance. With treatment by rosiglitazone, there is a significant improvement in insulin sensitivity for the ZDSD rats though less than that for the ZF rats (Fig. 5, [29]). When compared to the SD rats, the male ZDSD rats demonstrated age-dependent impaired glucose tolerance (Fig. 6, [30]). Thus, the ZDSD rats, with less fat content than the ZF rats (data not shown), appear to be an excellent

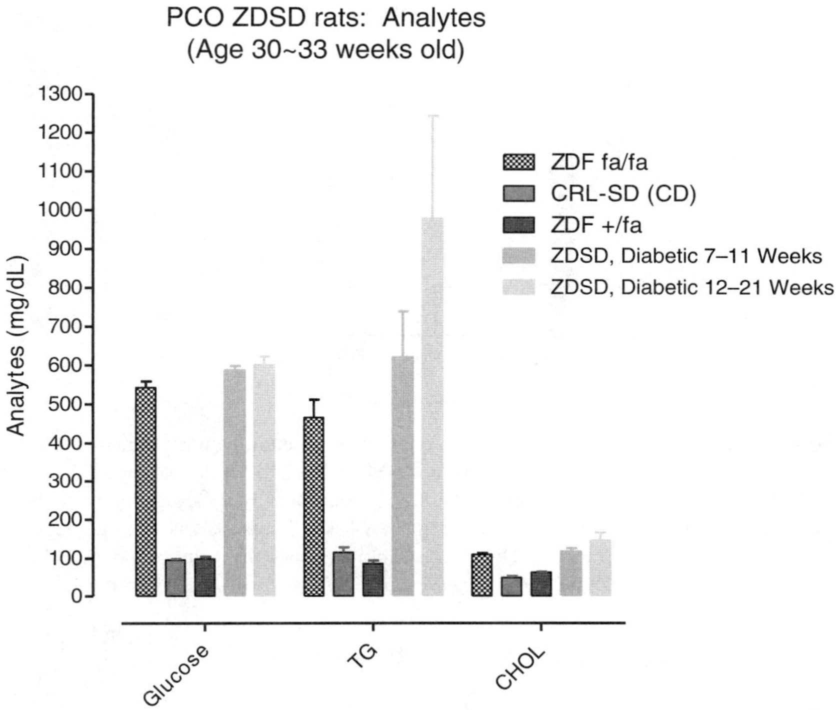

FIGURE 2 ZDSD rats: spontaneous diabetic development. Glucose, TG, and total CHOL were determined for five different 30–33-week-old rat models: ZDF fa/fa, CRL-SD (CD), ZDF + /fa, ZDSD diabetic from 7–11 weeks, and ZDSD diabetic from 12–21 weeks. For statistical analysis of this and following studies, as specified, *t*-test from GraphPad Prism 5 was used. ZDF and ZDSD diabetic from 7–11 weeks and from 12–21 weeks were analyzed with CRL-SD (CD) and ZDF + /fa; $P<0.0001$ for glucose. Similarly for TG, P values are <0.0001 with ZDF, $= 0.0104$ and $=0.0073$ with ZDSD diabetic 7–11 weeks, and $=0.0090$ and $=0.0075$ with ZDSD diabetic 12–21 weeks. For total CHOL, P values are <0.0001 with ZDF, <0.0001 and $=0.0003$ with ZDSD diabetic 7–11 weeks, and $=0.0012$ and $=0.0032$ with ZDSD diabetic 12–21 weeks. (*See insert for color representation of the figure.*)

alternative rat model in testing compounds for improving insulin sensitivity and glucose tolerance.

D. Identifying Biomarkers of Metabolic Diseases

Using Luminex multiplexing assays, collaboration between PreClinOmics (PCO) (Indianapolis, Indiana) and Rules-Based Medicine (Austin, Texas), the ZDSD rat has been shown as a highly useful model for multiple biomarkers in metabolic, renal, coagulation, and vascular diseases as well as for inflammation. Examples of these plasma biomarkers from an initial identification are shown in Table 4 [29].

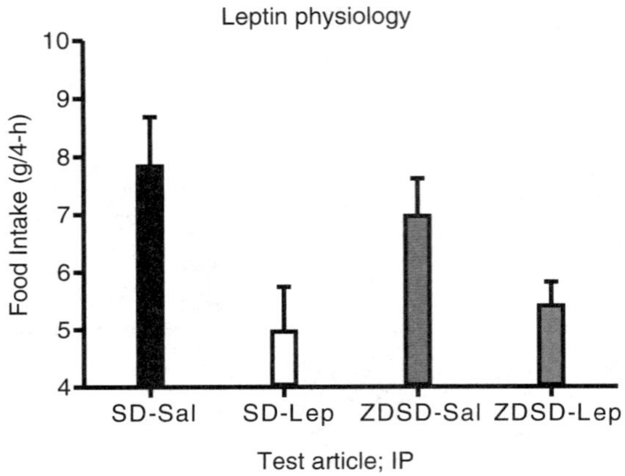

FIGURE 3 ZDSD rats: leptin physiology. Assessment of leptin pathway function was determined by feeding response to leptin injection. ZDSD rats of 15 weeks old and SD rats of 17 weeks old, $n = 9$, were dosed with leptin, 1 mg/kg intraperitoneally (IP), just before starting the dark cycle, and food intake was measured for the first 4 dark hours. Sal: saline; Lep: leptin. The food intake of this study was analyzed by t-test; in comparison to the saline group, $P = 0.0196$ for SD-Lep and $P = 0.0406$ for ZDSD-Lep. (*See insert for color representation of the figure.*)

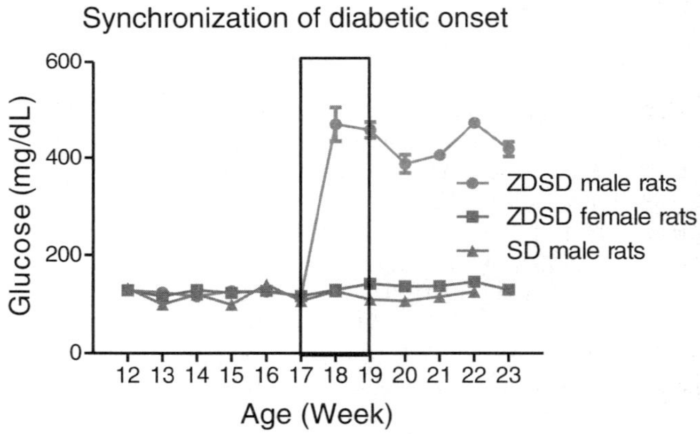

FIGURE 4 ZDSD rats: diet-induced diabetic synchronization. The ZDSD male rats, $n = 12$, can be placed on either D12468 (Research Diets) or 5SCA (LabDiet) at 17 weeks of age to synchronize the onset of diabetes. Two-way ANOVA from GraphPad Prism 5 was used in statistical analysis of this and following studies, as specified. Each P value is from interaction of column (effect) factor and row (time) factor, unless specified (i.e., from column factor only); in comparison to SD male rats, for analysis of glucose, $P < 0.0001$ for ZDSD male rats. (*See insert for color representation of the figure.*)

Also using the MSD platform, the ZDSD rat has been shown as a suitable model for evaluating several urine biomarkers of kidney injury related to nephropathy (Table 5, [30]). Neutrophil gelatinase-associated lipocalin (NGAL), the common plasma and kidney biomarker, has been shown recently as a biomarker of acute

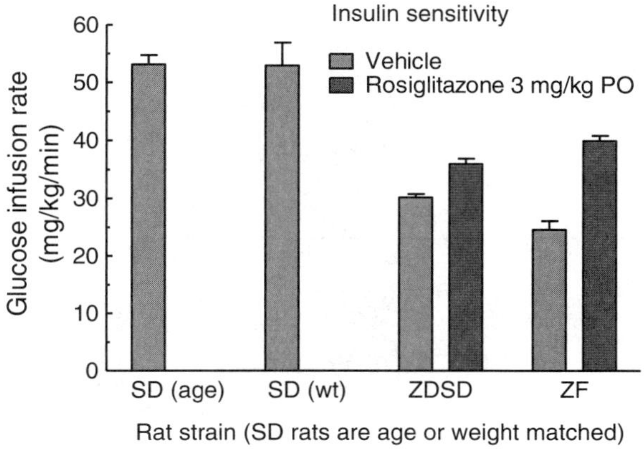

FIGURE 5 ZDSD rats: insulin resistance and improved sensitivity with rosiglitazone. The insulin sensitivity (or "Clamp") study was initiated with 8-week-old ZDSD, ZF, and SD rats, $n = 6$–8. Using t-test for insulin sensitivity (glucose infusion rate), for ZDSD rats, $P < 0.0001$, $= 0.0002$ and $= 0.0003$ when analyzed with SD (age) and SD (weight) control rats and with ZDSD rats treated with rosiglitazone 3 mg/kg, respectively, and for ZF rats, $P < 0.0001$ when analyzed with SD (age) and SD (weight) control rats and with ZF rats treated with rosiglitazone 3 mg/kg.

kidney injury and nephrotoxicity [31]. Selective use of these biomarkers, particularly upon further confirmation, should facilitate preclinical drug discovery in metabolic diseases.

IV. ZDSD RAT: EVALUATING CURRENT DRUGS

Since the ZDSD rat does not have a leptin or leptin receptor mutation and with apparent polygenic diseases that mimic type II diabetes in human population, this rat model appears more appropriate than other rodent models for use in antidiabetic drug discovery. One important question is whether current antidiabetic drugs are effective for the ZDSD rat. Tested in a prevention mode, metformin, rosiglitazone, and exenatide could significantly delay the development of diabetes in the ZDSD rats (Fig. 7). Though the modes of action for the antidiabetic drugs have not been completely elucidated, they appear to affect different molecular targets and metabolic pathways. Metformin acts through AMP-activated protein kinase (AMPK) pathway [32–35]. Rosiglitazone is peroxisome proliferator-activated receptor-γ (PPARγ) agonist [36] and inhibits obesity-linked phosphorylation of PPARγ by cyclin-dependent kinase 5 [37]. Exenatide is a GLP-1 mimetic for treatment of type 2 diabetes [38,39]. Showing preventive glucose-lowering effects by these antidiabetic drugs correlates with polygenic diseases of the ZDSD rats and validates its usefulness in drug discovery for metabolic diseases.

Rimonabant is a commonly used positive control for testing anti-obesity compounds in rodent models; it decreases insulin resistance via both adiponectin-dependent and

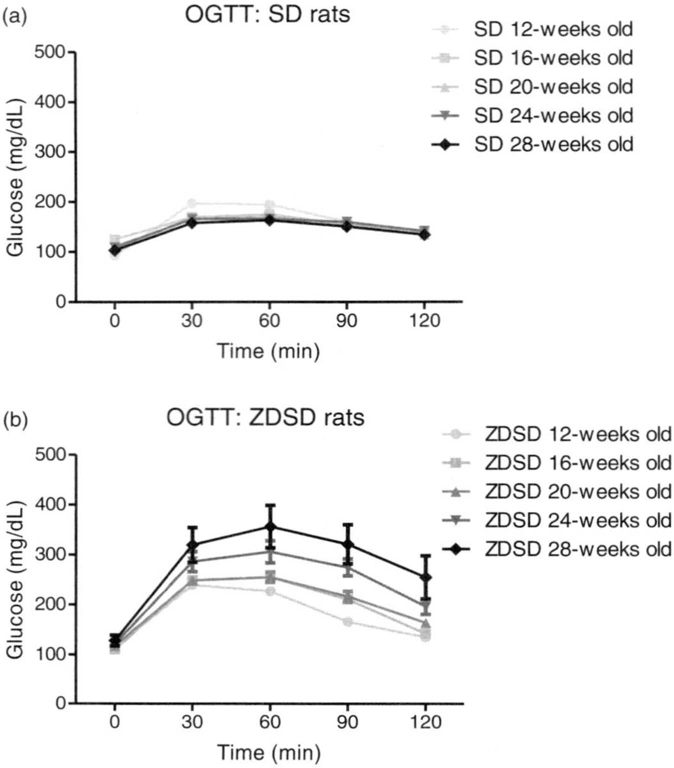

FIGURE 6 ZDSD rats: age-dependent impaired glucose tolerance. SD rats (a), $n=10$, and ZDSD rats (b), $n=10$, were used in an oral glucose tolerance test (OGTT) starting 12 weeks old. From 12 to 28 weeks old, for every 2 weeks, the rats were dosed with 2 g/kg glucose orally, and the OGTT data are shown for every 4 weeks only. Two-way ANOVA was used in analyzing glucose of this study; in comparison to SD rats, $P<0.0001$ for ZDSD rats from four age groups (12 to 24 weeks old) and $P=0.0083$ from 28 weeks old. (*See insert for color representation of the figure.*)

TABLE 4 ZDSD Rats: Serum Biomarkers[a] for Metabolic Disorders

Metabolic disease	Renal disease	Coagulation and vascular disease	Inflammation
Insulin	NGAL	VEGF	MCP-3
Leptin	β-2-microglobulin	Von Willebrand factor	Lymphotactin
GLP-1	Kim-1	Thrombopoietin	IL-11
Resistin	GST-α	Factor VII	MIP-1α
Adiponectin	Clusterin	PAI-1	MPO
			Eotoxin
			CD40 ligand

NGAL, neutrophil gelatinase-associated lipocalin; Kim-1, kidney-injury molecule-1; GST-α, glutathione S-transferase alpha; VEGF, vascular endothelial growth factor; PAI-1, plasminogen activator inhibitor-1; MCP-3, monocyte chemotactic protein-3; IL-11, interleukin-11; MIP-1α, macrophage inflammatory protein-1α; MPO, myeloperoxidase.

[a]Statistically significant elevation of the listed analytes of ZDSD versus SD rats at 14, 18, and/or 20 weeks old, analyzed by Rules-Based Medicine. Rats were on Purina 5SCA from 17 to 19 weeks of age.

TABLE 5 ZDSD Rats: Urinary Excretion of Kidney Biomarkers for Renal Disease

Biomarker	ZDSD-8/SD (Fold)	ZDSD-16/SD (Fold)
Albumin	243[a]	842[a]
Clusterin	4.9	17.6[a]
GSTYb-1	1.4	2.6[a]
NGAL	3.6	10.7[a]
Osteopontin	17.3	80[a]
RPA-1	5.5[a]	4.5[a]
Tim-1	4.1[a]	9.5[a]

The urinary biomarkers were compared from 33-week-old ZDSD rats to SD(CD) rats. ZDSD-8/SD is for the biomarker ratio that the ZDSD rats had been 8 weeks diabetic, and ZDSD-16/SD is for the biomarker ratio that the ZDSD rats had been 16 weeks diabetic.

GSTYb-1, glutathione S-transferase Yb1; RPA-1, renal papillary antigen 1; Tim-1, T cell immunoglobulin-mucin; analyzed by MSD.

[a]Statistically significant elevation of the analyte for the ZDSD rats.

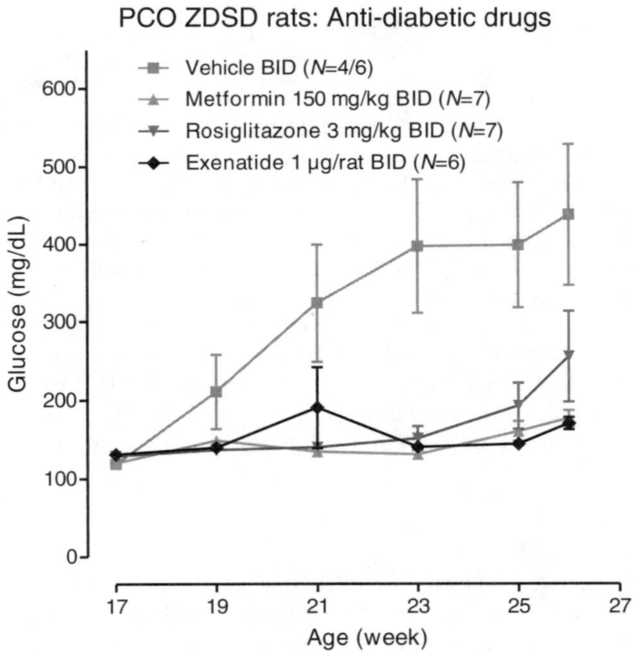

FIGURE 7 ZDSD rats: antidiabetic drug treatment with metformin, rosiglitazone, and exenatide. Using two-way ANOVA for glucose of the study, in comparison to the vehicle group, $P<0.0001$ for metformin, $P=0.0024$ for rosiglitazone, and $P=0.0002$ for exenatide. (*See insert for color representation of the figure.*)

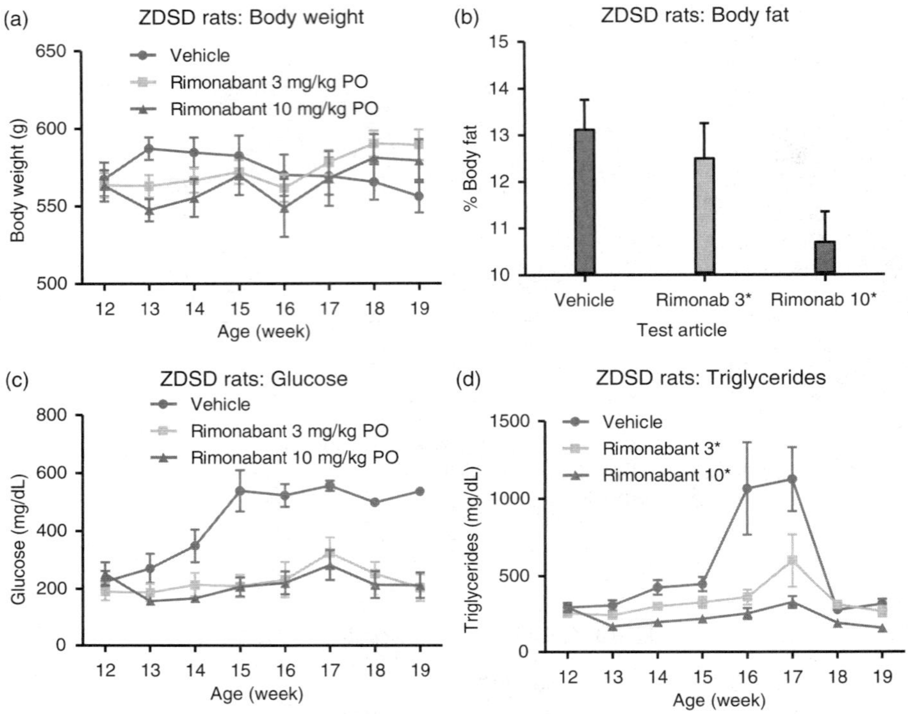

FIGURE 8 ZDSD rats: anti-obesity drug treatment with rimonabant. The drug treatment started with 12-week-old rats, $n=8$, that were on a standard diet for 3 weeks, followed by a high-fat diet for 2 weeks (15–17 weeks of age) and returned to the standard diet for 2 weeks. Body weight (a), fed plasma glucose (c), and TG (d) were monitored weekly; 3* is 3 mg/kg PO and 10* is 10 mg/kg PO. Body fat (b) was determined at 27-week-old rats by Quantitative Nuclear Magnetic Resonance (QNMR); Rimonab 3* is rimonabant 3 mg/kg and Rimonab 10* is rimonabant 10 mg/kg. Using two-way ANOVA, in comparison to the vehicle group, $P<0.0001$ for glucose at rimonabant 3 and 10 mg/kg and $P=0.0018$ and $P<0.0001$ for TG at rimonabant 3 and 10 mg/kg, respectively. For 12–15-week-old rats on the standard diet, $P=0.2517$ and $P=0.0087$ for body weight at rimonabant 3 and 10 mg/kg, respectively. Using t-test, in comparison to the vehicle group, $P=0.5419$ and $P<0.0202$ for body fat at rimonabant 3 and 10 mg/kg, respectively. (*See insert for color representation of the figure.*)

adiponectin-independent pathways [40]. Rimonabant lowered body weight increase of 12- to 15-week-old ZDSD rats on the standard diet (Fig. 8a). When tested with rimonabant at 10 mg/kg, the ZDSD rat showed a significant decrease in body fat (Fig. 8b). Interestingly, decreased body fat of the ZDSD rat also prevented spontaneous development of diabetes (Fig. 8c) and lowered TG (Fig. 8d). Thus, the ZDSD rat appears useful for both anti-obesity and antidiabetic drug discovery.

V. EXISTING RODENT MODELS: ESTABLISHING POSITIVE CONTROLS

Two types of existing rodent models are typically used in drug discovery for diabetes, metabolic syndrome, and obesity: genetic and diet induced. Two genetic models, db/db mice and ZDF rats, both with a leptin receptor mutation, as well as the DIO mice are probably the most used rodent models for antidiabetic drug discovery. Rosiglitazone is frequently used as a positive control in rodent efficacy studies for testing antidiabetic compounds. Rosiglitazone shows a significant efficacy in lowering plasma glucose in db/db mice and ZDF rats at low dose levels (Fig. 9a and b); similar glucose-lowering effects can be observed from dosing the

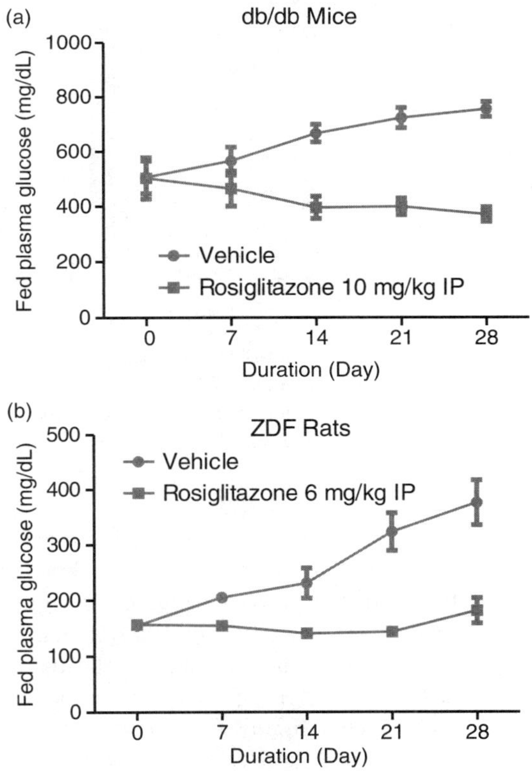

FIGURE 9 Rosiglitazone as a positive control for lowering plasma glucose in db/db mice (A) and in ZDF rats (B). The db/db mice, $n = 6$, were dosed with rosiglitazone IP at 10 mg/kg starting 5 weeks old for 4 weeks, and the ZDF rats, $n = 6$, were dosed with rosiglitazone IP at 6 mg/kg starting 6 weeks old for 4 weeks. Using two-way ANOVA for glucose of the study, $P < 0.0001$ for both the db/db mice and the ZDF rats.

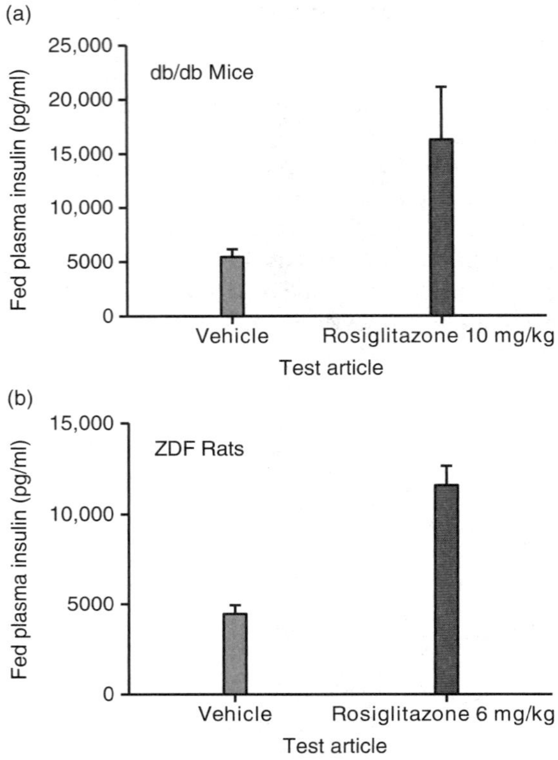

FIGURE 10 Rosiglitazone as a positive control for increasing plasma insulin in db/db mice (a) and in ZDF rats (b). The db/db mice and the ZDF rats were dosed with rosiglitazone as described in Figure 9. Plasma insulin was determined after the 4-week dosing with rosiglitazone. Using *t*-test for insulin of the study, $P=0.0524$ for the db/db mice and $P=0.0001$ for the ZDF rats.

drug by oral, subcutaneous, or intraperitoneal route of administration. Other positive controls for lowering glucose in antidiabetic drug discovery include metformin, exenatide, and GLP-1. In addition to lowering glucose, rosiglitazone serves as a positive control for increasing plasma insulin (Fig. 10a and b), though with marginal statistical significance for db/db mice, and for lowering plasma TG (Fig. 11a and b) in db/db mice and ZDF rats. Rosiglitazone appears to raise the ratio of high-density lipoprotein (HDL) to total CHOL slightly (Fig. 12); however, not statistically significant, the ratio was not a good positive control for this study. Rosiglitazone also increased insulin extracted from pancreas (Fig. 13), particularly for the 6- to 9-week-old mice, and increased TG extracted from liver (Fig. 14) in db/db mice; thus, rosiglitazone is likely the most used positive control for both antidiabetic and anti-obesity drug discovery. For reminder, as a therapeutic agent from several years of concern, rosiglitazone (trade name Avandia) increases the risk of heart failure [41]. Both the United States and Europe have greatly restricted the use of the

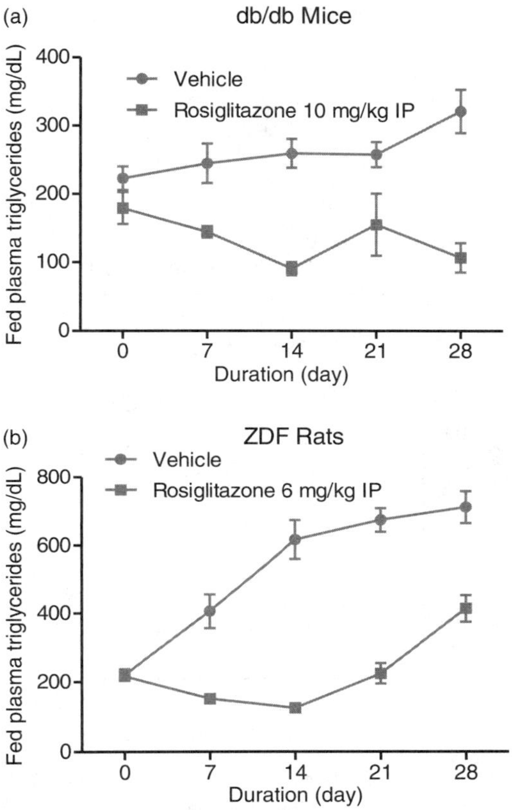

FIGURE 11 Rosiglitazone as a positive control for lowering plasma TG in db/db mice (a) and in ZDF rats (b). The db/db mice and the ZDF rats were dosed with rosiglitazone as described in Figure 9. Using two-way ANOVA for TG of the study, $P=0.0074$ for the db/db mice and $P<0.0001$ for the ZDF rats.

diabetes drug. For anti-obesity drug discovery, rimonabant is frequently used as a positive control in rodent efficacy studies for testing anti-obesity compounds. Using diet-induced obesity for Long–Evans rats at a low dose level, rimonabant shows a significant efficacy in lowering body weight (Fig. 15a) and body fat (Fig. 15b). Another genetic model, ob/ob mice with a leptin mutation, is also used for antidiabetic and anti-obesity studies. In addition to use in the treatment mode as described earlier, both rosiglitazone and rimonabant can be used in a preventive mode in rodent efficacy studies for metabolic syndrome, which is defined as abdominal obesity, high glucose and TG, low HDL, and high blood pressure [42]. Thus, rosiglitazone and rimonabant, among others, have been established as effective positive controls for these and other rodent models in drug discovery for diabetes, metabolic syndrome, and obesity.

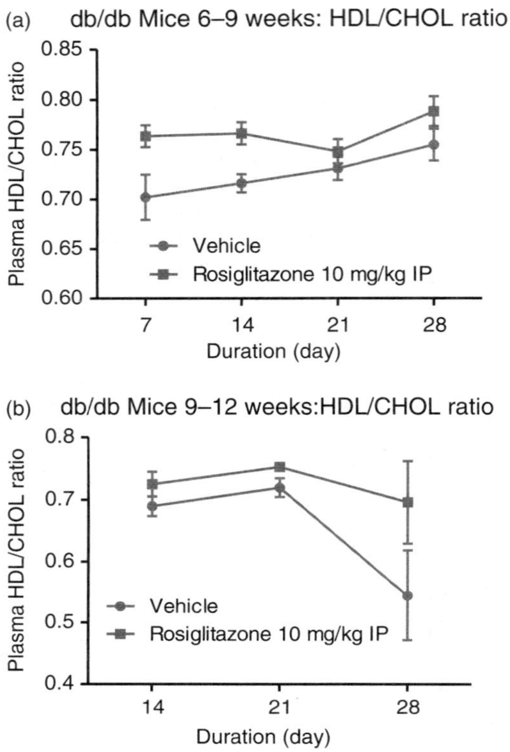

FIGURE 12 Rosiglitazone as a positive control for raising HDL/CHOL ratio in db/db mice. Starting 5 weeks (a) or 8 weeks (b) old, the db/db mice were dosed with 10 mg/kg rosiglitazone IP for 4 weeks. Using two-way ANOVA for HDL/CHOL ratio of the study, $P = 0.4305$ for the db/db mice 6–9 weeks and $P = 0.3050$ for the db/db mice 9–12 weeks.

VI. EXISTING RODENT MODELS AND ONE CANCER PATIENT: TESTING NUTRITIONAL SUPPLEMENT (ALKA VITA)

As this chapter goes to print, health-care reform will likely be a reality for the first time in the US history. Though there is a continuous debate about the economic impact of such health-care reform, unanimous scientific consensus is that obesity, diabetes, and metabolic diseases have a very significant role in human health and economic expenditure in the United States and worldwide. About 72 million Americans or one-third of adults were obese in 2006, according to the US Centers for Disease Control and Prevention. Approximately 300 000 deaths per year in the United States are associated with obesity in 2008 [43]. Obesity-related medical costs reached $147 billion in 2008, or about 10% of US medical spending [43]. Since fat overload is considered as the major factor for diabetic development [24], many of the obese individuals are also diabetic; both the United States and China have an alarming and equal diabetic rate of 11% in 2010. To improve or maintain good health, common knowledge and

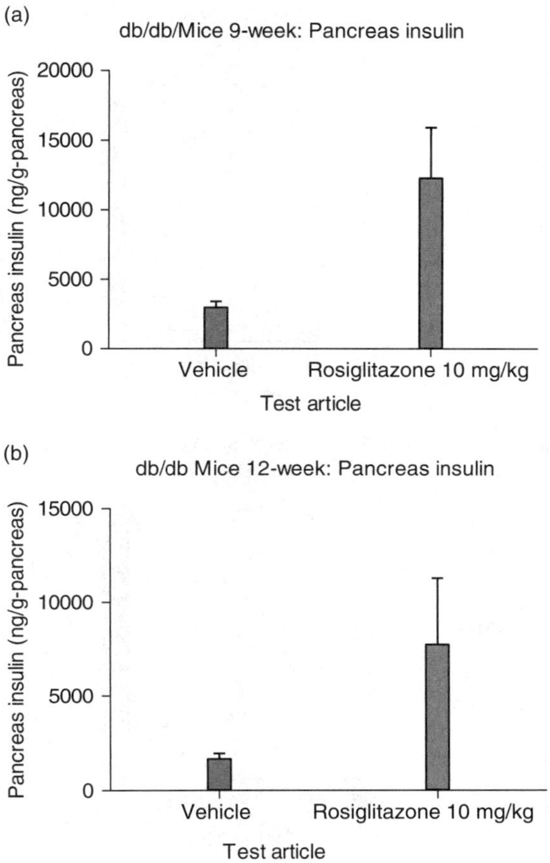

FIGURE 13 Rosiglitazone as a positive control for raising pancreas insulin in db/db mice. Starting 5 weeks (a) or 8 weeks (b) old, the db/db mice were dosed with 10 mg/kg rosiglitazone IP for 4 weeks. The mice were fasted for 12 h and an OGTT was conducted on day 29. Following the OGTT, the pancreas was removed, and insulin was extracted from the pancreas and analyzed by an MSD assay. Using *t*-test for pancreas insulin of the study, $P=0.0301$ for the db/db mice 6–9 weeks and $P=0.0940$ for the db/db mice 9–12 weeks.

experience recommend a balanced diet, with abundant fruits and vegetables and moderate meat consumption. Regular exercise, such as walking and running, is as important as balanced food for good health. Exercise is known to improve good hormone production and normal physiological function, the latter, for example, in raising AMPK-linked autophagy [44]. In addition, a recent "Blockbuster Study" finds that vigorous long-term exercise, as that from a distance runner, may contribute to the prevention of telomere shortening [45], thus reducing the risk of many diseases. The enzymes responsible for maintenance of telomeres, found on the ends of DNA structures, are called telomerases, the studies of which produced the 2009 Nobel Prize in Physiology or Medicine [46]. Since there is deficiency for vitamins and minerals in

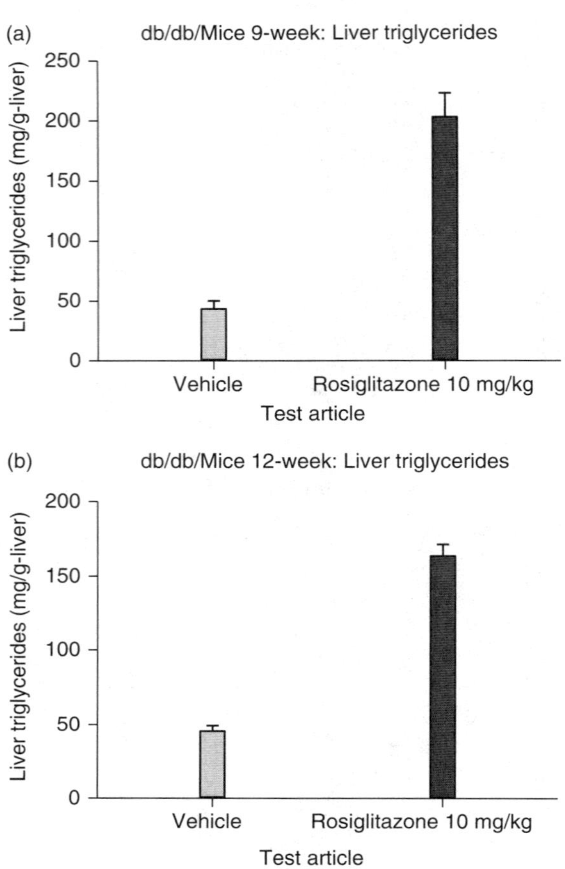

FIGURE 14 Rosiglitazone as a positive control for increasing liver TG in db/db mice. The db/db mice were dosed with rosiglitazone, and an OGTT was conducted as described in Figure 13. Following the OGTT, the liver was removed and TG were determined from 1% Tween-80 homogenates (25) using a kit from Wako Diagnostics. Using *t*-test for liver TG of the study, $P<0.0001$ for both db/db mice 6–9 weeks and 9–12 weeks.

western food, much of which is from mass commercial production, taking nutritional supplements (nutraceuticals) is often needed for maintaining healthy human physiology. For prevention of metabolic and other diseases, one needs balanced diet, moderate physical activity, and suitable nutritional supplements.

To maintain good human health, there are two predominant yet not proven basic biochemical hypotheses. One is originated from German scientists, suggesting that adequate oxygenation is the basis of normal physiological processes [47]. The other was hypothesized from Japan researchers, implicating that alkalinity plays an important role in human health [47,48]. Acid waste accumulation in the body, which can be reduced by alkaline water and fruit/vegetable diet, may cause oxygen deficiency and related symptoms for a variety of adult diseases [47]. A recent report describes phosphoric acid acting as a pH biosensor that links membrane biogenesis to metabolism [49]. To test the second pH-related hypothesis, an alkaline nutritional supplement has been extensively evaluated in preclinical antibacterial and antidiabetic applications.

Alka Vita (renamed as Neutrevive and LipH, of Orizon International and Cisne Enterprises, Odessa, Texas) is a uniquely modified liquid silicon containing a mixture of trimeric sodium silicate and sodium silicate pentahydrate. At a low concentration, it can raise pH and is suitable for human consumption [48,50]. It shows a strong inhibition of common pathogenic bacteria, including *Staphylococcus aureus* [51], a major infectious agent at hospitals. The effect of Alka Vita is attributed to both alkalinity and liquid silicon.

When tested in the db/db mice alone, Alka Vita moderately lowered glucose as well as AST and ALT (not shown), the liver toxicity biomarkers. When evaluated in

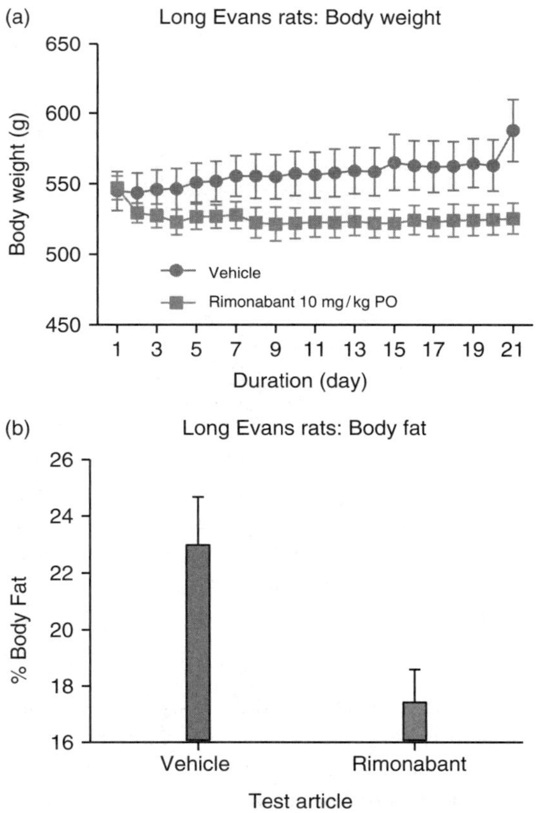

FIGURE 15 Rimonabant as a positive control for reducing body weight (a) and fat (b) in Long–Evans rats. DIO Long–Evans rats, $n=6$, at 14 weeks old were dosed orally with 10 mg/kg rimonabant (PO); body weight was measured daily for 21 days and % fat was determined by QNMR on day 21. Using two-way ANOVA by column (effect) factor only, in comparison to the vehicle group, $P<0.0001$ for body weight. Using t-test, in comparison to the vehicle group, $P=0.0272$ for body fat.

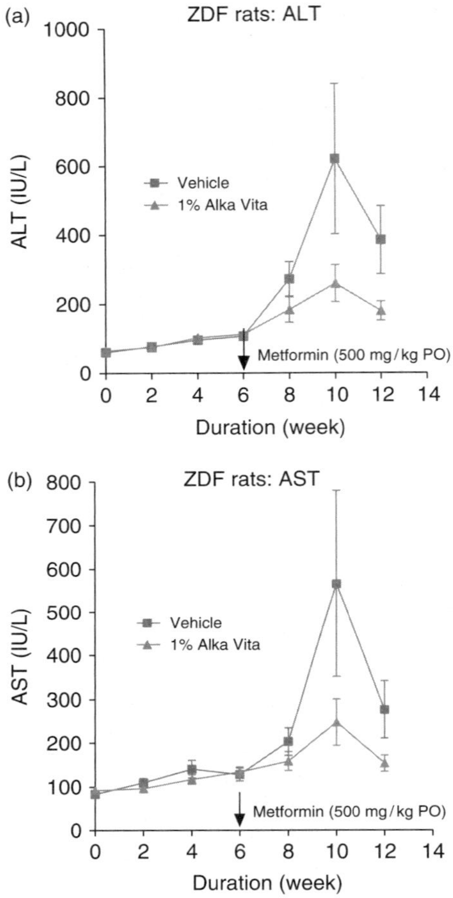

FIGURE 16 Effects of Alka Vita and metformin on ALT (a) and AST (b) of ZDF rats. ZDF obese male rats, $n = 15$, were dosed orally with 1% Alka Vita daily from 11 weeks old and dosed orally with 500 mg/kg metformin daily from 6 weeks of the study for 6 weeks; and ALT and AST determined as indicated in Scheme 3. Using two-way ANOVA for the study, $P = 0.0192$ for ALT and $P = 0.0590$ for AST.

the ZDF rats with metformin, it decreased ALT significantly and AST moderately (Fig. 16a and b). It also lowered glucose significantly when coadministered with metformin in the ZDF rats. The glucose-lowering effect appears to be additive (Fig. 17a). It decreased HbA1c and increased fasting insulin significantly (not shown). Alka Vita with metformin also significantly reduced cataract (Fig. 17b); the latter is a diabetic complication. From available scientific literature, including online publications, Alka Vita may serve as an effective and preventive nutritional supplement for bacterial and viral infections, diabetic development, and other pathological progression such as that of some cancer types [52]. The statement for Neutrevive to

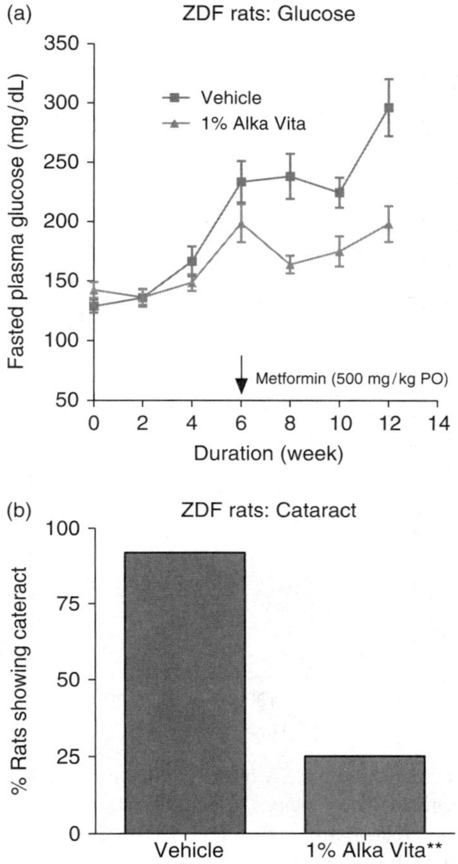

FIGURE 17 Effects of Alka Vita and metformin on glucose (a) and cataract (b) of ZDE rats. ZDF obese male rats were dosed with 1% Alka Vita daily from 11 weeks old and with 500 mg/kg metformin daily from 6 weeks of the study for 6 weeks as described in Figure 16; glucose determined from 16-h fasted rats; and cataract measured at a scale of 0 (none) to 5 (all) from both eyes of each rat at 17 weeks of the study (i.e., 5 weeks after termination of metformin treatment at 12 weeks). 1% Alka Vita**: with 500 mg/kg metformin for 6 weeks. Using two-way ANOVA for the study, $P=0.0001$ for glucose. Cataract was observed in both eyes from six of eight rats for the control group and from two of eight rats for the Alka Vita group.

revive the immune system has not been confirmed by the Food and Drug Administration [48].

More recently, Alka Vita was shown to have cancer chemotherapeutic properties *in vitro*, including an antioxidant response to glutathione, catalase, and superoxide dismutase [50]. *In vivo* tests are imperative to determine its true effectiveness, and an extreme *in vivo* case is described briefly here. After being informed by her physician that she had one and a half year to live and with the conventional chemotherapies, the 49-year-old female patient having terminal colon cancer survived for almost 2 years

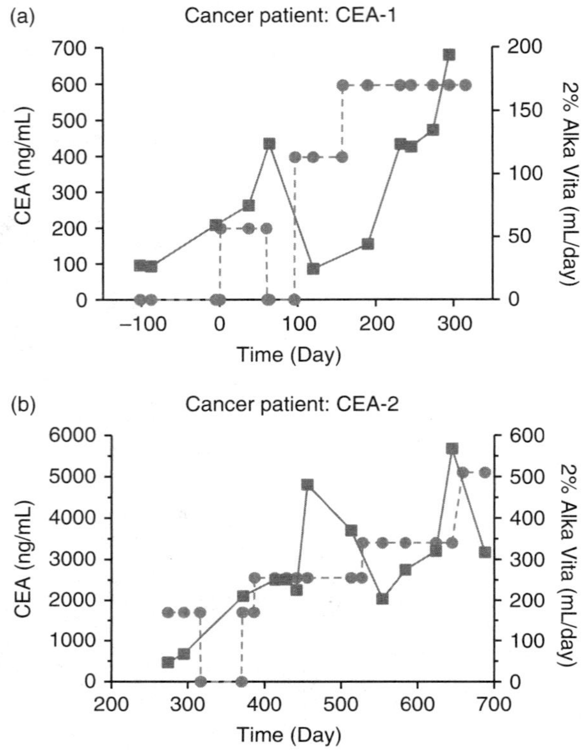

FIGURE 18 Effect of Alka Vita on CEA of the colon cancer patient. A 49-year-old female was diagnosed for a terminal colon cancer in March of 2008, and after surgeries she was informed by the physician that, with conventional chemotherapy, she would likely live for another 18 months. With the biweekly chemotherapies, the patient lived for 23 months, and in February of 2010, she continued the biweekly chemotherapies and additionally started taking 2% Alka Vita (LipH as the new name). From day 1 to day 584, except for two extended times (36 days in Fig. 18a and 54 days in Fig. 18b) without taking Alka Vita due to its unavailability, she took two ounces of 2% Alka Vita once a day initially and increased 2% Alka Vita stepwise as shown in the following text: to two ounces twice a day, to two ounces three times a day, to three ounces three times a day, to three ounces four times a day, and to three ounces six times a day (dashed line). CEA (solid line) was determined from the serum of the non-fasting patient by a clinical laboratory, and normal values of CEA are 0.5–2.5 ng/mL.

before she started taking Alka Vita. With the additional use of Alka Vita, though interrupted twice due to unavailability of Alka Vita, the patient remains alive today for more than 24 months. Incrementally elevating the use of Alka Vita either lowered the level or slowed the increase of carcinoma embryonic antigen (CEA; Fig. 18a and b), a proposed biomarker for colon cancer [53]. Also, Alka Vita appears to function by maintaining lower levels of AST (Fig. 19a) and glucose (Fig. 19b). It is important to share with the readers that, after being informed by another physician that she had only 3–6 months of life by the regular chemotherapy, the patient has been surviving

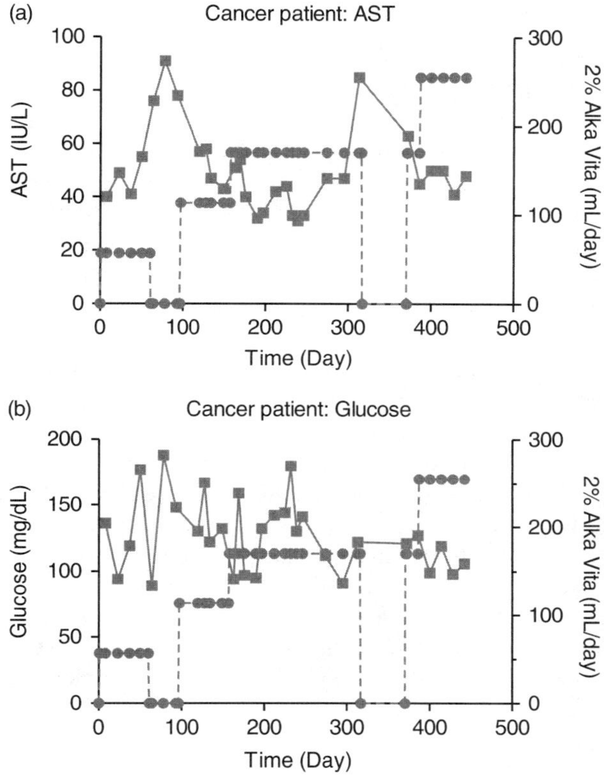

FIGURE 19 Effects of Alka Vita on AST and glucose of the colon cancer patient. See the colon cancer history with conventional chemotherapy of the patient and additional use of Alka Vita in Figure 18. AST (solid line) and glucose (solid line) were determined as indicated in Figure 18.

more than 16 months and is still looking forward to more quality life with continuous use of Alka Vita. Probably by modulating CEA, AST, glucose, and other biomarkers such as ALT and AP (not shown), Alka Vita boosts the immune system, as suggested for the product but not yet approved by FDA, and importantly it extends the life span and its quality for the patient.

VII. CONCLUDING REMARKS

The central theme for this chapter involves using preclinical enzyme assays and rodent models for drug discovery in metabolic diseases, especially diabetes, metabolic syndrome, and obesity. The chapter content derives from the research experience by the authors over three decades. Selective application of the enzyme assays as well as the rodent models (new and established) included in the chapter and others can significantly facilitate discovery of therapeutic agents for unmet medical needs.

A. Enzyme Assays

Application of the enzyme assays has been a continuously evolving process, for example, from theoretical (e.g., metabolic evolution) to practical (e.g., metabolic engineering, combinatorial biosynthesis, and metabolic diseases) and from low to high throughput. The enzyme assays may be classified by the role of the enzyme or by the detection method. Based on the role of enzyme, in addition to the drug target, the enzyme can either serve as the biocatalyst or as the signal reporter for the assay. From the detection method, the common signal is either by absorbance (UV or color), fluorescence, or product separation and mass for the assay. Appropriate utilization of equipment, such as HPLC and LC/MS, is often needed in separation and analysis of an enzymatic reaction product. For lead optimization in early drug discovery process, both HPLC and LC/MS can play a significant role in determining, in addition to the potency, the kinetic mechanism of an enzyme inhibitor, as competitive, noncompetitive, or uncompetitive binding mode that is useful in evaluating how the inhibitor may function in biological system. Toward discovery of an enzyme inhibitor as therapeutic agent, typically tens of thousands of compounds have been screened in the *in vitro* assays. Thus, in the early phase for drug discovery, high-throughput screening of enzyme inhibitors (or activators) with robotic automation is the standard practice by the pharmaceutical industry for the last two decades. Druggable enzyme inhibitors are identified through testing in multiple *in vitro* assays. A series of structure–activity relationship iterations is applied to develop a lead. During the lead optimization stage, compounds are evaluated using *in vivo* preclinical models to identify the clinical candidate.

B. Rodent Models

Mice or rats need smaller amount of compound than large animals, for example, dog or monkey. Thus, rodent models have been commonly used in preclinical drug discovery. For testing antidiabetic compounds, the db/db and ob/ob mice as well as the ZDF obese rats are the established rodent models and frequently are used. These rodent models have monogenic mutations, in either leptin or its receptor, and rosiglitazone is commonly used as the positive control. After a decade of selective breeding and extensive characterization, the ZDSD rat model without a leptin or leptin receptor mutation has been developed by PCO (Indianapolis, Indiana) for commercialization. Since the ZDSD rat has polygenic diseases that mimic human conditions, this rat model may become a useful preclinical model for drug discovery in diabetes, metabolic syndrome, and obesity. Metabolic syndrome can be defined by abdominal obesity, high glucose and TG, low HDL CHOL, and hypertension. It is exciting to expect that, by using the new ZDSD rat model as well as the established rodent models, drug discovery for metabolic diseases can accelerate significantly in the coming decade.

C. Applying Enzyme Assays and Rodent Models

Since a rodent model can be treated as a complex enzyme system, the enzyme assays become an integral part in evaluating compound efficacy *in vivo*. Samples for the enzyme assays can come from whole blood, serum or plasma, urine, and tissue extract. For antidiabetic and anti-obesity drug discovery, the following enzyme assays are commonly used: glucose, insulin, leptin, GLP-1, glucagon, amylin, adiponectin, CHOL, TG, free fatty acids, glycerol, and ketone bodies (Tables 1, 2 and 3). For the ZDSD rat, many serum biomarkers for metabolic disorders (Table 4, [29]) and several urine biomarkers for kidney injury and nephropathy (Table 5, [30]) have now become available for drug discovery in metabolic diseases.

D. Toward Preventive Health and Treatment

To improve and maintain good health, keeping proper diet and physical activity as well as taking needed supplements (nutraceuticals) have been the most common practices for human well-being. The importance of these preventive practices is easily understood but appears quite difficult for a majority of human population to follow through, due to a combination of both genetic and environmental factors. When prevention of metabolic diseases is not adequate for a person, then a therapeutic treatment becomes one of the best options to restore health or extend life.

There is increasing research for supplements during the last decade worldwide. Improved understanding in preventive health of supplements, for example, for good benefit/risk ratio, appears leading to better acceptance and wider use. As a unique mineral supplement, Alka Vita showed a glucose-lowering effect in both the ZDF rats and db/db mice, and, with metformin, it significantly decreased cataract formation. Since diabetes involves fat overload and inflammation [24], Alka Vita has been used [48], alone and in combination with other drugs such as metformin or insulin, in antidiabetic management of a growing population. Though the manufacturer's suggestion that it restores the immune system needs further scientific evaluation, Alka Vita may play a role in anti-infection [49] and other human disorders such as diabetes and cancer [48,51]. By regulating CEA, AST, glucose, and other biomarkers, Alka Vita has proved beneficial in extending the life extensively of the patient with terminal colon cancer (after a 3–6-month life projection, more than 16 months still living today); probably it can also be beneficial to other cancer types, but they need to be individually tested. There is excessive consumption of meat as well as carbonated and sugary drinks by a significant fraction of human population, likely leading to disease-causing acid formation. For the health-conscious readers of this chapter, since a major component for beneficial effects of Alka Vita is attributed to alkalinity, judicious use of alkaline water in combination with regular consumption of alkaline-forming fruits and vegetables can prove helpful as a simple and inexpensive step toward preventive health of modern acidic life.

Acknowledgments

We are grateful to Dr. Milt Zmijewski for the critical review of this chapter.

REFERENCES

1. MS Boguski, KD Mandl, VP Sukhatme. Repurposing with a difference. Science 324:1394–1395, 2009.
2. CP Adams, VV Brantner. Estimating the cost of new drug development: is it really $802 million? Health Aff 25:420–428, 2006.
3. JA DiMasi, RW Hansen, HG Grabowski. The price of innovation: new estimates of drug development costs. J Health Econ 22:151–185, 2003.
4. LN Ornston, WK Yeh. Origins of metabolic diversity: evolutionary divergence by sequence repetition. Proc Natl Acad Sci USA 76:3996–4000, 1979.
5. WK Yeh, JE Dotzlaf, GW Huffman. Biochemical characterization and evolutionary implication of β-lactam expandase/hydroxylase, expandase and hydroxylase. In: H Kleikauf, H von Dohren, eds. *50 Years of Penicillin Applications; History and Trends*. Prague: Public, 1994, pp. 208–223.
6. JE Dotzlaf, SW Queener, WK Yeh. Metabolic engineering for cephalosporin C yield improvement and production of intermediates. In: HA Kirst, WK Yeh, MJ Zmijewski, Jr., eds. *Enzyme Technologies for Pharmaceutical and Biotechnological Applications*. New York: Marcel Dekker, 2001, pp. 39–60.
7. MA DeSieno, CA Denard, H Zhao. Drug discovery and development by combinatorial biosynthesis. In: WK Yeh, HC Yang, JR McCarthy, eds. *Enzyme Technologies: Metagenomics, Evolution, Biocatalysis and Biosynthesis*. Hoboken, NJ: John Wiley & Sons, Inc., 2010, pp. 253–283.
8. MA Cornier, D Dabelea, TL Hernandez, RC Lindstrom, AJ Steig, NR Stob, RE Van Pelt, H Wang, RH Eckel. The metabolic syndrome. Endocr Rev 29:777–822, 2008.
9. A Meister. Reduction of α,γ-diketo and α-keto acids catalyzed by muscle preparations and by crystalline lactic dehydrogenase. J Biol Chem 184:117–129, 1950.
10. CF Feist, GD Hegeman. Phenol and benzoate metabolism by *Pseudomonas putida*: regulation of tangential pathways. J Bacteriol 100:869–877, 1969.
11. WK Yeh, N Bauer, JE Dotzlaf. High-performance liquid chromatographic assay for S-adenosyl-L-methionine: macrocin O-methyltransferase. J Chromatogr 288:157–165, 1984.
12. NJ Bauer, AJ Kreuzman, JE Dotzlaf, WK Yeh. Purification, characterization, and kinetic mechanism of S-adenosyl-L-methionine: macrocin O-methyltransferase from *Streptomyces fradiae*. J Biol Chem 263:15619–15625, 1988.
13. ET Seno, K Cox. Eli Lilly & Co., Indianapolis, IN. Unpublished results. 1990.
14. JE Dotzlaf, WK Yeh. Copurification and characterization of deacetoxycephalosporin C synthetase/hydroxylase from *Cephalosporium acremonium*. J Bacteriol 169:1611–1618, 1987.
15. PL Skatrud, AJ Teitz, TD Ingolia, CA Cantwell, DL Fisher, JL Chapman, SW Queener. Use of recombinant DNA to improve production of cephalosporin C by *Cephalosporium acremonium*. Bio/Technology 7:477–485, 1989.
16. AK Ghosh, S Gemma, J Tang. β-Secretase as a therapeutic target for Alzheimer's disease. Neurotherapeutics 5:399–408, 2008.
17. HC Yang, X Chai, M Mosior, W Kohn, LN Boggs, JA Erickson, DB McClure, WK Yeh, L Zhang, P Gonzales-DeWhitt, et al. Biochemical and kinetic characterization of BACE1: investigation into the putative species-specificity for β- and β′-cleavage sites by human and murine BACE1. J Neurochem 91:1249–1259, 2004.

18. KA Hughes, SP Webster, BR Walker. 11-Beta-hydroxysteroid dehydrogenase type 1 (11 β-HSD1) inhibitors in Type 2 diabetes mellitus and obesity. Expert Opin Investig Drugs 17:481–496, 2008.

19. JL Link. Pharmacological regulation of hepatic glucose production. Curr Opin Investig Drugs 4:421–429, 2003.

20. A Blum, A Raum, E Maser. Functional characterization of human 11β-hydroxysteroid dehydrogenase 1B (11β-HSD 1B) variant. Biochemistry 35:7715–7730, 2003.

21. SJ Thibodeaux, KDW Roth, WK Yeh, DA Briere, MD Michael, J Dage. Determination of K_i and *in vivo* efficacy for 11β-hydroxysteroid dehydrogenases using LC/MS/MS. *Proceedings of the 52nd ASMS Conference on Mass Spectrometry and Allied Topics*. Nashville, TN; May 23–27, 2004.

22. TD Aicher, Z Chen, RJ Hinklin, GA Hite, AP Krasutsky, R Li, JR McCowan, A Saeed, NJ Snyder, JL Toth, et al. Biphenyl amide lactam derivatives as inhibitors of 11-beta-hydroxysteroid dehydrogenase 1. Patent US20090156571A1, June 18, 2009.

23. CD Boyle, TJ Kowalski. 11β-hydroxysteroid dehydrogenase type 1 inhibitors: a review of recent patents. Expert Opin Ther Pat 19:801–825, 2009.

24. G Taubes. Prosperity's plague. Science 325:256–260, 2009.

25. WK Yeh, CR Stauss, K Zimmerman, RG Peterson. Determining total cholesterol and triglycerides from rat liver and muscle. *Third Annual Diabetes Drug Discovery: Targeting Diabetes with Novel Therapeutics*. Boston, MA, November 3–4, 2010.

26. Luminex Multiplexing: see Biomarkers from http://www.millipore.com/bmia/flx4/multiplexing-instruments (accessed July 22, 2013).

27. Meso Scale Discovery: see www.mesoscale.com (accessed July 22, 2013).

28. RG Peterson. The Zucker diabetic fatty (ZDF) rat- lessons from a leptin receptor defect diabetic model: In: E Shafrir, ed. *Animal Models of Diabetes, Frontiers in Research*. Boca Raton, FL: CRC Press, 2007.

29. RG Peterson. Use of new rodent models in discovery research: the ZDSD rat, a model for diabetes, metabolic syndrome and obesity without leptin or leptin receptor mutations. *2009 American Diabetes Association Meeting*. New Orleans, LA, June 5–9, 2009.

30. RG Peterson, VH Gattone, K Zimmerman. The ZDSD/Pco rat, a type II diabetic model for diabetic nephropathy without leptin or leptin receptor mutations. *2009 American Society of Nephropathy Meeting*. San Diego, CA, October 27–November 1, 2009.

31. P Devarajan. NGAL as a biomarker of acute kidney injury and nephrotoxicity. *2010 Cambridge Healthtech Institute's Ninth Annual World Pharmaceutical Congress: New Tools for Detecting Nephrotoxicity*. Philadelphia, PA, June 15–17, 2010.

32. RJ Shaw, KA Lamia, D Vasquez, SH Koo, N Bardeesy, RA DePinho, M Montminy, LC Cantley. The kinase LKB1 mediates glucose homeostasis in liver and therapeutic effects of Metformin. Science 310:1642–1646, 2005.

33. JM Lee, WY Seo, KH Song, D Chanda, YD Kim, DK Kim, MW Lee, D Ryu, YH Kim, JR Noh, et al. AMPK-dependent repression of hepatic gluconeogenesis via disruption of CREB-CRTC2 complex by orphan nuclear receptor small heterodimer partner. J Biol Chem 285:32182–32191, 2010.

34. JH Um, S Yang, S Yamazaki, H Kang, B Viollet, M Foretz, JH Chung. Metformin: activation of 5′-AMP-activated kinase with diabetes drug metformin induces casein kinase I-epsilon (CKI-epsilon)-dependent degradation of clock protein mPer2. J Biol Chem 282:20794–20798, 2007.

35. D Bungard, BJ Fuerth, PY Zeng, B Faubert, NL Maas, B Viollet, D Carling, CB Thompson, RG Jones, SL Berger. Signaling kinase AMPK activates stress-promoted transcription via histone H2B phosphorylation. Science 329:1201–1205, 2010.
36. T Takazawa, T Yamauchi, A Tsuchida, M Takada, Y Hada, M Iwabu, M Okada-Iwabu, K Ueki, T Kadowaki. Peroxisome proliferator-activated receptor γ agonist rosiglitazone increases expression of very low density lipoprotein receptor gene in adipocytes. J Biol Chem 284:30049–30057, 2009.
37. JH Choi, AS Banks, JL Estall, S Kajimura, P Bostrom, D Laznik, JL Ruas, MJ Chalmers, TM Kamenecka, M Bluher, et al. Anti-diabetic drugs inhibit obesity-linked phosphorylation of PPARγ by Cdk5. Nature 466:451–456, 2010.
38. A Cervera, E Wajcberg, A Sriwijitkamol, M Fernandez, P Zuo, C Triplitt, N Musi, RA DeFronzo, E Cersosimo. Mechanism of action of exenatide to reduce postprandial hyperglycemia in type 2 diabetes. Am J Physiol Endocrinol Metab 294:E846–E852, 2008.
39. OG Kolterman, DD Kim, LShen, JA Ruggles, LL Nielsen, MS Fineman, AD Baron. Pharmacokinetics, pharmacodynamics, and safety of exenatide in patients with type 2 diabetes mellitus. Am J Health Syst Pharm 62:173–181, 2005.
40. T Watanabe, N Kubota, M Ohsugi, T Kubota, I Takamoto, M Iwabu, M Awazawa, H Katsuyama, C Hasegawa, K Tokuyama, et al. Rimonabant ameliorates insulin resistance via both adiponectin-dependent and adiponectin-independent pathways. J Biol Chem 284:1803–1812, 2009.
41. J Cousin-Frankel. FDA, Europe curb risky diabetes drug. ScienceInsider, September 23, 2010.
42. Metabolic Syndrome: From Third Report of National Cholesterol Education Program Expert Panel on Detection, Evaluation and Treatment of High Blood Cholesterol in Adults, NIH Publication No. 01-3670, 2001.
43. See www.orexigen.com (accessed July 3, 2013).
44. K Gaber. Autophagy explaining exercise. Science 335:281, 2012.
45. C Werner, T Furster, T Widmann, J Poss, C Roggia, M Hanhoun, J Scharhag, N Buchner, T Meyer, W Kindermann, et al. Physical exercise prevents cellular senescence in circulating leucocytes and in the vessel wall. Circulation 120:2438–2447, 2009.
46. G Vogel, E Pennisi. Telomerases. Physiology Nobel: US researchers recognized for work on Telemeres. Science 326:212–213. 2009.
47. S Whang. *Reverse Aging*. Twenty first printing, 2008. Miami, FL: JSP Publishing. Also see: www.alkalife.com (accessed July 3, 2013).
48. Alka Vita: see www.liphsolutions.com (accessed July 3, 2013).
49. BP Young, JJH Shin, R Orij, JT Chao, SC Li, XL Guan, A Khong, E Jan, MR Wenk, WA Prinz, et al. Phosphatidic acid is a pH biosensor that links membrane biogenesis to metabolism. Science 329:1085–1088, 2010.
50. DL Townsend, LM White, I Cisneros, CR Richardson, DA Vattem. Cancer chemotherapeutic effects of modified sodium silicate (Alkahydroxy/Alka V6). FASEB J 24:537.16, 2010.
51. RM Jeter. Antibacterial susceptibility test results for Alka Vita product; July 17, 2006. See http://synergygroupnz.com/alka-vita-3 (accessed July 3, 2013).
52. CR Richardson, DA Vattem. Anti-cancer and anti-HIV effects of Alka Hydroxy – Alka Vita – Alka-V6: see http://www.orizonresearch.com (accessed July 3, 2013).
53. GY Locker, S Hamilton, J Harris, JM Jessup, N Kemeny, JS Macdonald, MR Somerfield, DF Hayes, RC Bast, Jr. ASCO 2006 update of recommendations for the use of tumor markers in gastrointestinal cancer. J Clin Oncol 24:5313–5327, 2006.

PART B

ENZYMES – INDISPENSABLE TOOLS FOR IMPROVING DRUGGABILITY

5

ENZYMES AND TARGETED ACTIVATION OF PRODRUGS

YANHUI YANG, YU CHEN, HERVE ALOYSIUS, DAIGO INOYAMA, AND LONGQIN HU

Department of Medicinal Chemistry, Ernest Mario School of Pharmacy, Rutgers, The State University of New Jersey, Piscataway, New Jersey

I. INTRODUCTION

Most drugs, in order to produce their desired pharmacological action, have to overcome many hurdles before reaching the desired site of action. These hurdles include the intestinal barrier, the blood–brain barrier (BBB), and metabolic reactions that could render them inactive. Most drugs distribute randomly throughout the body, and the amount of drugs reaching the site of action is relatively small. For an effective amount to reach the site of action and not cause severe systemic side effects, a drug must possess certain physicochemical properties that make it conducive to penetrate through various biological membranes, to avoid metabolic inactivation by various enzymes, and to avoid retention in body depot tissues that could lead to undesirable long-lasting effects. These desired physicochemical properties are not always present in pharmacologically active compounds.

With the advance of new technologies such as combinatorial and computational chemistry, more and more compounds are being identified with extremely potent *in vitro* activity but are found to be inactive *in vivo*. They may have the optimal configuration and conformation needed to interact with their target receptor or enzyme, but they do not necessarily possess the best molecular form and physicochemical properties needed for their delivery to the site of action.

Enzyme Technologies: Pluripotent Players in Discovering Therapeutic Agents, First Edition.
Edited by Hsiu-Chiung Yang, Wu-Kuang Yeh, and James R. McCarthy.

Some of the problems often encountered include (1) limited solubility and poor chemical stability preventing the drug from being adequately formulated, (2) low or variable bioavailability due to incomplete absorption across biological membranes or extensive first-pass metabolism, and (3) lack of site specificity. Prodrugs have been designed and used clinically to solve these problems. A prodrug is inactive or much less active and has to undergo activation *in vivo* to form the active drug molecule.

Prodrug design has proven to work for many pharmacologically active compounds or drugs in improving their physicochemical and biological properties and their target selectivity. Currently, 5–7% of the drugs approved worldwide can be classified as prodrugs, and approximately 15% of all new drugs approved each year were prodrugs [1]. Prodrugs have been used in a wide variety of therapeutic areas including anti-influenza, antihypertension, antibiotics, antiulcer, anticoagulant, anti-inflammatory, antifungal, anesthetic, and anticancer (Table 1). Many prodrugs have gained huge clinical success. For example, oseltamivir (**1**) significantly enhances oral bioavailability of GS4071, a selective inhibitor of influenza A and B viral neuraminidase, to approximately 80% [2,3]; enalapril (**2**) improves the oral bioavailability of enalaprilat, an angiotensin-converting enzyme (ACE) inhibitor, to 53–74% [1,4]; pivampicillin (**3**), a β-lactam antibiotic prodrug, improves the lipophilicity of ampicillin, and, as a result, the oral bioavailability is increased from 32–55% to 87–94% [5,6]; miproxifene phosphate (**4**), an anticancer prodrug, has increased aqueous solubility at pH 7.4 by 1000-fold as compared to the parent [7]; and capecitabine (**5**), another anticancer prodrug, has less systemic toxicity due to the increased expression, in tumor tissues, of thymidine phosphorylase (TP), one of three enzymes responsible for the prodrug's activation to 5-fluorouracil (5-FU) [8].

In vivo biotransformation of prodrugs to active drugs determines the ultimate usefulness since prodrugs themselves have no pharmacological activities. In most cases, prodrugs are simple derivatives that require only one to two steps of chemical or enzymatic transformation to yield the active parent drugs. For example, antiulcer prodrug omeprazole (**6**) is activated by acid to form a cyclic sulfenamide, which irreversibly binds to the gastric H^+,K^+-ATPase and inhibits the secretion of gastric acid (Fig. 1a) [9]. ACE inhibitor prodrug enalapril (**2**), with improved oral bioavailability, was rapidly hydrolyzed by carboxylesterase (CE) to the active form enalaprilat (Fig. 1b) [1]. Capecitabine (Xeloda, **5**) is a prodrug that has reduced gastrointestinal toxicity and high tumor selectivity. The enzymatic bioconversion as shown in Figure 1c is initiated in the liver, where human carboxylesterases (hCEs) cleave the ester bond of the carbamate. This cleavage is followed by a fast, spontaneous decarboxylation leading to the formation of 5′-deoxy-5-fluorocytidine (5′-dFCyd). Generation of the parent drug continues in the liver, and to some extent in tumors, by cytidine deaminase, which converts 5′-dFCyd to 5′-deoxyuridine (5′-dFUrd). Finally, TP liberates the active drug 5-fluorouracil in tumors. Besides enzyme-mediated and acid/base-catalyzed transformation, prodrugs can also be activated by photolysis [15,16,18] and radiation [17–19].

TABLE 1 Examples of Prodrugs in Clinical Use or Development

Prodrug	Therapeutic area	Structure	Activation	References
Oseltamivir (**1**)	Anti-influenza	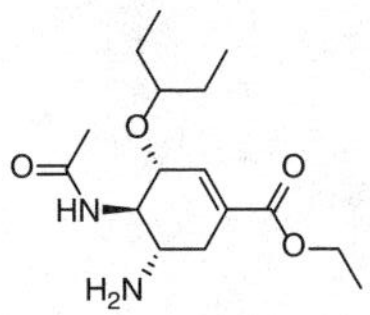	Carboxylesterase	2,3
Enalapril (**2**)	Antihypertension	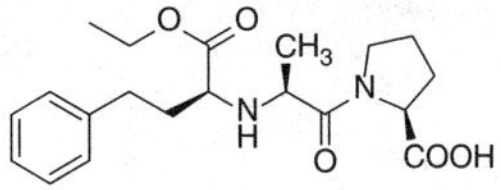	Carboxylesterase	1,4
Pivampicillin (**3**)	β-lactam antibiotic	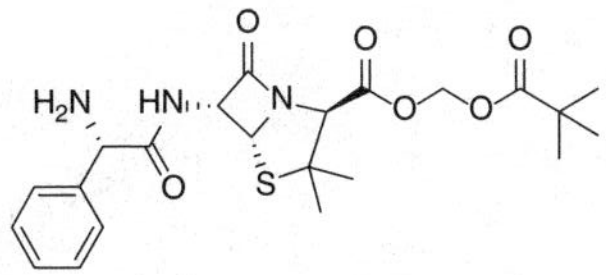	Esterases	5,6
Miproxifene phosphate (**4**)	Anticancer	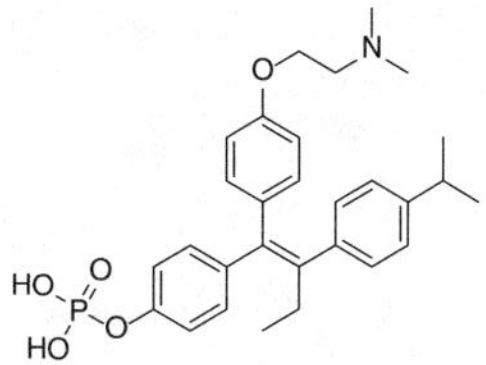	AP	7
Capecitabine (**5**)	Anticancer	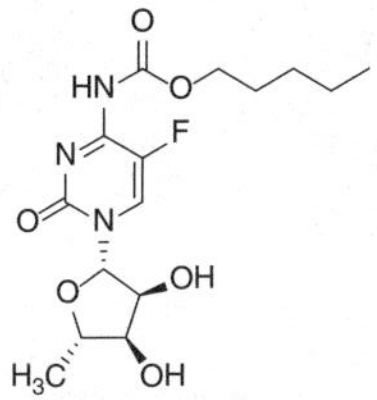	CE CD TP	1,8

(*Continued overleaf*)

TABLE 1 (*Continued*)

Prodrug	Therapeutic area	Structure	Activation	References
Omeprazole (**6**)	Antiulcer	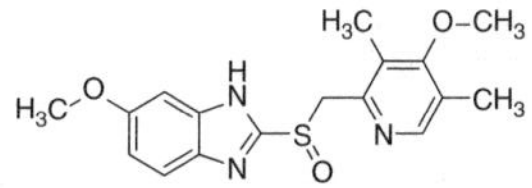	Acid (proton)	9
cBR96-Dox (**7**)	Anticancer	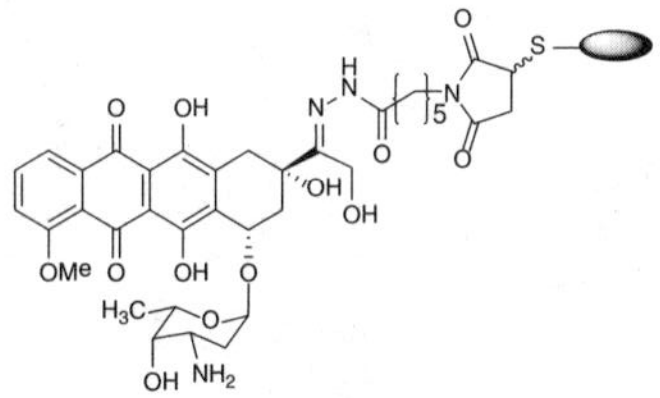	Acidic hydrolysis	10
Ximelagatran (**8**)	Anticoagulant	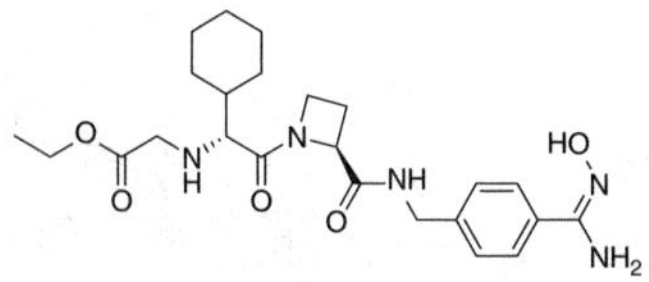	CE P450 Red	11
Sulindac (**9**)	Nonsteroidal anti-inflammatory	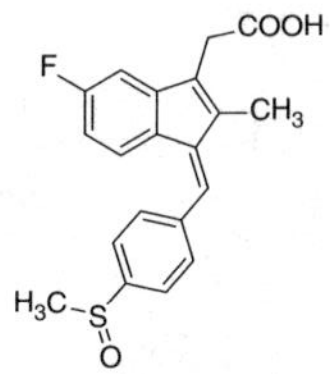	Methionine sulfoxide reductases	12
Fosfluconazole (**10**)	Antifungal	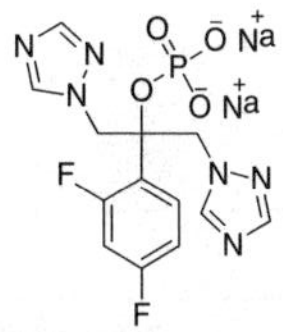	AP	13

Phosphonooxymethyl propofol (**11**)	Anesthetic	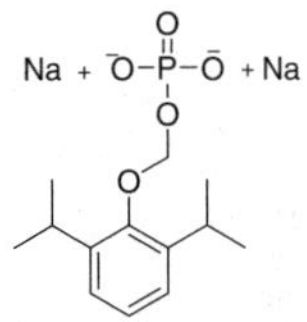	AP	14
5-FdU carbonate (**12**)	Anticancer	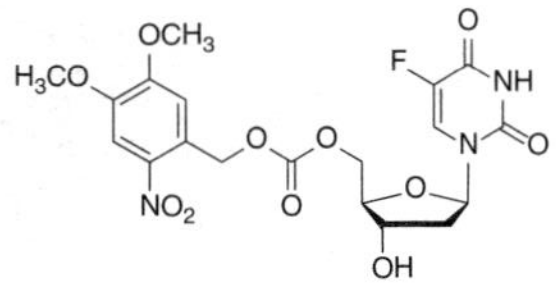	UV irradiation	15
Nitrobenzyl 5-FU (**13**)	Anticancer	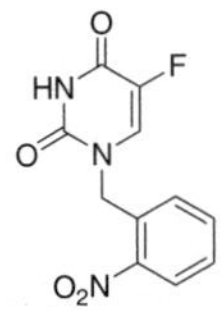	UV irradiation	16
Co(III)(cyclen)(azaCBI) (**14**)	Anticancer	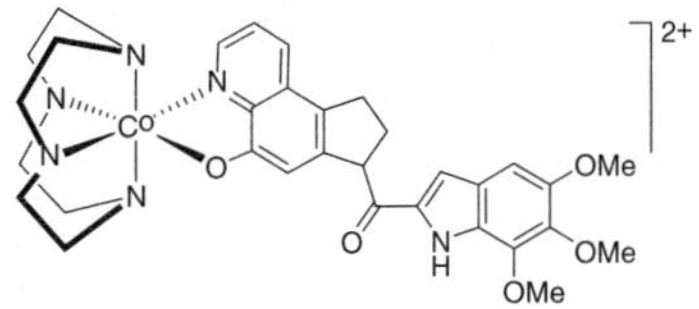	Radiation	17

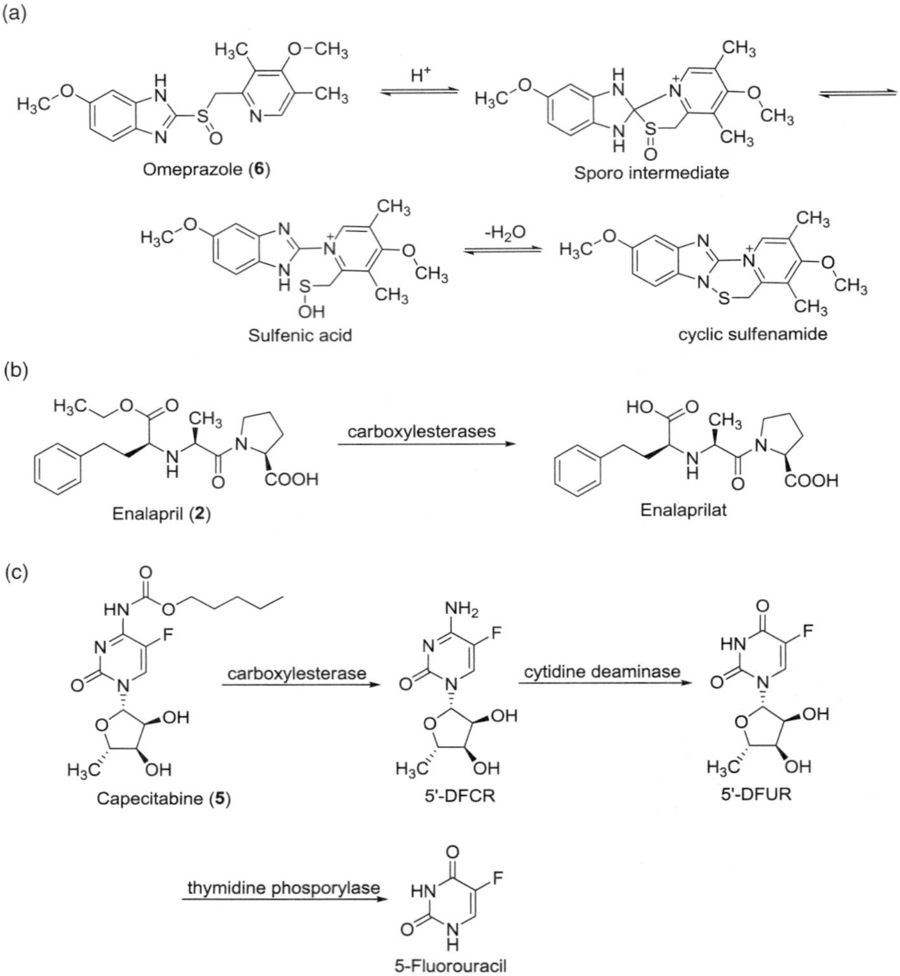

FIGURE 1 The mechanisms of activation of representative prodrugs. (a) Activation of antiulcer omeprazole by acids. (b) Activation of ACE inhibitor prodrug enalapril by carboxylesterases. (c) Activation of anticancer prodrug capecitabine by three enzymes.

Activation of prodrugs can involve many endogenous enzymes including CYP450, DT-diaphorase, tyrosinase, CE, and β-glucuronidase. Exogenous enzymes can also be used to activate prodrugs as in antibody-directed enzyme prodrug therapy (ADEPT) [20,21], gene-directed enzyme prodrug therapy (GDEPT) [22–24], virus-directed enzyme prodrug therapy (VDEPT) [25], and polymer-directed enzyme prodrug therapy (PDEPT) [26–28]. This chapter will give an overview of the enzymes used in prodrug activation with emphasis to their structure, distribution, and mechanism of activation for typical prodrugs.

II. ENDOGENOUS ENZYMES

Plenty of endogenous enzymes have been used in the activation of prodrugs. Since endogenous enzymes are naturally present in the human body, there are no immunogenic issues in using them to activate prodrugs, whereas most nonendogenous enzymes would raise problems of antigenic response when they are delivered or expressed *in vivo* for the activation of prodrugs. Endogenous enzymes used in the activation of prodrugs belong to following four categories: oxidoreductases, transferases, hydrolases, and lyases.

A. Oxidoreductases

Oxidoreductases are common enzymes that exist in most organisms. Oxidoreductases used in prodrug activation include aldehyde oxidases, amino acid oxidases, CYP450 monooxygenases, CYP450 reductase (P450 Red), DT-diaphorase, and tyrosinase.

Aldehyde Oxidases Aldehyde oxidases (EC 1.2.3.1) are a small group of proteins belonging to the larger family of molybdo-flavoenzymes (MFEs) [29]. Similar to other MFEs, aldehyde oxidases require flavin adenine dinucleotide (FAD), molybdenum, and two distinct [2Fe-2S] clusters as cofactors. Aldehyde oxidases catalyze the oxidation of a variety of organic aldehydes and N-heterocyclic compounds to corresponding carboxylic acids and hydroxylated N-heterocycles by using molecular oxygen.

The enzyme consists of two identical subunits of 140–150 kDa. Each subunit has a typical tripartite structure consisting of three binding sites: iron–sulfur binding site, FAD-binding site, and substrate-binding pocket, each in a separate domain [29]. The N-terminal 20-kDa domain is characterized by two nonidentical iron–sulfur [2Fe-2S] centers; the 40-kDa central domain contains the FAD-binding site, whereas the large 85-kDa C-terminal domain has the substrate-binding pocket lying in close proximity to the MoCo-binding site [29,30]. Unlike other molybdoproteins, such as sulfite oxidases and nitrate reductase, aldehyde oxidases require posttranslational sulfuration of MoCo to become catalytically active.

Aldehyde oxidases are widely distributed and are mainly in the cytosol [31]. In humans, aldehyde oxidases are expressed predominantly in the liver, lung, and kidney. Several anticancer prodrugs have been reported to be activated by aldehyde oxidases. As shown in Figure 2, 5-ethynyl-2(1H)-pyrimidinone (**15**) was oxidized by

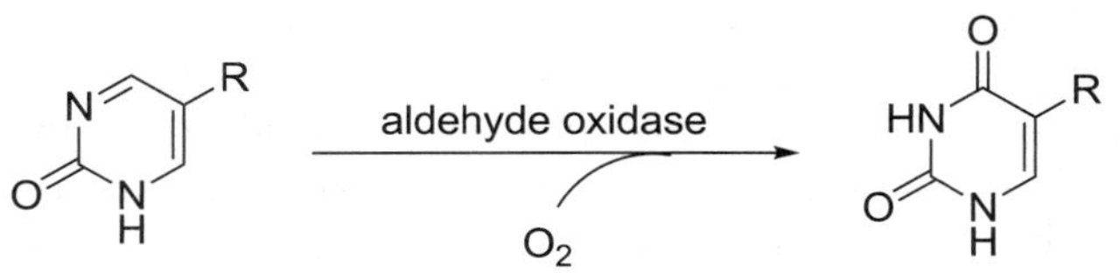

5-ethynyl-2(*1H*)-pyrimidinone (**15**) R = C≡CH 5-ethynyluracil R = C≡CH

5-fluoro-2-pyrimidinone (**16**, 5-FP) R = F 5-fluorouracil (5-FU) R = F

FIGURE 2 Activation of pyrimidinone prodrugs by aldehyde oxidases.

rabbit liver aldehyde oxidase to 5-ethynyluracil, which is a mechanism-based inhibitor of dihydropyrimidine dehydrogenase and prevents the rapid breakdown of 5-FU [31,32]. Similarly, 5-fluoro-2-pyrimidinone (**16**, 5-FP) was designed as a prodrug activated by aldehyde oxidases to avoid the rapid breakdown of 5-FU in the gastrointestinal tract, and 5-FP was shown to be as effective as and about twofold less toxic than 5-FU [31,33].

Amino Acid Oxidases Amino acid oxidases are classified as D-amino acid oxidase (DAAO) and L-amino acid oxidase (LAAO) based on their substrate. DAAO (EC 1.4.3.3) is a FAD-dependent enzyme that catalyzes the oxidation of D-amino acids to their corresponding imino acids with concomitant reduction of FAD. The reduced flavin is subsequently reoxidized by molecular oxygen generating H_2O_2, and the imino acid is released into the solvent where it nonenzymatically hydrolyzes to yield the corresponding α-keto acid and ammonia [34]. As illustrated recently in several reviews [35–39], D-amino acid oxidase plays an important role in microbial metabolism, utilization of endogenous D-amino acids, regulation of the nervous system, and aging in mammals.

D-Amino acid oxidase has 320–376 amino acid residues with a molecular weight of 38–125 kDa. Sequences of D-amino acid oxidase from different sources (yeast, algae, microscopic fungi, bivalve mollusk, fish, mammals) exhibit low homology of less than 30–40% [38]. The human kidney D-amino acid oxidase has 347 amino acid residues with a molecular weight of 39 kDa. This enzyme has an 84.4% sequence similarity with the pig kidney enzyme [40]. The crystal structure of the human D-amino acid oxidase with the inhibitor benzoate bound was reported recently [34]. The asymmetric unit contained four molecules of human D-amino acid oxidase in the form of two homodimers. Each of the four molecules has the same conformation, and the two molecules in a homodimer are in a "head-to-head" arrangement. Every asymmetric subunit has one molecule of noncovalently bound FAD as a cofactor and one molecule of benzoate as an inhibitory substrate analog. There are a total of 11 α-helices and 14 β-strands in each subunit, which fold into two domains, the FAD-binding domain and the interface domain [34].

LAAO (EC 1.4.3.2) catalyzes the oxidative deamination of L-amino acid in a similar fashion to D-amino acid oxidase [41]. This enzyme is involved in eight metabolic pathways: (1) alanine and aspartate metabolism; (2) methionine metabolism; (3) valine, leucine, and isoleucine degradation; (4) tyrosine metabolism; (5) phenylalanine metabolism; (6) tryptophan metabolism; (7) phenylalanine, tyrosine, and tryptophan biosynthesis; and (8) alkaloid biosynthesis. Snake venom L-amino acid oxidase is the best characterized in this family of enzymes [41].

Currently amino acid oxidases are known to mainly locate in peroxisomes and present in many species. In mammals, amino acid oxidases are mainly present in the kidneys, less in the liver. Significant levels of amino acid oxidases have also been found in the brain, nerve, leukocytes, adrenal cortex, intestines, heart, lungs, tongue, skin, stomach, spleen, muscle, and fat tissue [31]. In humans, the enzyme is less widely distributed, majorly in the kidney and liver, low in the brain. Interestingly, it is not found in the human lungs, spleen, heart, and serum. Thus, it might be possible to design prodrugs to target specific organs where amino acid oxidases are present.

Selenocysteine (SeCys, **17**) *Se*-conjugates was shown to have chemopreventive activity in animal tumor models for chemical carcinogenesis [42]; SeCys conjugates can be activated by amino acid oxidase *in vivo* [43]. As showed in Figure 3, SeCys conjugates can undergo the usual oxidative deamination reaction catalyzed by amino acid oxidase and are also activated by an amino acid oxidase-catalyzed β-elimination to form pyruvate, ammonia, and the chemopreventive selenols [31,43]. The β-elimination route has been proposed as a possible mechanism for the activation of SeCys conjugates.

D-Alanine is another potential prodrug activated by amino acid oxidase; it was proposed that D-alanine can be used in combination with amino acid oxidase from the red yeast *Rhodotorula gracilis* in the GDEPT approach and that the metabolite hydrogen peroxide can readily cross cellular membranes to damage biologically important molecules (including DNA, proteins, and lipids) and induce apoptosis [31,44].

Cytochrome P450 Monooxygenases Cytochrome P450s (CYP450s, CYPs, EC 1.14.14.1) are a large group of ubiquitous proteins containing a single iron protoporphyrin IX prosthetic heme group [45]. This superfamily is comprised of 18 families and 57 members. Among them, classes I and II are the major enzymes, which act as versatile monooxygenases. These enzymes catalyze a multitude of reactions, including the hydroxylation of alkanes to alcohols; conversion of alkenes to epoxides, arenes to phenols, and sulfides to sulfoxides and sulfones; and the oxidative cleavage of C–N, C–O, C–C, or C–S bonds [45].

CYP enzymes have been identified from all lineages of life, including mammals, birds, fish, insects, worms, sea squirts, sea urchins, plants, fungi, slime molds, bacteria, and archaea. More than 7700 distinct CYP sequences are known. Mammalian CYP450 enzymes are mainly expressed in the liver and only at low levels in other tissues or tumors.

CYP families 1, 2, and 3 are involved in the metabolism of xenobiotic substances such as carcinogens, pro-carcinogens, and chemotherapeutics [45]. The CYP1 family

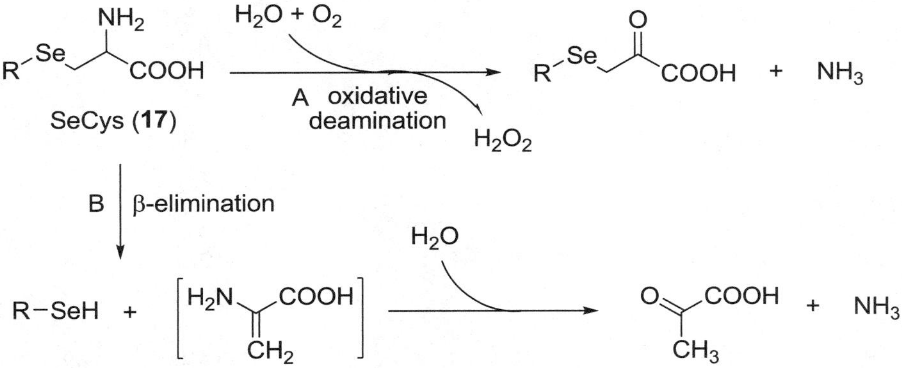

FIGURE 3 The bioactivation of SeCys conjugates by amino acid oxidases. Shown are the usual reaction of oxidative deamination (a) and β-elimination reaction (b) of SeCys conjugates [31].

of enzymes plays an important role in the bioactivation of anticancer drugs [46]. Although several CYP450s are overexpressed in tumors, the most notable of these is CYP1B1. CYP1B1 was detected in many tumor tissue types including breast, colon, lung, skin, lymph node, esophagus, brain, and testis, while the corresponding normal tissues were shown to have no significant expression of CYP1B1. CYP2B6 and CYP3A4 are also known to activate several clinically used prodrugs such as cyclophosphamide and ifosfamide. Table 2 shows several prodrugs that are known to be activated by CYP450.

Phortress (**18**) is the water-soluble L-lysyl prodrug of 2-(4-amino-3-methylphenyl)-5-fluorobenzothiazole (5 F-203) for parental administration; 5 F-203 is one of a series of benzothiazoles that form DNA adducts following activation and oxidation by P450 CYP1A1 [46,47]. DMU-135 (**19**) is a prodrug activated by CYP1B1, and the dealkylated metabolite DMU-117 with methylenedioxy cleaved is a potent nonselective tyrosine kinase inhibitor [45]. Cyclophosphamide (**20**), a commonly used antitumor agent for a variety of human cancers, including breast cancer, endometrial cancer, lung cancer, and various leukemias and lymphomas [48], is metabolized in the liver to the 4-hydroxylated metabolite by CYP2B6 and CYP3A4. The 4-hydroxyl cyclophosphamide metabolite undergoes spontaneous decomposition to phosphoramide mustard and acrolein as shown in Figure 4. Phosphoramide mustard is the ultimate cytotoxic agent that is responsible for the cross-linking of DNA, whereas acrolein is a reactive Michael acceptor that is capable of alkylating proteins [46].

TABLE 2 Prodrugs Activated by CYP450 Enzymes

Name	Structure	CYP isozyme	References
Phortress (**18**)	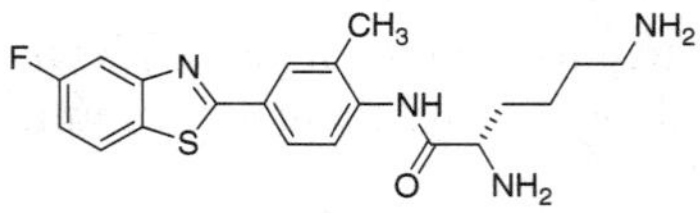	CYP1A1	46,47
DMU-135 (**19**)	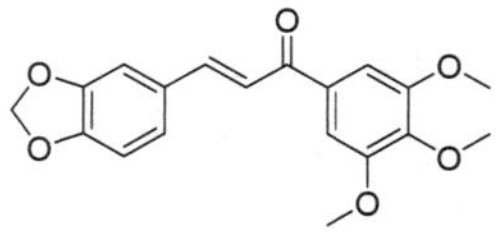	CYP1B1	45
Cyclophosphamide (**20**)	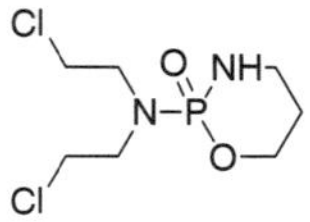	CYP2B6	46,48
Ifosfamide (**21**)	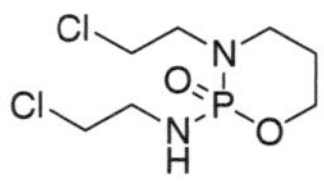	CYP3A4	48
Procarbazine (**22**)	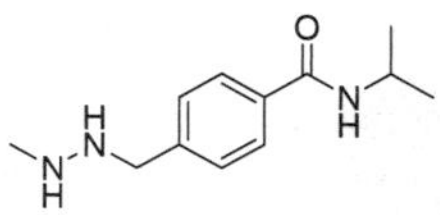	CYP450s	49

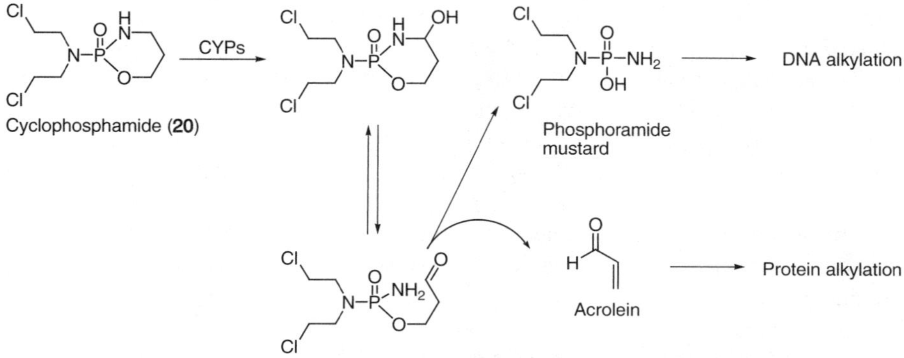

FIGURE 4 Mechanism of cyclophosphamide activation by CYP450s [50].

Ifosfamide (**21**), an isomer of cyclophosphamide, displays high activity against soft tissue sarcomas, testicular cancer, and ovarian and breast cancer, among others [48]. The activation of ifosfamide is similar to that of cyclophosphamide. While the human CYP2B6 is the major activator of cyclophosphamide, CYP3A4 is the primary activator of ifosfamide. Procarbazine (**22**) is an orally active alkylating anticancer prodrug that undergoes a complex metabolic transformation into its active intermediates [49]. Oxidation of procarbazine to azo-procarbazine occurs by microsomal CYP450 oxido-reductase and mitochondrial monoamine oxidase, and then azo-procarbazine is microsomally oxidized to azoxyderivatives. Of these, methylazoxy-procarbazine is the major cytotoxic compound, showing severalfold greater cytotoxicity than procarbazine and other metabolites [49].

Cytochrome P450 Reductase CYP450 Red (CPR, EC 1.6.2.4) is an FAD- and flavin mononucleotide (FMN)-containing flavoenzyme that catalyzes the transfer of electrons to the multiple microsomal CYP450s present in a given cell. A total of two electrons from NADPH are transferred, first to the reductase's bound FAD and then to its FMN, before being transferred by the reductase one at a time to the CYP450 heme protein [51]. CYP450, in turn, utilizes these reducing equivalents for various monooxygenase reactions, including the hydroxylation reactions associated with the activation of cyclophosphamide and other anticancer prodrugs. P450 Red enhances the chemosensitivity of these anticancer drugs on CYP450-expressing tumor cell lines by increasing the CYP450 metabolic activity, which is directly dependent on the extent of the increase in P450 Red expression [52].

P450 Red has two functional domains, a hydrophobic N-terminal membrane binding domain (6 kDa) which serves to anchor the protein molecule to the endoplasmic reticulum and nuclear envelope, and a hydrophilic C-terminal catalytic domain (72 kDa) that is comprised of several structural domains [53]. The N-terminal domain, consisting of approximately 170 amino acid residues, binds FMN, and the C-terminal region binds FAD and NADPH. Therefore, the 3-D structure of P450 Red is composed of four structural domains. From N- to C-terminus, they are the FMN-binding domain, the connecting domain, and the FAD- and the NADP-binding domains [53]. The FMN-binding domain consists of a five-stranded parallel β-sheet flanked by five α-helices with the FMN positioned at the tip of the C-terminal side of the β-sheet. The core of the FAD-binding domain is an antiparallel flattened β-barrel, and that of the NADP(H)-binding domain is another parallel five-stranded β-sheet sandwiched by α-helices. The connecting domain, situated between the two flavin domains, consists of residues 240–270 and 325–450 and is comprised mainly of α-helices. This domain appears to be responsible for bringing the two flavins together and, therefore, for modulating electron transfer between the two flavins. Residues located at the interface between the FMN-binding and the connecting domains are mainly hydrophilic, and, therefore, the interaction between them is largely electrostatic [53].

P450 Red is present in a variety of human tissues, high in the liver, lung, and small intestines and low in stomach and colon. P450 Red is also present in a variety of tumor cell lines including cells from leukemia and melanoma and central nervous system, breast, colon, lung, ovarian, prostate, and renal tumors [31]. However, P450 Red activity is lower in tumor tissue than in the corresponding normal tissue and correlates with CYP450 activity [31].

Mitomycin C (**23**), an antitumor antibiotic alkylating agent, is activated by P450 Red as shown in Figure 5 [54,55]. This process can occur through a one-electron pathway, producing the semiquinone free-radical intermediate, or a two-electron pathway, yielding the hydroquinone intermediate. In the presence of oxygen, the semiquinone will enter into a redox cycle, generating reactive oxygen species [54]. After reductive activation of the quinone moiety, a series of spontaneous rearrangements results in the opening of the aziridine ring and the formation of a quinone methide intermediate, which alkylates DNA [31]. Menadione (**24**), a synthetic derivative of vitamin K1, displays antitumor activity against several tumor types, such as leukemia and tumors from the liver, colon, lung, stomach, and prostate [31]. Menadione is activated by P450 Red via a one-electron reduction resulting in the formation of a semiquinone radical, which subsequently may reduce molecular oxygen to superoxide anion, induce the depletion of glutathione (GSH), and cause single-stranded DNA breaks and apoptosis [31]. Tirapazamine (**25**), a novel bioreductive prodrug exhibiting a high specificity towards hypoxic tumor cells, can be activated by P450 Red, by a one-electron reduction, yielding the tirapazamine nitroxide radical, which causes DNA single- and double-strand breaks [56].

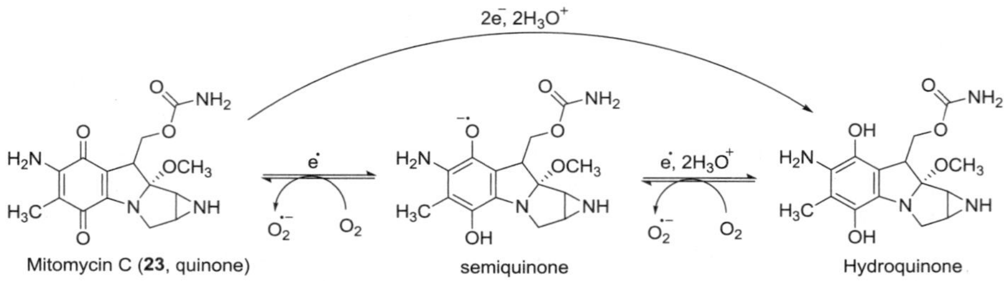

FIGURE 5 Activation of mitomycin C by P450 Red.

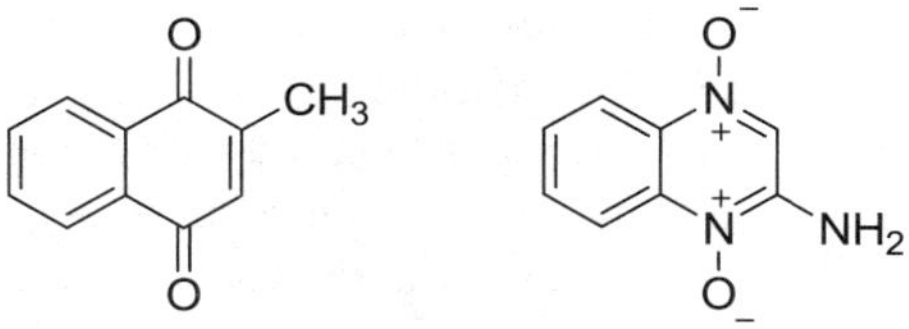

Menadione (**24**) Tirapazamine (**25**)

SCHEME 1

DT-Diaphorase DT-Diaphorase (EC 1.6.5.2), also called NAD(P)H quinone oxidoreductase 1 (NQO1), is a flavoprotein that catalyzes two-electron reduction of quinones and quinonoid compounds to hydroquinones, using either NADH or NADPH as the electron donor. DT-diaphorase is a tightly associated physiological dimer of identical subunits, each comprising 273 amino acids with a molecular weight of 30 kDa [57–60]. Each subunit contains two separate domains: a major, catalytic domain (residues 1–220) folded in a predominantly α/β structure and a small, C-terminal domain (residues 221–273) [61]. Each monomer contains a non-covalently bound FAD prosthetic group that is essential for catalytic activity. Based on the crystal structure of a tetramer [62], the four monomers are arranged as two noncrystallographic physiological dimers, with the active site being at each dimer interface. All four chains have essentially identical structures. Residues Trp^{105}, Phe^{106}, and Gln^{104} form a pocket around the isoalloxazine moiety. Together with the backbone carbonyl of Leu^{103}, these residues form the FAD-binding site.

DT-Diaphorase has become important in cancer pharmacology because its activity varies with tissues and its expression is elevated in tumors [63]. Because of the over-expression of DT-diaphorase in many tumors, including those of the lung, colon, liver, and breast, DT-diaphorase has been used as an activating enzyme for the development of anticancer prodrugs [59]. Several cytotoxic antitumor quinones, including the mitomycins, anthracyclines, and aziridinylbenzoquinones, are believed to be activated by DT-diaphorase to generate reactive intermediates which can undergo nucleophilic additions with DNA, resulting in single-strand breaks.

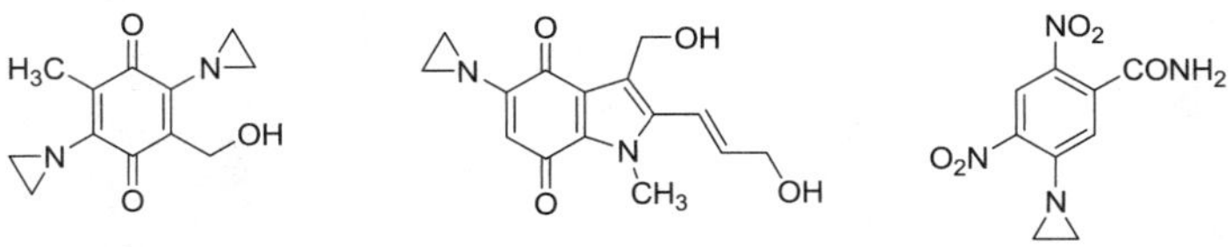

RH1 (**26**) EO9 (**27**) CB1954 (**28**)

SCHEME 2

RH1 (**26**) is a very efficient substrate for DT-diaphorase, and reduction by DT-diaphorase results in activation of the aziridine groups and subsequent DNA alkylation with interstrand cross-links formed preferentially at 5-GCC-3′ sequences [64]. RH1 was considered as a prodrug in an enzyme-directed prodrug therapy to treat tumors, like non-small cell lung cancer, that have high DT-diaphorase activity. EO9 (**27**), an aziridinylindolequinone, is a better DT-diaphorase substrate than the clinically proven mitomycin C, and it shows high toxicity in DT-diaphorase-rich cell lines [65].

Another bioreductive prodrug that appears to be selectively activated by DT-diaphorase is CB 1954 (**28**) [5-(1-aziridinyl)-2,4-dinitrobenzamide], which was shown to have high potency and specificity against Walker tumor cells both *in vivo* and *in vitro*. As showed in Figure 6, the compound can be activated by reducing the 4-nitro group to generate 5-(1-aziridinyl)-4-hydroxylamino-2-nitrobenzamide [66].

Azoreductase Azoreductase (EC 1.7.1.6) belongs to the family of oxidoreductases, and it catalyzes the reductive cleavage of azo groups (–N=N–) utilizing FMN as the cofactor and NADH, but not NADPH, as an electron donor [67].

Azoreductases are produced by the microflora in the colon. Therefore, azoreductases are responsible for the activation of several clinically used prodrugs for colon-specific delivery. Active drug molecules have been linked through an azo group to a nontherapeutic moiety, another active drug molecule, or a polymer to form prodrugs. After the prodrugs are ingested, the azoreductases from colonic bacteria cleave the azo bonds, releasing the drugs in their active forms. For example, several prodrugs of the anti-inflammatory agent mesalazine are marketed drugs including balsalazine (**29**), ipsalazide (**30**), olsalazine (**31**), and sulfasalazine (**32**). These prodrugs can pass intact and unabsorbed from the upper gastrointestinal tract, but because they are azo-containing compounds, they can be activated by bacterial azoreductases in the colon to release the active 5-amino salicylic acid (5-ASA or mesalazine) as shown in Figure 7 [68,69].

Tyrosinase Tyrosinase (EC 1.14.18.1) is an enzyme with a binuclear copper center that catalyzes two different oxidative reactions by molecular oxygen. These reactions are the *ortho*-hydroxylation of monophenols to *o*-diphenols (monophenolase activity) and the oxidation of *o*-diphenols to their *o*-quinones (diphenolase activity) [70].

Tyrosinase is present in vegetables, fruits, and mushrooms and is a key enzyme responsible for the browning that occurs upon bruising or long-term storage. In mammals, the enzyme is responsible for skin pigmentation abnormalities, such as flecks and defects. Recently, the enzyme was reported to be linked to Parkinson's disease and other neurodegenerative diseases [71].

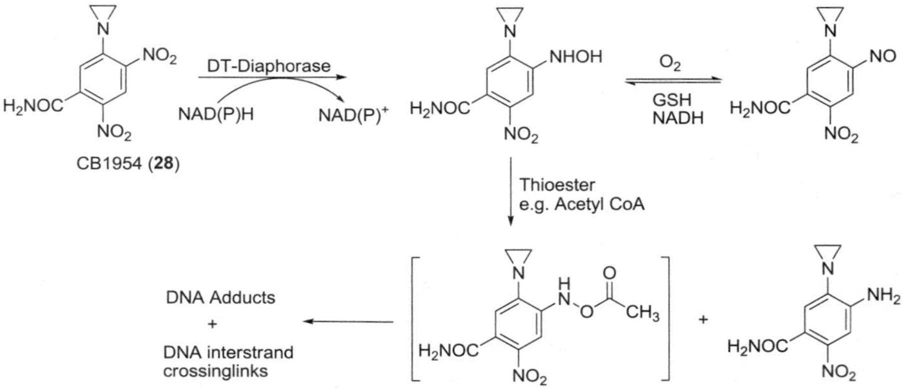

FIGURE 6 Mechanism of activation of CB 1954 by DT-diaphorase [66].

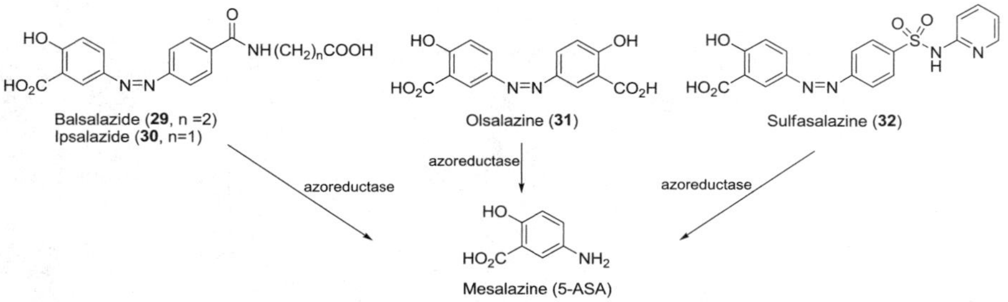

FIGURE 7 Activation of azo-containing prodrugs to mesalazine by azoreductase.

The crystal structure of tyrosinase was recently published as a complex with the caddie protein ORF378 [71]. Tyrosinase takes α-helical structures with the core of the enzyme formed by a four-helix bundle (α2, α3, α6, and α7 helices). The catalytic dinuclear copper center is lodged in the helical bundle. One copper ion (designated Cu^A) is coordinated by His^{38}, His^{54}, and His^{63}, while the second copper ion (Cu^B) is coordinated by His^{190}, His^{194}, and His^{216}. This dicopper center is located at the bottom of the large concavity as a putative substrate-binding pocket that is formed by hydrophobic residues. In addition to the helical structure, tyrosinase has a few β-structures, as judged from the backbone torsion angles. Among them, only the N- and C-terminal β-strands form a sheet structure [71]. The active site of the enzyme contains a pair of antiferromagnetically coupled copper (II) ions or T3 center.

It has been reported that when melanocytes become malignant, the expression of tyrosinase is upregulated, resulting in high levels of tyrosinase within the melanoma cells [72]. Based on this observation, a melanocyte-directed enzyme prodrug therapy (MDEPT) was proposed as a potential new treatment of malignant melanoma using prodrugs designed to function as tyrosinase substrates [72,73]. A number of tyrosinase-activated prodrugs have been investigated for the treatment of melanoma. 2,4-Dihydroxyphenyl alanine (**33**) is one of the first tyrosinase-activated prodrugs reported. Activation of 2,4-dihydroxyphenyl alanine gives 6-hydroxydopa, which reacts rapidly with molecular oxygen to generate a number of cytotoxic species including superoxide radical, hydroxyl radical, hydrogen peroxide, and quinones [31,74]. Recently, a series of phenyl carbamate mustards (**34**) and phenyl urea mustards (**35**) were synthesized as substrates of tyrosinase [72,73]. These compounds can be activated by tyrosinase to release a cytotoxic mustard via a cyclization-release mechanism as represented by the activation of compound **34** in Figure 8 [73].

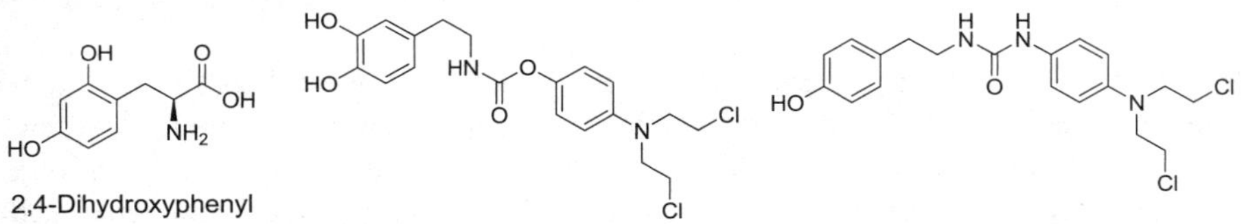

SCHEME 3

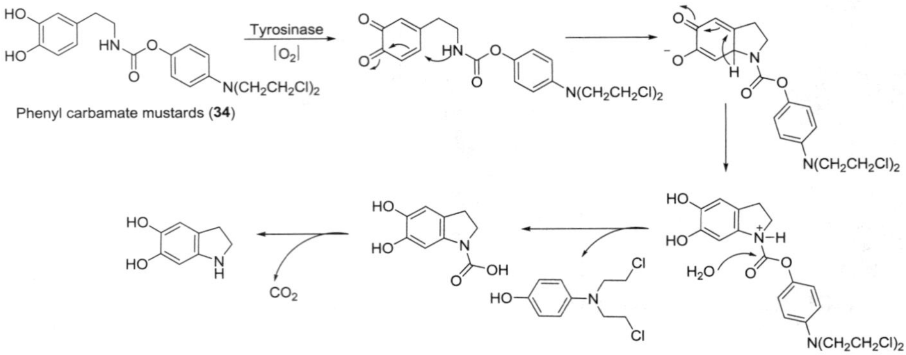

FIGURE 8 Proposed activation of a phenyl carbamate mustard prodrug (**34**) by tyrosinase [73].

B. Transferase

Transferases involved in prodrug activation include thymidylate synthase (TS), TP, glutathione S-transferase (GST), and deoxycytidine kinase (DCK).

Thymidylate Synthase TS (EC 2.1.1.148) is one of the best-known targets in anticancer chemotherapy; it has both catalytic and regulatory functions. As a catalytic protein, TS catalyzes the methylation reaction of 2′-deoxyuridine-5′-monophosphate (dUMP) to 2′-deoxythymidine-5′-monophosphate (dTMP). As a regulatory protein, it interacts with its own mRNA in a feedback mechanism and interacts with other sequences to regulate the expression of proteins involved in apoptosis such as bcl2, c-myc, and p53 [75].

TS is overexpressed in actively proliferating cells such as tumor cells and has been used to activate anticancer prodrugs in tumor cells and help overcome drug resistance [76–78]. Development of chemical agents for the inhibition of TS has led to several FDA-approved anticancer drugs, such as fluorouracil, raltitrexed, pemetrexed, and nolatrexed [79]. However, most tumors are resistant to the therapy by TS inhibitors due to their high levels of TS. Continuous exposure of tumors to these TS inhibitors can also increase the level of TS, thus resulting in decreased clinical response and more rapid progressing disease [31]. Toxicity related to TS inhibitors is another problem that may develop because of high-level expression and activity of TS in some normal cells. Instead of inhibiting TS, prodrugs can be designed as substrates of TS to release cytotoxic species upon TS-catalyzed reaction [76,78,80].

As shown in Figure 9, the less cytotoxic 2′-deoxyuridine analogs like 2′-F-ara-deoxyuridine (FAU) (**35**) can be converted into the more cytotoxic thymidylate analogs through phosphorylation by thymidine kinase (TK) and methylation by TS [80]. As a prototype, FAU readily entered cells and was phosphorylated, methylated, and subsequently incorporated into cellular DNA. FAU produced varying degrees of growth inhibition against several cancer cell lines [80].

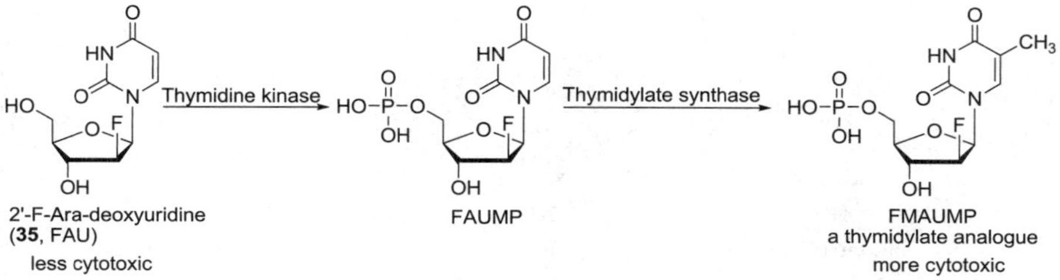

FIGURE 9 Activation of FAU by TK and TS.

NB1011 (**36**) is a phosphoramidate prodrug of E-5-(2-bromovinyl)-2′-deoxyuridine monophosphate (BVdUMP), which is a substrate for TS and is charged under physiological pH making it difficult to permeate through cellular membranes. The phosphoramidate NB1011 facilitates the intracellular delivery of BVdUMP. Within cells, NB1011 is converted through hydrolysis to the corresponding monophosphate, BVdUMP, which was subsequently activated by TS to form the cytotoxic allyl bromide product as shown in Figure 10. The cytotoxicity of NB1011 is more pronounced on high TS-expressing tumor cells [76,78].

Thymidine Phosphorylase TP (EC 2.4.2.4) is a highly selective enzyme that in its normal intracellular environment catalyzes analogous reactions with a variety of 5-substituted uracils and their β-D-2-deoxyribofuranosyl derivatives [82]. It is strictly enantioselective, accepting in its active site and acting only on the naturally occurring D-thymidine and some of its close analogs, but not on their L-counterparts. TP functions principally as a catabolic enzyme, generating 2-deoxyribose 1-phosphate (dR1P) during the process of maintaining adequate intra- and extracellular levels of thymine. Therefore, TP plays a key role in maintaining the balance of the nucleotide pool and controlling nucleic acid homeostasis, by ensuring the correct supply of deoxyribonucleoside triphosphates (dNTPs) for DNA replication and repair [83].

TP is present in a variety of tissues. Highest levels are present in the liver, lung, and lymph node in humans. Other tissues contain lower levels of TP. TP activity is elevated up to 10-fold in tumors compared with normal tissues. Overexpression of TP is mainly found in the hypoxic areas of solid tumors [31].

TP protein exists as a dimer, but dimerization seems to have no clear relevance to its function. Crystal structures indicate that the binding sites for phosphate/sulfate and thymine are exposed to bulk water in an open cleft and are too far apart for the reaction to occur. A large domain movement, closing the cleft and bringing the substrates together, has been suggested as a necessary step towards an active complex [84].

The role of TP as a prodrug-activation enzyme is highlighted in the metabolic activation of FUra and 5′-dFUrd to anticancer drugs as shown in Figure 11. Both FUra (base) and 5′-dFUrd (nucleoside) are substrates of human TP. 5′-dFUrd is converted to FUra through the actions of TP. FUra is the precursor of 5-fluoro-2′-deoxyuridine monophosphate (F-dUMP) and F-UTP. On one hand, F-dUMP forms a ternary covalent complex with TS and methylenetetrahydrofolate leading to the inhibition of

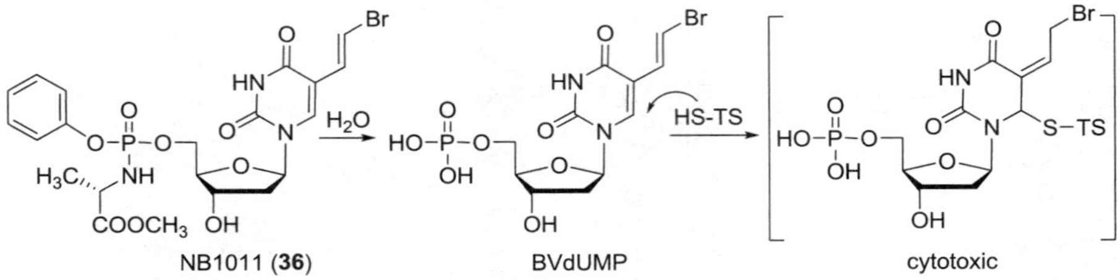

FIGURE 10 The mechanism of activation of NB1011 by intracellular hydrolysis and TS [77,81].

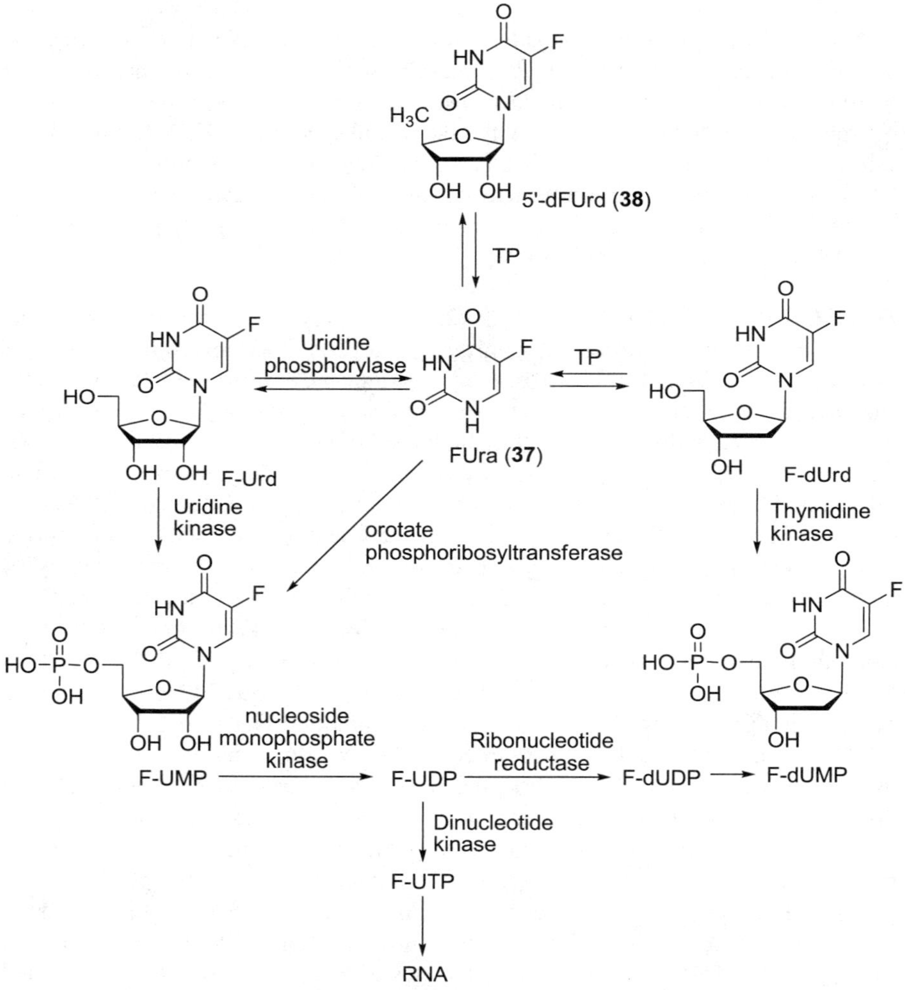

FIGURE 11 Activation of 5-fluorouracil (**37**, FUra) and 5′-dFUrd (**38**) to (1) F-dUMP, a specific inhibitor of DNA synthesis (by blocking TK activity), and (2) F-UTP, an antimetabolite of RNA synthesis through its incorporation into nascent mRNA [83].

TS enzyme activity and an increase in DNA fragmentation. These changes affect DNA replication and homeostasis. On the other hand, FUra, once activated to F-UTP, could be incorporated into RNA, thus interfering with its synthesis and its nuclear processing. Therefore, TP plays a key role in mediating the sensitivity of tumor cells to FUra as an inhibitor of TS activity, DNA replication, and RNA synthesis [83].

Glutathione S-Transferases The cellular GSH system is a critical component of detoxification of cytostatics in the cell [85]. As key enzymes in GSH detoxication system, GSTs (EC 2.5.1.18) are a superfamily of detoxication enzymes with members in several structural classes [86]. Human cytosolic GSTs are highly polymorphic and can be divided into six classes: α, μ, ω, π, θ, and ξ. The GST isozyme π is a widely studied member of this family, because its expression is frequently elevated in many human tumors and it is involved in the development of resistance to several anticancer drugs [87,88]. A major task of GSTs is to catalyze the conjugation of GSH, a thiol-containing tripeptide with a wide variety of electrophilic compounds, and thereby facilitate their elimination from the body by transporter protein such as multidrug resistance (MDR)-related protein or P-glycoprotein (P-gp). GSTs have also been implicated in cellular resistance to chemotherapeutic agents, since exposure to these drugs is often associated with induction of GSTs, especially GST π [89]. Activation and overexpression of GSTs could provide protection of the targeted cells from a variety of endogenous and exogenous compounds, including anticancer drugs, and can thereby decrease the effectiveness of chemotherapy leading to drug resistance. In addition to the development of drug resistance, recent studies have demonstrated a regulatory role for the π and μ classes of GSTs in the mitogen-activated protein (MAP) kinase pathway involved in cellular survival and death signaling. GSTs are shown to be an inhibitor of one of the MAP kinases, JNK, by forming protein–protein complex. Many anticancer drugs induce apoptosis via the activation of JNK pathways; overexpression of GST π and the resulting inhibition of JNK could lead to reduced activity of these anticancer agents.

In order to solve the drug resistance problem related to GSTs, different GST inhibitors have been designed and tested [90]. An alternative strategy is to take advantage of the GST overexpression and develop prodrugs that are activated by GSTs, and thus also gain the benefit of delivering the effect where it is desired and thereby attenuate adverse side effects [91]. Several prodrugs have been developed to be activated by GST P1–1.They are metabolized to form nitrogen mustards [92,93], alkylating exocyclic enones [94], cytolytic NO [95], or thiopurine [96].

Because numerous solid tumors and a number of drug-resistant cell lines express high levels of GST π, higher selectivity could be achieved [97]. For example, primary ovarian cancer biopsies [98] and cell lines established from drug-resistant cancer patients [99] have been shown to express high levels of GST π. TLK286 (**39**) is one of the first reported as GST π-activated prodrug. As a novel prodrug activated by GST π through a β-elimination reaction releasing the alkylating phosphorodiamidate (Fig. 12) [100], TLK286 was predicted to have improved therapeutic index and is

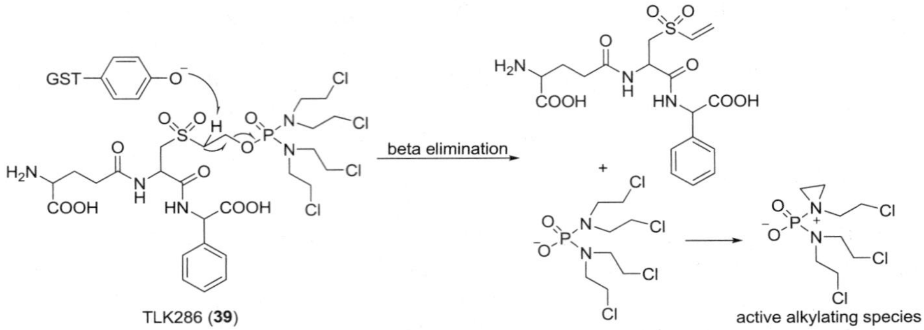

TLK286 (**39**)

FIGURE 12 The mechanism of activation of TLK286 by GST π.

currently under phase 3 clinical trial for the treatment of several tumors including breast, colon, lung, and ovarian cancer.

cAVTP (**40**) and tAVTG (**41**) are thiopurine prodrugs with an α,β-unsaturated substituent acetylvinyl (butanone) attached to the sulfur of 6-mercaptopurine (6-MP) and 6-thioguanine (6-TG), respectively [101]. 6-MP and 6-TG are cytotoxic agents commonly used in the treatment of acute lymphoblastic leukemia (ALL) [102]. Activation of these prodrugs is believed to go through a GST-catalyzed nucleophilic attack of the cellular thiol GSH on the butenone moiety, followed by an elimination reaction releasing the active parent drug 6-MP or 6-TG (Fig. 13) [96]. The prodrugs were shown to deliver the parent thiopurines more efficiently into the cells and exhibit less systemic toxicity than the parent thiopurines.

GST has also been used to activate prodrugs designed to release nitric oxide (NO) [95,103]. NO plays an important role in various physiological processes such as smooth muscle relaxation, platelet aggregation, vasodilation, neurotransmission, and immune response [104]. Although the relationship between NO and cancer is complex, it has been demonstrated that NO induces apoptosis and initiates differentiation in certain types of cancer cells via a unique novel mechanism of action [105]. Such agents with unique mechanisms of action are particularly desirable as new cancer drugs are needed to overcome the drug resistance problems and improve the therapeutic outcome in the treatment of cancer [106,107].

NO prodrugs activated by GSTs have been designed based on 1-chloro-2,4-dinitrobenzene (CDNB), a known substrate of GSTs. Figure 14 shows one typical GST-activated NO-releasing prodrug, O^2-(2,4-dinitrophenyl) 1-[(4-ethoxycarbonyl) piperazin-1-yl]diazen-1-ium-1,2-diolate (JS-K, **42**), and its proposed mechanism of activation [103]. NO release from JS-K is proposed to occur through a nucleophilic aromatic substitution by GSH catalyzed by GST to form a diazeniumdiolate anion that then spontaneously releases NO. Structure–activity studies indicate that piperazine ring can be replaced by the seven-membered homopiperazine without loss of antiproliferative activity and with a diminished reactivity towards both GSH and GSH/GST. Both JS-K and its homolog displayed no cytotoxic activity towards normal renal epithelial cell line at concentrations that effectively inhibited the proliferation of a panel of renal cancer cell lines. These properties may prove advantageous

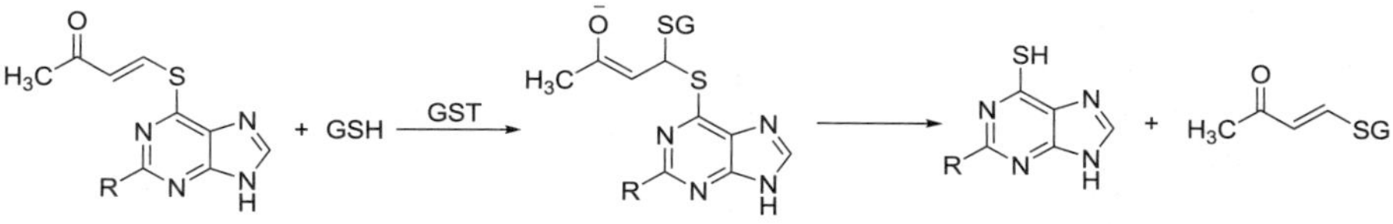

FIGURE 13 The activation of acetylvinylthiopurine prodrugs by GSH in the presence of GST.

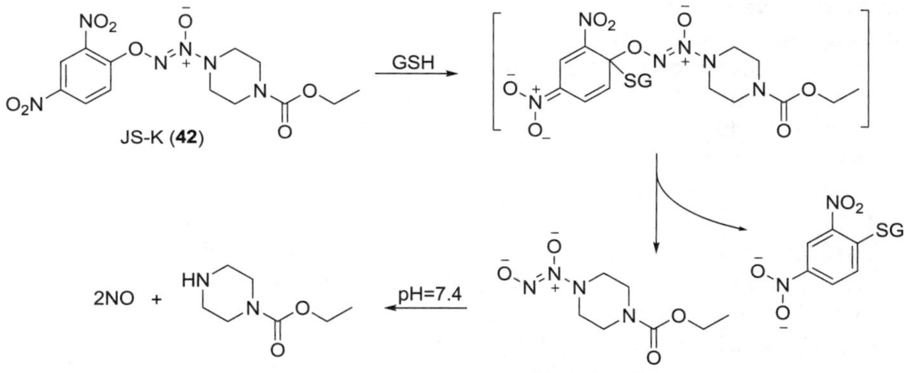

FIGURE 14 The proposed mechanism of activation of JS-K by GSH.

in the further development of this class of NO prodrugs as potential cancer therapeutic agents. JS-K has recently been shown to suppress the growth of human leukemia and prostate cancer xenografts in mice [108,109].

JS-K, however, is reported to be metabolized by GST α one magnitude faster than GST π; this could be problematic because GST α is the prevalent isoform required for the maintenance of cellular homeostasis in normal tissues [110]. PABA/NO (**43**) was thus designed using a structure-based approach to target NO more specifically to tumor cells by targeting to the less prevalent GST π to prevent collateral damage to normal tissues where the prevalent isoform is required for the maintenance of cellular homeostasis. PABA/NO is metabolized to release NO more efficiently by GST π than by GST α [95]. As a substrate of multidrug resistance-associated protein 1 (MRP1), a key component of the cellular efflux pump system, the resistance to PABA/NO and its metabolites is provided by MRP1. Therefore, the cells with high levels of GST π, but low MRP1 content, may be ideal targets for induction of cytotoxicity of PANA/NO [95]. In a recent study, it was found that homopiperazine analog of JS-K (**44**) was identical to JS-K in activity against a number of cancer cell lines, and it did not display any cytotoxicity against a normal renal epithelial cell line at the inhibited concentrations [111].

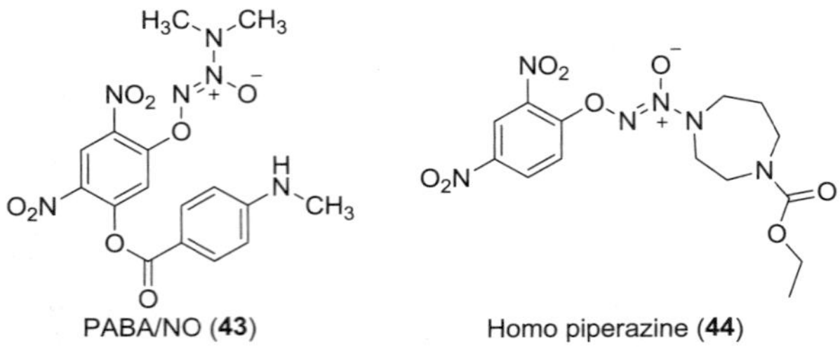

SCHEME 4

Deoxycytidine Kinase DCK (EC 2.7.1.74) is an enzyme that transfers phosphate to deoxycytidine, and it catalyzes the phosphorylation of 2′-deoxycytidine (dC), 2′-deoxyadenosine (dA), and 2′-deoxyguanosine (dG) to their corresponding monophosphates using nucleoside triphosphates as phosphoryl [112].

In humans, high levels of DCK have been reported in lymphocytes and infant thymus. Lower levels were found in the brain, muscle, ventricle, lymphoid, and colon, whereas the enzyme appeared to be absent in the liver [31]. DCK is expressed in tumor cells, although in low and varying levels.

In addition to recycling natural 2′-deoxyribonucleosides, DCK catalyzes the initial, often rate-determining phosphorylation of several antiviral and anticancer nucleoside analog prodrugs, such as gemcitabine (**45**), fludarabine (**46**), and pentostatin (**47**), which undergo phosphorylation by DCK before their incorporation into DNA causing cell death. Gemcitabine (2′-deoxy-2′,2′-difluorocytidine, dFdC) is a pyrimidine nucleoside analog with therapeutic activity against solid tumors and hematological malignancies [113]. It is clinically effective in the treatment of advanced or metastatic pancreatic cancer and non-small cell lung cancer. Gemcitabine exerts its antiproliferative activity via multiple mechanisms of action. It is phosphorylated intracellularly by DCK to form its corresponding monophosphate, which is the rate-limiting step in the formation of its active triphosphate metabolite (Fig. 15). Gemcitabine triphosphate is incorporated into DNA, thereby inhibiting DNA synthesis and DNA repair [114]. Fludarabine (arabinosyl-2-fluoroadenine phosphate, F-araAMP) shows a wide range of activity against hematological malignancies, whereas pentostatin (2′-deoxycoformycin, DCF) has been shown to have clinical activity against lymphoid malignancies [115]. The triphosphate metabolites of fludarabine and pentostatin formed via DCK-medicated phosphorylation can incorporate

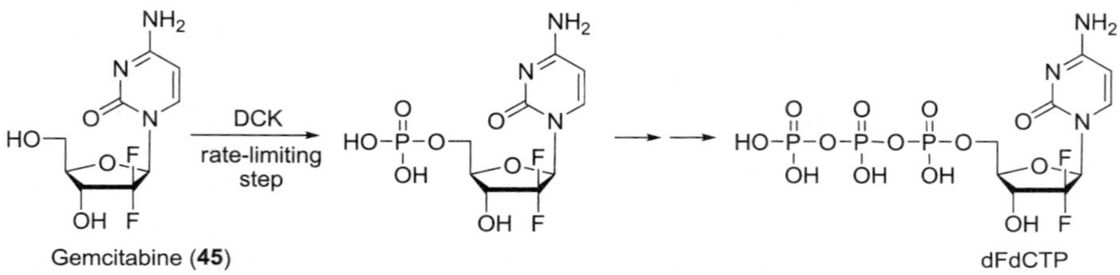

FIGURE 15 The mechanism of activation of gemcitabine by DCK.

into DNA during the process in which strand breaks are repaired and are therefore cytotoxic to slowly replicating cells [115]. Triphosphate of pentostatin also induces the accumulation of deoxyadenosine, which is in turn phosphorylated by DCK to dATP, resulting in interference with the normal adenosine metabolism.

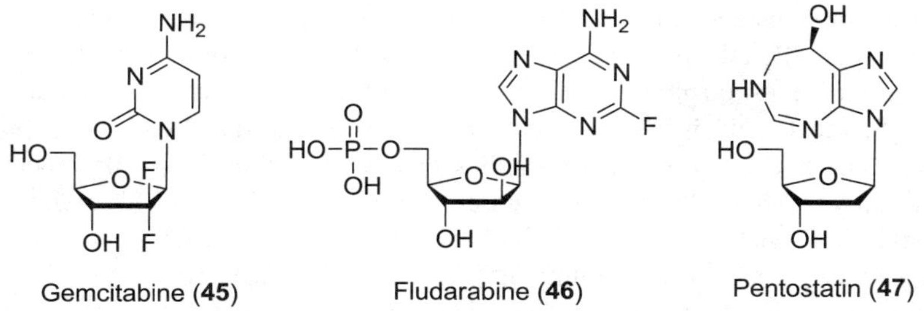

SCHEME 5

C. Hydrolases

Many prodrugs contain an ester bond, which is formed by derivatizing a phenol, alcohol, or carboxylic acid group present in the drug molecule [116]. Typically the activation of these prodrugs is through esterase-catalyzed hydrolysis. Although less common as compared to ester prodrugs, there are prodrugs that contain amides and can be activated by proteases. The most important hydrolases involved in prodrug activation include CE, acetylcholinesterase (AChE), butyrylcholinesterase (BChE), paraoxonase (PON), arylesterase, matrix metalloproteinase (MMP), alkaline phosphatase (AP), β-glucuronidase, human valacyclovirase (VACVase), and transpeptidase.

Carboxylesterase CEs (EC 3.1.1.1) can efficiently catalyze the hydrolysis of a variety of ester- and amide-containing xenobiotics to their respective free acids. They act as an effective biological barrier to limit the distribution of substrates that might be toxic and facilitate their elimination by turning them into more polar molecules.

Although CEs show varying activity and substrate specificity, depending on species and location, they are present virtually throughout the body, including the intestines, blood, brain, skin, and tumor. There are several different isozymes of carboxylesterases; based on sequence homology and similarity of characteristics, the isozymes have recently been classified into five subfamilies: CES1, CES2, CES3, CES4, and CES5. However, other names such as hCEs that are classified as human liver carboxylesterase 1 (hCE1), human carboxylesterase 2 (hCE2) and intestinal carboxylesterase (hiCE), and human brain carboxylesterase (hBr2) continue to be used in the literature.

CE is a homotrimer with each 60-kDa subunit that is composed of 550 amino acid residues. Each hCE1 monomer consists of 19 β-strands and 15 α-helices, forming a central catalytic domain and a regulatory domain. Two disulfide bonds are present (Cys^{87}–Cys^{116} and Cys^{274}–Cys^{285}), and these cysteine bridges probably play an important role in maintaining the proteins in an enzymatically active conformation [117].

The active site is buried at the bottom of a long gorge approximately 15–20 Å from the surface of the enzyme [117–120]. The gorge is lined with aromatic amino acids, creating a highly hydrophobic pocket and an electrostatic gradient that essentially "sucks" substrate molecules towards the catalytic residues [117]. At the base of the active site gorge, the catalytic serine, histidine, and glutamic acid residues are juxtaposed within a distance of several Å, such that nucleophilic attack of esterified substrates can readily take place. Adjacent to these amino acids is an oxyanion hole that is thought to be responsible for the stabilization of the tetrahedral intermediate [117].

Biochemical analysis indicated that CEs have substrate specificity. These proteins are efficient at the hydrolysis of small substrates but show vast differences with increasingly larger substrates. For example, the hCE1 was 100- to 1000-fold less efficient at CPT-11 metabolism than rabbit liver carboxylesterase (rCE), even though the two enzymes have over 80% amino acid identity [119,121]. Analysis of the crystal structures of these two proteins indicated that they were almost identical, except for two loops, which formed the entrance to the active site gorge. In rCE, the structure of these loops is considerably more flexible. In contrast, the same regions in hCE1 were considerably rigid. Analysis of the catalysis of a series of substrates of increasing size indicated that rCE was much more efficient in the hydrolysis of larger molecules (e.g., CPT-11, 1-naphthyl butyrate, *p*-nitrophenyl butyrate) than hCE1. The flexibility of the loops in rCE allows large substrates to be accommodated within the active site, while a similar process cannot occur with hCE1 due to the rigidity of these loops. Therefore, the substrate specificity of CEs is, in part, determined by constraints enforced by the dimensions of, and access to, the active site [117].

The CEs are involved in the activation of various antiviral, anticancer, and antibiotic prodrugs. Several representative esterase-activated prodrugs are shown in Figure 16. Irinotecan (CPT-11, **48**) is converted to its active metabolite, 7-ethyl-10-hydroxycamptothecin, which is an inhibitor of topoisomerase I [119,122]; temocapril (**49**), a long-acting ACE inhibitor, is quickly metabolized to the active dicarboxylic acid [123]; dipivefrin (**50**), available as an ophthalmic solution for the treatment of glaucoma, is hydrolyzed to epinephrine on penetrating cornea [124]; and lovastatin (**51**) and simvastatin (**52**), used to lower cholesterol in patients with hypercholesterolemia, were hydrolyzed by hCE1 to form the hydroxy acid forms, which inhibit the activity of HMG-CoA reductase [125].

Acetylcholinesterase AChE (EC 3.1.1.7) can rapidly hydrolyze the neurotransmitter acetylcholine (ACh) at cholinergic synapses and neuromuscular junctions. AChE is an extrinsic membrane-bound enzyme that projects into the synapse. It is responsible for the elimination of ACh released from the presynaptic nerve process in response to an action potential and is critical for maintaining normal cholinergic function. ACh diffuses across the synapse and binds to the ACh receptor, which among other functions serves as an ion gate for the entry of K^+ into the postsynaptic nerve process or muscle cell [116,126]. AChE consists of about 550 amino acid residues and has a molecule mass of about 70–80 kDa. The AChE active site consists of three major domains: (1) an ester-binding region containing the catalytic triad, serine, histidine, and glutamate; (2) an anionic region that is more than 14.7 Å from the

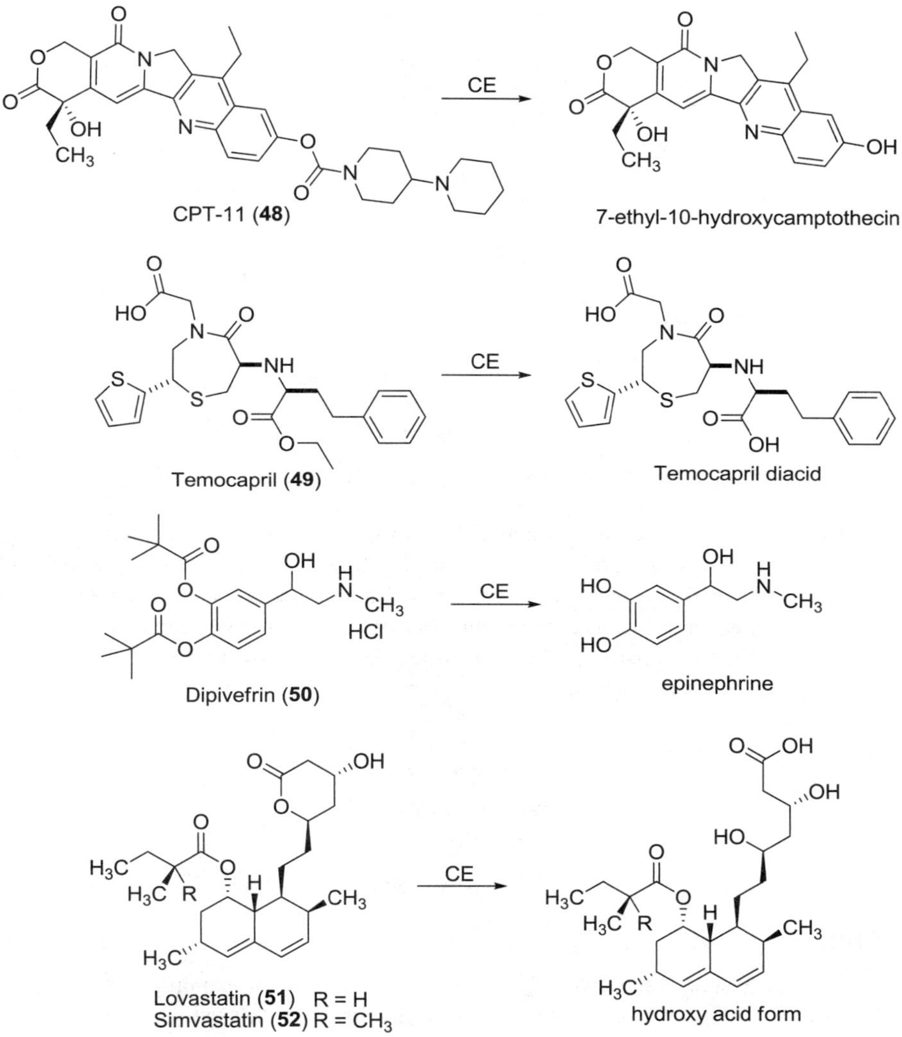

FIGURE 16 CE-mediated activation of representative ester and carbamate prodrugs.

catalytic serine and where the choline quaternary ammonium binds to; and (3) a peripheral hydrophobic region that is contiguous with or near the ester-binding and anionic choline-binding regions and that is important in binding aryl substrates and other active site ligands [126].

The nucleophilic nature of the carboxylate is transferred through the imidazole ring of histidine to the hydroxyl group of serine, allowing it to displace the choline moiety from the substrate, forming an acetyl–enzyme intermediate. A subsequent hydrolysis step frees the acetate group (Fig. 17) [127].

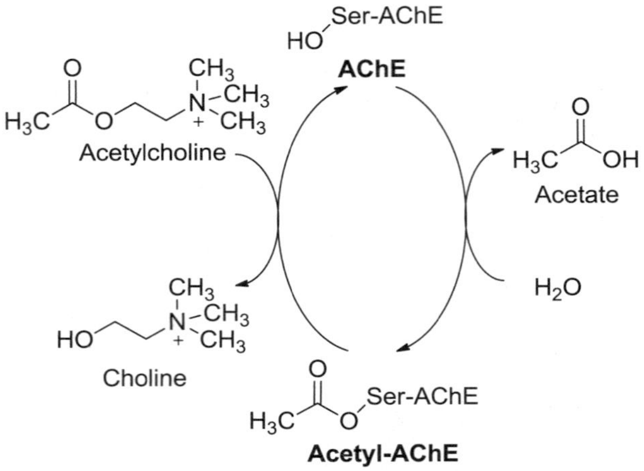

FIGURE 17 The mechanism of hydrolysis of ACh by AChE.

AChE is widely expressed in many tissues throughout the body, including skeletal muscle, many types of neurons in the central and peripheral nervous systems, and numerous other cell types that are targets of cholinergic nerves such as endocrine and exocrine glands, smooth and cardiac muscle cells, and sensory organs embedded in epithelia [128]. In humans, AChE is also expressed in nonattached cells such as erythrocytes and lymphocytes.

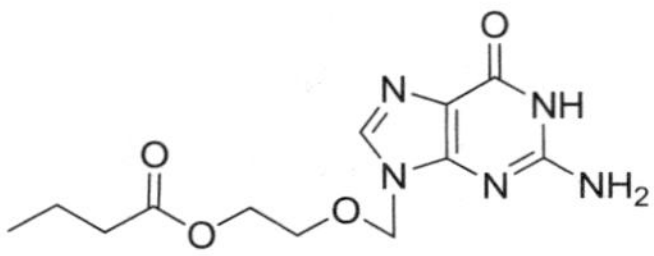

Butyryl ACV (**53**)

SCHEME 6

Several prodrugs activated by AChE have been reported. For example, butyryl ACV (**53**) is an AChE-activated ester prodrug of acyclovir (ACV), which is a synthetic purine nucleoside analog of guanine and clinically used in the treatment of herpes simplex virus infections because of its affinity for the viral TK [129].

Butyrylcholinesterase BChE (EC 3.1.1.8) is relatively abundant in plasma and is sometimes referred to as the serum cholinesterase. However, its physiological role remains unclear. BChE can degrade a large number of ester-containing compounds, including ester prodrugs [130].

The overall 3-D structure of BChE is very similar to that of AChEs. However, BChE does not form the same kind of dimer as observed in structures of *Torpedo californica* AChE, mouse AChE, and human AChE. In AChEs, the four helices involved in the subunit–subunit interaction are antiparallel, and the active site openings are located on opposite sides of the dimer. Although homologous helices are found at the

BChE dimer interface (residues 362–375 and 514–529), they are not antiparallel but are at 45° angle relative to each other, and the active site openings are located on the same side of the dimer. Constraints resulting from the crystal packing could be responsible for these differences [116,130].

BChE has a broader substrate specificity than AChE, and this difference in specificity can be explained by BChE's more spacious acyl-binding pocket at the active site and differences in amino acid residues [130]. Similar to AChE, the active site gorge of BChE consists of triad residues, and this gorge is surprisingly deep, of 20 Å [116,127]. This gorge attracts substrates and reduces their mobility, thus enabling enormously speedy delivery of the substrate to the active site where it is then hydrolyzed in a unique "dry" environment that allows only limited solvent–solute interactions [116,130].

BChE exists predominantly in plasma as well as in various tissues such as the brain, muscle, kidney, intestine, retina, and placenta. Several BChE-activated prodrugs have been developed. Bambuterol (**54**) is slowly converted by BChE to terbutaline, a β_2-adrenoceptor agonist [131,132]. Methylprednisolone acetate (**55**) can be hydrolyzed by BChE to its active form – methylprednisolone, a steroid anti-inflammatory agent [133]. Isosorbide diaspirinate (**56**), the aspirin diester of isosorbide, is stable towards aqueous hydrolysis and in the presence of α-chymotrypsin [134]. However, it undergoes rapid hydrolysis in the presence of BChE as shown in Figure 18.

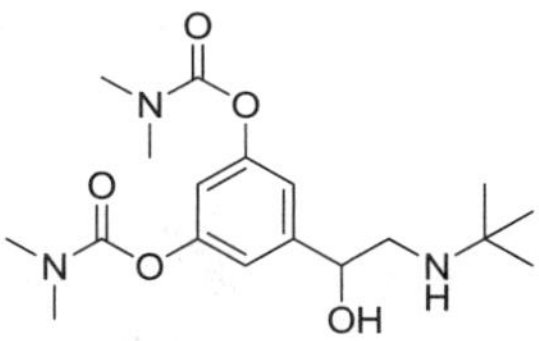

Bambuterol (**54**)

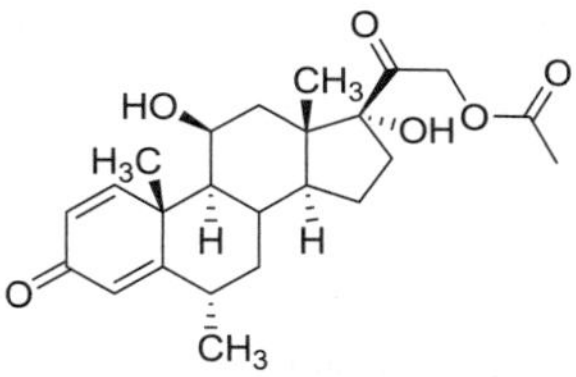

Methylprednisolone Acetate (**55**)

SCHEME 7

Paraoxonase The PON (EC 3.1.8.1) family is a hydrolase family, which includes PON1, PON2, and PON3. PON1 is an important esterase, which can catalyze the hydrolysis of various organophosphates. PON2 and PON3 share ~60% sequence identity with PON1; however, they lack or have very limited PON and arylesterase activities. Comparing with PON1's proficient esterase activity towards several

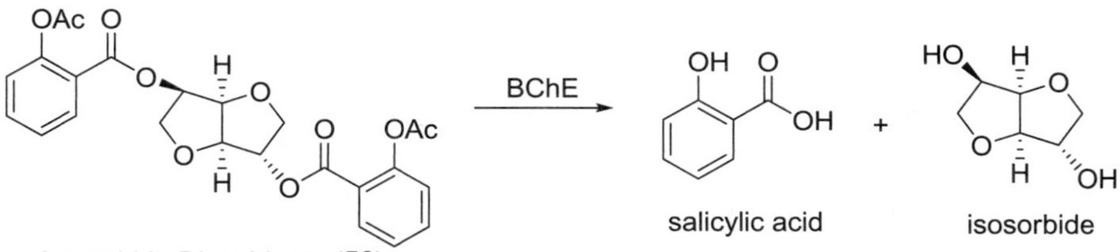

FIGURE 18 Activation of isosorbide diaspirinate by BChE.

synthetic substrates, PON2 and PON3 exhibit high lactonase activity. Human PON1 (hPON1) is synthesized in the liver and secreted into the blood, and similarly PON3 is expressed mostly in the liver and at low levels in the kidney [135]. In recent years, PON1 and its two known Q and R isoenzymes have been shown to possess both arylesterase and organophosphatase activities and play important roles in drug metabolism and in the prevention of atherosclerosis [136].

PON1 is a 43-kDa glycoprotein consisting of 554 amino acid residues. The crystal structure studies of recombinant PON1 (rePON1) variants indicate that PON1 is a six-bladed β-propeller, and each blade contains four strands [136]. A disulfide bridge (Cys^{42}–Cys^{353}) and two calcium ions (7.4 Å apart), which are located in the central tunnel of the propeller, are found in the crystal. The calcium ion at the top of the propeller, assigned as the "catalytic calcium," seems to interact with five protein residues (the side-chain oxygens of Asn^{224}, Asn^{270}, Asn^{168}, Asp^{269}, and Glu^{53}) about 2.2–2.5 Å away. Two other potential ligands are a water molecule and one of the oxygens of a phosphate ion [136].

Prulifloxacin (**57**), a lipophilic prodrug of ulifloxacin, is a new oral fluoroquinolone with a broad spectrum of *in vitro* activity against various Gram-positive and Gram-negative microorganisms. Its structure contains the skeletal quinolone with a four-membered ring in the 1,2-position including a sulfur atom to increase antibacterial activity and an oxodioxolenylmethyl group in the 7-piperazine ring to improve its oral absorption; it is immediately and quantitatively transformed into the active metabolite ulifloxacin (Fig. 19) [137]. Lactone-containing drugs lovastatin and simvastatin are all substrates of hPON3 [116,135,138].

Matrix Metalloproteinase MMPs, also known as matrixins, are zinc-dependent endopeptidases consisting of more than 25 family members classified based on domain structure and substrate specificity (Table 3). They play important roles in the degradation of extracellular matrix components, such as collagen, laminin, fibronectin, and elastin. MMPs also cleave a variety of proteins, such as growth factor receptors and cell adhesion molecules, which may be important for tumor growth and survival. Synthesized initially as inactive proenzymes, MMPs are processed by proteolysis to their active forms which generally contain three main domains, namely, a propeptide domain, a catalytic domain, and a hemopexin/vitronectin-like domain. Several MMPs, including MMP-1, MMP-2, MMP-8, MMP-13, MMP-14, MMP-18, and MMP-22, catalyze the first committed step in extracellular matrix degradation

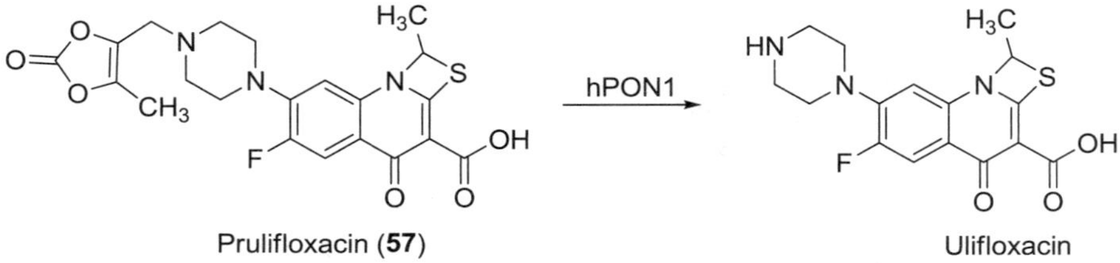

FIGURE 19 The activation of prulifloxacin by hPON1.

TABLE 3 MMP and Their Substrates

Enzyme	Alternative name	EC number	Substrates
MMP-1	Interstitial collagenase	3.4.24.7	Collagens (I, II, III, VII, VIII, and X); gelatin; aggrecan; L-selectin; IL-1beta; proteoglycans; entactin; ovostatin; MMP-2; MMP-9
MMP-2	Gelatinase A	3.4.24.24	Collagens (IV, V, VII, and X); gelatin
MMP-3	Stromelysin-1, proteoglycanase	3.4.24.17	Collagens (III, IV, V, and IX); gelatin; aggrecan; perlecan; decorin; laminin; elastin; casein; osteonectin; ovostatin; entactin; PLG; MBP; IL-1beta; MMP-2/TIMP-2; MMP-7; MMP-8; MMP-9; MMP-13
MMP-8	Neutrophil collagenase	3.4.24.34	Collagens (I, II, III, V, VII, VIII, and X); gelatin; aggrecan; fibronectin
MMP-9	Gelatinase B	3.4.24.35	Collagens (IV, V, VII, X, and XIV); gelatin; entactin; aggrecan; elastin; fibronectin; osteonectin; PLG; MBP; IL-1beta
MMP-13	Collagenase-3	[a]	Collagens (I, II, III, IV, IX, X, and XIV); gelatin; PLG; aggrecan; perlecan; fibronectin; osteonectin; MMP-9
MMP-14	MT1-MMP	3.4.24.80	Collagens (I–III); gelatin; casein; fibronectin; laminin; vitronectin; entactin; proteoglycans; MMP-2; MMP-13
MMP-18	Xenopus collagenase-4	[a]	Not known
MMP-22	Chicken MMP (C-MMP)	[a]	Not known

[a]Not assigned

by unwinding and cleaving the triple helix of interstitial collagens (types I–III) [139–144]. On the other hand, MMP-3 and MMP-9 only possess the ability to bind to single-strand type I collagen. The collagen cleavage site is similar for most MMPs, and the general motif preceding the scissile bond (between Gly and Ile/Leu) has been characterized as a four-triplet region rich in proline with one always at P3 position. It is important to recognize the ability to unwind and cleave interstitial collagen in the triple helix form as the main determinant of substrate specificity, rather than the primary structure [143,144].

Because of their significant role in tumor angiogenesis, invasion, and metastasis, MMPs have been implicated in several pathologies mediated by extracellular matrix degradation including various cancers (e.g., multiple myeloma), arthritis, glomerulonephritis, periodontal disease, and ulcer [144]. Among them, MMP-2 and MMP-9 represent important drug targets due to their significant involvement in angiogenesis and metastatic growth. A more recent and promising strategy involves targeted prodrug design for selective proteolytic activation by MMP-2 and MMP-9 in tumor cells, although the body of literature pertaining to this approach is still in its nascent and exploratory phase [145–148].

Model prodrugs have been reported to release, upon activation by MMP-9, anthraquinone-type cytotoxic agents that are often used in combination with conventional therapies for the treatment of multiple myeloma [147]. Although not curative, coadministration of liposomal doxorubicin, lenalidomide, thalidomide, and bortezomib showed clinical benefits in patients with multiple myeloma. L-Prolyl-*N*-(3-aminopropyl)-1-aminoanthraquinone, an adduct known to possess dual topoisomerase inhibitory activity *in vivo*, was coupled to an optimized substrate of MMP-9 as shown in EV1 (**58**) [147]. Cleavage of EV1 by MMP-9 occurs between glycine and norvaline, resulting in the liberation of the active inhibitor of topoisomerases.

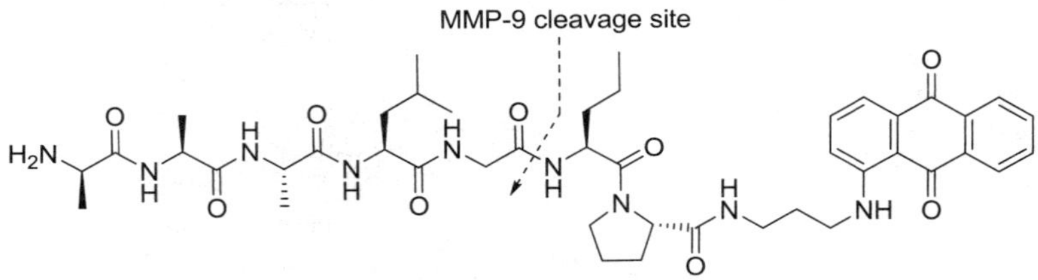

EV1 (**58**)

SCHEME 8

Selective delivery of doxorubicin to tumor tissues was investigated using its peptide conjugates as prodrugs activated by the combined proteolytic action of MMP-2, MMP-9, and MMP-14. Doxorubicin is a potent anthracycline used for the treatment of various types of cancer. It causes cytotoxicity through topoisomerase II-mediated DNA breaks. Coupling of doxorubicin to several MMP-cleavable peptides generated a series of prodrugs that were readily cleaved by MMP-2, MMP-9, and MMP-14. For example, the prodrug Ac–Glu–Pro–Cit–Gly–homo–Phe–Tyr–Leu–Dox was cleaved by MMP-2, MMP-9, and MMP-14 to generate a tripeptide–Dox conjugate, which was subsequently degraded by extracellular proteases to release Leu–doxorubicin as shown in Figure 20. Leu–doxorubicin was known to release doxorubicin upon further proteolytic cleavage. Prodrug testing in the fibrosarcoma cell line HT1080 expressing multiple MMPs and HT1080 xenografts in mice demonstrated the selective conversion of the doxorubicin prodrug to free doxorubicin [146]. The prodrug also exhibited a 10-fold increase in tumor/heart distribution ratio, greater efficacy, and less toxicity as compared to doxorubicin.

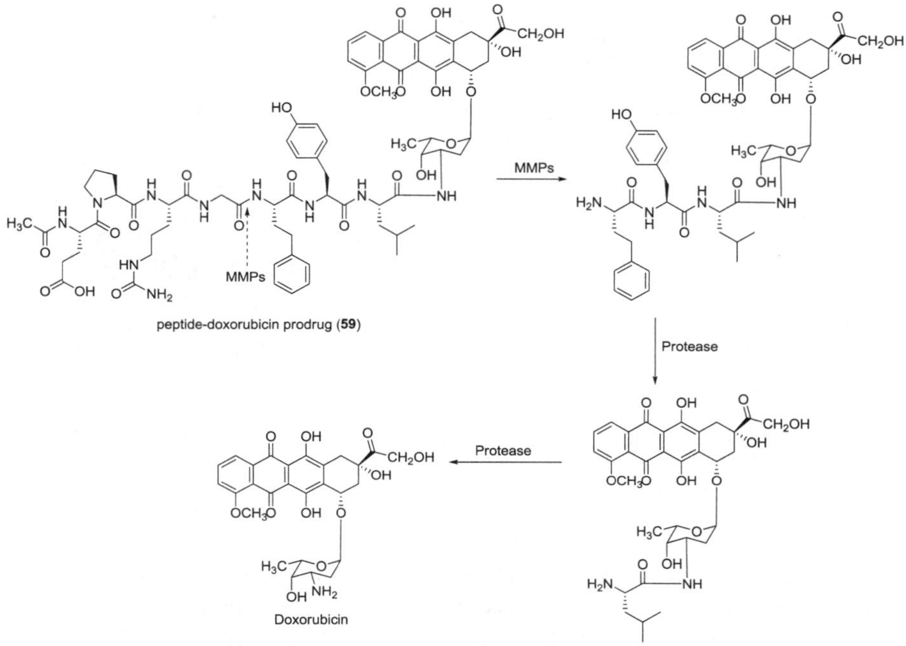

FIGURE 20 Activation of a peptide–doxorubicin prodrug by MMPs.

Alkaline Phosphatase AP (EC 3.1.3.1) is a metalloenzyme (contains zinc and magnesium) with broad substrate specificity. It catalyzes the removal of phosphate groups from a diverse class of compounds, including nucleotides, proteins, and alkaloids. AP is ubiquitously expressed in several species and tissues with the highest expression in the liver. The differential expression of various AP isozymes in liver and bone has been used as a diagnostic marker for various cancers and renal dysfunction [31].

AP is believed to be responsible for the activation of several clinically used phosphate prodrugs including fosfluconazole, fosphenytoin, fosaprepitant, amifostine, clindamycin phosphate, estramustine phosphate, and etoposide phosphate. Most of these prodrugs were developed to overcome the solubility problems of parent drugs and are for parental administration, while some are for the targeted treatment of cancer. Amifostine (WR-2721, S-2 [3-aminopropylamino]-ethylphosphorothioic acid, **60**) was developed as a protective agent against radiation- and chemotherapy-induced cellular injury [149]. It is selectively activated in normal cells by AP via dephosphorylation to afford the free thiol WR-1065, which acts a free-radical scavenger (Fig. 21). In addition to free-radical scavenging, the protective effect of amifostine has been postulated to originate from DNA repair through hydrogen donation from the free thiol group and inhibition of DNA damage. The protective effects of amifostine have been demonstrated in preclinical species in the presence of platinum compounds such as cisplatin. Following incubation of cisplatin in salmon sperm DNA in the presence of 50-fold molar excess of amifostine, WR-1065, and related compounds, a maximal

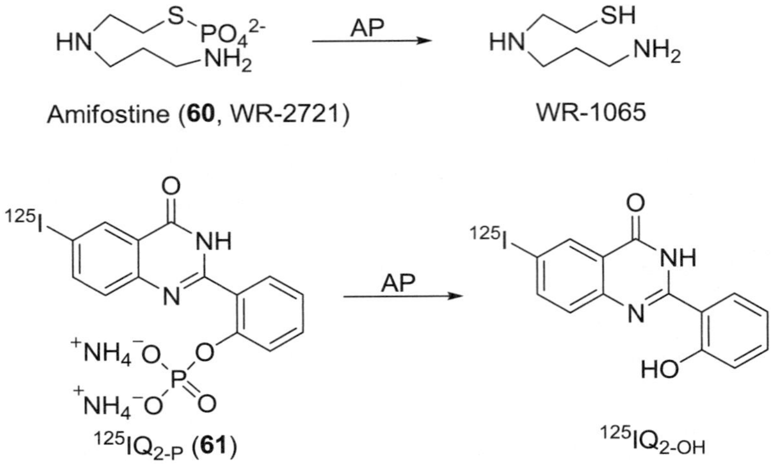

FIGURE 21 Activation of amifostine and $^{125}IQ_{2\text{-}P}$ by AP.

49% reduction in platinum–DNA adduct formation was observed with WR-1065 [150]. Clinically, the protective effects of amifostine administration 30 min prior to chemotherapy have proven effective although no proof of concept could be obtained in *in vivo* animal studies.

AP-mediated prodrug activation has been applied to cancer imaging and therapy. Such radiodetection of solid tumors *in vivo* was shown using the radioactive quinazolinone prodrug $^{125}IQ_{2\text{-}P}$, the ammonium salt of 2-(2′-phosphoryloxyphenyl)-6-[^{125}I]iodo-4-(*3H*)-quinazolinone (**61**). As shown in Figure 21, the prodrug is hydrolyzed to a water-insoluble, radiolabeled drug $^{125}IQ_{2\text{-}OH}$ by AP, which is specifically overexpressed on the exterior surfaces of tumor cell plasma membranes [151]. The quinazolinone molecule precipitates and is concentrated and permanently trapped within the extracellular spaces of targeted solid tumors. And the prolonged residence time of the drug should permit the noninvasive detection and therapy of tumors.

β-Glucuronidase Human β-glucuronidase (EC 3.2.1.31) is a lysosomal enzyme that plays an important role in the degradation of glucuronic acid containing glycosaminoglycans, such as heparin sulfate, chondroitin sulfate, and dermatan sulfate [152]. The function of human β-glucuronidase is essential for the restructuring of the extracellular matrix components, and mutation in the β-glucuronidase gene has been shown to cause a rare lysosomal storage disease called mucopolysaccharidosis type VII (Sly syndrome) [153]. β-Glucuronidase is a tetrameric glycoprotein with four identical subunits of 77 kDa and is localized in lysosomes of the cell. The optimal pH for the activity of human β-glucuronidase is around 4, while the physiological pH of 7.4 leads to decreased enzyme activity [152].

The tetrameric β-glucuronidase is an ellipsoid with approximate dihedral symmetry [154]. Each monomer is arranged into three structural domains comprised of AA22–223, AA224–342, and AA343–632, respectively. The first domain is a highly distorted barrel-like structure with a jelly roll motif and two β-hairpin insertions. The second domain is similar in topology to the immunoglobulin constant domain, while the C-terminal domain forms an α/β or TIM-barrel motif containing the active site of the enzyme [154]. The active site cavity of human β-glucuronidase consists of three amino acid residues: Glu^{540}, Glu^{451}, and Tyr^{504} [154,155]. Among them, Glu^{540} acts as a nucleophile, while Glu^{451} acts as an acid/base catalyst or the proton donor. Tyr^{504} is also important, although its role is unclear. β-Glucuronidase has a broader substrate specificity; almost any aglycone in a β-linkage can be hydrolyzed to glucuronic acid [152]. One possible mechanism proposed for the hydrolytic action by β-glucuronidase is illustrated in Figure 22 [156].

Tumor tissues including breast, lung, and gastrointestinal tract carcinomas and melanomas have been shown to have increased expression of β-glucuronidase as compared to normal tissues. The high expression of the β-glucuronidase enzyme is localized mainly in necrotic areas of the malignancies [152]. The liberation of lysosomal β-glucuronidase extracellularly leads to the accumulation of the enzyme in the extracellular matrix of the malignant tissues. Cancer tissues have been targeted by a wide variety of glucuronide prodrugs exploiting the presence of β-glucuronidase in high concentrations as the enzyme is not present in the general circulation. BHAMG (**62**), a synthetic glucuronide prodrug of aniline mustard, can be activated by *Escherichia coli* β-glucuronidase [157]. Pyrrolo[2,1-c][1,4]-benzodiazepine-β-glucuronide (**63**), a prodrug of DNA minor groove binder, has been shown to be activated by β-glucuronidase [158]. HMR1826 (**64**), where the glucuronide group is

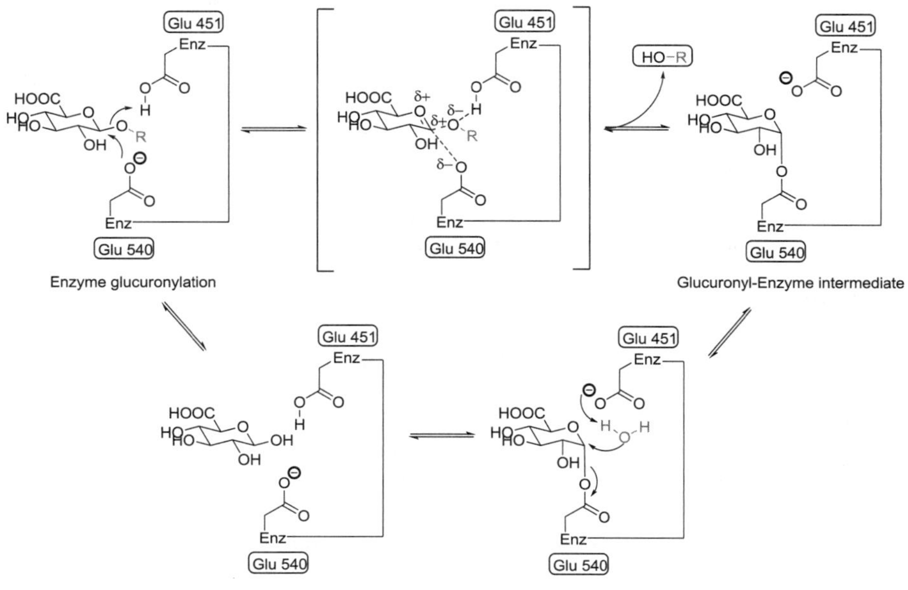

FIGURE 22 The catalytic cycle of β-glucuronidase. The R group shown in red is released upon hydrolysis of the glycosidic bond. (*See insert for color representation of the figure.*)

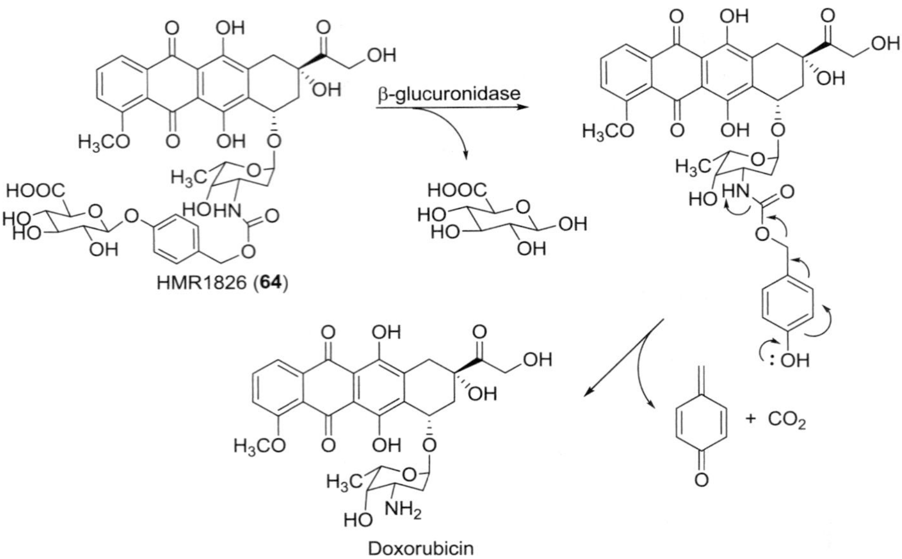

FIGURE 23 Activation of doxorubicin–glucuronide conjugate (HMR1826) by β-glucuronidase followed by 1,6-elimination of the self-immolative linker.

linked to doxorubicin as showed in Figure 23, has a V_{max} of 635 μmol min^{-1} mg^{-1} and a K_m of 1.3 mM as a substrate of β-glucuronidase [152].

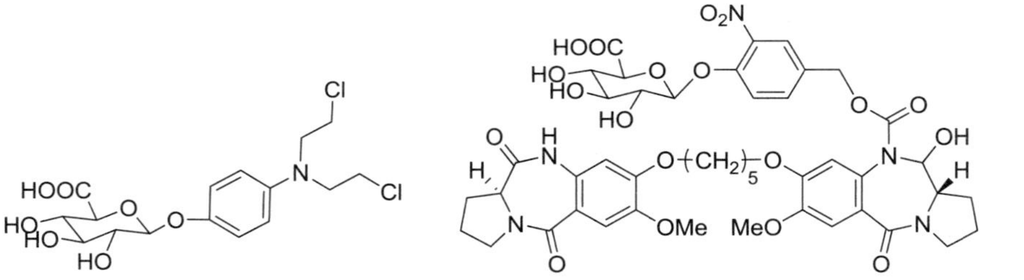

BHAMG (**62**) Pyrrolo[2,1-c][1,4]-benzodiazepine-β-glucuronide (**63**)

SCHEME 9

Human Valacyclovirase VACVase is a highly specific α-amino acid ester hydrolase. VACVase catalyzes the hydrolytic activation of two clinically important antiviral nucleoside prodrugs, valacyclovir (VACV) and valganciclovir (VGCV), to their corresponding active drugs, ACV and ganciclovir [159].

Structurally, VACVase contains 8 β-strands and 13 α-helices, which form two domains, an α/β hydrolase domain and a cap domain. Between the two domains lies a cavity containing a cave-like pocket with a single opening and a large open groove extending from the pocket mouth [159]. The catalytic triad of VACVase consists of three residues: Ser^{122}, His^{255}, and Asp^{227} [160]. The substrate-binding site of VACVase is believed to be in a flexible and mostly hydrophobic acyl pocket, possessing a charge–charge interaction site for the α-amino group and a relatively generous spatial binding site for the nucleoside leaving group with relatively limited space around the site for the cleavable ester group [161].

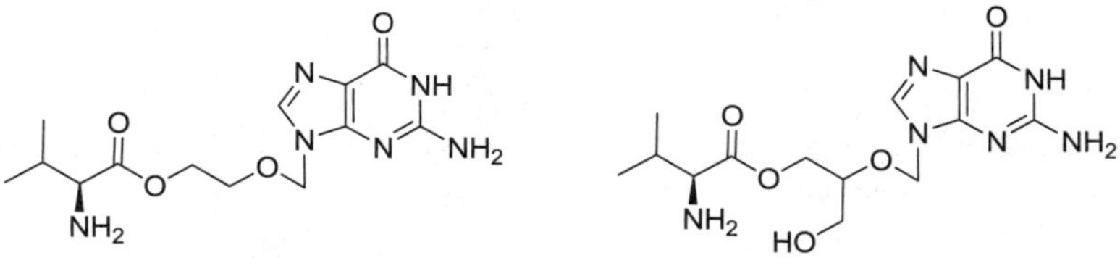

Valacyclovir (**65**) Valganciclovir (**66**)

SCHEME 10

VACVase is expressed at a high level in human liver and kidney and lower levels in the intestines, heart, and skeletal muscle [161]. Nucleoside analogs VACV (**65**) and VGCV (**66**) are two typical prodrugs activated by VACVase. VACV, the valyl ester prodrug of ACV, increases the oral bioavailability of ACV three- to fivefold [162]. In addition to activating VACV and VGCV, VACVase has been shown to hydrolyze the prodrugs of a broad range of antiviral and anticancer nucleoside analogs such as zidovudine, floxuridine, and gemcitabine [161].

Plasmin Plasmin (EC 3.4.21.7) is an important serine protease present in blood that degrades many blood plasma proteins, most notably fibrin clots. In humans, the plasmin protein is encoded by the plasminogen (PLG) gene. Plasmin is derived from PLG, a circulating glycoprotein (91 kDa) present in plasma at a concentration of ~1.5 μM [163]. PLG is a single-chain protein containing 791 amino acid residues, while plasmin molecule is a two-chain disulfide-linked serine protease, which is activated from PLG by a single cleavage of the Arg^{561}–Val^{562} peptide bond in the zymogen. The N-terminal heavy chain of plasmin (residues 1–561, ~60 kDa) is composed of five kringle domains, each containing approximately 80 amino acid residues [163]. These kringle domains are responsible for many properties of PLG, including interaction with activation inhibitors, activation stimulators, mammalian and bacterial cells, and its physiological substrate fibrin and its primary inhibitor α_2-antiplasmin [163]. The C-terminal light chain of plasmin (residues 562–791, ~25 kDa) is a typical serine protease homologous to trypsin and contains the classic catalytic triad, composed of His^{603}, Asp^{646}, and Ser^{741} [163].

Plasmin is believed to be involved in migration, invasion, and metastasis of tumor cells by catalyzing the breakdown of extracellular matrix proteins [164,165]. Plasmin is predominantly present in its inactive proenzyme form PLG in the body. Active plasmin is formed locally at or near the surface of tumor cells from cell-bound PLG activated by cell-bound urokinase-type plasminogen activator (u-PA), produced by cancer or stroma cells [166]. In the general circulation, urokinase and plasmin are inactivated by protease inhibitors such as PAI-1 and α_2-antiplasmin present in the blood. Many tumor cell lines and tumor tissues have a significantly higher u-PA level than their normal counterparts [167], and u-PA is a strong prognostic factor for reduced survival and increased relapse in many types of tumors [168,169]. Therefore, plasmin is considered a promising enzyme for the tumor-specific activation of anticancer prodrugs.

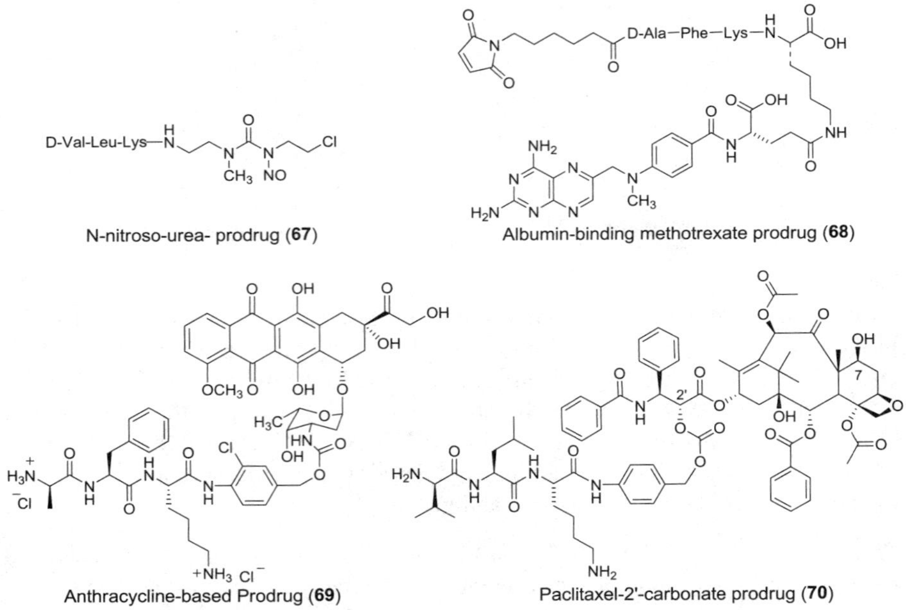

SCHEME 11

Characteristic substrate tripeptides, D-Ala–Phe–Lys and D-Val–Leu–Lys, with an N-terminal D-amino acid have been used in the design of potential prodrugs for activation by plasmin. Examples include *N*-nitrosourea-based prodrugs with a substrate tripeptide coupled via a cyclization linker [170]. As shown in Figure 24, an ethylenediamine and a mono-methylated ethylenediamine serve to link the tripeptide substrates to *N*-nitrosourea. In the presence of PLG, the tripeptide segment is cleaved off by plasmin, and subsequent intramolecular cyclization leads to the formation of imidazolidin-2-one with concomitant expulsion of the free parent drug.

Albumin-binding methotrexate (MTX) prodrug (**68**) was reported by conjugation of albumin with the maleimide-linked peptide MTX [171]. Plasmin cleavage between

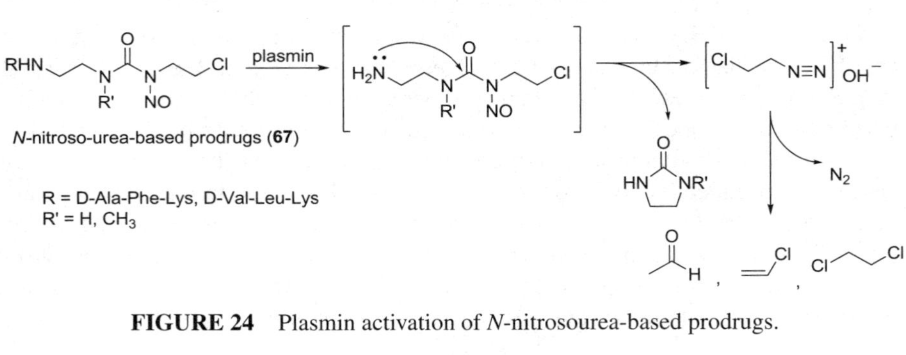

FIGURE 24 Plasmin activation of *N*-nitrosourea-based prodrugs.

D-Ala–Phe–Lys tripeptide and the lysine linker releases γMTX-εLys-OH, which is an active cytotoxic agent. Anthracycline-based prodrug (**69**), formed by insertion of a cleavable linker between the tripeptide substrate and doxorubicin, was shown to be stable in Tris buffer and bovine serum and is readily cleaved by plasmin [172]. Paclitaxel prodrug (**70**) with a carbonate linker was also designed to be selectively activated by plasmin [173]. Paclitaxel-2′-carbonate prodrug is stable in Tris buffer, and the release of free paclitaxel through 1,6-elimination process was shown to be instantaneous upon proteolysis by plasmin.

Prostate-Specific Antigen Prostate-specific antigen (PSA, EC 3.4.21.77) was first identified in human seminal plasma in 1969. PSA is a glycoprotein composed of a single polypeptide chain with a total mass of 33–34 kDa containing approximately 7% carbohydrate [174]. It belongs to the human kallikrein family of serine protease with both antigenic and enzymatic activity [175]. The three members of this kallikrein gene family include tissue kallikrein, human glandular kallikrein (hk), and PSA. Although PSA has ~70% sequence homology with tissue kallikrein and ~80% homology with hk2, it has chymotrypsin-like activity, while the other two kallikreins have trypsin-like activity [176]. PSA is primarily produced by prostate ductal and acinar epithelium and is secreted into the lumen, where it cleaves semenogelins I and II in the seminal coagulum [177]. PSA has been shown to activate u-PA, which is thought to be involved in cancer invasion and metastasis [178]. PSA could also affect tumor spread by proteolytic modulation of cell adhesion receptors [179]. Furthermore, PSA is able to cleave insulin-like growth factor-binding protein 3 (IGFBP-3) causing the release of active IGF-I, which could enhance tumor growth [180]. However, other studies indicated that PSA may inhibit tumor growth by generating antiangiogenic angiostatin from PLG [181,182].

The concentration of PSA in the prostate extracellular fluid is around 1.6–2.1 μM (50–68 μg/mL) in normal prostate and primary prostate cancer, where 80–90% of PSA is enzymatically active [183]. However, intracellular PSA is enzymatically inactive because of the presence of high concentration of zinc ion. Normally, PSA is present in blood at very low levels (<4.0 ng/mL). However, increased serum PSA levels are often seen in prostate cancer patients, which are believed to be the result of cellular PSA leak caused by the absence of a basement membrane barrier that normally separates the prostate gland from the surrounding stoma; this lack of confinement is the hallmark of invasive prostate cancer [184,185]. There is evidence that serum PSA levels correlate well with the number of malignant prostate cells, and high serum PSA levels suggest a high risk of metastatic prostate cancer [186]. PSA has been approved by FDA as a clinically important serological biomarker for the screening of prostate cancer. Most importantly, PSA present in bloodstream lacks enzymatic activity because it forms an inactive complex with protease inhibitors such as α_1-antichymotrypsin and α_2-macroglobulin [175,176,187]. Serum concentrations of these protease inhibitors are 10^5–10^6-fold higher than serum PSA levels and, therefore, are able to trap any active PSA present [185,188]. The loss of enzymatic activity of serum PSA does not affect the use of PSA as a serological marker because its immunoreactivity is still maintained. These characteristics of PSA, including specific production in the prostate tissue, enzymatically active form only in extracellular environment, enzymatically inactive form in serum and intracellular environment, and high concentration of PSA in malignant prostate cancer cells, suggest that PSA can be used as a prodrug-converting enzyme to active anticancer prodrugs to achieve selective therapeutic effects in the prostate-derived cancer tissues.

Peptide-based prodrugs linking a substrate peptide sequence to an anticancer drug have been reported for proteolytic activation by PSA. Upon proteolytic cleavage by PSA, the anticancer agent is then released locally in the microenvironment that surrounds prostate cancer cells. The anticancer selectivity is, therefore, achieved through the site-specific activation of PSA. The peptide should be cleaved specifically by PSA and resist the hydrolytic actions of other enzymes; thus, activation of the prodrugs will be localized in the prostate and prostate-derived cancer tissues where 80–90% of PSA is enzymatically active. Several short peptide sequences have been identified as PSA specific, including Ser–Lys–Leu–Gln–, His–Ser–Ser–Lys–Leu–Gln–, and glutaryl–Hyp–Ala–Ser–Chg–Gln–, all with a unique glutamine residue at P_1 site (hydroxyproline (Hyp), cyclohexylglycine (Chg)) [189,190]. These unique PSA substrate sequences were attached to anticancer drugs with the dipeptide Ser–Leu or Leu as the linker to improve the rate of PSA cleavage (e.g., glutaryl–Hyp–Ala–Ser–Chg–Gln–Ser–Leu–Dox (Fig. 25), *N*-morpholinocarbonyl–His–Ser–Ser–Lys–Leu–Gln–Leu–Drug). The peptide–drug conjugates were shown to have increased therapeutic index in preclinical prostate cancer models, validating the use of PSA as a prodrug-converting enzyme [191].

Various anticancer agents, including doxorubicin [190,192], thapsigargin [193,194], vinblastine [195,196], 5-fluoro-2′-deoxyuridine [197], 3-hydroxylapachone, diazeniumdiolate [198], nitrogen mustard [199], protoxin [200], and paclitaxel [201], have been used in prodrugs that showed selective cytotoxicity against PSA-expressing cells. The peptide–doxorubicin conjugate (**71**) was evaluated in phase I clinical trials [191].

Another design of PSA substrate peptide conjugates of anticancer drugs is to incorporate cleavable linkers between PSA substrate peptide and anticancer drugs [193,195,196,202]. As shown in Figure 26a, the parent drug 5-fluorodeoxyuridine

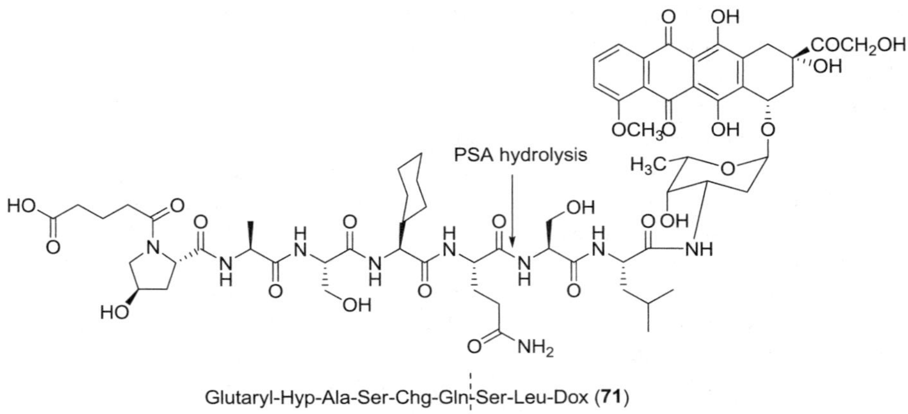

FIGURE 25 Structure of glutaryl–Hyp–Ala–Ser–Chg–Gln–Ser–Leu–Dox, and the arrow indicates the site of PSA hydrolysis.

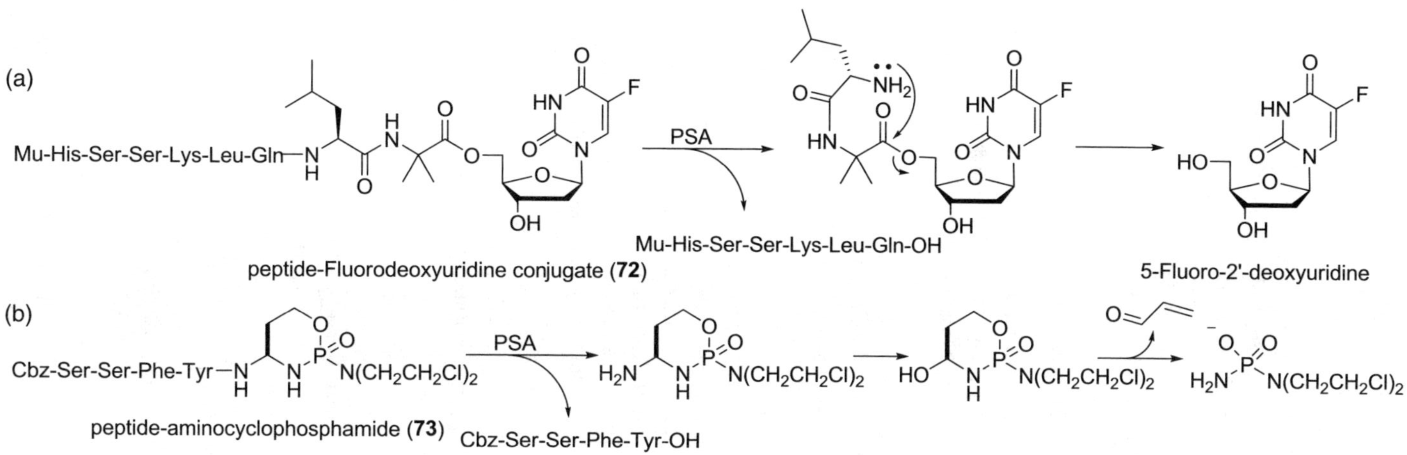

FIGURE 26 Cleavage of peptide–fluorodeoxyuridine (a) and peptide–aminocyclophosphamide conjugates (b) by PSA and the mechanisms of drug release.

(FUDR) could be released through cyclization upon proteolytic cleavage of prodrug (**72**), thus maintaining the potency and physiological property of the parent drug. Spontaneous conversion of 4-aminocyclophosphamide to phosphoramide mustard under physiological conditions was demonstrated and exploited towards the design of PSA-activated prodrugs. A series of 4-aminocyclophosphamide prodrugs (**73**) were synthesized by conjugating 4-aminocyclophosphamides to the tetrapeptide (Cbz–Ser–Ser–Phe–Tyr) sequence, a PSA substrate [193,195,196,202]. The stereoselectivity of PSA-mediated cleavage of the prodrugs was found to favor the *cis*-(2R, 4R)-conjugate, which has a $t_{1/2}$ of 12 min in the presence of PSA (Fig. 26b) [193,195,196,202].

D. Lyases: Cysteine *S*-Conjugate β-Lyases

The only lyases used in prodrug activation are cysteine *S*-conjugate β-lyases (CCBLs) (EC 4.4.1.13). CCBLs use pyridoxal 5′-phosphate as a coenzyme and catalyze β-elimination reactions with L-cysteine *S*-conjugates [$RSCH_2CH(NH_3^+)CO_2^-$] [203]. The reaction generates an RSH fragment and aminoacrylate [$CH_2=CH(NH_3^+)CO_2^-$]. The aminoacrylate is an enamine that rapidly tautomerizes to the more stable ketimine tautomer, α-iminopropionate [$CH_3CH(=NH_2^+)CO_2^-$], followed by hydrolysis to pyruvate [$CH_3C(O)CO_2^-$] and ammonium. Most of the CCBLs discovered thus far are aminotransferases, and an aminotransferase reaction may therefore compete with the β-elimination reaction. Human CCBL activity is present within the microsomes in addition to the cytosol and mitochondria.

CCBLs are believed to be responsible for the activation of cisplatin to a nephrotoxin in proximal tubule cells as shown in Figure 27a. Cisplatin is one of the most effective anticancer agents in the treatment of many solid tumors, including testicular and ovarian cancers. Cisplatin enters cells and binds to DNA, which is toxic to dividing cells but not to quiescent cells such as the proximal tubules in the kidney.

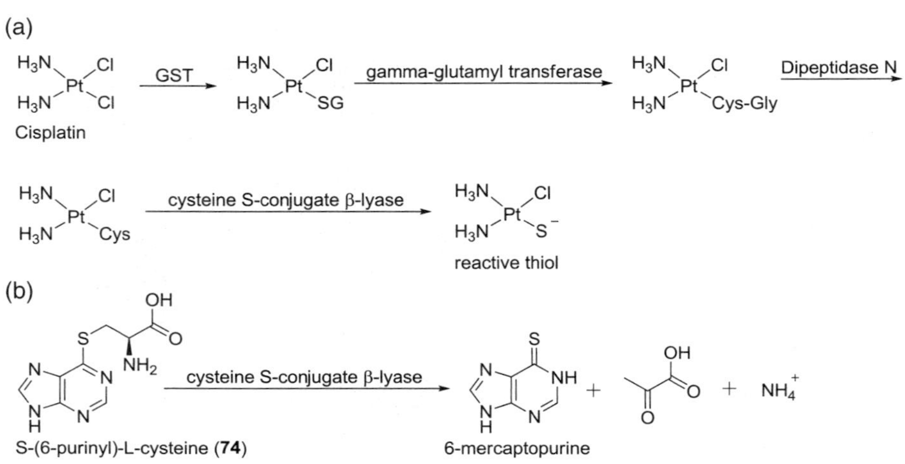

FIGURE 27 (a) The pathway of cisplatin activation proposed based on metabolism of the glutathione conjugates of the halogenated alkenes to nephrotoxins [204] (b) The activation of *S*-(6-purinyl)-L-cysteine to 6-MP by cysteine S-conjugate β-lyase.

Formation of cisplatin–glutathione conjugates reduces the amount of platinum bound to DNA and protects dividing cells from cisplatin toxicity. Then the conjugates are transported out of the cell, followed by cleaving to a platinum–cysteinyl–glycine conjugate by gamma-glutamyl transpeptidase on the cell surface. The platinum conjugate is further cleaved to a platinum–cysteine conjugate by aminodipeptidase N on the cell surface. The platinum–cysteine conjugate is taken up into the cell where it is converted to a highly reactive thiol by CCBL. Binding of the reactive thiol to essential proteins within the cell is toxic [204]. Elucidation of this pathway could help design ways to mitigate cisplatin nephrotoxicity. *S*-(6-Purinyl)-L-cysteine (**74**) was designed as a prodrug of the anticancer agent 6-MP activated by CCBL(s) in the kidney as shown in Figure 27b [205].

III. NONENDOGENOUS ENZYMES

Although utilizing overexpressed endogenous enzymes in tumors to activate prodrugs has been a recurring theme in anticancer drug design, there is evidence of varying enzyme regulation in tumors, and consistent patterns of expression in clinical tumors have usually proved elusive [206]. Recently, ADEPT, GDEPT, and VDEPT have also been proposed as a solution to the problem [207]. In these approaches, an enzyme is delivered site specifically by chemical conjugation or genetic fusion to a tumor-specific antibody or by an enzyme gene delivery system into tumor cells [208]. This is then followed by the administration of a prodrug, which is selectively activated by the delivered enzyme at the tumor cells. There are a wide variety of enzymes (primarily from bacterial sources), such as carboxypeptidase G2 (CPG2), β-glucuronidase, and nitroreductase (NTR), that have been used in these approaches.

These are approaches to deliver cytotoxic agents to tumor cells in selective and specific manners. Differ from ADEPT, which uses a conjugate or a fusion protein to target tumor cells, GDEPT and VDEPT deliver, specifically to the tumor cells, the gene that encodes for a prodrug-activating enzyme and expresses the enzyme intracellularly. This means that prodrugs need to penetrate across tumor cell membranes in the GDEPT and VDEPT approaches, while prodrugs are most likely activated in the extracellular space by enzymes localized on the cell surface in the ADEPT approach.

A. Nitroreductases

NTR are a family of homodimeric flavoenzymes of 24-kDa subunits with tightly bound FMN. They catalyze the reduction of a broad spectrum of nitroaromatic compounds by converting them to their corresponding hydroxylamines with the aid of nicotinamide cofactors such as NADH and NADPH. This catalysis proceeds via a Bi Bi ping-pong mechanism involving two-electron transfers.

NTR have a potential advantage in suicide gene therapy, since metabolism of an aromatic nitro group to the corresponding hydroxylamine results in a very large electronic change, providing an effective "switch" mechanism [209]. The most common and well-characterized NTR used in cancer prodrug design is the oxygen-insensitive form derived from *E. coli* NTR (*nfsB* gene product). The classic example

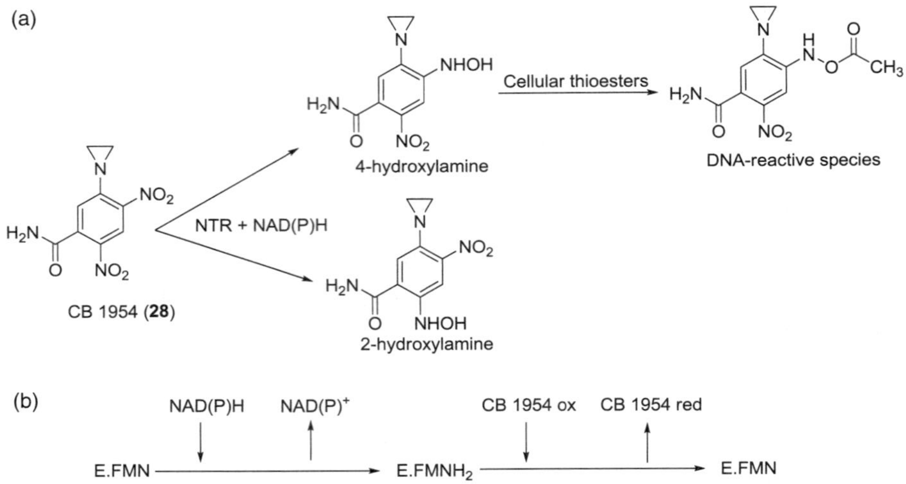

FIGURE 28 (a) Mechanism of CB 1954 reduction by NTR and the subsequent activation of the 4-hydroxylamine metabolite by acetyl CoA and (b) schematic of the Bi Bi ping-pong mechanism of NTR [212].

of NTR-activated prodrug is the 5-aziridinyl-2,4-dinitrobenzamide, CB 1954 (**28**), delivered by GDEPT [210]. Initially, CB 1954 was shown to be highly cytotoxic in cancer lines and efficacious in the Walker 256 rat carcinoma model *in vivo*. In its mode of action, CB 1954 is specifically activated by rat DT-diaphorase which reduces its 4-nitro group to the hydroxylamine (Fig. 28). In addition to the alkylating aziridine functional group, the hydroxylamine metabolite of CB 1954 can be further activated through a thioester to yield a reactive intermediate to alkylate DNA. However, because CB 1954 was found to be a poorer substrate for human DT-diaphorase ($k_{cat} = 0.64\,min^{-1}$) than rat ($k_{cat} = 4.1\,min^{-1}$), and a highly specific substrate for *E. coli* NTR, it is used as a prodrug in combination with *E. coli* NTR in a GDEPT approach and is currently in phase I clinical trials [211].

Since the introduction of the *E. coli* NTR for the delivery of CB 1954 by GDEPT, several NTR family members and mutants with improved kinetics and stability profile have been explored. The NTR YwrO (Bam YwrO) from *Bacillus amyloliquefaciens* has also been proposed for the activation of CB 1954 [213]. It is a 174-amino acid-long 19.7-kDa flavoenzyme expressed from the *ywrO* gene with no sequence similarity to *E. coli* NTR. The catalytic efficiency of YwrO was $0.013\,\mu M^{-1}\,s^{-1}$, which is somewhat comparable to that of *E. coli* NTR ($0.007\,\mu M^{-1}\,s^{-1}$). Another NTR from the thermophilic bacterium *Bacillus licheniformis* has been proposed as an improved substitute for GDEPT-assisted activation of CB 1954 [214]. In contrast to *E. coli* NTR, the *B. licheniformis* enzyme only reduces CB 1954 to the desired 4-hydroxylamine metabolite. Despite its preference for NADPH, the *B. licheniformis* NTR uses pyridinium cofactors and can reduce a broad range of CB 1954 derivatives.

In parallel to enzyme optimization efforts, several CB 1954 derivatives as well as nitrogen mustard prodrugs have been developed. Among the introduced nitrogen

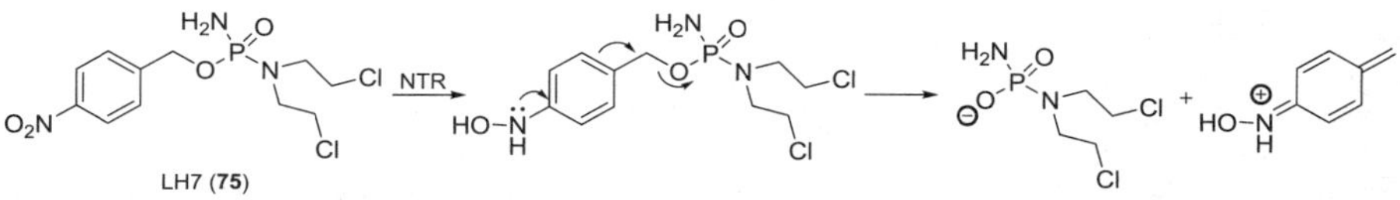

FIGURE 29 Proposed mechanism of activation of prodrug LH7.

mustard prodrugs, the novel 4-nitrobenzyl phosphoramide mustard LH7 (**75**) exhibited improved efficacy and bystander effects in V79 cells and SKOV3 ovarian cells expressing *E. coli* NTR as compared to CB 1954 [215]. Because of its improved catalytic efficiency, the acyclic prodrug LH7 exhibited 167 500-fold selectivity in its cytotoxic activity against NTR-expressing V79 cells and an IC_{50} value of 0.4 nM, which represents a 100- and 10-fold increase in selectivity and potency as compared to CB 1954. Somewhat lower selectivity and potency were observed in SKOV3 cells (IC_{50} value of 1.2 nM, selectivity greater than 833). The activation mechanism of LH7 is illustrated in Figure 29.

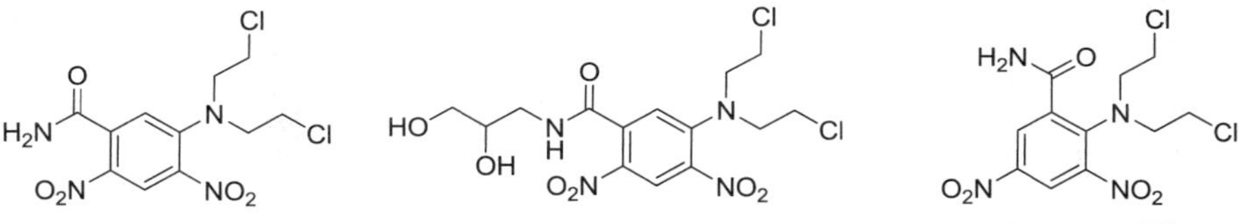

SCHEME 12

SN 23862 (**76**), a dinitrobenzamide mustard, is rapidly activated by NTR with the release of electrons to the mustard providing a potential DNA cross-linking species [23]. The metabolic activation and cytotoxicity of SN 23862 in mammalian cells are inhibited by oxygen much more effectively than is the case with CB 1954 [216]. Therefore, several derivatives of SN 23862 were developed. For example, SN 25261 (**77**), featuring a dihydroxypropylcarboxamide ring substitution, was 10 times more soluble than SN 23862 and retained the kinetics and cytotoxicity profile of SN 23862. Alteration of the carboxamide group position generated SN 25084 (**78**) with improved selectivity and k_{cat}.

Results from the aforementioned studies can be explained by X-ray analyses of reduced NTR in complex with dinitrobenzamide prodrugs [209]. NTR crystals revealed a rigid enzyme structure with the FMN cofactor undergoing most of the changes associated with hydride transfer during catalysis. The binding of CB 1954 and SN 23862 was found to be markedly different. In its binding mode, CB 1954 was found to occupy the two active sites A and B of NTR in nonidentical ways with both nitro groups stacking above the FMN, thereby allowing reduction of both groups, which explains why equal amounts of the 2-hydroxylamine and 4-hydroxylamine metabolites are generated from CB 1954. By contrast, SN 23862U forcefully binds to both active sites with only the 2-nitro group stacked on top of the FMN for ready hydride transfer.

B. CYP450s

Due to the low level of CYP450 in mammalian tumor cells, exogenous CYP450 genes are delivered in GDEPT for anticancer therapy. The flavoprotein NADPH P450 Red interacts with P450 as an electron donor and catalyzes P450 monooxygenase reactions [52]. Therefore, P450/P450 Red combination system is being explored in GDEPT. CYP450-based GDEPT shows specific advantage, that is, the use of human CYP genes involved in prodrug activation enhances the clinical development by avoiding complications associated with immune responses elicited by the expression of CYP genes from heterologous species [217].

Several CYP450-activated prodrugs, including cyclophosphamide, ifosfamide, 4-ipomeanol (**79**), and AQ4N (**80**), are used in GDEPT. Cyclophosphamide and its isomer – ifosfamide – are alkylating prodrugs of the oxazaphosphorine class, and both undergo bioactivation in the liver via the CYP450-catalyzed 4-hydroxylation reaction. 4-Ipomeanol, a naturally occurring furane, is activated by the bovine CYP4B1 to a furane epoxide [218]. AQ4N, an alkylaminoanthraquinone *N*-oxide, is a novel prodrug of a topoisomerase II inhibitor that is reductively metabolized in hypoxic cells to the active agent AQ4 [219].

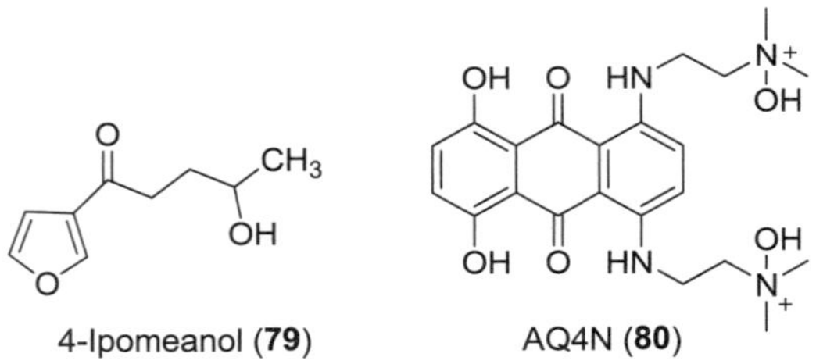

4-Ipomeanol (**79**) AQ4N (**80**)

SCHEME 13

C. Purine Nucleoside Phosphorylase

Purine nucleoside phosphorylase (PNP) (EC 2.4.2.1) catalyzes the cleavage of the glycosidic bond of ribo- and deoxyribonucleosides, in the presence of inorganic orthophosphate (Pi) as a second substrate, to generate the purine base and ribose(deoxyribose)-1-phosphate. For the natural substrates, the reaction is reversible [220]. PNPs are found in a broad range of organisms and divided into two main categories based on size, low molecule mass homotrimers (about 80–100 kDa) and high molecular mass homohexamers (about 110–160 kDa) [220]. Low molecular mass homotrimers, mainly isolated from mammals, are specific for catalysis of 6-oxopurines and their nucleosides, while high molecular mass homohexamers, mainly found in bacteria, accept as substrates both 6-oxo- and/or 6-aminopurines and their nucleosides [220]. In mammals it is located especially in the liver, brain, thyroid, kidney, and spleen. The highest concentrations of the enzyme are found in the kidney, peripheral lymphocytes, and granulocytes [31].

Differences in specificity between PNPs from *E. coli* and humans have been investigated mainly for the development of GDEPT. From the structure of *E. coli* PNP [221], the PNP hexamer is a disc 60 Å thick and 100 Å in diameter, with D3 symmetry

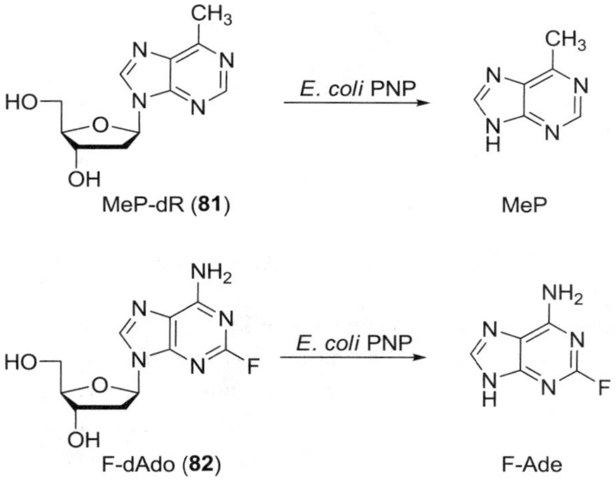

FIGURE 30 The bioconversion of MeP-dR and F-dAdo by *E. coli* PNP in GDEPT.

and a noncrystallographic threefold axis in the crystal form utilized for these studies. The monomers alternate in an up/down fashion around the disc, with three of the active sites near the top and three near the bottom. Each of the six active sites utilizes residues from a pair of twofold related monomers. Each monomer consists of a mixed β-sheet core flanked by eight α-helices. This enzyme's active site consists of Ala^{156}, Phe^{159}, Val^{178}, Met^{180}, Ile^{206}, and Asp^{204}. Among them, the first four of these residues form a hydrophobic pocket around the purine base; Phe^{159} is located between the purine base and the hydrophobic face of the sugar and makes an angle of −60 with the plane of the purine ring; Met^{180} also lies between the purine base and the hydrophobic face of the ribosyl group; and protonated Asp^{204} stabilizes the transition state, in which electron density from the weakening bond is transferred to the purine ring [221].

The selective cleavage of purine nucleoside analogs by *E. coli* PNP is being evaluated as a possible suicide gene therapy strategy for the treatment of solid tumors. Excellent antitumor activity has been demonstrated with 9-(2-deoxy-β-D-ribofuranosyl)-6- methylpurine (MeP-dR) against tumors that express *E. coli* PNP [222]. The active metabolite 6-methylpurine (MeP) can readily diffuse across cell membranes and kill adjacent tumor cells that do not express *E. coli* PNP (Fig. 30) [222]. 2-Fluoro-2′-deoxyadenosine (F-dAdo) is another nucleoside analog that is an excellent substrate for *E. coli* PNP [223]. It is better tolerated by mice than MeP-dR, and the active form is F-Ade, which is active against human tumor cells that do not express *E. coli* PNP due to its conversion to toxic metabolites and the subsequent inhibition of DNA synthesis [223].

D. Thymidine Kinase

TK (EC 2.7.1.21) is a deoxyribonucleoside kinase involved in the pyrimidine salvage pathway by catalyzing the phosphorylation of 2′-deoxythymidine to thymidine monophosphate. Four types of deoxyribonucleoside kinases exist in mammals; they are cytosolic deoxycytidine kinase and thymidine kinase (TK1) and the mitochondrial

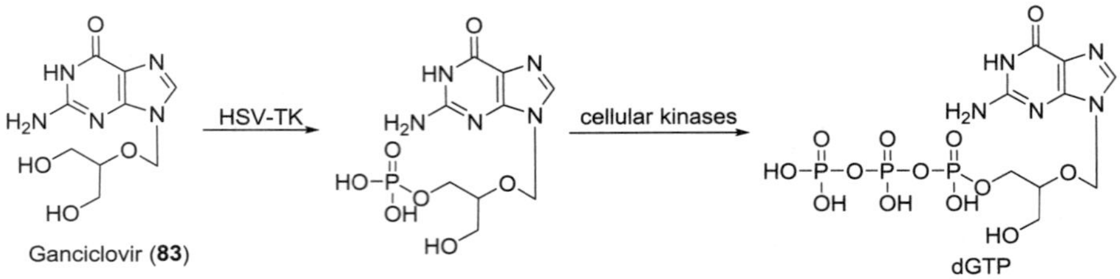

FIGURE 31 The activation of ganciclovir by TK in GDEPT.

thymidine kinase (TK2) and deoxyguanosine kinase. Despite sequence similarities with deoxyguanosine kinases, and with the TK-1 like group of kinases, herpes simplex virus type I thymidine kinase (HSV-TK) is not classified as a TK-1 like enzyme and is structurally different from type II cytosolic human thymidine kinase, as confirmed by crystal structures of the latter enzyme with its feedback inhibitor 2′-deoxythymine-5′-triphosphate.

TK enzyme from the HSV-TK in conjunction with the antiviral drug ganciclovir was one of the first GDEPT protocols developed and has been tested in numerous clinical trials, primarily in gliomas by intratumoral injection [206]. Due to its unique ability to phosphorylate nucleoside analog prodrugs such as ganciclovir (**83**), HSV-TK has been widely used as a suicide gene therapy enzyme. Once delivered to tumor cells, HSV-TK phosphorylates ganciclovir to its monophosphate derivative. Further phosphorylation by other cellular kinases results in the conversion of ganciclovir monophosphate to ganciclovir triphosphate which ultimately interferes with DNA elongation by intercalating between DNA strands (Fig. 31) [224–228].

E. Alkaline Phosphatase

Due to the wide distribution of AP, they cannot be used to locally activate prodrugs, except by using ADEPT, GDEPT, and VDEPT strategies [31]. Phosphorylated prodrug of etoposide (**84**) was evaluated in a ADEPT strategy, and a high degree of *in vivo* antitumor activity was obtained through the combination of MAb–AP conjugate and prodrug [229].

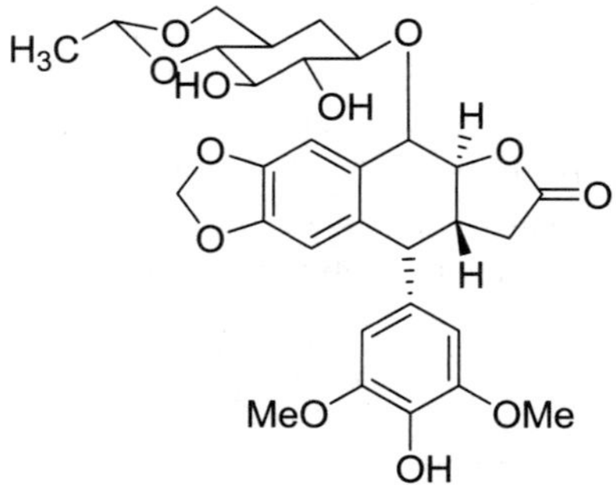

Etoposide (**84**)

SCHEME 14

F. β-Glucuronidase

Since endogenous extracellular β-glucuronidase is present at high levels in necrotic tumors, glucuronide prodrugs could potentially be used as a monotherapy. Moreover, in the case of non-necrotic areas of tumors where the enzyme concentration is low, exogenous β-glucuronidases can be delivered to these tumor tissues using ADEPT and GDEPT strategies. There are several advantages in these new strategies, for example, plenty of glucuronide prodrugs (doxorubicin, analog of CPT, paclitaxel, and so on) are available; glucuronide prodrugs cannot readily enter cells, which minimizes the interactions with endogenous β-glucuronidase [230]. In ADEPT strategy, introduction of exogenous human β-glucuronidase via a fusion protein consisting of an anti-CEA antibody linked to a β-glucuronidase has already been assessed [231]. The bacterial enzyme β-glucuronidase, encoded by the *E. coli* gusA gene, has also been used.

Many glucuronide antitumor prodrugs have been developed for application in ADEPT [232]. An example is the docetaxel prodrug **85** in which a glucuronic acid moiety is linked to a double spacer containing para-hydroxybenzyl alcohol connected to a diamine tether via a carbamate linkage [231,233]. This system was shown to be efficient in releasing drugs in the presence of β-glucuronidase (Fig. 32a). 9-Aminocamptothecin prodrug (**86**) is much more water soluble at neutral and acidic pH than the parent drug (80 times at pH 4 and 1800 times at pH 7), the potassium salt

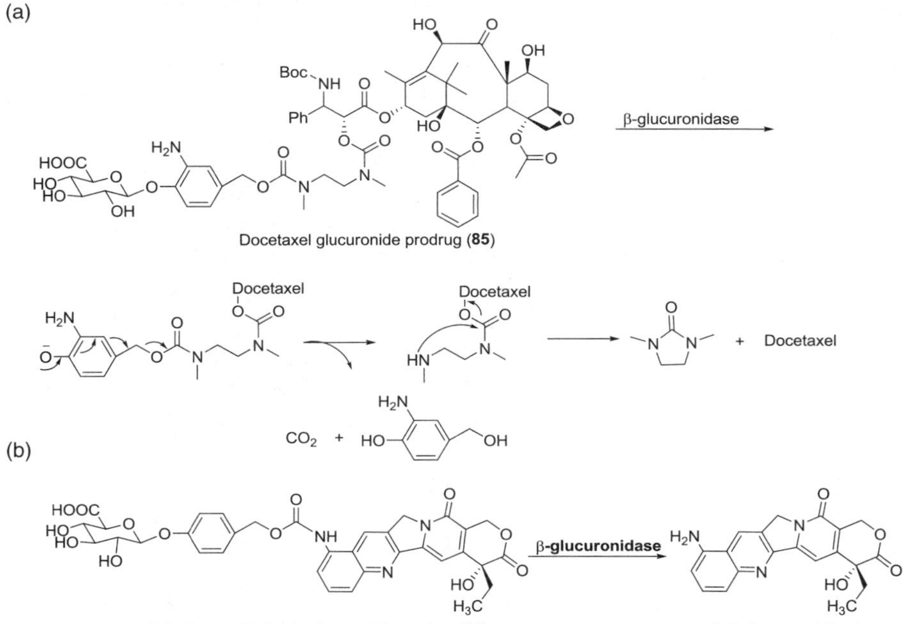

FIGURE 32 The activation of glucuronide prodrugs of docetaxel (a) and 9-aminocamptothecin (b) by β-glucuronidase in ADEPT [233].

displaying excellent solubility even at pH4. The prodrug was cleaved by β-glucuronidase at pH 7 to afford the parent drug (Fig. 32b) [234].

G. Carboxypeptidase G2

Glucarpidase under development as a chemoprotective agent to neutralize the toxic effects of MTX is a recombinant bacterial CPG2 (EC 3.4.17.11). This CPG2 from *Pseudomonas* RS-16 is a 42-kDa exopeptidase in the aminoacylase-1/M20 family of enzymes that is not found in mammals. The active conformation of CPG2 requires homodimerization and the presence of two atoms of zinc per subunit. Each subunit of the active enzyme contains a zinc-binding domain and a dimerization domain, both of which participate in catalysis [235]. The natural substrate of CPG2 is folic acid, but it is capable to hydrolyze glutamate from amidic [236,237], urethanic, and ureidic bonds [238,239], which allows it to be used to activate synthetic prodrugs in GDEPT and ADEPT as represented by the reaction shown in Figure 33.

A series of CPG2-activated nitrogen mustard prodrugs has been developed with various cytotoxic nitrogen mustards linked to glutamate directly [236,237, 240–242] or through a self-immolative linker [243,244]. CPG2 catalyzes the hydrolysis of these prodrugs, releasing glutamic acid and the parent cytotoxins including benzoic acid nitrogen mustards (BANM), aniline nitrogen mustards (ANM), and phenol nitrogen mustards (PNM). 4-[(2-Mesyloxylethyl)(2-chloroethyl)amino]benzoyl-L-glutamic acid (CMDA, **87**) was the first ADEPT prodrug evaluated clinically [245]. A phase I ADEPT clinical trial of CMDA showed promising results, with one patient showing a partial response and six with stable disease for 4 months. However, the activated drug, 4-[(2-mesyloxylethyl)-(2-chloroethyl)amino]benzoic acid, had a long half-life that resulted in dose-limiting myelosuppression due to diffusion of the activated drug back into the circulation.

The pyrrolo[2,1-c][1,4]benzodiazepines (PBDs) are a family of antitumor antibiotics produced by *Streptomyces* species [246]. They exert their cytotoxicity by covalently bonding to the exocyclic C2–NH_2 group of guanine residues in the minor groove of DNA through their N10–C11 imine functionality. The more stable carbamate prodrugs are good substrates for CPG2. The prodrug (**88**) was completely converted into the cytotoxic parent PBD within 50 min of incubation with the enzyme *in vitro*.

MTX conjugates with amino acids or peptides covalently linked to the α-carboxyl group of MTX are prodrug forms of the parent antifolate [247–249]. When tested

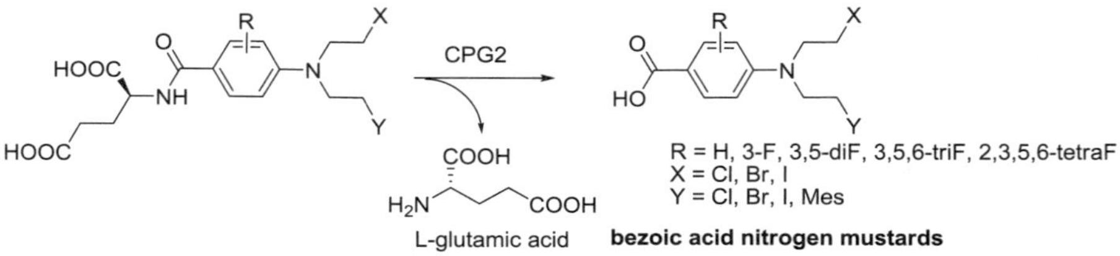

FIGURE 33 The mechanism of mustard prodrugs activated by CPG2.

against UCLA-P3 human lung carcinoma cells *in vitro*, the IC_{50} values for MTX–amino acid conjugates in the absence of CPG2 was in the micromolar range (about 7.0 μM). This value was much higher than the IC_{50} for MTX, which was in nanomolar range (28 nM). When incubated with UCLA-P3 human lung adenocarcinoma cells treated with KS1/4–CPG2 conjugate (KS1/4 is a carcinoma-specific monoclonal antibody), MTX–Phe (**89**) produced an IC_{50} of 63 nM. KS1/4–CPG2 conjugate treatment alone produced no cytotoxicity, and, in the absence of KS1/4–CPG2 conjugate, cells were much less susceptible to MTX–Phe (IC_{50} = 2.2 μM).

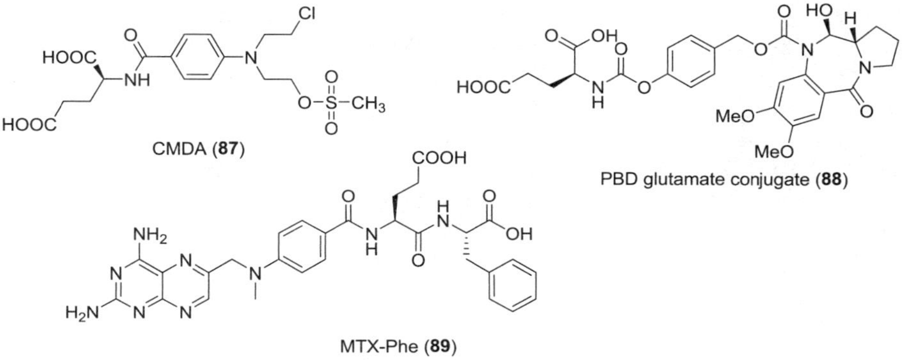

SCHEME 15

H. Penicillin Amidase

Penicillin amidase (PA), also known as penicillin acylase, is widely expressed in various microorganisms including yeast, fungi, and bacteria. It is a member of the N-terminal-nucleophile hydrolase family, a group of enzymes sharing a characteristic fold proximal to their active sites as well as a nucleophilic residue at their N-terminus. The PA family members hydrolyze a series of penicillins and are subdivided into three subtypes based on their substrate specificity, namely, penicillin V amidases (PVA) (type I; substrate: penicillin V, phenoxymethylpenicillin), penicillin G amidase (PGA) (type II; substrate: penicillin G, benzylpenicillin), and ampicillin amidases (type III; substrate: ampicillin). The *E. coli* isozyme (EC 3.5.1.11) is a well-characterized, inexpensive, and commercially available type II amidase commonly used for the production of semisynthetic penicillins. It is heterodimeric and catalyzes the hydrolysis of penicillin G to phenylacetic acid and 6-aminopenicillanic acid (6-APA), which is the building block for the natural penicillins (Fig. 34). The catalytic mechanism of PGA has been shown from crystal structures of PGA–penicillin sulfoxide complex to proceed via nucleophilic attack of active site Ser B1 on the carbonyl of the scissile amide bond; in addition, Asn B241 plays a key role in stabilizing optimal active site geometry [250–252].

PGA is known to have broad substrate specificity; several targeted prodrugs activated by PGA have been reported. The potent cytotoxin palytoxin exhibits activity against a wide spectrum of cell types including fibroblasts, lymphocytes,

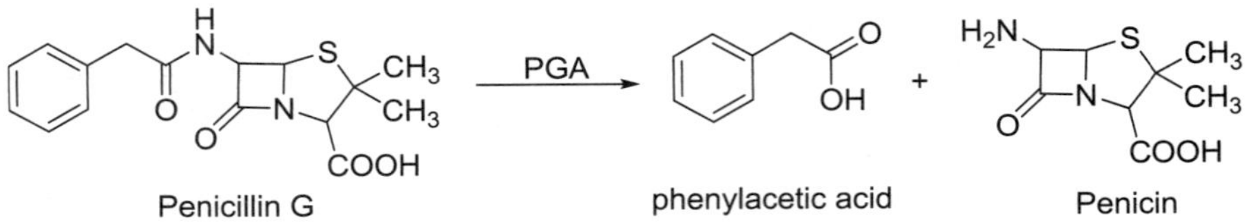

FIGURE 34 Catalysis of penicillin G hydrolysis by PGA.

erythrocytes, and epithelial cells [253]. It is cytotoxic against cells expressing the Na^+-K^+-ATPase-associated toxin receptor. The *N*-acyl palytoxin derivative, *N*-(4′-hydroxyphenylacetyl)palytoxin (NHPAP, **90**) was developed as a targeted prodrug for PGA activation via ADEPT. PGA was covalently linked to the monoclonal antibody L6 mAb. Compared to palytoxin, the prodrug was 1000 times less toxic to carcinoma and lymphoma cells lacking PGA expression. NHPAP was efficiently converted to palytoxin in H2981 cells expressing the L6 antigen and exhibited comparable cytotoxic activity to the free drug (IC_{50} values were 0.06 and 0.01 nM for NHPAP and palytoxin, respectively).

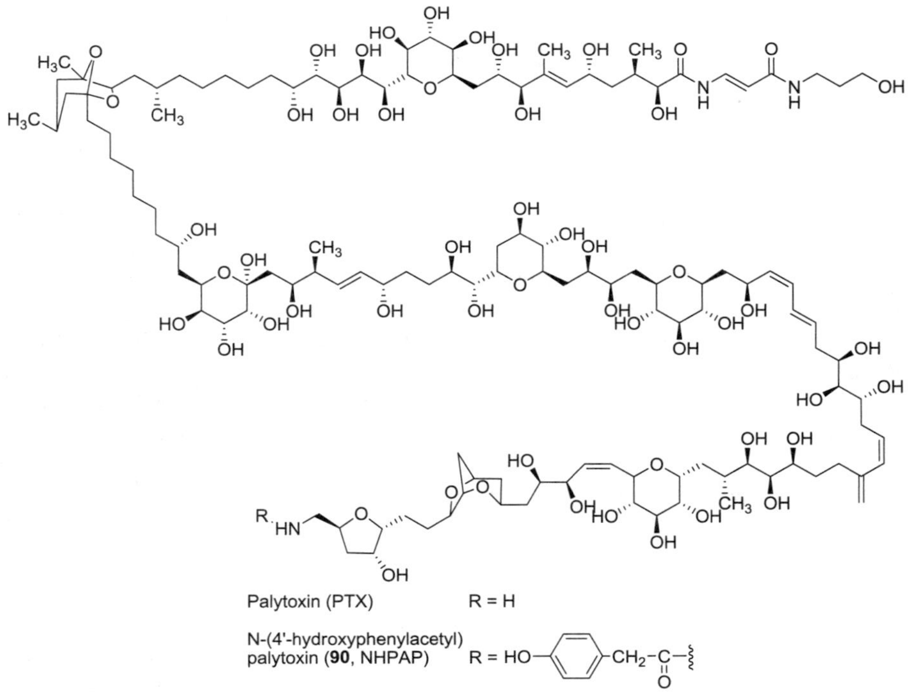

SCHEME 16

The L6–PGA conjugate has also been used in the ADEPT-assisted activation of doxorubicin and melphalan [254]. The prodrugs *N*-(phenylacetamide)doxorubicin (**91**) and *N*-(phenylacetyl)melphalan (**92**) were found to be 10- and 20-fold less cytotoxic than free drugs in H2981 cancer lines lacking PGA. Co-incubation with PGA increased the cytotoxicity of both prodrugs due to PGA activation. For example, the IC_{50} values of prodrug (**91**) were 2 μM in the absence of PGA and 0.2 μM in the presence of PGA, whereas the IC_{50} value of doxorubicin was 0.2 μM in the same assay.

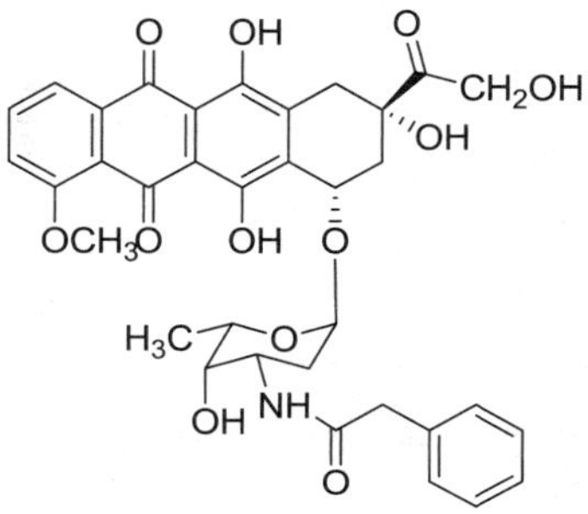

N-(phenylactemido)doxorubicin (**91**)

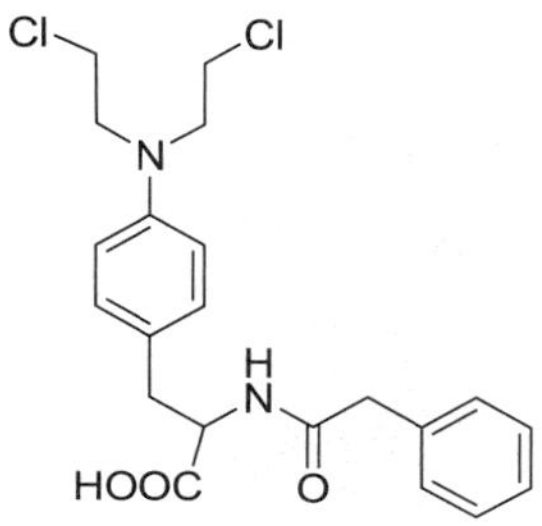

N-(phenylactemido)melphalan (**92**)

SCHEME 17

I. β-Lactamase

β-Lactamase (EC 3.5.2.6) is a serine protease produced by various bacteria; it catalyzes the hydrolysis of the β-lactam moiety in penicillin and other similar β-lactam antibiotics to β-amino acid. Based on primary sequence and metal ion requirement, β-lactamases have been grouped into four classes. Classes A, C, and D enzymes utilize a serine residue in the active site, whereas class B enzymes require an additional catalytic zinc atom for activity. All of these enzymes contain the active site serine residue as well as a conserved triad of KS(T)G between the active site serine and the C-terminus. The class A β-lactamases are largely plasmid encoded and are widely distributed in both Gram-positive and Gram-negative bacteria. β-Lactamases produced by Gram-positive organisms are usually secreted enzymes. β-Lactamase may be clinically beneficial when orally administered to preserve the natural intestinal flora during the parenteral administration of antibiotics.

The advantages of using β-lactamase in prodrug activation in ADEPT are that the enzyme is not present in human cells and that it is highly selective to the β-lactam-containing compounds. Cephalosporin, with the ability to release drug from C3′ position upon cleavage of the β-lactam ring by β-lactamase, has been widely utilized as the core of β-lactamase-activated prodrugs.

As shown in Figure 35, opening of the β-lactam ring in cephalosporin is initiated by the nucleophilic attack from the hydroxyl group of the Ser^{70} side chain at the carbonyl carbon (C=O) of the β-lactam ring. The drug attached to the C3′ position represented as a leaving group (LG) is released by the movement of electrons towards the thiazine ring of the cephem core. It has been noted that the propensity for the drug release depends on the leaving group ability of the drug molecule and the stability of the linkage to the cephalosporin.

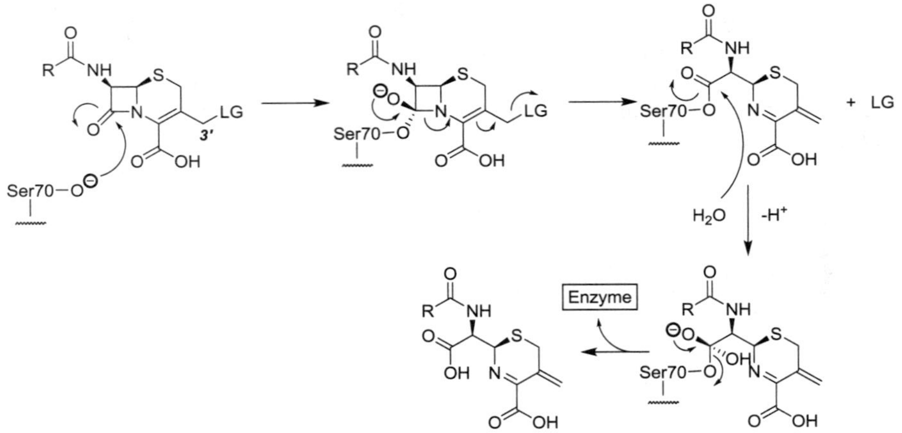

FIGURE 35 The mechanism of activation of cephalosporin prodrugs by β-lactamases [255].

Vinca alkaloid derivative (LY233425, **93**) was also conjugated to C3′ position of cephalosporin through an alkyl thioether, which would leave as a thiol upon hydrolysis of the β-lactam ring [256]. This vinca alkaloid prodrug was shown to be fivefold less toxic than the free drug, but it was equally potent as the parent drug in the presence of β-lactamase.

A carbamate linker was used a cephalosporin–taxol prodrug, Protax (**94**) [257]. Because the C2′ hydroxyl group of taxol did not readily react with the *p*-nitrophenyl carbonate of cephalosporin activated at C3′, a GABA linker was used to conjugate the C2′ hydroxyl group of taxol with cephalosporin. Since the C2′ hydroxyl group is required for biological activity, the prodrug will only be activated after the C2′ linker is removed.

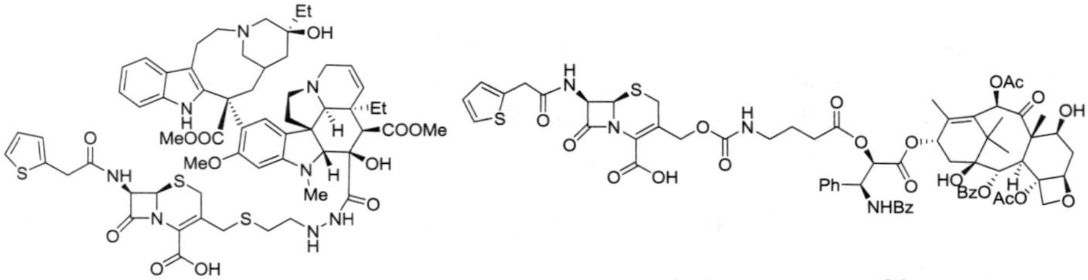

LY233425 cephalosporin prodrug (**93**)

Cephalosporin taxol (**94**, Protax)

SCHEME 18

J. Cytosine Deaminase

Cytosine deaminase (CD, EC 3.5.4.1) catalyzes the deamination of cytosine to uracil and 5-methylcytosine to thymine. The enzyme has been found in bacteria and fungi, where it plays an important role in pyrimidine salvage [258]. However, no CD is present in mammalian cells, which utilize the cytidine deaminase instead.

The protein structure is composed of a central five-stranded β-sheet (β1–β5) with the strand order 2, 1, 3, 4, and 5. The β-sheet is sandwiched by two α-helices (αA and αE) on one side and by three α-helices (αB–αD) on the other side. This enzyme forms a tightly packed dimer in the crystallographic asymmetric unit [258]. The active site contains one tightly bound zinc ion, which is tetrahedrally coordinated to His^{62} (2.06 Å), Cys^{91} (2.31 Å), Cys^{94} (2.25 Å), and the hydroxyl group of the bound inhibitor (2.08 Å). The protein ligands are located on the N termini of the αB and αC helices, and the helix dipoles may partially neutralize the negative charges of the cysteinate ligands to lower the p*Ka* value of the nucleophilic water molecule. Comparison of the active sites of CD with other cytidine and deoxycytidylate deaminase family reveals a very similar interaction network between the attacking water molecule, the zinc ion, the zinc ligands, and the proton shuttle, suggesting a similar zinc-assisted deamination mechanism in the family [258].

The yeast enzyme CD has also been widely studied for GDEPT in conjunction with the antifungal agent 5-fluorocytosine (**95**), which is converted by CD to the toxic TS inhibitor 5-FU as shown in Figure 36. This freely diffusible effector, already a drug of choice for treating colon cancer, has much better bystander effects. GDEPT model studies with CD/5-fluorocytosine have focused mainly on colon cancer models. Possible drawbacks of this protocol include the relatively low potency of the effector, coupled with its pronounced cell cycle selectivity [206,259].

K. Methionine γ-Lyase

Methionine γ-lyase (MGL, EC 4.4.1.11) is an enzyme that catalyzes the γ-elimination reaction of L-methionine to produce α-ketobutyric acid, methanethiol, and ammonia. In addition to its physiological reaction, the enzyme also catalyzes γ-displacement reactions of L-methionine and its derivatives as well as β-elimination and β-displacement reactions of L-cysteine and S-substituted L-cysteines [260].

MGL exists as a tetramer in solution, and the tetramer consists of catalytic dimers related by a twofold axis. The catalytic dimer consists of two subunits related by a twofold axis and includes two active sites. Each active site is formed by amino acid residues from both subunits and contains PLP. Each monomer consists of three structurally and functionally distinct regions called the N-terminal (AA1–62), PLP-binding (AA63–259), and C-terminal domains (AA260–398) [260,261].

Selenomethionine (**96**) is an amino acid containing selenium. The L-isomer of selenomethionine, known as Se-MET, is a common natural food source of selenium.

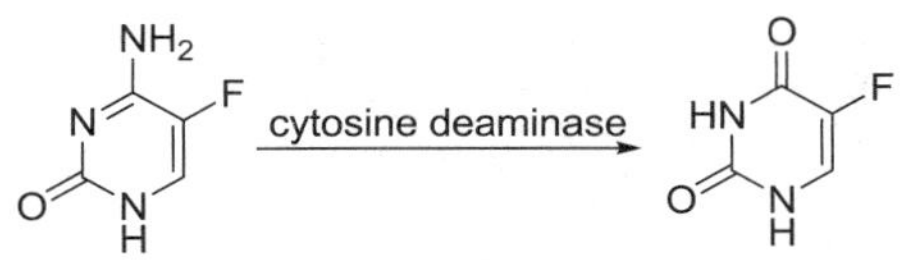

5-Fluorocytosine (**95**) 5-Fluorouracil

FIGURE 36 The activation of 5-fluorocytosine by CD in GEDPT.

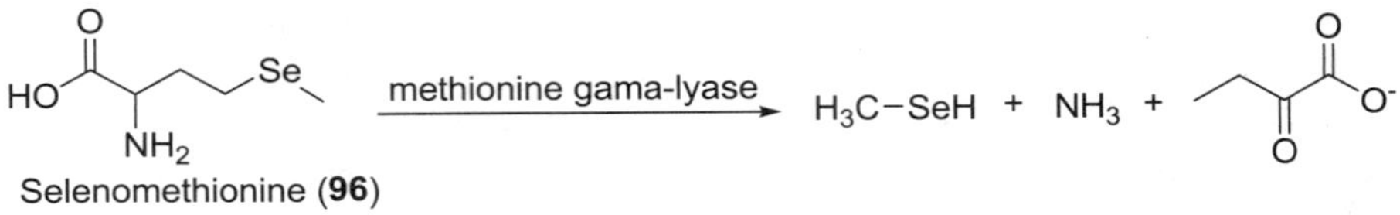

FIGURE 37 Activation of selenomethionine by methionine γ-lyase in GDEPT.

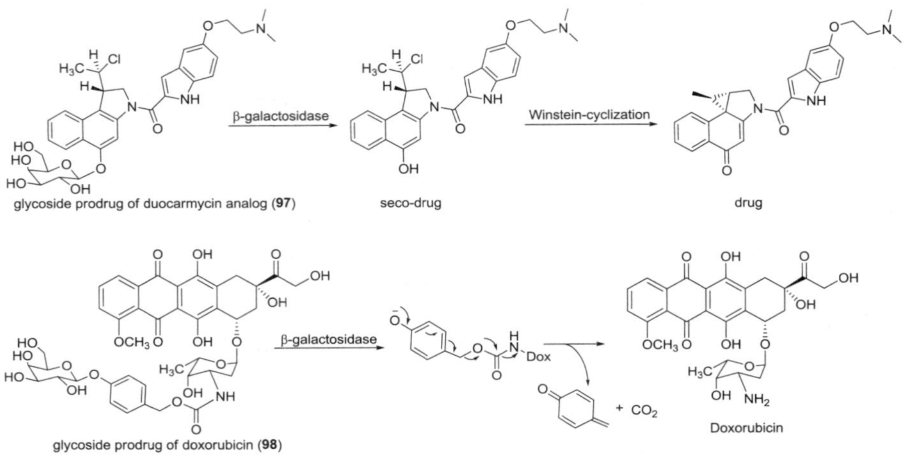

FIGURE 38 Activation of glycoside prodrugs by β-galactosidase.

As shown in Figure 37, Se-MET can be cleaved by the adenoviral-delivered MGL to methylselenol, which cause the release of cytochrome C into the cytosol and subsequent apoptosis in cancer cells [262,263].

L. β-Galactosidase

E. coli β-galactosidase (EC 3.2.1.23), encoded by lacZ, has been used in ADEPT. The enzyme is a 464-kDa homotetramer and each subunit is composed of five domains. The enzyme is often used for histological localization with 5-bromo-4-chloro-3-indolyl-beta-D-galactopyranoside (X-Gal) as the substrate.

A glycoside prodrug of duocarmycin analog (**97**) consists of the (1*S*,10*R*)-methyl-*seco*-CBI-skeleton as effector, a dimethylaminoethoxyindole (DMAI) side chain for binding to the minor groove of DNA and β-D-galactose as a promoiety [264]. This prodrug has a rather low cytotoxicity with an IC_{50} of 3600 nM but a very high cytotoxicity upon activation by β-galactosidase to give a highly cytotoxic duocarmycin analog (Fig. 38) [265]. Another example is the glycoside prodrug (**98**) of doxorubicin which was found to be stable in buffer and could be readily converted to doxorubicin by β-galactosidase [266].

IV. CONCLUDING REMARKS

Prodrugs have been very effective in solving problems of the parent molecules to provide the desired physicochemical and pharmaceutical properties for improved oral bioavailability, patient acceptability, pharmacokinetic profiles, and target specificity. There are many examples of prodrugs that are used clinically. Enzymes play an important role in the activation of many prodrugs although some prodrugs can be activated solely through chemical reactions without involving enzymes.

With the discovery of more compounds with desired activity but suboptimal pharmaceutical properties, prodrugs have become an increasingly important approach that can be used to solve some of these problems. Knowing the various enzymes that could potentially be used to activate potential prodrugs will definitely help with the design process.

Most often endogenous enzymes can be effectively utilized for the activation of prodrugs. With the recent shift of focus to deliver active drug molecules to a specific target tissue or organ, nonendogenous enzymes have been employed for site-specific activation of prodrugs. These novel approaches include ADEPT, GDEPT, and VDEPT strategies for the treatment of cancer. Enzymes used in these strategies often come from bacteria or fungi such as *E. coli* NTR and HSV-TK. These approaches are very promising but are also very challenging. There are still many problems to solve. With the advance of new technologies, it is very likely we will see some of these approaches become effective tools against various diseases.

REFERENCES

1. J Rautio, H Kumpulainen, T Heimbach, R Oliyai, D Oh, T Jarvinen, J Savolainen. Prodrugs: design and clinical applications. Nat Rev Drug Discov 7:255–270, 2008.
2. K McClellan, CM Perry. Oseltamivir: a review of its use in influenza. Drugs 61:263–283, 2001.
3. A Bardsley-Elliot, S Noble. Oseltamivir. Drugs 58:851–860; discussion 861–862, 1999.
4. PA Todd, RC Heel. Enalapril. A review of its pharmacodynamic and pharmacokinetic properties, and therapeutic use in hypertension and congestive heart failure. Drugs 31:198–248, 1986.
5. W Daehne, E Frederiksen, E Gundersen, F Lund, P Morch, HJ Petersen, K Roholt, L Tybring, WO Godtfredsen. Acyloxymethyl esters of ampicillin. J Med Chem 13:607–612, 1970.
6. H Chanteux, F Van Bambeke, MP Mingeot-Leclercq, PM Tulkens. Accumulation and oriented transport of ampicillin in Caco-2 cells from its pivaloyloxymethylester prodrug, pivampicillin. Antimicrob Agents Chemother 49:1279–1288, 2005.
7. J Shibata, T Toko, H Saito, AE Lykkesfeldt, A Fujioka, K Sato, A Hashimoto, K Wierzba, Y Yamada. Estrogen agonistic/antagonistic effects of miproxifene phosphate (TAT-59). Cancer Chemother Pharmacol 45:133–141, 2000.
8. F Kratz, IA Muller, C Ryppa, A Warnecke. Prodrug strategies in anticancer chemotherapy. ChemMedChem 3:20–53, 2008.

9. C Litalien, Y Theoret, C Faure. Pharmacokinetics of proton pump inhibitors in children. Clin Pharmacokinet 44:441–466, 2005.
10. J-C Florent, C Monneret. Doxorubicin conjugates for selective delivery to tumors. In: K Krohn, ed. *Anthracycline Chemistry and Biology II: Mode of Action, Clinical Aspects and New Drugs, Topics in Current Chemistry*. Berlin/Heidelberg: Springer, 2008, pp. 99–140.
11. SJ Ho, TA Brighton. Ximelagatran: direct thrombin inhibitor. Vasc Health Risk Manag 2:49–58, 2006.
12. F Etienne, L Resnick, D Sagher, N Brot, H Weissbach. Reduction of Sulindac to its active metabolite, sulindac sulfide: assay and role of the methionine sulfoxide reductase system. Biochem Biophys Res Commun 312:1005–1010, 2003.
13. H Yuan, N Li, Y Lai. Evaluation of *in vitro* models for screening alkaline phosphatase mediated bioconversion of phosphate ester prodrugs. Drug Metab Dispos 37:1443–1447, 2009.
14. H Kumpulainen, T Jarvinen, A Mannila, J Leppanen, T Nevalainen, A Mantyla, J Vepsalainen, J Rautio. Synthesis, *in vitro* and *in vivo* characterization of novel ethyl dioxy phosphate prodrug of propofol. Eur J Pharm Sci 34:110–117, 2008.
15. Y Wei, Y Yan, D Pei, B Gong. A photoactivated prodrug. Bioorg Med Chem Lett 8:2419–2422, 1998.
16. Z Zhang, H Hatta, T Ito, S Nishimoto. Synthesis and photochemical properties of photo-activated antitumor prodrugs releasing 5-fluorouracil. Org Biomol Chem 3:592–596, 2005.
17. GO Ahn, KJ Botting, AV Patterson, DC Ware, M Tercel, WR Wilson. Radiolytic and cellular reduction of a novel hypoxia-activated cobalt(III) prodrug of a chloromethylbenzindoline DNA minor groove alkylator. Biochem Pharmacol 71:1683–1694, 2006.
18. T Ito, K Tanabe, H Yamada, H Hatta, S Nishimoto. Radiation- and photo-induced activation of 5-fluorouracil prodrugs as a strategy for the selective treatment of solid tumors. Molecules 13:2370–2384, 2008.
19. K Tanabe, Y Mimasu, A Eto, Y Tachi, S Sakakibara, M Mori, H Hatta, S Nishimoto. One-electron reduction characteristics of N(3)-substituted 5-fluorodeoxyuridines synthesized as radiation-activated prodrugs. Bioorg Med Chem 11:4551–4556, 2003.
20. M Jung. Antibody directed enzyme prodrug therapy (ADEPT) and related approaches for anticancer therapy. Mini Rev Med Chem 1:399–407, 2001.
21. CJ Springer, II Niculescu-Duvaz. Antibody-directed enzyme prodrug therapy (ADEPT): a review. Adv Drug Deliv Rev 26:151–172, 1997.
22. CJ Springer, I Niculescu-Duvaz. Gene-directed enzyme prodrug therapy (GDEPT): choice of prodrugs. Adv Drug Deliv Rev 22:351–364, 1996.
23. WA Denny. Nitroreductase-based GDEPT. Curr Pharm Des 8:1349–1361, 2002.
24. A Nawa, T Tanino, C Luo, M Iwaki, H Kajiyama, K Shibata, E Yamamoto, K Ino, Y Nishiyama, F Kikkawa. Gene directed enzyme prodrug therapy for ovarian cancer: could GDEPT become a promising treatment against ovarian cancer? Anticancer Agents Med Chem 8:232–239, 2008.
25. RJ Knox, PJ Burke, S Chen, DJ Kerr. CB 1954: from the Walker tumor to NQO2 and VDEPT. Curr Pharm Des 9:2091–2104, 2003.
26. R Duncan, S Gac-Breton, R Keane, R Musila, YN Sat, R Satchi, F Searle. Polymer-drug conjugates, PDEPT and PELT: basic principles for design and transfer from the laboratory to clinic. J Control Release 74:135–146, 2001.

27. R Satchi-Fainaro, H Hailu, JW Davies, C Summerford, R Duncan. PDEPT: polymer-directed enzyme prodrug therapy. 2. HPMA copolymer-beta-lactamase and HPMA copolymer-C-Dox as a model combination. Bioconjug Chem 14:797–804, 2003.
28. R Satchi, TA Connors, R Duncan. PDEPT: polymer-directed enzyme prodrug therapy. I. HPMA copolymer-cathepsin B and PK1 as a model combination. Br J Cancer 85:1070–1076, 2001.
29. E Garattini, M Fratelli, M Terao. Mammalian aldehyde oxidases: genetics, evolution and biochemistry. Cell Mol Life Sci 65:1019–1048, 2008.
30. E Garattini, R Mendel, MJ Romao, R Wright, M Terao. Mammalian molybdo-flavoenzymes, an expanding family of proteins: structure, genetics, regulation, function and pathophysiology. Biochem J 372:15–32, 2003.
31. M Rooseboom, JN Commandeur, NP Vermeulen. Enzyme-catalyzed activation of anticancer prodrugs. Pharmacol Rev 56:53–102, 2004.
32. DJ Porter, JA Harrington, MR Almond, GT Lowen, TP Zimmerman, T Spector. 5-ethynyl-2(1H)-pyrimidinone: aldehyde oxidase-activation to 5-ethynyluracil, a mechanism-based inactivator of dihydropyrimidine dehydrogenase. Biochem Pharmacol 47:1165–1171, 1994.
33. X Guo, M Lerner-Tung, HX Chen, CN Chang, JL Zhu, CP Chang, G Pizzorno, TS Lin, YC Cheng. 5-Fluoro-2-pyrimidinone, a liver aldehyde oxidase-activated prodrug of 5-fluorouracil. Biochem Pharmacol 49:1111–1116, 1995.
34. T Kawazoe, H Tsuge, MS Pilone, K Fukui. Crystal structure of human D-amino acid oxidase: context-dependent variability of the backbone conformation of the VAAGL hydrophobic stretch located at the si-face of the flavin ring. Protein Sci 15:2708–2717, 2006.
35. L Pollegioni, G Molla, S Sacchi, E Rosini, R Verga, MS Pilone. Properties and applications of microbial D-amino acid oxidases: current state and perspectives. Appl Microbiol Biotechnol 78:1–16, 2008.
36. T Kawazoe, HK Park, S Iwana, H Tsuge, K Fukui. Human D-amino acid oxidase: an update and review. Chem Rec 7:305–315, 2007.
37. L Pollegioni, L Piubelli, S Sacchi, MS Pilone, G Molla. Physiological functions of D-amino acid oxidases: from yeast to humans. Cell Mol Life Sci 64:1373–1394, 2007.
38. VI Tishkov, SV Khoronenkova. D-Amino acid oxidase: structure, catalytic mechanism, and practical application. Biochemistry (Mosc) 70:40–54, 2005.
39. MS Pilone. D-Amino acid oxidase: new findings. Cell Mol Life Sci 57:1732–1747, 2000.
40. K Momoi, K Fukui, F Watanabe, Y Miyake. Molecular cloning and sequence analysis of cDNA encoding human kidney D-amino acid oxidase. FEBS Lett 238:180–184, 1988.
41. RG Stabeli, CD Sant'Ana, PH Ribeiro, TR Costa, FK Ticli, MG Pires, A Nomizo, S Albuquerque, NR Malta-Neto, M Marins, et al. Cytotoxic L-amino acid oxidase from Bothrops moojeni: biochemical and functional characterization. Int J Biol Macromol 41:132–140, 2007.
42. J Venhorst, M Rooseboom, NP Vermeulen, JN Commandeur. Studies on the inhibition of human cytochromes P450 by selenocysteine Se-conjugates. Xenobiotica 33:57–72, 2003.
43. M Rooseboom, NP Vermeulen, N van Hemert, JN Commandeur. Bioactivation of chemopreventive selenocysteine Se-conjugates and related amino acids by amino acid oxidases novel route of metabolism of selenoamino acids. Chem Res Toxicol 14:996–1005, 2001.
44. LD Stegman, H Zheng, ER Neal, O Ben-Yoseph, L Pollegioni, MS Pilone, BD Ross. Induction of cytotoxic oxidative stress by D-alanine in brain tumor cells expressing Rhodotorula gracilis D-amino acid oxidase: a cancer gene therapy strategy. Hum Gene Ther 9:185–193, 1998.

45. RD Bruno, VC Njar. Targeting cytochrome P450 enzymes: a new approach in anti-cancer drug development. Bioorg Med Chem 15:5047–5060, 2007.

46. D Downie, MCE McFadyen, GI Murray, PH Rooney. Pharmacogenomics of Cytochrome P450 Enzymes in Tumours. Curr Pharmacogenomics 2:243–254, 2004.

47. TD Bradshaw, AD Westwell. The development of the antitumour benzothiazole prodrug, Phortress, as a clinical candidate. Curr Med Chem 11:1009–1021, 2004.

48. L Chen, DJ Waxman. Cytochrome P450 gene-directed enzyme prodrug therapy (GDEPT) for cancer. Curr Pharm Des 8:1405–1416, 2002.

49. JP Armand, V Ribrag, JL Harrousseau, L Abrey. Reappraisal of the use of procarbazine in the treatment of lymphomas and brain tumors. Ther Clin Risk Manag 3:213–224, 2007.

50. L Chen, DJ Waxman. Intratumoral activation and enhanced chemotherapeutic effect of oxazaphosphorines following cytochrome P-450 gene transfer: development of a combined chemotherapy/cancer gene therapy strategy. Cancer Res 55:581–589, 1995.

51. DJ Waxman, L Chen, JE Hecht, Y Jounaidi. Cytochrome P450-based cancer gene therapy: recent advances and future prospects. Drug Metab Rev 31:503–522, 1999.

52. J Lengler, M Omann, D Duvier, H Holzmuller, W Gregor, B Salmons, WH Gunzburg, M Renner. Cytochrome P450 reductase dependent inhibition of cytochrome P450 2B1 activity: implications for gene directed enzyme prodrug therapy. Biochem Pharmacol 72:893–901, 2006.

53. M Wang, DL Roberts, R Paschke, TM Shea, BS Masters, JJ Kim. Three-dimensional structure of NADPH-cytochrome P450 reductase: prototype for FMN- and FAD-containing enzymes. Proc Natl Acad Sci USA 94:8411–8416, 1997.

54. VJ Spanswick, J Cummings, JF Smyth. Current issues in the enzymology of mitomycin C metabolic activation. Gen Pharmacol 31:539–544, 1998.

55. MF Belcourt, WF Hodnick, S Rockwell, AC Sartorelli. Exploring the mechanistic aspects of mitomycin antibiotic bioactivation in Chinese hamster ovary cells overexpressing NADPH:cytochrome C (P-450) reductase and DT-diaphorase. Adv Enzyme Regul 38:111–133, 1998.

56. Y Jounaidi, DJ Waxman. Combination of the bioreductive drug tirapazamine with the chemotherapeutic prodrug cyclophosphamide for P450/P450-reductase-based cancer gene therapy. Cancer Res 60:3761–3769, 2000.

57. JA Robertson, HC Chen, DW Nebert. NAD(P)H:menadione oxidoreductase. Novel purification of enzyme cDNA and complete amino acid sequence, and gene regulation. J Biol Chem 261:15794–15799, 1986.

58. S Chen, K Wu, R Knox. Structure-function studies of DT-diaphorase (NQO1) and NRH: quinone oxidoreductase (NQO2). Free Radic Biol Med 29:276–284, 2000.

59. M Faig, MA Bianchet, P Talalay, S Chen, S Winski, D Ross, LM Amzel. Structures of recombinant human and mouse NAD(P)H:quinone oxidoreductases: species comparison and structural changes with substrate binding and release. Proc Natl Acad Sci USA 97:3177–3182, 2000.

60. MA Adams, Z Jia. Modulator of drug activity B from *Escherichia coli*: crystal structure of a prokaryotic homologue of DT-diaphorase. J Mol Biol 359:455–465, 2006.

61. R Li, MA Bianchet, P Talalay, LM Amzel. The three-dimensional structure of NAD(P) H:quinone reductase, a flavoprotein involved in cancer chemoprotection and chemotherapy: mechanism of the two-electron reduction. Proc Natl Acad Sci USA 92:8846–8850, 1995.

62. JV Skelly, MR Sanderson, DA Suter, U Baumann, MA Read, DS Gregory, M Bennett, SM Hobbs, S Neidle. Crystal structure of human DT-diaphorase: a model for interaction with the cytotoxic prodrug 5-(aziridin-1-yl)-2,4-dinitrobenzamide (CB1954). J Med Chem 42:4325–4330, 1999.

63. PL Gutierrez. The role of NAD(P)H oxidoreductase (DT-Diaphorase) in the bioactivation of quinone-containing antitumor agents: a review. Free Radic Biol Med 29:263–275, 2000.

64. A Begleiter, MK Leith, D Patel, BB Hasinoff. Role of NADPH cytochrome P450 reductase in activation of RH1. Cancer Chemother Pharmacol 60:713–723, 2007.

65. M Faig, MA Bianchet, S Winski, R Hargreaves, CJ Moody, AR Hudnott, D Ross, LM Amzel. Structure-based development of anticancer drugs: complexes of NAD(P) H:quinone oxidoreductase 1 with chemotherapeutic quinones. Structure 9:659–667, 2001.

66. JV Skelly, RJ Knox, TC Jenkins. Aerobic nitroreduction by flavoproteins: enzyme structure, mechanisms and role in cancer chemotherapy. Mini Rev Med Chem 1:293–306, 2001.

67. K Ito, M Nakanishi, WC Lee, H Sasaki, S Zenno, K Saigo, Y Kitade, M Tanokura. Three-dimensional structure of AzoR from *Escherichia coli*. An oxidereductase conserved in microorganisms. J Biol Chem 281:20567–20576, 2006.

68. H Chen. Recent advances in azo dye degrading enzyme research. Curr Protein Pept Sci 7:101–111, 2006.

69. MK Chourasia, SK Jain. Pharmaceutical approaches to colon targeted drug delivery systems. J Pharm Pharm Sci 6:33–66, 2003.

70. F Garcia-Molina, JL Munoz, R Varon, JN Rodriguez-Lopez, F Garcia-Canovas, J Tudela. A review on spectrophotometric methods for measuring the monophenolase and diphenolase activities of tyrosinase. J Agric Food Chem 55:9739–9749, 2007.

71. Y Matoba, T Kumagai, A Yamamoto, H Yoshitsu, M Sugiyama. Crystallographic evidence that the dinuclear copper center of tyrosinase is flexible during catalysis. J Biol Chem 281:8981–8990, 2006.

72. S Knaggs, H Malkin, HM Osborn, NA Williams, P Yaqoob. New prodrugs derived from 6-aminodopamine and 4-aminophenol as candidates for melanocyte-directed enzyme prodrug therapy (MDEPT). Org Biomol Chem 3:4002–4010, 2005.

73. AM Jordan, TH Khan, H Malkin, HM Osborn. Synthesis and analysis of urea and carbamate prodrugs as candidates for melanocyte-directed enzyme prodrug therapy (MDEPT). Bioorg Med Chem 10:2625–2633, 2002.

74. ME Morrison, MJ Yagi, G Cohen. *In vitro* studies of 2,4-dihydroxyphenylalanine, a prodrug targeted against malignant melanoma cells. Proc Natl Acad Sci USA 82:2960–2964, 1985.

75. MP Costi, S Ferrari, A Venturelli, S Calo, D Tondi, D Barlocco. Thymidylate synthase structure, function and implication in drug discovery. Curr Med Chem 12:2241–2258, 2005.

76. DB Lackey, MP Groziak, M Sergeeva, M Beryt, C Boyer, RM Stroud, P Sayre, JW Park, P Johnston, D Slamon, et al. Enzyme-catalyzed therapeutic agent (ECTA) design: activation of the antitumor ECTA compound NB1011 by thymidylate synthase. Biochem Pharmacol 61:179–189, 2001.

77. OA Sergeeva, HG Khambatta, BE Cathers, MV Sergeeva. Kinetic properties of human thymidylate synthase, an anticancer drug target. Biochem Biophys Res Commun 307: 297–300, 2003.

78. Q Li, C Boyer, JY Lee, HM Shepard. A novel approach to thymidylate synthase as a target for cancer chemotherapy. Mol Pharmacol 59:446–452, 2001.

79. YM Rustum, A Harstrick, S Cao, U Vanhoefer, MB Yin, H Wilke, S Seeber. Thymidylate synthase inhibitors in cancer therapy: direct and indirect inhibitors. J Clin Oncol 15: 389–400, 1997.
80. JM Collins, RW Klecker, AG Katki. Suicide prodrugs activated by thymidylate synthase: rationale for treatment and noninvasive imaging of tumors with deoxyuridine analogues. Clin Cancer Res 5:1976–1981, 1999.
81. PJ Barr, NJ Oppenheimer, DV Santi. Thymidylate synthetase-catalyzed conversions of E-5-(2-bromovinyl)-2′-deoxyuridylate. J Biol Chem 258:13627–13631, 1983.
82. PN Edwards. A kinetic, modeling and mechanistic re-analysis of thymidine phosphorylase and some related enzymes. J Enzyme Inhib Med Chem 21:483–499, 2006.
83. F Focher, S Spadari. Thymidine phosphorylase: a two-face Janus in anticancer chemotherapy. Curr Cancer Drug Targets 1:141–153, 2001.
84. MJ Pugmire, WJ Cook, A Jasanoff, MR Walter, SE Ealick. Structural and theoretical studies suggest domain movement produces an active conformation of thymidine phosphorylase. J Mol Biol 281:285–299, 1998.
85. JD Hayes, JU Flanagan, IR Jowsey. Glutathione transferases. Annu Rev Pharmacol Toxicol 45:51–88, 2005.
86. B Mannervik, PG Board, JD Hayes, I Listowsky, WR Pearson. Nomenclature for mammalian soluble glutathione transferases. Methods Enzymol 401:1–8, 2005.
87. D Townsend, K Tew. Cancer drugs, genetic variation and the glutathione-S-transferase gene family. Am J Pharmacogenomics 3:157–172, 2003.
88. ML O'Brien, KD Tew. Glutathione and related enzymes in multidrug resistance. Eur J Cancer 32A:967–978, 1996.
89. K Sato. Glutathione transferases as markers of preneoplasia and neoplasia. Adv Cancer Res 52:205–255, 1989.
90. KD Tew, S Dutta, M Schultz. Inhibitors of glutathione S-transferases as therapeutic agents. Adv Drug Deliv Rev 26:91–104, 1997.
91. S Mahajan, WM Atkins. The chemistry and biology of inhibitors and pro-drugs targeted to glutathione S-transferases. Cell Mol Life Sci 62:1221–1233, 2005.
92. A Satyam, MD Hocker, KA Kane-Maguire, AS Morgan, HO Villar, MH Lyttle. Design, synthesis, and evaluation of latent alkylating agents activated by glutathione S-transferase. J Med Chem 39:1736–1747, 1996.
93. AS Morgan, PE Sanderson, RF Borch, KD Tew, Y Niitsu, T Takayama, DD Von Hoff, E Izbicka, G Mangold, C Paul, et al. Tumor efficacy and bone marrow-sparing properties of TER286, a cytotoxin activated by glutathione S-transferase. Cancer Res 58:2568–2575, 1998.
94. DS Hamilton, X Zhang, Z Ding, I Hubatsch, B Mannervik, KN Houk, B Ganem, DJ Creighton. Mechanism of the glutathione transferase-catalyzed conversion of antitumor 2-crotonyloxymethyl-2-cycloalkenones to GSH adducts. J Am Chem Soc 125: 15049–15058, 2003.
95. VJ Findlay, DM Townsend, JE Saavedra, GS Buzard, ML Citro, LK Keefer, X Ji, KD Tew. Tumor cell responses to a novel glutathione S-transferase-activated nitric oxide-releasing prodrug. Mol Pharmacol 65:1070–1079, 2004.
96. BI Eklund, S Gunnarsdottir, AA Elfarra, B Mannervik. Human glutathione transferases catalyzing the bioactivation of anticancer thiopurine prodrugs. Biochem Pharmacol 73:1829–1841, 2007.

97. KD Tew. Glutathione-associated enzymes in anticancer drug resistance. Cancer Res 54:4313–4320, 1994.
98. JC Schisselbauer, WM Hogan, KH Buetow, KD Tew. Heterogeneity of glutathione S-transferase enzyme and gene expression in ovarian carcinoma. Pharmacogenetics 2:63–72, 1992.
99. AD Lewis, JD Hayes, CR Wolf. Glutathione and glutathione-dependent enzymes in ovarian adenocarcinoma cell lines derived from a patient before and after the onset of drug resistance: intrinsic differences and cell cycle effects. Carcinogenesis 9:1283–1287, 1988.
100. MH Lyttle, A Satyam, MD Hocker, KE Bauer, CG Caldwell, HC Hui, AS Morgan, A Mergia, LM Kauvar. Glutathione-S-transferase activates novel alkylating agents. J Med Chem 37:1501–1507, 1994.
101. S Gunnarsdottir, M Rucki, AA Elfarra. Novel glutathione-dependent thiopurine prodrugs: evidence for enhanced cytotoxicity in tumor cells and for decreased bone marrow toxicity in mice. J Pharmacol Exp Ther 301:77–86, 2002.
102. S Coulthard, L Hogarth. The thiopurines: an update. Invest New Drugs 23:523–532, 2005.
103. H Chakrapani, RC Kalathur, AE Maciag, ML Citro, X Ji, LK Keefer, JE Saavedra. Synthesis, mechanistic studies, and anti-proliferative activity of glutathione/glutathione S-transferase-activated nitric oxide prodrugs. Bioorg Med Chem 16:9764–9771, 2008.
104. DA Wink, JB Mitchell. Chemical biology of nitric oxide: insights into regulatory, cytotoxic, and cytoprotective mechanisms of nitric oxide. Free Radic Biol Med 25:434–456, 1998.
105. DA Wink, MB Grisham, JB Mitchell, PC Ford. Direct and indirect effects of nitric oxide in chemical reactions relevant to biology. Methods Enzymol 268:12–31, 1996.
106. B Bonavida, S Khineche, S Huerta-Yepez, H Garban. Therapeutic potential of nitric oxide in cancer. Drug Resist Updat 9:157–173, 2006.
107. LK Keefer. Progress toward clinical application of the nitric oxide-releasing diazeniumdiolates. Annu Rev Pharmacol Toxicol 43:585–607, 2003.
108. PJ Shami, JE Saavedra, CL Bonifant, J Chu, V Udupi, S Malaviya, BI Carr, S Kar, M Wang, L Jia, et al. Antitumor activity of JS-K [O^2-(2,4-dinitrophenyl) 1-[(4-ethoxycarbonyl)piperazin-1-yl]diazen-1-ium-1,2-diolate] and related O^2-aryl diazeniumdiolates *in vitro* and *in vivo*. J Med Chem 49:4356–4366, 2006.
109. V Udupi, M Yu, S Malaviya, JE Saavedra, PJ Shami. JS-K, a nitric oxide prodrug, induces cytochrome c release and caspase activation in HL-60 myeloid leukemia cells. Leuk Res 30:1279–1283, 2006.
110. PJ Shami, JE Saavedra, LY Wang, CL Bonifant, BA Diwan, SV Singh, Y Gu, SD Fox, GS Buzard, ML Citro, et al. JS-K, a glutathione/glutathione S-transferase-activated nitric oxide donor of the diazeniumdiolate class with potent antineoplastic activity. Mol Cancer Ther 2:409–417, 2003.
111. RS Nandurdikar, AE Maciag, ML Citro, PJ Shami, LK Keefer, JE Saavedra, H Chakrapani. Synthesis and evaluation of piperazine and homopiperazine analogues of JS-K, an anti-cancer lead compound. Bioorg Med Chem Lett 19:2760–2762, 2009.
112. P Iyidogan, S Lutz. Systematic exploration of active site mutations on human deoxycytidine kinase substrate specificity. Biochemistry 47:4711–4720, 2008.
113. X Song, PL Lorenzi, CP Landowski, BS Vig, JM Hilfinger, GL Amidon. Amino acid ester prodrugs of the anticancer agent gemcitabine: synthesis, bioconversion, metabolic bioevasion, and hPEPT1-mediated transport. Mol Pharm 2:157–167, 2005.

114. SA Veltkamp, JH Beijnen, JH Schellens. Prolonged versus standard gemcitabine infusion: translation of molecular pharmacology to new treatment strategy. Oncologist 13:261–276, 2008.
115. SA Johnson. Clinical pharmacokinetics of nucleoside analogues: focus on haematological malignancies. Clin Pharmacokinet 39:5–26, 2000.
116. BM Liederer, RT Borchardt. Enzymes involved in the bioconversion of ester-based prodrugs. J Pharm Sci 95:1177–1195, 2006.
117. PM Potter, RM Wadkins. Carboxylesterases – detoxifying enzymes and targets for drug therapy. Curr Med Chem 13:1045–1054, 2006.
118. S Bencharit, CL Morton, JL Hyatt, P Kuhn, MK Danks, PM Potter, MR Redinbo. Crystal structure of human carboxylesterase 1 complexed with the Alzheimer's drug tacrine: from binding promiscuity to selective inhibition. Chem Biol 10:341–349, 2003.
119. S Bencharit, CL Morton, EL Howard-Williams, MK Danks, PM Potter, MR Redinbo. Structural insights into CPT-11 activation by mammalian carboxylesterases. Nat Struct Biol 9:337–342, 2002.
120. S Bencharit, CL Morton, Y Xue, PM Potter, MR Redinbo. Structural basis of heroin and cocaine metabolism by a promiscuous human drug-processing enzyme. Nat Struct Biol 10:349–356, 2003.
121. RM Wadkins, CL Morton, JK Weeks, L Oliver, M Wierdl, MK Danks, PM Potter. Structural constraints affect the metabolism of 7-ethyl-10-[4-(1-piperidino)-1-piperidino]carbonyloxycamptothecin (CPT-11) by carboxylesterases. Mol Pharmacol 60:355–362, 2001.
122. LP Rivory, MR Bowles, J Robert, SM Pond. Conversion of irinotecan (CPT-11) to its active metabolite, 7-ethyl-10-hydroxycamptothecin (SN-38), by human liver carboxylesterase. Biochem Pharmacol 52:1103–1111, 1996.
123. K Yasunari, K Maeda, M Nakamura, T Watanabe, J Yoshikawa, A Asada. Pharmacological and clinical studies with temocapril, an angiotensin converting enzyme inhibitor that is excreted in the bile. Cardiovasc Drug Rev 22:189–198, 2004.
124. M Nakamura, E Shirasawa, M Hikida. Characterization of esterases involved in the hydrolysis of dipivefrin hydrochloride. Ophthalmic Res 25:46–51, 1993.
125. CD Fleming, S Bencharit, CC Edwards, JL Hyatt, L Tsurkan, F Bai, C Fraga, CL Morton, EL Howard-Williams, PM Potter, et al. Structural insights into drug processing by human carboxylesterase 1: tamoxifen, mevastatin, and inhibition by benzil. J Mol Biol 352:165–177, 2005.
126. DM Quinn. Acetylcholinesterase: enzyme structure, reaction dynamics, and virtual transition states. Chem Rev 87:955–979, 1987.
127. H Soreq, S Seidman. Acetylcholinesterase – new roles for an old actor. Nat Rev Neurosci 2:294–302, 2001.
128. RL Rotundo. Expression and localization of acetylcholinesterase at the neuromuscular junction. J Neurocytol 32:743–766, 2003.
129. RV Tak, D Pal, H Gao, S Dey, AK Mitra. Transport of acyclovir ester prodrugs through rabbit cornea and SIRC-rabbit corneal epithelial cell line. J Pharm Sci 90:1505–1515, 2001.
130. Y Nicolet, O Lockridge, P Masson, JC Fontecilla-Camps, F Nachon. Crystal structure of human butyrylcholinesterase and of its complexes with substrate and products. J Biol Chem 278:41141–41147, 2003.

131. A Tunek, LA Svensson. Bambuterol, a carbamate ester prodrug of terbutaline, as inhibitor of cholinesterases in human blood. Drug Metab Dispos 16:759–764, 1988.

132. B Waldeck. Beta-adrenoceptor agonists and asthma – 100 years of development. Eur J Pharmacol 445:1–12, 2002.

133. C Meyers, O Lockridge, BN La Du. Hydrolysis of methylprednisolone acetate by human serum cholinesterase. Drug Metab Dispos 10:279–280, 1982.

134. JF Gilmer, LM Moriarty, MN Lally, JM Clancy. Isosorbide-based aspirin prodrugs. II. Hydrolysis kinetics of isosorbide diaspirinate. Eur J Pharm Sci 16:297–304, 2002.

135. DI Draganov, BN La Du. Pharmacogenetics of paraoxonases: a brief review. Naunyn Schmiedebergs Arch Pharmacol 369:78–88, 2004.

136. M Harel, A Aharoni, L Gaidukov, B Brumshtein, O Khersonsky, R Meged, H Dvir, RB Ravelli, A McCarthy, L Toker, et al. Structure and evolution of the serum paraoxonase family of detoxifying and anti-atherosclerotic enzymes. Nat Struct Mol Biol 11:412–419, 2004.

137. MG Matera. Pharmacologic characteristics of prulifloxacin. Pulm Pharmacol Ther 19 (Suppl. 1):20–29, 2006.

138. S Billecke, D Draganov, R Counsell, P Stetson, C Watson, C Hsu, BN La Du. Human serum paraoxonase (PON1) isozymes Q and R hydrolyze lactones and cyclic carbonate esters. Drug Metab Dispos 28:1335–1342, 2000.

139. M Farr, M Pieper, J Calvete, H Tschesche. The N-terminus of collagenase MMP-8 determines superactivity and inhibition: a relation of structure and function analyzed by biomolecular interaction analysis. Biochemistry 38:7332–7338, 1999.

140. N Diaz, D Suarez. Peptide hydrolysis catalyzed by matrix metalloproteinase 2: a computational study. J Phys Chem B 112:8412–8424, 2008.

141. I Botos, E Meyer, SM Swanson, V Lemaitre, Y Eeckhout, EF Meyer. Structure of recombinant mouse collagenase-3 (MMP-13). J Mol Biol 292:837–844, 1999.

142. K Briknarova, M Gehrmann, L Banyai, H Tordai, L Patthy, M Llinas. Gelatin-binding region of human matrix metalloproteinase-2: solution structure, dynamics, and function of the COL-23 two-domain construct. J Biol Chem 276:27613–27621, 2001.

143. GB Fields. A model for interstitial collagen catabolism by mammalian collagenases. J Theor Biol 153:585–602, 1991.

144. JL Lauer-Fields, D Juska, GB Fields. Matrix metalloproteinases and collagen catabolism. Biopolymers 66:19–32, 2002.

145. I Watermann, J Gerspach, M Lehne, J Seufert, B Schneider, K Pfizenmaier, H Wajant. Activation of CD95L fusion protein prodrugs by tumor-associated proteases. Cell Death Differ 14:765–774, 2007.

146. CF Albright, N Graciani, W Han, E Yue, R Stein, Z Lai, M Diamond, R Dowling, L Grimminger, SY Zhang, et al. Matrix metalloproteinase-activated doxorubicin prodrugs inhibit HT1080 xenograft growth better than doxorubicin with less toxicity. Mol Cancer Ther 4:751–760, 2005.

147. E Van Valckenborgh, D Mincher, A Di Salvo, I Van Riet, L Young, B Van Camp, K Vanderkerken. Targeting an MMP-9-activated prodrug to multiple myeloma-diseased bone marrow: a proof of principle in the 5T33MM mouse model. Leukemia 19:1628–1633, 2005.

148. SJ Kridel, E Chen, LP Kotra, EW Howard, S Mobashery, JW Smith. Substrate hydrolysis by matrix metalloproteinase-9. J Biol Chem 276:20572–20578, 2001.

149. WJ van der Vijgh, GJ Peters. Protection of normal tissues from the cytotoxic effects of chemotherapy and radiation by amifostine (Ethyol): preclinical aspects. Semin Oncol 21:2–7, 1994.

150. M Treskes, LG Nijtmans, AM Fichtinger-Schepman, WJ van der Vijgh. Effects of the modulating agent WR2721 and its main metabolites on the formation and stability of cisplatin-DNA adducts *in vitro* in comparison to the effects of thiosulphate and diethyldithiocarbamate. Biochem Pharmacol 43:1013–1019, 1992.

151. K Chen, K Wang, AM Kirichian, AF Al Aowad, LK Iyer, SJ Adelstein, AI Kassis. In silico design, synthesis, and biological evaluation of radioiodinated quinazolinone derivatives for alkaline phosphatase-mediated cancer diagnosis and therapy. Mol Cancer Ther 5:3001–3013, 2006.

152. M de Graaf, E Boven, HW Scheeren, HJ Haisma, HM Pinedo. Beta-glucuronidase-mediated drug release. Curr Pharm Des 8:1391–1403, 2002.

153. C Vogler, J Barker, MS Sands, B Levy, N Galvin, WS Sly. Murine mucopolysaccharidosis VII: impact of therapies on the phenotype, clinical course, and pathology in a model of a lysosomal storage disease. Pediatr Dev Pathol 4:421–433, 2001.

154. S Jain, WB Drendel, ZW Chen, FS Mathews, WS Sly, JH Grubb. Structure of human beta-glucuronidase reveals candidate lysosomal targeting and active-site motifs. Nat Struct Biol 3:375–381, 1996.

155. MR Islam, S Tomatsu, GN Shah, JH Grubb, S Jain, WS Sly. Active site residues of human beta-glucuronidase. Evidence for Glu(540) as the nucleophile and Glu(451) as the acid-base residue. J Biol Chem 274:23451–23455, 1999.

156. AW Wong, S He, JH Grubb, WS Sly, SG Withers. Identification of Glu-540 as the catalytic nucleophile of human beta-glucuronidase using electrospray mass spectrometry. J Biol Chem 273:34057–34062, 1998.

157. BM Chen, LY Chan, SM Wang, MF Wu, JW Chern, SR Roffler. Cure of malignant ascites and generation of protective immunity by monoclonal antibody-targeted activation of a glucuronide prodrug in rats. Int J Cancer 73:392–402, 1997.

158. A Kamal, V Tekumalla, P Raju, VG Naidu, PV Diwan, R Sistla. Pyrrolo[2,1-c][1,4]benzodiazepine-beta-glucuronide prodrugs with a potential for selective therapy of solid tumors by PMT and ADEPT strategies. Bioorg Med Chem Lett 18:3769–3773, 2008.

159. L Lai, Z Xu, J Zhou, KD Lee, GL Amidon. Molecular basis of prodrug activation by human valacyclovirase, an alpha-amino acid ester hydrolase. J Biol Chem 283:9318–9327, 2008.

160. I Kim, GM Crippen, GL Amidon. Structure and specificity of a human valacyclovir activating enzyme: a homology model of BPHL. Mol Pharm 1:434–446, 2004.

161. I Kim, X Song, BS Vig, S Mittal, HC Shin, PJ Lorenzi, GL Amidon. A novel nucleoside prodrug-activating enzyme: substrate specificity of biphenyl hydrolase-like protein. Mol Pharm 1:117–127, 2004.

162. I Kim, XY Chu, S Kim, CJ Provoda, KD Lee, GL Amidon. Identification of a human valacyclovirase: biphenyl hydrolase-like protein as valacyclovir hydrolase. J Biol Chem 278:25348–25356, 2003.

163. V Novokhatny. Structure and activity of plasmin and other direct thrombolytic agents. Thromb Res 122(Suppl. 3):S3–S8, 2008.

164. PH Quax, AC de Bart, JA Schalken, JH Verheijen. Plasminogen activator and matrix metalloproteinase production and extracellular matrix degradation by rat prostate cancer cells *in vitro*: correlation with metastatic behavior *in vivo*. Prostate 32:196–204, 1997.

165. PH Quax, GN van Muijen, EJ Weening-Verhoeff, LR Lund, K Dano, DJ Ruiter, JH Verheijen. Metastatic behavior of human melanoma cell lines in nude mice correlates with urokinase-type plasminogen activator, its type-1 inhibitor, and urokinase-mediated matrix degradation. J Cell Biol 115:191–199, 1991.

166. R Hewitt, K Dano. Stromal cell expression of components of matrix-degrading protease systems in human cancer. Enzyme Protein 49:163–173, 1996.

167. PH Quax, RT van Leeuwen, HW Verspaget, JH Verheijen. Protein and messenger RNA levels of plasminogen activators and inhibitors analyzed in 22 human tumor cell lines. Cancer Res 50:1488–1494, 1990.

168. M Schmitt, N Harbeck, C Thomssen, O Wilhelm, V Magdolen, U Reuning, K Ulm, H Hofler, F Janicke, H Graeff. Clinical impact of the plasminogen activation system in tumor invasion and metastasis: prognostic relevance and target for therapy. Thromb Haemost 78:285–296, 1997.

169. S Ganesh, CF Sier, G Griffioen, HJ Vloedgraven, A de Boer, K Welvaart, CJ van de Velde, JH van Krieken, JH Verheijen, CB Lamers, et al. Prognostic relevance of plasminogen activators and their inhibitors in colorectal cancer. Cancer Res 54:4065–4071, 1994.

170. G Eisenbrand, S Lauck-Birkel, WC Tang. An approach towards more selective anticancer agents. Synthesis 1996:1246–1258, 1996.

171. A Warnecke, I Fichtner, G Sass, F Kratz. Synthesis, cleavage profile, and antitumor efficacy of an albumin-binding prodrug of methotrexate that is cleaved by plasmin and cathepsin B. Arch Pharm 340:389–395, 2007.

172. FM de Groot, AC de Bart, JH Verheijen, HW Scheeren. Synthesis and biological evaluation of novel prodrugs of anthracyclines for selective activation by the tumor-associated protease plasmin. J Med Chem 42:5277–5283, 1999.

173. FM de Groot, LW van Berkom, HW Scheeren. Synthesis and biological evaluation of 2′-carbamate-linked and 2′-carbonate-linked prodrugs of paclitaxel: selective activation by the tumor-associated protease plasmin. J Med Chem 43:3093–3102, 2000.

174. MC Wang, LA Valenzuela, GP Murphy, TM Chu. Purification of a human prostate specific antigen. Invest Urol 17:159–163, 1979.

175. H Lilja. A kallikrein-like serine protease in prostatic fluid cleaves the predominant seminal vesicle protein. J Clin Invest 76:1899–1903, 1985.

176. A Christensson, CB Laurell, H Lilja. Enzymatic activity of prostate-specific antigen and its reactions with extracellular serine proteinase inhibitors. Eur J Biochem 194:755–763, 1990.

177. H Lilja, J Oldbring, G Rannevik, CB Laurell. Seminal vesicle-secreted proteins and their reactions during gelation and liquefaction of human semen. J Clin Invest 80:281–285, 1987.

178. E Yoshida, S Ohmura, M Sugiki, M Maruyama, H Mihara. Prostate-specific antigen activates single-chain urokinase-type plasminogen activator. Int J Cancer 63:863–865, 1995.

179. CS Killian, DA Corral, E Kawinski, RI Constantine. Mitogenic response of osteoblast cells to prostate-specific antigen suggests an activation of latent TGF-beta and a proteolytic modulation of cell adhesion receptors. Biochem Biophys Res Commun 192:940–947, 1993.

180. P Cohen, HC Graves, DM Peehl, M Kamarei, LC Giudice, RG Rosenfeld. Prostate-specific antigen (PSA) is an insulin-like growth factor binding protein-3 protease found in seminal plasma. J Clin Endocrinol Metab 75:1046–1053, 1992.

181. AH Fortier, BJ Nelson, DK Grella, JW Holaday. Antiangiogenic activity of prostate-specific antigen. J Natl Cancer Inst 91:1635–1640, 1999.
182. HH Heidtmann, DM Nettelbeck, A Mingels, R Jager, HG Welker, RE Kontermann. Generation of angiostatin-like fragments from plasminogen by prostate-specific antigen. Br J Cancer 81:1269–1273, 1999.
183. SR Denmeade, LJ Sokoll, DW Chan, SR Khan, JT Isaacs. Concentration of enzymatically active prostate-specific antigen (PSA) in the extracellular fluid of primary human prostate cancers and human prostate cancer xenograft models. Prostate 48:1–6, 2001.
184. WJ Castellani. Cancer (prostate-specific antigen). Anal Chem 67:399–403, 1995.
185. RT McCormack, HG Rittenhouse, JA Finlay, RL Sokoloff, TJ Wang, RL Wolfert, H Lilja, JE Oesterling. Molecular forms of prostate-specific antigen and the human kallikrein gene family: a new era. Urology 45:729–744, 1995.
186. WJ Catalona, DS Smith, TL Ratliff, KM Dodds, DE Coplen, JJ Yuan, JA Petros, GL Andriole. Measurement of prostate-specific antigen in serum as a screening test for prostate cancer. N Engl J Med 324:1156–1161, 1991.
187. A Otto, J Bar, G Birkenmeier. Prostate-specific antigen forms complexes with human alpha 2-macroglobulin and binds to the alpha 2-macroglobulin receptor/LDL receptor-related protein. J Urol 159:297–303, 1998.
188. H Lilja, A Christensson, U Dahlen, MT Matikainen, O Nilsson, K Pettersson, T Lovgren. Prostate-specific antigen in serum occurs predominantly in complex with alpha 1-antichymotrypsin. Clin Chem 37:1618–1625, 1991.
189. SR Denmeade, W Lou, J Lovgren, J Malm, H Lilja, JT Isaacs. Specific and efficient peptide substrates for assaying the proteolytic activity of prostate-specific antigen. Cancer Res 57:4924–4930, 1997.
190. VM Garsky, PK Lumma, DM Feng, J Wai, HG Ramjit, MK Sardana, A Oliff, RE Jones, D DeFeo-Jones, RM Freidinger. The synthesis of a prodrug of doxorubicin designed to provide reduced systemic toxicity and greater target efficacy. J Med Chem 44:4216–4224, 2001.
191. RS DiPaola, J Rinehart, J Nemunaitis, S Ebbinghaus, E Rubin, T Capanna, M Ciardella, S Doyle-Lindrud, S Goodwin, M Fontaine, et al. Characterization of a novel prostate-specific antigen-activated peptide-doxorubicin conjugate in patients with prostate cancer. J Clin Oncol 20:1874–1879, 2002.
192. D DeFeo-Jones, VM Garsky, BK Wong, DM Feng, T Bolyar, K Haskell, DM Kiefer, K Leander, E McAvoy, P Lumma, et al. A peptide-doxorubicin "prodrug" activated by prostate-specific antigen selectively kills prostate tumor cells positive for prostate-specific antigen *in vivo*. Nat Med 6:1248–1252, 2000.
193. SR Denmeade, CM Jakobsen, S Janssen, SR Khan, ES Garrett, H Lilja, SB Christensen, JT Isaacs. Prostate-specific antigen-activated thapsigargin prodrug as targeted therapy for prostate cancer. J Natl Cancer Inst 95:990–1000, 2003.
194. CM Jakobsen, SR Denmeade, JT Isaacs, A Gady, CE Olsen, SB Christensen. Design, synthesis, and pharmacological evaluation of thapsigargin analogues for targeting apoptosis to prostatic cancer cells. J Med Chem 44:4696–4703, 2001.
195. SF Brady, JM Pawluczyk, PK Lumma, DM Feng, JM Wai, R Jones, D DeFeo-Jones, BK Wong, C Miller-Stein, JH Lin, et al. Design and synthesis of a pro-drug of vinblastine targeted at treatment of prostate cancer with enhanced efficacy and reduced systemic toxicity. J Med Chem 45:4706–4715, 2002.

196. D DeFeo-Jones, SF Brady, DM Feng, BK Wong, T Bolyar, K Haskell, DM Kiefer, K Leander, E McAvoy, P Lumma, et al. A prostate-specific antigen (PSA)-activated vinblastine prodrug selectively kills PSA-secreting cells *in vivo*. Mol Cancer Ther 1:451–459, 2002.

197. A Mhaka, SR Denmeade, W Yao, JT Isaacs, SR Khan. A 5-fluorodeoxyuridine prodrug as targeted therapy for prostate cancer. Bioorg Med Chem Lett 12:2459–2461, 2002.

198. X Tang, M Xian, M Trikha, KV Honn, PG Wang. Synthesis of peptide-diazeniumdiolate conjugates: towards enzyme activated antitumor agents. Tetrahedron Lett 42:2625–2629, 2001.

199. GB Jones, MO Mitchell, JS Weinberg, AV D'Amico, GJ Bubley. Towards enzyme activated antiprostatic agents. Bioorg Med Chem Lett 10:1987–1989, 2000.

200. SA Williams, RF Merchant, E Garrett-Mayer, JT Isaacs, JT Buckley, SR Denmeade. A prostate-specific antigen-activated channel-forming toxin as therapy for prostatic disease. J Natl Cancer Inst 99:376–385, 2007.

201. SK Kumar, SA Williams, JT Isaacs, SR Denmeade, SR Khan. Modulating paclitaxel bioavailability for targeting prostate cancer. Bioorg Med Chem 15:4973–4984, 2007.

202. Y Jiang, RS Dipaola, L Hu. Synthesis and stereochemical preference of peptide 4-aminocyclophosphamide conjugates as potential prodrugs of phosphoramide mustard for activation by prostate-specific antigen (PSA). Bioorg Med Chem Lett 19:2587–2590, 2009.

203. AJ Cooper, JT Pinto. Cysteine S-conjugate beta-lyases. Amino Acids 30:1–15, 2006.

204. L Zhang, MH Hanigan. Role of cysteine S-conjugate beta-lyase in the metabolism of cisplatin. J Pharmacol Exp Ther 306:988–994, 2003.

205. LH Lash, A Shivnani, J Mai, P Chinnaiyan, RJ Krause, AA Elfarra. Renal cellular transport, metabolism, and cytotoxicity of S-(6-purinyl)glutathione, a prodrug of 6-mercaptopurine, and analogues. Biochem Pharmacol 54:1341–1349, 1997.

206. WA Denny. Tumor-activated prodrugs – a new approach to cancer therapy. Cancer Invest 22:604–619, 2004.

207. G Xu, HL McLeod. Strategies for enzyme/prodrug cancer therapy. Clin Cancer Res 7:3314–3324, 2001.

208. L Hu, B Liu, DR Hacking. 5′-[2-(2-Nitrophenyl)-2-methylpropionyl]-2′-deoxy-5-fluorouridine as a potential bioreductively activated prodrug of FUDR: synthesis, stability and reductive activation. Bioorg Med Chem Lett 10:797–800, 2000.

209. E Johansson, GN Parkinson, WA Denny, S Neidle. Studies on the nitroreductase prodrug-activating system. Crystal structures of complexes with the inhibitor dicoumarol and dinitrobenzamide prodrugs and of the enzyme active form. J Med Chem 46:4009–4020, 2003.

210. NK Green, DJ Youngs, JP Neoptolemos, F Friedlos, RJ Knox, CJ Springer, GM Anlezark, NP Michael, RG Melton, MJ Ford, et al. Sensitization of colorectal and pancreatic cancer cell lines to the prodrug 5-(aziridin-1-yl)-2,4-dinitrobenzamide (CB1954) by retroviral transduction and expression of the *E. coli* nitroreductase gene. Cancer Gene Ther 4:229–238, 1997.

211. RJ Knox, F Friedlos, M Jarman, LC Davies, P Goddard, GM Anlezark, RG Melton, RF Sherwood. Virtual cofactors for an *Escherichia coli* nitroreductase enzyme: relevance to reductively activated prodrugs in antibody directed enzyme prodrug therapy (ADEPT). Biochem Pharmacol 49:1641–1647, 1995.

212. PR Race, AL Lovering, SA White, JI Grove, PF Searle, CW Wrighton, EI Hyde. Kinetic and structural characterisation of *Escherichia coli* nitroreductase mutants showing improved efficacy for the prodrug substrate CB1954. J Mol Biol 368:481–492, 2007.

213. MM AbuKhader, J Heap, CI De Matteis, SW Doughty, N Minton, M Paoli. Crystallization and preliminary X-ray characterization of the Bacillus amyloliquefaciens YwrO enzyme. Acta Crystallogr Sect F Struct Biol Cryst Commun 63:746–750, 2007.

214. CD Emptage, RJ Knox, MJ Danson, DW Hough. Nitroreductase from *Bacillus licheniformis*: a stable enzyme for prodrug activation. Biochem Pharmacol 77:21–29, 2009.

215. PF Searle, MJ Chen, L Hu, PR Race, AL Lovering, JI Grove, C Guise, M Jaberipour, ND James, V Mautner, et al. Nitroreductase: a prodrug-activating enzyme for cancer gene therapy. Clin Exp Pharmacol Physiol 31:811–816, 2004.

216. GM Anlezark, RG Melton, RF Sherwood, WR Wilson, WA Denny, BD Palmer, RJ Knox, F Friedlos, A Williams. Bioactivation of dinitrobenzamide mustards by an *E. coli* B nitroreductase. Biochem Pharmacol 50:609–618, 1995.

217. Y Jounaidi. Cytochrome P450-based gene therapy for cancer treatment: from concept to the clinic. Curr Drug Metab 3:609–622, 2002.

218. M Czerwinski, TL McLemore, RM Philpot, PT Nhamburo, K Korzekwa, HV Gelboin, FJ Gonzalez. Metabolic activation of 4-ipomeanol by complementary DNA-expressed human cytochromes P-450: evidence for species-specific metabolism. Cancer Res 51:4636–4638, 1991.

219. HO McCarthy, A Yakkundi, V McErlane, CM Hughes, G Keilty, M Murray, LH Patterson, DG Hirst, SR McKeown, T Robson. Bioreductive GDEPT using cytochrome P450 3A4 in combination with AQ4N. Cancer Gene Ther 10:40–48, 2003.

220. A Bzowska, E Kulikowska, D Shugar. Purine nucleoside phosphorylases: properties, functions, and clinical aspects. Pharmacol Ther 88:349–425, 2000.

221. EM Bennett, C Li, PW Allan, WB Parker, SE Ealick. Structural basis for substrate specificity of *Escherichia coli* purine nucleoside phosphorylase. J Biol Chem 278:47110–47118, 2003.

222. WB Parker, SA King, PW Allan, LL Bennett, Jr., JA Secrist, 3rd, JA Montgomery, KS Gilbert, WR Waud, AH Wells, GY Gillespie, et al. *In vivo* gene therapy of cancer with *E. coli* purine nucleoside phosphorylase. Hum Gene Ther 8:1637–1644, 1997.

223. WB Parker, PW Allan, AE Hassan, JA Secrist, 3rd, EJ Sorscher, WR Waud. Antitumor activity of 2-fluoro-2′-deoxyadenosine against tumors that express *Escherichia coli* purine nucleoside phosphorylase. Cancer Gene Ther 10:23–29, 2003.

224. N Ostermann, I Schlichting, R Brundiers, M Konrad, J Reinstein, T Veit, RS Goody, A Lavie. Insights into the phosphoryltransfer mechanism of human thymidylate kinase gained from crystal structures of enzyme complexes along the reaction coordinate. Structure 8:629–642, 2000.

225. N Ostermann, D Segura-Pena, C Meier, T Veit, C Monnerjahn, M Konrad, A Lavie. Structures of human thymidylate kinase in complex with prodrugs: implications for the structure-based design of novel compounds. Biochemistry 42:2568–2577, 2003.

226. U Kosinska, C Carnrot, MP Sandrini, AR Clausen, L Wang, J Piskur, S Eriksson, H Eklund. Structural studies of thymidine kinases from Bacillus anthracis and Bacillus cereus provide insights into quaternary structure and conformational changes upon substrate binding. Febs J 274:727–737, 2007.

227. MS Birringer, MT Claus, G Folkers, DP Kloer, GE Schulz, L Scapozza. Structure of a type II thymidine kinase with bound dTTP. FEBS Lett 579:1376–1382, 2005.

228. N Janoschek, E van de Leur, AM Gressner, R Weiskirchen. Induction of cell death in activated hepatic stellate cells by targeted gene expression of the thymidine kinase/ganciclovir system. Biochem Biophys Res Commun 316:1107–1115, 2004.

229. PD Senter, GJ Schreiber, DL Hirschberg, SA Ashe, KE Hellstrom, I Hellstrom. Enhancement of the *in vitro* and *in vivo* antitumor activities of phosphorylated mitomycin C and etoposide derivatives by monoclonal antibody-alkaline phosphatase conjugates. Cancer Res 49:5789–5792, 1989.

230. KC Chen, TL Cheng, YL Leu, ZM Prijovich, CH Chuang, BM Chen, SR Roffler. Membrane-localized activation of glucuronide prodrugs by beta-glucuronidase enzymes. Cancer Gene Ther 14:187–200, 2007.

231. AE Alaoui, N Saha, F Schmidt, C Monneret, JC Florent. New Taxol (paclitaxel) prodrugs designed for ADEPT and PMT strategies in cancer chemotherapy. Bioorg Med Chem 14:5012–5019, 2006.

232. FM de Groot, EW Damen, HW Scheeren. Anticancer prodrugs for application in monotherapy: targeting hypoxia, tumor-associated enzymes, and receptors. Curr Med Chem 8:1093–1122, 2001.

233. E Bouvier, S Thirot, F Schmidt, C Monneret. First enzymatically activated Taxotere prodrugs designed for ADEPT and PMT. Bioorg Med Chem 12:969–977, 2004.

234. S Papot, I Tranoy, F Tillequin, JC Florent, JP Gesson. Design of selectively activated anticancer prodrugs: elimination and cyclization strategies. Curr Med Chem Anticancer Agents 2:155–185, 2002.

235. HA Lindner, VV Lunin, A Alary, R Hecker, M Cygler, R Menard. Essential roles of zinc ligation and enzyme dimerization for catalysis in the aminoacylase-1/M20 family. J Biol Chem 278:44496–44504, 2003.

236. CJ Springer, P Antoniw, KD Bagshawe, F Searle, GM Bisset, M Jarman. Novel prodrugs which are activated to cytotoxic alkylating agents by carboxypeptidase G2. J Med Chem 33:677–681, 1990.

237. CJ Springer, I Niculescu-Duvaz, RB Pedley. Novel prodrugs of alkylating agents derived from 2-fluoro- and 3-fluorobenzoic acids for antibody-directed enzyme prodrug therapy. J Med Chem 37:2361–2370, 1994.

238. CJ Springer, R Dowell, PJ Burke, E Hadley, DH Davis, DC Blakey, RG Melton, I Niculescu-Duvaz. Optimization of alkylating agent prodrugs derived from phenol and aniline mustards: a new clinical candidate prodrug (ZD2767) for antibody-directed enzyme prodrug therapy (ADEPT). J Med Chem 38:5051–5065, 1995.

239. RI Dowell, CJ Springer, DH Davies, EM Hadley, PJ Burke, FT Boyle, RG Melton, TA Connors, DC Blakey, AB Mauger. New mustard prodrugs for antibody-directed enzyme prodrug therapy: alternatives to the amide link. J Med Chem 39:1100–1105, 1996.

240. CJ Springer. CMDA, antineoplastic prodrug. Drug Future 18:212–215, 1993.

241. I Niculescu-Duvaz, I Scanlon, D Niculescu-Duvaz, F Friedlos, J Martin, R Marais, CJ Springer. Significant differences in biological parameters between prodrugs cleavable by carboxypeptidase G2 that generate 3,5-difluoro-phenol and -aniline nitrogen mustards in gene-directed enzyme prodrug therapy systems. J Med Chem 47:2651–2658, 2004.

242. LC Davies, F Friedlos, D Hedley, J Martin, LM Ogilvie, IJ Scanlon, CJ Springer. Novel fluorinated prodrugs for activation by carboxypeptidase G2 showing good *in vivo* antitumor activity in gene-directed enzyme prodrug therapy. J Med Chem 48:5321–5328, 2005.

243. D Niculescu-Duvaz, I Niculescu-Duvaz, F Friedlos, J Martin, R Spooner, L Davies, R Marais, CJ Springer. Self-immolative nitrogen mustard prodrugs for suicide gene therapy. J Med Chem 41:5297–5309, 1998.
244. D Niculescu-Duvaz, I Niculescu-Duvaz, F Friedlos, J Martin, P Lehouritis, R Marais, CJ Springer. Self-immolative nitrogen mustards prodrugs cleavable by carboxypeptidase G2 (CPG2) showing large cytotoxicity differentials in GDEPT. J Med Chem 46:1690–1705, 2003.
245. WA Denny. Prodrug strategies in cancer therapy. Eur J Med Chem 36:577–595, 2001.
246. DE Thurston. Advances in the study of Pyrrolo[2,1-c][1,4]-benzodiazepine (PBD) antitumor antibiotics. In: S Neidle, MJ Waring, eds. *Molecular Aspects of Anticancer Drug-DNA Interactions*. London: The Macmillan Press Ltd, 1993, pp. 54–88.
247. DH Palmer, AE Milner, DJ Kerr, LS Young. Mechanism of cell death induced by the novel enzyme-prodrug combination, nitroreductase/CB1954, and identification of synergism with 5-fluorouracil. Br J Cancer 89:944–950, 2003.
248. FM Sirotnak, PL Chello, JR Piper, JA Montgomery, JI DeGraw. Structural specificity of folate analog transport and binding to dihydropholate reductase in murine tumor and normal cells: relevance to therapeutic efficacy. In: RL Kisliuk, GM Brown, eds. *Chemistry and Biology of Pteridines*. New York: Elsevier, 1979, pp. 597–608.
249. U Kuefner, U Lohrmann, YD Montejano, KS Vitols, FM Huennekens. Carboxypeptidase-mediated release of methotrexate from methotrexate alpha-peptides. Biochemistry 28:2288–2297, 1989.
250. CA Claridge, A Gourevitch, J Lein. Bacterial penicillin amidase. Nature 187:237–238, 1960.
251. WB Alkema, CM Hensgens, EH Kroezinga, E de Vries, R Floris, JM van der Laan, BW Dijkstra, DB Janssen. Characterization of the beta-lactam binding site of penicillin acylase of *Escherichia coli* by structural and site-directed mutagenesis studies. Protein Eng 13:857–863, 2000.
252. CE McVey, MA Walsh, GG Dodson, KS Wilson, JA Brannigan. Crystal structures of penicillin acylase enzyme-substrate complexes: structural insights into the catalytic mechanism. J Mol Biol 313:139–150, 2001.
253. GS Bignami, PD Senter, PG Grothaus, KJ Fischer, T Humphreys, PM Wallace. N-(4′-hydroxyphenylacetyl)palytoxin: a palytoxin prodrug that can be activated by a monoclonal antibody-penicillin G amidase conjugate. Cancer Res 52:5759–5764, 1992.
254. VM Vrudhula, PD Senter, KJ Fischer, PM Wallace. Prodrugs of doxorubicin and melphalan and their activation by a monoclonal antibody-penicillin-G amidase conjugate. J Med Chem 36:919–923, 1993.
255. G Minasov, X Wang, BK Shoichet. An ultrahigh resolution structure of TEM-1 beta-lactamase suggests a role for Glu166 as the general base in acylation. J Am Chem Soc 124:5333–5340, 2002.
256. LN Jungheim, TA Shepherd. Design of antitumor prodrugs: substrates for antibody targeted enzymes. Chem Rev 94:1553–1566, 1994.
257. ML Rodrigues, P Carter, C Wirth, S Mullins, A Lee, BK Blackburn. Synthesis and beta-lactamase-mediated activation of a cephalosporin-taxol prodrug. Chem Biol 2:223–227, 1995.
258. TP Ko, JJ Lin, CY Hu, YH Hsu, AH Wang, SH Liaw. Crystal structure of yeast cytosine deaminase. Insights into enzyme mechanism and evolution. J Biol Chem 278:19111–19117, 2003.

259. BE Huber, EA Austin, CA Richards, ST Davis, SS Good. Metabolism of 5-fluorocytosine to 5-fluorouracil in human colorectal tumor cells transduced with the cytosine deaminase gene: significant antitumor effects when only a small percentage of tumor cells express cytosine deaminase. Proc Natl Acad Sci USA 91:8302–8306, 1994.

260. A Nikulin, S Revtovich, E Morozova, N Nevskaya, S Nikonov, M Garber, T Demidkina. High-resolution structure of methionine gamma-lyase from Citrobacter freundii. Acta Crystallogr D Biol Crystallogr 64:211–218, 2008.

261. DV Mamaeva, EA Morozova, AD Nikulin, SV Revtovich, SV Nikonov, MB Garber, TV Demidkina. Structure of Citrobacter freundii L-methionine gamma-lyase. Acta Crystallogr Sect F Struct Biol Cryst Commun 61:546–549, 2005.

262. K Miki, M Xu, A Gupta, Y Ba, Y Tan, W Al-Refaie, M Bouvet, M Makuuchi, AR Moossa, RM Hoffman. Methioninase cancer gene therapy with selenomethionine as suicide prodrug substrate. Cancer Res 61:6805–6810, 2001.

263. N Yamamoto, A Gupta, M Xu, K Miki, Y Tsujimoto, H Tsuchiya, K Tomita, AR Moossa, RM Hoffman. Methioninase gene therapy with selenomethionine induces apoptosis in bcl-2-overproducing lung cancer cells. Cancer Gene Ther 10:445–450, 2003.

264. LF Tietze, T Herzig, A Fecher, F Haunert, I Schuberth. Highly selective glycosylated prodrugs of cytostatic CC-1065 analogues for antibody-directed enzyme tumor therapy. Chembiochem 2:758–765, 2001.

265. LF Tietze, HJ Schuster, B Krewer, I Schuberth. Synthesis and biological studies of different duocarmycin based glycosidic prodrugs for their use in the antibody-directed enzyme prodrug therapy. J Med Chem 52:537–543, 2009.

266. H Devalapally, RS Navath, V Yenamandra, RR Akkinepally, RK Devarakonda. Beta-galactoside prodrugs of doxorubicin for application in antibody directed enzyme prodrug therapy/prodrug monotherapy. Arch Pharm Res 30:723–732, 2007.

6

EVOLUTION OF AN ORALLY ACTIVE PRODRUG OF GEMCITABINE

JAMES R. MCCARTHY

Department of Chemistry and Chemical Biology, Indiana University Purdue University Indianapolis, Indianapolis, Indiana

I. INTRODUCTION

Gemcitabine hydrochloride (**1**, Fig. 1) is a pyrimidine nucleoside analog with proven anticancer efficacy for a variety of solid tumor types in patients including pancreatic, NSCL, and breast cancer [1–4] that is marketed as Gemzar™ in more than 70 countries. The drug is administered as a parenteral formulation with a 30 min intravenous infusion of 1000–1250 mg/m^2 given once weekly for a 3- or 4-week cycle. Gemcitabine is activated within cells by sequential phosphorylation by deoxycytidine kinase to mono-, di-, and triphosphate nucleotides [5–7] (Fig. 2). The triphosphate (**6**) is believed to be the predominant cause for the observed apoptosis in treated malignant cells by incorporating into DNA, causing DNA chain termination and thus diminishing DNA synthesis [5]. These effects on DNA are potentiated since the diphosphate (**5**) inhibits human ribonucleotide reductase (hRR), the enzyme that catalyzes the conversion of nucleoside diphosphates to deoxynucleotide diphosphates [8–10]. This inhibition depletes levels of all deoxynucleoside diphosphates and as a result depletes deoxynucleotide diphosphates in the cell and enhances the ability of the triphosphate (**6**) to compete with deoxycytidine triphosphate (dCTP) for incorporation into DNA [11,12]. In addition, depletion of dCTP releases the tight feedback inhibition of human deoxycytidine kinase (hdCK) by dCTP and results in elevated levels of the diphosphate (**5**) and triphosphate (**6**) [13]. Alternatively, gemcitabine

Enzyme Technologies: Pluripotent Players in Discovering Therapeutic Agents, First Edition.
Edited by Hsiu-Chiung Yang, Wu-Kuang Yeh, and James R. McCarthy.

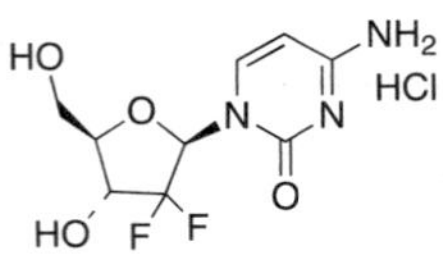

1, gemcitabine HCl
(dFdC HCl)

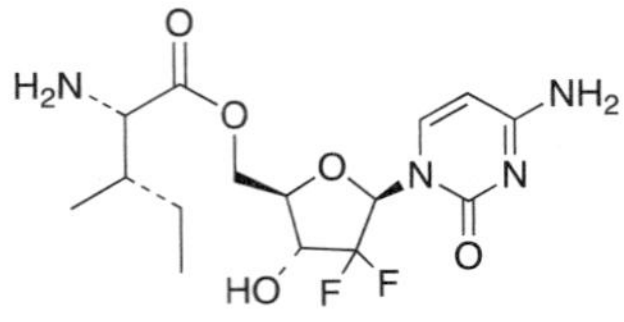

2, 5'-O-L-Ile-gemcitabine

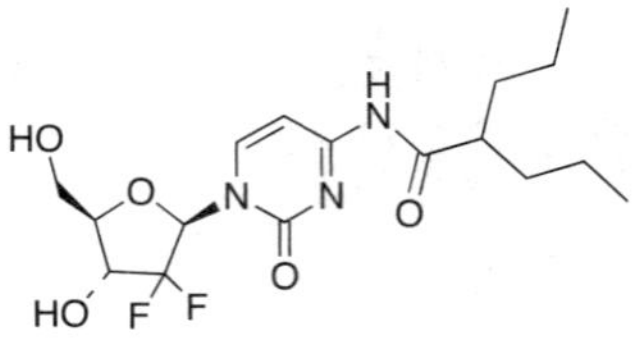

3, LY2334737

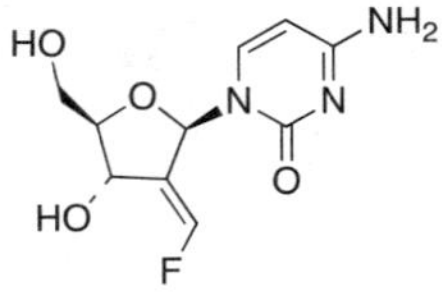

4, tezacitabine

FIGURE 1 Chemical structures of gemcitabine, prodrugs of gemcitabine, and tezacitabine.

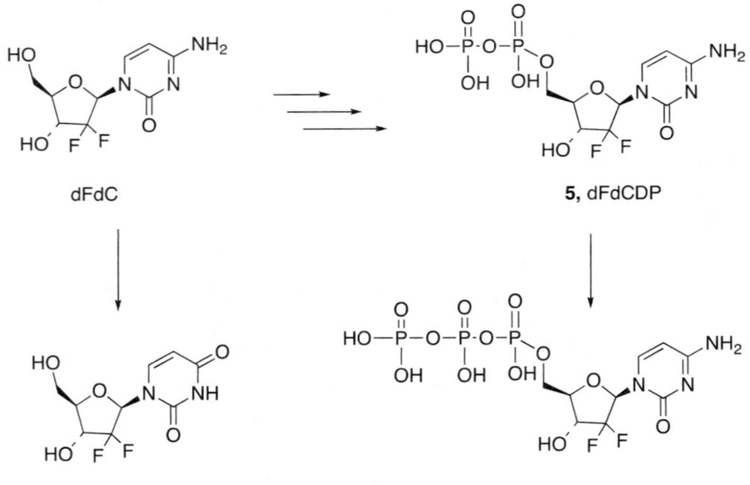

FIGURE 2 Activation and inactivation pathways for gemcitabine (dFdC).

may be metabolized by deamination to its inactive uridine analog 2,2′-difluorodeoxyuridine (**7**) (dFdU) by deoxycytidine deaminase, which is present at high levels in both human plasma and liver [14,15] (Fig. 2). Deamination occurs rapidly in plasma with a half-life of approximately 70 min. Despite this rapid inactivation, gemcitabine is clinically efficacious in a number of solid tumors.

In contrast to gemcitabine, tezacitabine (**4**) [16] (Fig. 1), which is also a mechanism-based inhibitor of hRR and a DNA chain terminator [17,18], has a half-life in man of 3–4 h, and 23% of the administered drug was recovered in the urine as parent molecule [19,20]. Even though tezacitabine (**4**) has a more favorable PK profile than gemcitabine in man, was well tolerated, and demonstrated objective antitumor activity in patients with refractory solid tumors [19,20], the drug has still not been introduced on the market. As a result, a major effort was initiated to obtain a prodrug of gemcitabine with a substantially improved PK profile that could be administered orally.

Prodrugs have proven to be an effective method for the oral delivery of several nucleoside drugs such as capecitabine [21], a prodrug of 5-fluorouracil (5-FU), and valacyclovir, the valine ester of acyclovir [22,23]. Modification of acyclovir to valacyclovir resulted in a fivefold increase in oral bioavailability via recognition and active transport by the amino acid transporter PEPT-1. Amidon and coworkers [24] have prepared a series of natural amino acid ester prodrugs of gemcitabine targeting the PEPT-1 transporter and reported that the L-isoleucine derivative (**2**) (Fig. 1) was relatively stable to hydrolysis in pH 7.4 buffer and exhibited a very high degree of stability when incubated with crude cytidine deaminase. They reported that (**2**) was approximately 10-fold more stable in Caco-2 homogenates than the L-valine analog (**8**) (Table 1), with an estimated half-life of 75 min.

II. PREPARATION OF GEMCITABINE PRODRUGS

We prepared a number of 5′-amino acid esters of gemcitabine and observed that the 5′-cyclopropylamino acid esters (**9**) and the *N*-methyl cyclopropyl amino acid ester (**10**) were substantially more stable than the valine analog (**8**), the cyclobutyl analog (**11**), or the α-methyl-α-ethyl analog (**12**) (Table 1). After a 30 min incubation in a pH 8 buffer at 40 °C, the *N*-methyl cyclopropyl amino acid ester (**10**) underwent only 4% hydrolysis and the desmethyl analog (**9**) underwent 13% hydrolysis. In contrast, the cyclobutyl analog (**11**) was 100% hydrolyzed, while the valine analog (**8**) was 65% hydrolyzed. This was an exciting observation and we believed of significance for designing substantially more stable prodrugs than the classical acetyl ester analogs. We studied this enhanced stability of the cyclopropyl esters in more detail [25] and found that esters of cyclopropanecarboxylic acid demonstrate a substantial increase in stability under both acid- and base-catalyzed hydrolytic conditions relative to related esters. Table 2 shows the half-lives for benzyl cyclopropanecarboxylic esters (**13**), compound **22**, and compound **23** in 0.1 N HCl and pH 10 buffer at 40 °C compared to several other esters. Benzyl cyclopropanecarboxylate (**13**) is almost six times more stable than the open-chain isopropyl analog (**15**) in 0.1 N HCl at 40 °C ($t_{1/2}$ = 180.2 vs. 32.1 h) and twice as stable at pH 10 buffer ($t_{1/2}$ = 78.7 vs. 40.0 h). This cyclopropanecarboxylic acid ester is similarly more stable than cyclopentanecarboxylic acid ester (**17**). The contrast is even greater for the cyclobutane analog (**16**), similar to compound **10**, where the $t_{1/2}$ is only 14.1 h in 0.1 N HCl and 16.0 h in pH 10 buffer. While the benzoate ester **19** is less stable in pH 10 buffer than the cyclopropanecarboxylic

TABLE 1 Solution Stability of 5′-Esters of Gemcitabine at 40 °C for pH 6 and pH 8 (% Degradation) at 30 min, Inhibition of [³H]Gly-Sar Uptake in HeLa/hPEPT1 Cells by 5′-Esters of Gemcitabine and the Rate of Hydrolysis in Mouse and Human SIH Measured by Formation of Gemcitabine. Compound 11 Not Tested for Inhibition or Hydrolysis

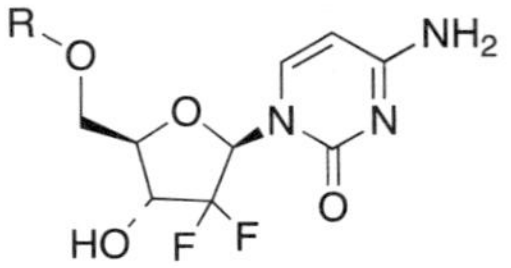

Cpd#	R	% degradation		Gly-SAR inhibition (mM)	Mouse SIH hydrolysis (uM/min)	Human SIH hydrolysis (uM/min)
		pH 6	pH 8			
8	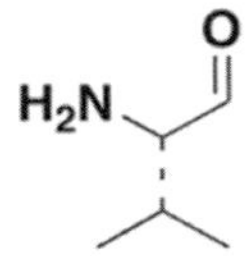	15	65	0.47	15.8	16.2
9	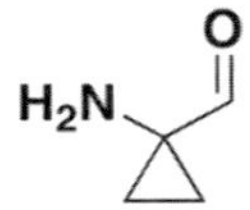	5	13	2.85	2.85	12.9
10	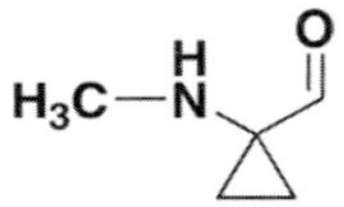	2	4	1.25	1.25	0.1
11	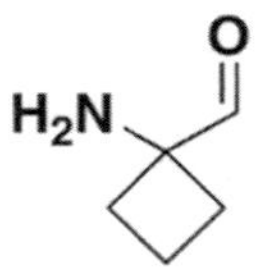	89	100			
12	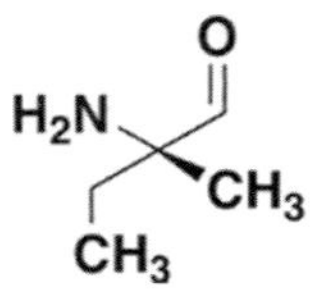	11	64	3.46	3.46	16.2

acid ester (**13**) ($t_{1/2}$ = 55.4 vs. 78.7 h, respectively), it is more stable in 0.1 N HCl. The cyclohexanecarboxylate ester (**18**) is slightly more stable than compound **13** in pH 10 buffer (87 vs. 78.7 h). Comparing the more sterically hindered pivalic acid ester (**20**), with 1-methylcyclopropanecarboxylic acid benzyl ester (**22**), the cyclopropane ester is more than twice as stable in acid ($t_{1/2}$ = 267.5 vs. 125.7 h) and

TABLE 2 Half-Lives for the Hydrolysis of Benzyl Esters 13–18

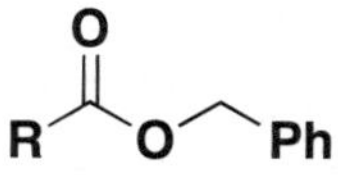

$t_{1/2}$ at 40 °C (h)

Ester	R	0.1N HCl	pH 10
13		180.2	78.7
14	CH2	14.2	8.7
15	i-Pr	32.1	40
16		14.1	16
17		26.3	43.1
18		52.4	87
19	Ph	>300	55.4
20	t-Bu	125.7	142.4
21	$PhCH_2$	36.1	12
22		267.5	192
23	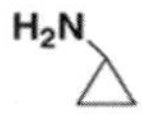	>300	43.1
24	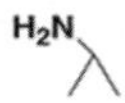	>300	7.8

1.3 times more stable in pH 10 buffer. This trend holds for amino acid derivatives (**23**) versus compound **24** as well. Both amino acids are stable under acidic conditions, but the cyclopropane analog (**23**) is almost six times more stable ($t_{1/2}$ = 7.8 vs. 43.1 h) in pH 10 buffer.

III. ADVANTAGE OF CYCLOPROPYL ESTER PRODRUGS

The advantage of the increased hydrolytic stability of cyclopropane derivatives of amino acid esters, for the design of a suitable prodrug, is highlighted by comparing the stability of valacyclovir (**25**) with the cyclopropane analog (**26**) (Table 3). At 40°C, the half-life of the cyclopropane amino acid ester at pH 6 is >300 h, while the value for valacyclovir is 69.7 h. This trend holds at pH 8 and pH 10 as well. The enhancement in chemical stability is very important in terms of drug development potential. Anand and coworkers [26] note that acyclovir would be a suitable drug for the treatment of herpes simplex keratitis, the leading cause of blindness in the United States [26], but the corneal permeability is very low. Valacyclovir has much better permeability, but the half-life in aqueous buffer is unsatisfactory for an ophthalmic

TABLE 3 Hydrolysis Rates of 25 and Cyclopropane Derivative 26

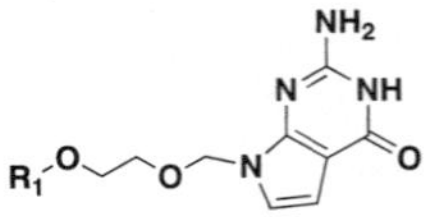

$t_{1/2}$ @ 40 °C (hrs)

Ester	R_1	0.1 N HCl	pH 2	pH 4	pH 6	pH 8	pH 10
25	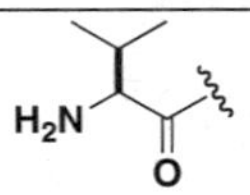	>300	>300	>300	69.7	7.8	6.8
26	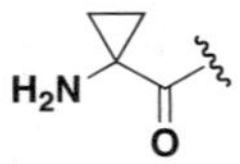	>300	>300	>300	>300	90.1	23.8

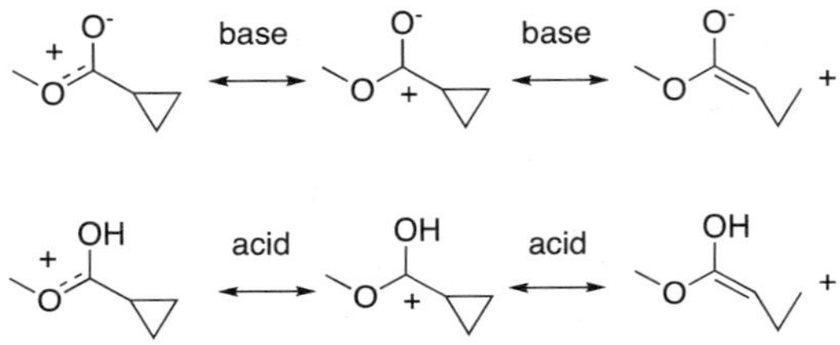

FIGURE 3 Increased stability of cyclopropane amino acid esters attributed to hyperconjugative effect of the cyclopropane ring.

solution. This issue could be overcome with the cyclopropane prodrug (**26**), where the half-life in pH 6 buffer at 5 °C is 352 days.

The cyclopropyl group provides hyperconjugative stabilization (see Fig. 3) [25].

IV. ISSUES WITH CYCLOPROPYL ESTERS OF GEMCITABINE

Interestingly, the L-valine ester (**8**) and the cyclopropyl glycine ester (**9**) showed equivalent activity to gemcitabine in preliminary *in vivo* efficacy studies at one-third the dose of gemcitabine. Both compounds **8** and **9** were good substrates for PEPT-1. However, evaluation of both compounds in a 4-day toxicological model for gastrointestinal (GI) and liver toxicity showed unacceptable toxicity. This was attributed to the ease of hydrolysis in the mouse small intestine hydrolysis assay (Table 1). In summary, the cyclopropyl ester analog of gemcitabine (**9**) showed good aqueous stability at pH 2 and pH 8, in contrast to the valine analog (**8**), and was a good substrate for hPEPT1. However, the compound demonstrated poor stability against enzymatic hydrolysis in intestinal cells.

V. ALTERNATIVE PRODRUGS OF GEMCITABINE

A. Aminoacyl Amides

As a result of the stability issues associated with amino acid esters of gemcitabine and the associated toxicity observed, we targeted aminoacyl amides as a way to provide recognition by PEPT-1 for oral activity and for increased stability in the stomach and intestines (see Scheme 1). It was of interest that Stella and coworkers [27] reported that aminoacyl amide prodrugs of prazosin undergo an intramolecular rearrangement involving the α-amine of the amino acid, forming a spiro-intermediate that opens to the α-alkyl amides. No citations of this report had appeared in the literature at the time we targeted the aminoacyl amides. However, two publications [28,29] and three patents [30–32] on aminoacyl amides of cytosine nucleosides isolated as HCl or TFA salts had appeared. Because of these potentially conflicting reports, we felt it was imperative to utilize X-ray data to confirm structures. The aminoacyl amides of gemcitabine (**29a–g**) were prepared through a PyBOP/HOAt-mediated selective acylation at the N^4-position of gemcitabine using Boc-protected amino acids (**27**). Removal of the Boc-protecting group with TFA/CH_2Cl_2 (1:1) afforded the TFA salts of the corresponding N4-(α-aminoacyl)gemcitabines (**29a–g**). The L-proline derivative (**30h**) was prepared in a similar fashion [33].

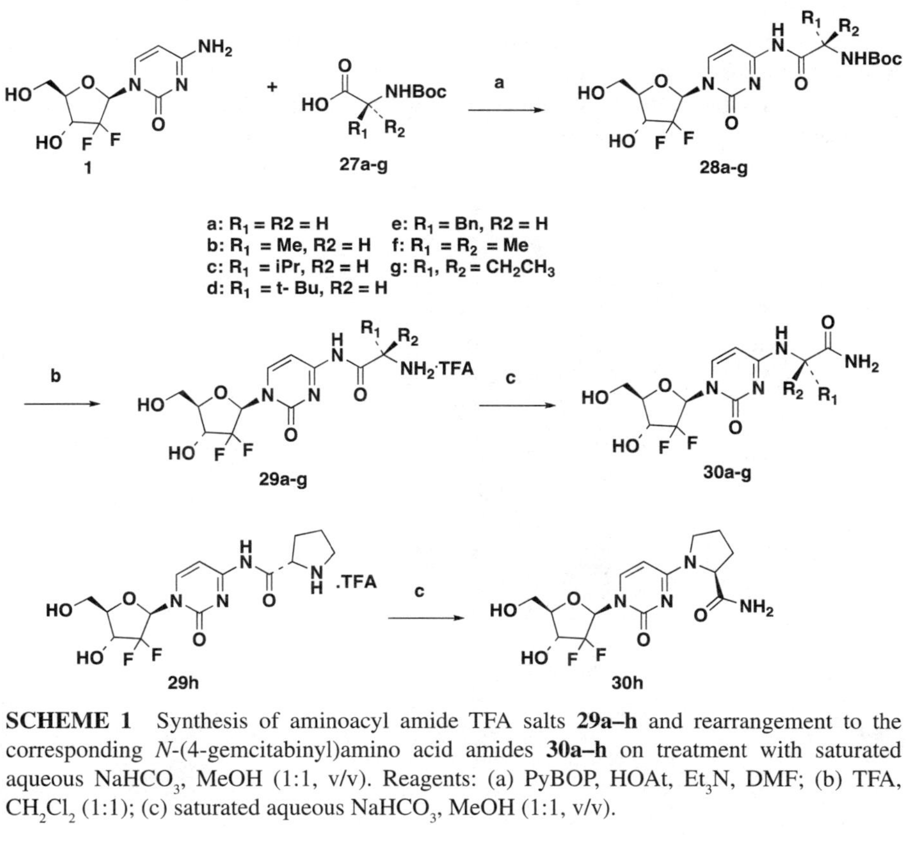

SCHEME 1 Synthesis of aminoacyl amide TFA salts **29a–h** and rearrangement to the corresponding *N*-(4-gemcitabinyl)amino acid amides **30a–h** on treatment with saturated aqueous $NaHCO_3$, MeOH (1:1, v/v). Reagents: (a) PyBOP, HOAt, Et_3N, DMF; (b) TFA, CH_2Cl_2 (1:1); (c) saturated aqueous $NaHCO_3$, MeOH (1:1, v/v).

Interestingly, compound **29d** with the bulky *t*-butyl group was 100% degraded at pH 8 at 40°C in 4 h to a water-insoluble product. While some amide hydrolysis to gemcitabine occurred under acidic conditions (pH 1–2), the observed degradation product at pH 8 was inconsistent with simple hydrolysis to gemcitabine. Comparison of the UV spectra of this product and compound **29d** indicated a change in electron density in the cytosine moiety. Treatment of compound **29d** with a mixture of saturated aqueous $NaHCO_3$ and CH_3OH (Scheme 1) provided a single new compound (**30d**) with the same molecular weight as compound **29d** that exhibited a bathochromic shift in the UV spectrum. The structure of compound **30d** was assigned based on both 1-D 1H and 2-D homonuclear 1H and heteronuclear 1H–13C NMR experiments. A single crystal X-ray on the HCl salt of compound **30d** confirmed the structure (Fig. 4). Similar studies using aqueous $NaHCO_3$–CH_3OH were performed on additional N4-(aminoacyl)gemcitabines, and the results are summarized in Table 4. Compounds **29a–h** were found to undergo rearrangement to compounds **30a–h**. The amide bond on the glycyl analog (**29a**) was very labile. Products arising from both hydrolysis to gemcitabine and rearrangement to compound **30a** were observed. Increasing the steric bulk at the α-position (**29b–g**) improved the hydrolytic stability of the aminoacyl derivatives and afforded the rearrangement product in good yields (64–81%) with the exception of α,α-dimethy glycyl amide (**29f**) that provided **30f** in 52% yield.

It is interesting to note that the proline derivative (**29h**) underwent rearrangement similar to the open-chain analogs providing compound **30h**, albeit in 38% yield. These results indicate the involvement of the amino group in the *N*4-(α-aminoacyl)-gemcitabine degradation process.

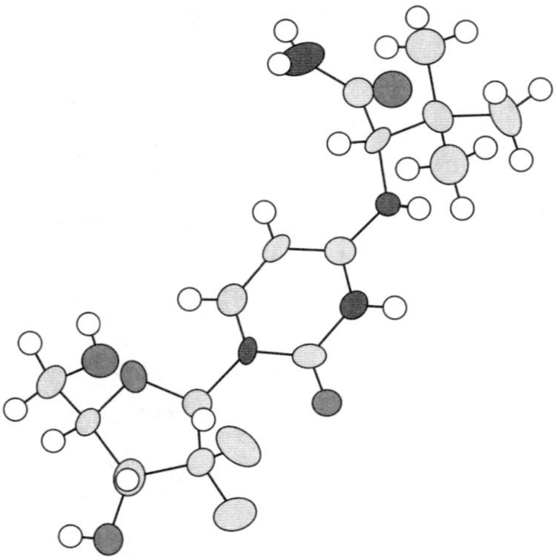

FIGURE 4 ORTEP drawing of **30d**.

TABLE 4 Results from Rearrangement Reactions

Product	Amino acid moiety	Rearrangement yield (%)
30a	L-glycyl	51
30b	L-alanyl	75
30c	L-valyl	64
30d	L-t-leucyl	78
30c	L-phenylalanyl	81
30f	α,α-dimethylglycyl	52
30g	α-aminocyclopropylcarboxyl	69
30h	L-prolyl	38

B. Instability of Aminoacyl Amides and Conflicting Literature

A mechanism for this rearrangement similar to the one proposed by Stella [27] for the aminoacyl prazosin analogs is in Scheme 2. The reaction involves a nucleophilic attack of the free α-amino group in compound (**29**) at the 4-position of the cytidine ring, resulting in the displacement of the amide nitrogen.

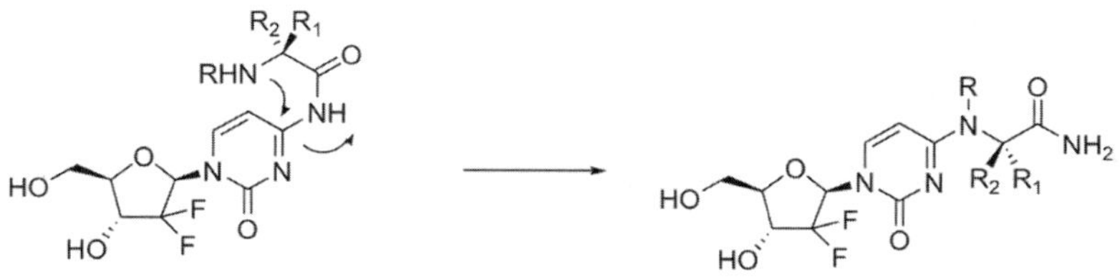

4 5

SCHEME 2 Proposed mechanism for the rearrangement of **4** to **5**.

It is important to note that *N*4-(α-aminoacyl)gemcitabines (**29**) are not stable in solution and undergo rapid pH-dependent degradation. Under near neutral and basic conditions, primary *N*4-(α-aminoacyl)-gemcitabines (**29**) rearrange to novel *N*-(4-gemcitabinyl)amino acid amides (**30**) that are no longer of interest as soluble prodrugs of gemcitabine.

C. Valproate Amide Prodrug of Gemcitabine

In order to further probe the utility of this type of moiety, a series of additional amide prodrugs were prepared, culminating in the discovery of valproate amide **3**.

Valproic acid was coupled directly to gemcitabine using peptide-coupling conditions (EDC/HOBt/NMM) to give compound **3** in excellent yield (95%) (Scheme 3). Compound **3** was not crystalline as the free base.

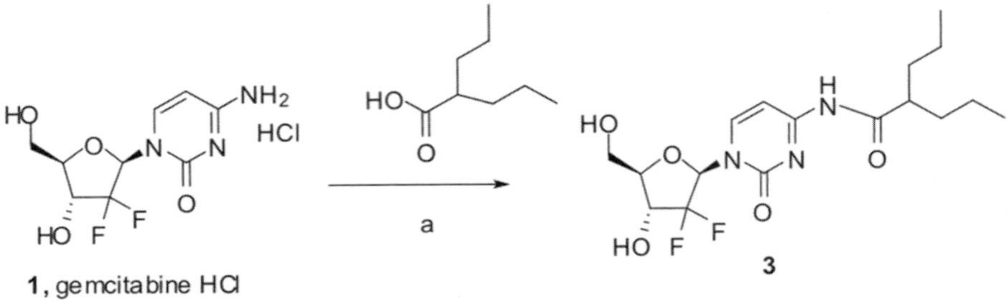

SCHEME 3 Reaction conditions for the conversion of gemcitabine hydrochloride (**1**) to the valproate amide prodrug of gemcitabine (**3**). Reaction conditions: (a) EDC, HOBt, NMM, DMF/DMSO (3:1), 55 °C.

VI. CO-CRYSTALS OF VALPROATE AMIDE OF GEMCITABINE

To provide a means of purification (via crystallization) and oral drug delivery, a screen for salts and co-crystals of compound **3** was conducted. The crystallization screen yielded viable salts with *p*-toluenesulfonic acid (TSA) and benzenesulfonic acid. Ultimately, a novel TSA salt co-crystal (**31**) comprised of two equivalents of compound **3**, one equivalent of TSA, and one equivalent of water was chosen as the crystalline form for future development based on its superior physicochemical properties and ease of preparation. X-ray crystal structure analysis of the salt co-crystal confirmed a single proton transfer between TSA and one equivalent of compound **3** (Fig. 5).

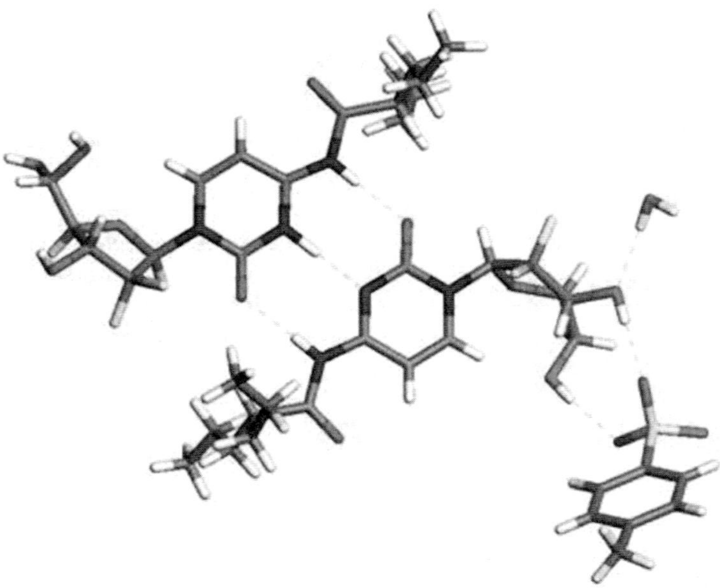

FIGURE 5 X-ray structure of the 2:1:1 gemcitabine prodrug hemi-tosylate hemihydrate **31**. (*See insert for color representation of the figure.*)

Protonation of the cytosine ring creates a complementary structure to the neutral molecule, resulting in a second equivalent of compound **3** being sequestered in the crystal structure by triple hydrogen bonding like that observed for the guanine–cytosine (GC) base pair in DNA. Interestingly, hydrogen bonding in the cytosine–cytosinium ion dimer appears to be stronger (based on shorter intermolecular contacts) than in the GC base pair [34].

VII. STABILITY OF VALPROATE AMIDE PRODRUG

In order to deliver the intact prodrug to the systemic circulation, the compound must first be stable at the various pH environments encountered in the GI tract following oral dosing. The solution stability of the compound was studied in a series of buffer solutions ranging from pH 1 to pH 8 (Table 5). The stability of **3** was found to be pH dependent across the entire pH range. Approximately 21% degradation (hydrolysis to gemcitabine) occurred at pH 1 and none at neutral pH 6–8 after 4 h at 40 °C. To assess the enzymatic stability of the prodrug, the rate of hydrolysis was determined in small intestine homogenates (SIH). Crude homogenates of small intestinal epithelial cells were prepared from sections of the upper small intestine from CD-1 mice and humans. The rate of hydrolysis of **3** (at 96 μM) in SIH incubations (2.2 to 4.4 mg/mL protein) was slow (<10 pmol/mg/min) in both species. In comparison, turnover of the 5′-O-isoleucine ester **2** was found to be substantially more rapid at 2.3 and 3.7 nmol/mg/min in mouse and human, respectively. Over a 30 min incubation period, these rates correspond to 100% hydrolysis of the O-Ile ester **2**, compared to approximately 1% of the prodrug **3**. The *in vitro* hydrolysis profile of the compound was further evaluated in liver S9 incubations to determine the extent of hepatic activation. The hydrolysis rate in hepatic S9 (100 μM) was also relatively slow, at 27 pmol/min/mg in human S9 and 11 pmol/mg/min in mouse. These results suggest **3** would be orally absorbed as the intact prodrug, with minimal hydrolysis occurring in the GI tract due to the molecule's chemical and enzymatic stability. Slow hydrolysis in the liver, however, should result in sustained delivery of gemcitabine to the systemic circulation.

TABLE 5 Solution Stability and Solubility of 3 in Buffer Solutions at 40 °C for 4 h[a]

pH	% hydrolysis to gemcitabine	Solubility (mg/ml)
1	21	1.12
2	12	1.79
4	2	>2.0
6	0	>2.0
8	0	>2.0

[a]100 μg/ml of compound **3** was dissolved in each buffer solution and incubated at 40 C for 4 h.

VIII. BIOAVAILABILITY OF AMIDE PRODRUG

Compound **3** was administered orally to CD-1 mice at a dose of 14.3 mg/kg, and systemic plasma concentrations of both **3** and gemcitabine were analyzed. A single sample was also collected from the hepatic portal vein at 5 min post-dose. The prodrug was rapidly absorbed and passed the intestinal epithelium largely intact, with a mean concentration of 653 ng/mL in portal blood at 5 min. The mean gemcitabine concentration was 99 ng/mL in the same samples, suggesting that a large percentage of the absorbed fraction is stable during absorption. Systemic plasma samples collected at 5 min post-dose had approximately threefold lower concentrations of compound **3** compared to the corresponding portal sample, consistent with hepatic first-pass hydrolysis. Systemic exposure (AUC) to gemcitabine was 778 ng*hr/mL and Cmax was 373 ng/mL. These values compare to 536 ng*hr/mL and 535 ng/mL in mice dosed with a molar-equivalent dose of oral gemcitabine. The Tmax values were 1 h for compound **3** and 0.5 h for gemcitabine. Oral dosing of the prodrug also resulted in prolonged gemcitabine exposure and a change in the shape of the gemcitabine concentration–time profile, suggesting formation rate-limited kinetics.

In a separate study, compound **3** was administered orally to CD-1 mice at doses of 6,12, and 24 mg/kg, and concentrations of compound **3** and gemcitabine were measured, as well as the concentration of dFdU produced via subsequent deamination of gemcitabine. The ratio of dFdU to gemcitabine was found to be less than half of that observed in mice who received equivalent doses of gemcitabine alone, which provides evidence for a reduction in first-pass metabolism through blocking the site of deamination with the prodrug moiety.

IX. ANTITUMOR ACTIVITY OF PRODRUG

The *in vivo* antitumor activity of **3** was evaluated in a HCT-116 human colon tumor xenograft using CD-1 nude mice (Table 6). Initial *in vivo* 3-day cytotoxicity studies indicated that the prodrug itself was not cytotoxic to HCT-116 cells and therefore

TABLE 6 Comparison of Efficacy of 3 (LY2334737) to Gemcitabine (1) when Dosed Orally in a Human Colon HCT-116 Xenograft Model

Dose[a] (mg/kg)	Gemcitabine free base equivalent dose (mg/kg)	Route and schedule	Maximum tumor reduction[b] (% vehicle)
3.77(3)	2.5	oral, qd × 14	56.3 ± 6.0[c]
7.55 (3)	5.1	oral, qd × 14	67.1 ± 4.9[c]
160 (1)	140	ip, q3d × 4	71.5 ± 3.8[c]

[a]Mice were treated with either the vehicle or the indicated drug. LY2334737 or gemcitabine was dosed as an oral gavage daily for 14 days (QD×14). Alternatively, gemcitabine was administered intraperitoneally every 3 days for four doses (Q3D × 4). Tumor volumes and animal's weight were measured three times a week for 6 weeks.
[b]Maximum reduction in tumor volume observed with drug treatment versus the vehicle treatment group.
[c]Statistically significantly different ($p<0.0001$) from the vehicle control group.

needed to be cleaved to release gemcitabine. The compound displayed significant antitumor activity at the 7.55 and 3.77 mg/kg doses when given daily for 14 days. These data also indicated that the 7.55 mg/kg dose of the prodrug was equally efficacious to gemcitabine given intraperitoneally at 160 mg/kg for a total of four doses. The treatment was well tolerated with 2% or less weight loss observed for all groups.

X. SUMMARY

In summary, an orally active prodrug of gemcitabine was developed that possesses excellent chemical and enzymatic stability and delivers intact prodrug to the systemic circulation, thereby reducing intestinal exposure to gemcitabine. The compound achieved oral efficacy on a daily schedule that is equivalent to a clinically relevant i.p. dose of gemcitabine in HCT-116 human colon cancer xenografts. Based on these results, **3** was selected for further clinical development and has advanced into Phase I clinical trial [35].

REFERENCES

1. LW Hertel, GB Boder, JS Kroin, SM Rinzel, GA Poore, GC Todd, GB Grindey. Evaluation of the antitumor-activity of gemcitabine (2,2′-difluorodeoxycytidine). Cancer Res 50:4417–4422, 1990.
2. F Eckel, G Schneider, RM Schmid. Pancreatic cancer: a review of recent advances. Expert Opin Invest Drugs 15:1395–1410, 2006.
3. C Manegold. Gemcitabine (Gemzar) in non cell lung cancer. Expert Rev Anticancer Ther 4:345–360, 2004.
4. V Heinemann. Gemcitabine in metastatic breast cancer. Expert Rev Anticancer Ther 5:429–443, 2005.
5. P Huang, S Chubb, LW Hertel, GB Grindey, W Plunkett. Action of 2,2′- difluorodeoxycytidine on DNA synthesis. Cancer Res 51:6110–6117, 1991.
6. V Ruiz van Haperen, G Vefreman, J Vermorken, GG Peters. 2,2′- Difluorodeoxycytidine (gemcitabine) incorporation into RNA and DNA from tumour cell lines. Biochem Pharmacol 46:762–766, 1993.
7. V Heinemann, Y Xu, S Chubb, A Sen, LW Hertel, GB Grindey, W Plunkett. Cellular elimination of 2,2′-difluorodeoxycytidine 5′-triphosphate: a mechanism of self-potentiation. Cancer Res 52:533–539, 1992.
8. J Stubbe, WA van der Donk. Ribonucleotide reductases: radical enzymes with suicidal tendencies. Chem Biol 12:793–801, 1995.
9. J Stubbe, WA van der Donk. Protein radicals in enzyme catalysis. Chem Rev 98:705–762, 1998.
10. E Artin, J Wang, GJS Lohman, K Yokoyama, G Yu, RG Griffin, G Bar, J Stubbe. Insight into the mechanism of inactivation of ribonucleotide reductase by gemcitabine 5′-diphosphate in the presence or absence of reductant. Biochemistry 48:11622–11629, 2009.
11. V Gandhi, W Plunkett. Modulatory activity of 2,2′-difluorodeoxycytidine on the phosphorylation and cytotoxicity of arabinosyl nucleosides. Cancer Res 50:3675–3680, 1990.

12. V Heinemann, YZ Xu, S Chubb, S Sen, L Hertel, GB Grindey, W Plunkett. Inhibition of ribonucleotide reduction in cerf-cem cells by 2,2′-difluorodeoxycytidine. Mol Pharmacol 38:567–572, 1990.
13. V Gandhi, P Huang, YZ Xu, V Heinemann, W Plunkett. Metabolism and action of 2,2′-difluorodeoxycytidine: self-potentiation of cytotoxicity. Adv Exp Med Biol 309A: 125–130, 1991.
14. L Shipley, T Brown, J Cornpropst, M Hamilton, W Daniels, H Culp. Metabolism and disposition of gemcitabine, and oncolytic deoxycytidine analog in mice, rats, and dogs. Drug Metab Dispos 20:G849–G855, 1992.
15. J Kroep, C van Moorsel, G Veerman, D Voorn, R Schultz, J Worzallia, L Tanzer, R Merriman, H Pinedo, G Peters. Role of deoxycytidine kinase (dCK), thymidine kinase 2 (TK2) and deoxycytidine deaminase (dCDA) in the antitumor activity of gemcitabine (dFdC). Adv Exp Med Biol 431:657–660, 1998.
16. JR McCarthy, DP Matthews, DM Stemerick, EW Huber, P Bey, BJ Lippert, RD Snyder, PS Sunkara. Stereospecific method to (E) and (Z) terminal fluoroolefins and its application to the synthesis of 2′-deoxy-2′-fluoromethylenenucleosides as inhibitors of ribonucleotide reductase. J Am Chem Soc 113:7439–7440, 1991.
17. GJ Gerfen, WA van der Donk, G Yu, JR McCarthy, E Jarvi, DP Matthews, C Farrar, RG Griffin, J Stubbe. Characterization of a substrate-derived radical detected during the inactivation of ribonucleotide reductase from *Escherichia coli* by 2′fluoromethylene-2′deoxycytidine 5′diphosphate. J Am Chem Soc 120:3823–3835, 1998.
18. Y Yonetani, T Mizukami. Kinetic analysis of inhibition of human ribonucleotide reductase and DNA polymerase a by a new anticancer agent, (E)-2′-deoxy-2′-(fluoromethylene) cytidine. Proc Am Assoc Cancer Res 37:405, 1996.
19. GI Rodriguez, RE Jones, EK Orenberg, ML Stoltz, DJ Brooks. Phase I clinical trials of tezacitabine [(E)-2′-deoxy-2′-(fluoromethylene)cytidine] in patients with refractory solid tumors. Clin Cancer Res 8:2828–2834, 2002.
20. N Masuda, S Negoro, K Takeda, N Takifuji, T Hirashima, T Yana, N Kurata, T Kuwabara, S Kobayashi, S Kudoh, et al. Phase I and pharmacologic study of oral (E)-2′-deoxy-2′-(fluoromethylene)cytidine on a daily × 5-day schedule. Invest New Drugs 16:245–254, 1999.
21. N Shimma, I Umeda, M Arasaki, C Murasaki, K Masubuchi, Y Kochu, M Miwa, M Ura, N Sawada, H Tahara, et al. The design and synthesis of a new tumor-selective fluoropyrimidine carbamate, capecitabine. Bioorg Med Chem 8:1697–1706, 2000.
22. P Ettmayer, GI Amidon, B Clement, B Testa. Lessons learned from marketed and investigational prodrugs. J Med Chem 47:2393–2404, 2004.
23. C MacDougall, J Guglielmo. Pharmacokinetics of valaciclovir. J Antimicrob Chemother 53:899–901, 2004.
24. X Song, PL Lorenzi, CP Landowski, BS Vig, JM Hilfinger, GL Amidon. Amino acid ester prodrugs of the anticancer agent gemcitabine: synthesis, bioconversion, metabolic bioevasion, and hPEPT-1 mediated transport. Mol Pharm 2:157–167, 2005.
25. DM Bender, JA Peterson, JR McCarthy, H Gunaydin, Y Takano, KN Houk. Cyclopropanecarboxyl acid esters as potential prodrugs with enhanced hydrolytic stability. Org Lett 10:509–511, 2008.
26. BS Anand, YE Nashed, AK Mitra. Novel dipeptide prodrugs of acyclovir for ocular herpes infections: bioconversion, antiviral activity and transport across rabbit cornea. Curr Eye Res 26:151–163, 2003.

27. NL Pochopin, WN Charman, VJ Stella. Kinetics of rearrangement and hydrolysis of amino acid derivatives of prazosin. Int J Pharm 105:169–176, 1994.

28. YH Choe, CD Conover, D Wu, M Royzen, RB Greenwald. Anticancer drug delivery systems: N4-acyl poly(ethyleneglycol) prodrugs of ara-C l. Efficacy in solid tumors. J Control Release 79:41–53, 2002.

29. SC Mauldin, CJ Paget, Jr., CD Jones, JM Colacino, AJ Baxter, KA Staschke, NG Johansson, L Vrang. Synthesis and antiviral activity of prodrugs of the nucleoside 1-[2′,3′-dideoxy-3′-C-(hydroxymethyl)-β-D-erythropentofuranosyl] cytosine. Bioorg Med Chem 6:577–585, 1998.

30. RB Greenwald, YH Choe. Terminally branched polymeric linkers containing extension moieties for prodrug conjugates. US Patent 20020015691, Feb 7, 2002.

31. YH Choe, RB Greenwald. Terminally-branched polymeric linkers and polymeric conjugates as prodrugs. WO 2002065988, Aug 29, 2002.

32. D Standring, JP Sommadossi, AL Patty, M Seiffer. β-L-2′-Deoxynucleosides for the treatment of resistant hepatitis B virus (HBV) strains, and combination therapies. WO 2004024095, Mar 25, 2004.

33. D Zhang, DM Bender, F Victor, JA Peterson, RD Boyer, GA Stephenson, A Azman, JR McCarthy. Facile rearrangement of N^4-(α-aminoacyl)cytidines to N-(4-cytidinyl) amino acid amides. J Med Chem 52:6958–6961, 2009.

34. J Šponer, J Leszczynski, V Vetterl, P Hobza. Base stacking and hydrogen bonding in protonated cytosine dimer: The role of molecular ion-dipole and induction interactions. J Biomol Struct Dyn 13:695–706, 1996.

35. K Koolen, PO Witteveen, I Garcia-Ribas, S Callies, V Andre, RH Kronemeijer, A Noj, JH Beijnen, EE Voest, JH Schellens. Phase I study of oral gemcitabine prodrug (LY2334737) alone and in combination with erlotinib in patients (pts) with advanced solid tumors. J Clin Oncol 27 (Suppl):15s; Abstr 2576, 2009; http://www.asco.org/ascov2/Meetings/Abstracts?&vmview=abst_detail_view &confID=65&abstractID=33804 (accessed July 3, 2013).

7

ENZYMATICALLY ACTIVATED PHOSPHATE AND PHOSPHONATE PRODRUGS

IVAN S. KRYLOV AND CHARLES E. MCKENNA
Department of Chemistry, University of Southern California, Los Angeles, California

I. INTRODUCTION

To be successful, therapeutic agents must not only exhibit high affinity for their biological targets but once administered *in vivo* must also be able to reach the target site at the effective concentration and sufficiently sustained intact (or as an active metabolite). In addition they must be eliminated. These requirements describing the journey of the drug inside the patient are applicable to all drugs and are traditionally categorized as administration, distribution, metabolism, and excretion (ADME), which constitute a pharmacokinetic (PK) profile of the drug [1]. Additionally, drugs and their metabolites should ideally be nontoxic to the treated patient. The ADME profile is defined by a drug's physical and chemical properties, such as solubility, polarity, and chemical and enzymatic stabilities. In order to establish optimal ("drug-like") ADME properties for drug candidates, multiple approaches analyzing physicochemical properties, mechanisms of action, and administration routes of numerous existing drugs and drug candidates have been utilized [2–4]. One of the first measures of drug-like properties was proposed by Lipinski et al. ("Lipinski's Rule of 5"), which claimed to predict poor absorption or permeation of the drug candidate if two out of the following parameters apply: the MW > 500 Da, number of H-bond donors > 5, number of H-bond acceptors > 10, and the calculated LogP (CLogP) > 5

Enzyme Technologies: Pluripotent Players in Discovering Therapeutic Agents, First Edition.
Edited by Hsiu-Chiung Yang, Wu-Kuang Yeh, and James R. McCarthy.

[5]. This rule especially applies to orally administered drugs because absorption, along with delivery and metabolism, determines overall oral bioavailability – the fraction of the drug that reaches the plasma compartment after oral administration. Indeed, reduced bioavailability is considered to be the main obstacle to further development of orally administered drugs as their pharmacological performance depends on concentration of the drug in plasma cells or targeted tissues [6]. Drug's ADME profile can be altered utilizing various prodrug approaches or pharmaceutical formulations of the drug candidate with agents improving certain physicochemical properties.

Historically, the idea of targeted delivery of pharmacologically active agents has its origins in Paul Ehrlich's postulate of an ideal drug that would function as a "magic bullet," uniquely affecting its therapeutic target. Recognition that the form of a drug at administration may need to be different from the effective molecule acting *in situ* led to the concept of the prodrug, which implies chemical or preferably specific enzymatic activation. Prodrug may exhibit little or no pharmacological activity itself while releasing a therapeutically active metabolite upon transformation *in vivo* [7]. The prodrug strategy for optimization of a drug's therapeutic potential gained increasing popularity in the early 1990s–2000s, as evidenced by an exponential increase in the number of prodrug patents (>2000% over the years 1993–2003) [8]. A similar trend is revealed by analysis of the number of publications including "prodrugs" as a key word during 1984–2003 (Fig. 1, left). Since 2003, however, the number of publications per year through 2010 has grown by only 20% (Fig. 1, right) [9].

Prodrug's physicochemical properties that may be targeted for improvement include water or lipid membrane solubility, brain permeability, local delivery, presystemic metabolism, toxicity, or unacceptable taste of the parent drug. In order to achieve clinical objectives, the designed prodrug should meet several basic criteria:

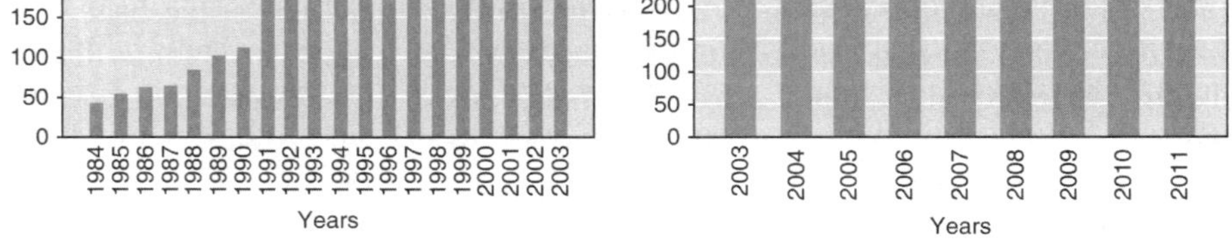

FIGURE 1 Number of publications with "prodrugs" as a key word published from 1984 until 2003 (left) and from 2003 through 2011 (right) [9].

(1) ready absorption from the site of administration, (2) stability under delivery conditions and adequate transport to the target site, and (3) conversion to the active parent drug or metabolite *in vivo* by enzymatic or chemical reaction at a rate consistent with pharmacological efficacy, preferably at a higher conversion rate than elsewhere in the body. Additionally, both the prodrug and the metabolized promoiety should be nontoxic [10]. The release of the parent drug via a purely chemical process has the advantage of being independent from variability in levels of enzyme expression in various tissues and species, but examples of successful prodrugs selectively activated via chemical reactions at the target site are limited [11–13], in part due to the difficulty in reaching an appropriate balance between chemical stability and an efficient activation rate under physiological conditions (e.g., pH).

On the other hand, enzymatic activation of chemically stable prodrugs not only allows targeted and efficient release of the parent drug or metabolite but also simplifies their handling and development [14]. A drawback is that the expression pattern, selectivity, and structure–function dependence of the enzymes participating in prodrug activation are likely to differ in various tissues, individuals, and species. A potential prodrug activating enzyme should meet the following criteria: (1) be a member of a well-characterized enzyme family that plays a known role in the target tissue or in the progression of disease and (2) be unique to the target tissue or at least be present in significantly higher concentrations compared to healthy tissues. Finally, the enzyme concentration and the rate it selectively liberates the parent drug should be sufficient to generate the desired pharmacological effect. Unfortunately, enzymes utilized for prodrug activation (e.g., esterases) often demonstrate overlapping substrate selectivities and ubiquitous expression throughout the body, thus diminishing the possibility for site-selective activation (targeted delivery) and making assessment, which enzymes are responsible for conversion of the prodrug, *in vivo* difficult [15]. As a result, these important parameters have rarely been studied before a particular prodrug becomes a viable candidate for preclinical/clinical development, except in cases where the prodrug is purpose-designed for targeted drug delivery to tissues (e.g., tumor) or organs (e.g., liver). The fact that relatively few prodrugs have reached the market despite hundreds of reports on the design of prodrugs showing proof of concept emphasizes this missing or underdeveloped mechanistic component in prodrug development. In this chapter, we will survey several representative examples of current prodrug approaches which explicitly depend on enzyme-mediated activation.

Modifications designed to improve the ADME properties of phosphate or phosphonate group-containing drugs are an important area of current prodrug research. The hydrophilic phosphoric or phosphonic acid groups are almost completely ionized at physiological pH, dramatically affecting absorption and delivery of the pharmaceutical product by increasing aqueous solubility but concurrently decreasing cell membrane permeability and thus oral bioavailability. Phosphate and phosphonate groups have been long recognized as targets for prodrug design aimed either to mask the negative charges of these groups in order to increase the drug's ability to penetrate cell membrane (Table 1, No 1–3) or to incorporate the phosphate group into a hydrophobic drug molecule in order to increase its aqueous solubility (Table 1, No 4–26).

TABLE 1 Selected Examples of Drugs Incorporating a Phosphate Group to Increase Aqueous Solubility (No 1–3) and of Nucleotide Analogs Incorporating Phosphate and Phosphonate Groups Targeted for Prodrug Design (No 4–26)

No	Structures	No	Structures
1	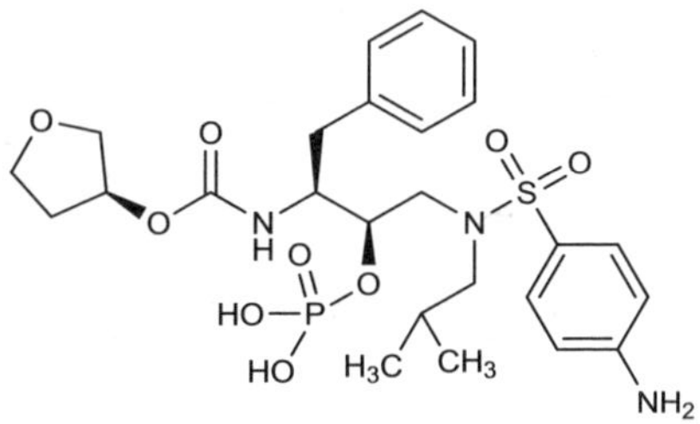 Fosamprenavir	2	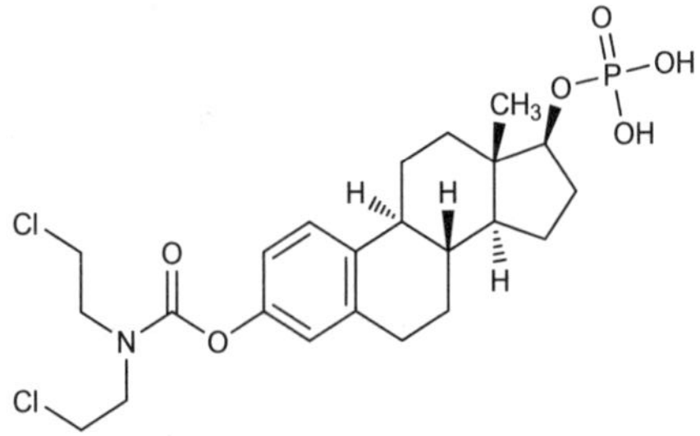 Estramustine phosphate
3	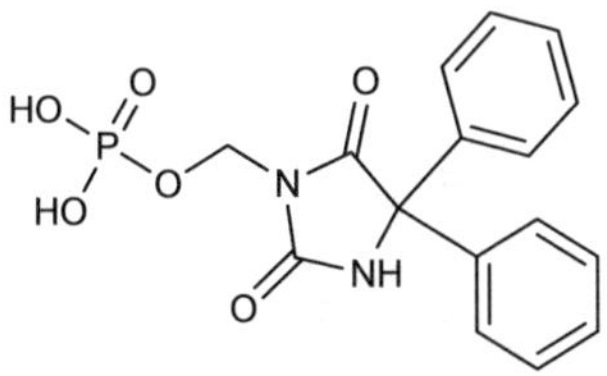 Fosphenytoin		
4	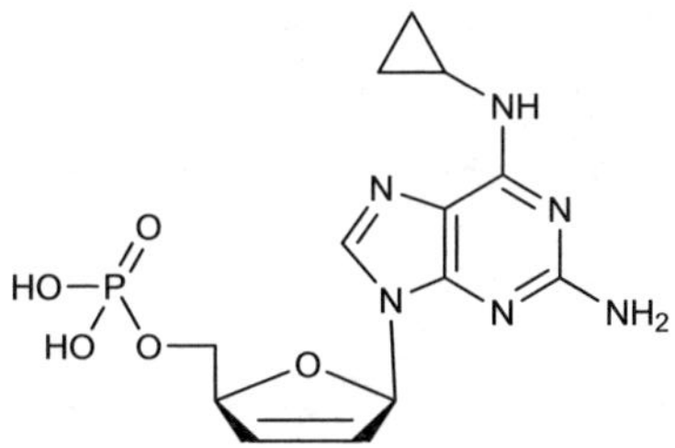 Abacavir (ABC) monophosphate	5	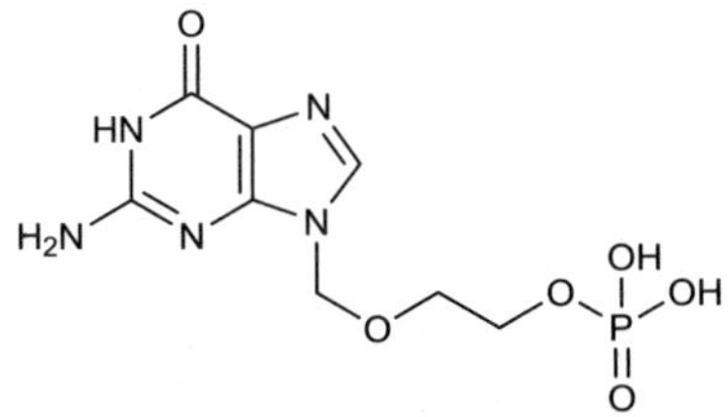 Acyclovir monophosphate

6 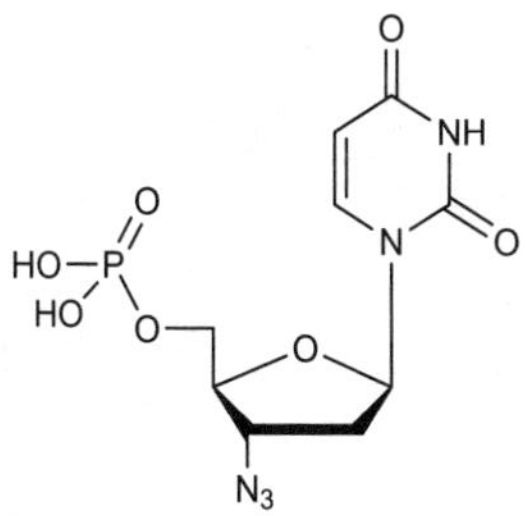 3-Azido-thymidine (AZT) monophosphate	7 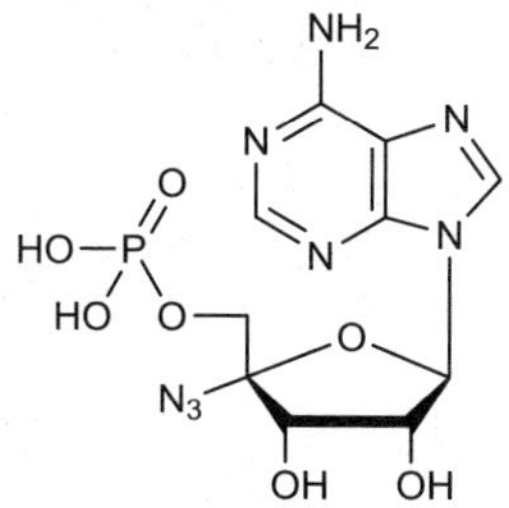 4'-Azidoadenosine monophosphate
8 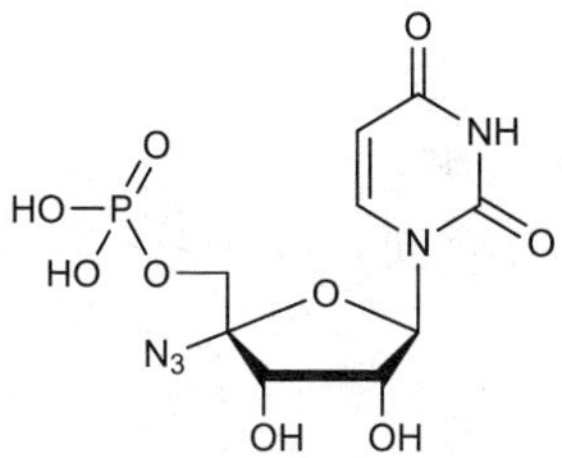 4'-Azidouridine monophosphate	9 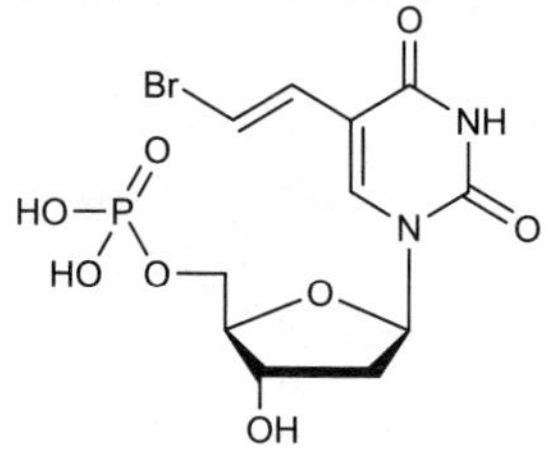 (E)-5-(2-Bromovinyl)-2'-deoxy-uridine (BVDU, Brivudine) monophosphate
10 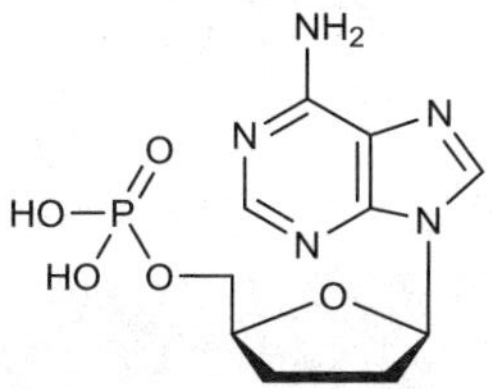 2',3'-Dideoxyadenosine (ddA) monophosphate	11 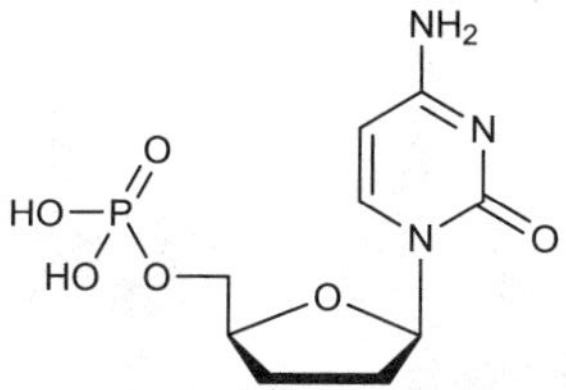 2',3'-Dideoxycytidine (ddC) monophosphate

(Continued overleaf)

TABLE 1 (*Continued*)

No	Structures	No	Structures
12	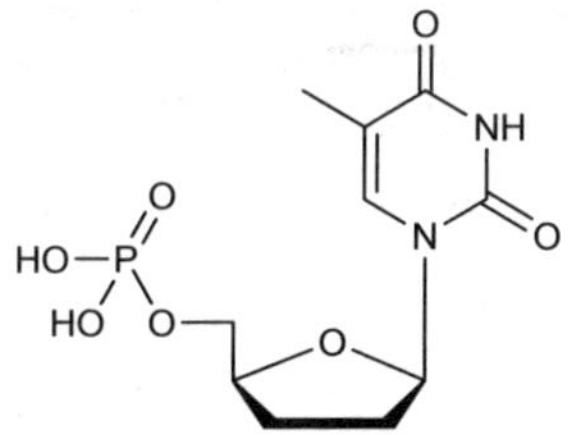 2',3'-Dideoxythymidine (ddT) monophosphate	13	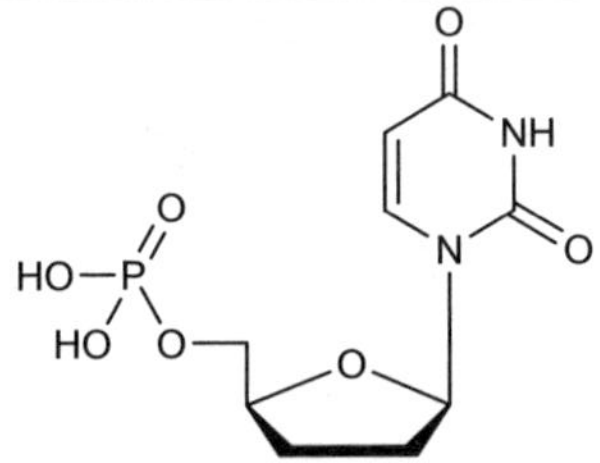 2',3'-Dideoxyuridine (ddU) monophosphate
14	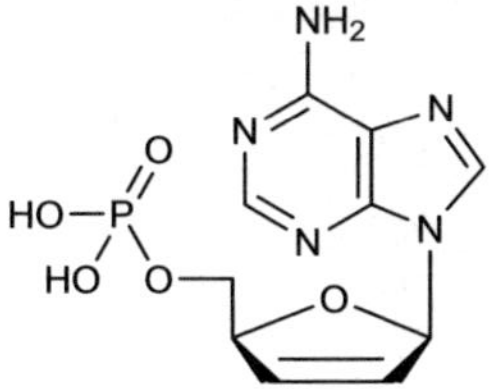 2',3'-Didehydro-2',3'-dideoxy-adenosine (d4A) monophosphate	15	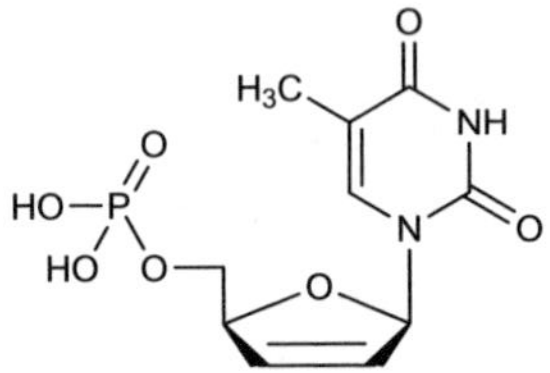 2',3'-Didehydro-2',3'-dideoxy-thymidine (d4T) monophosphate
16	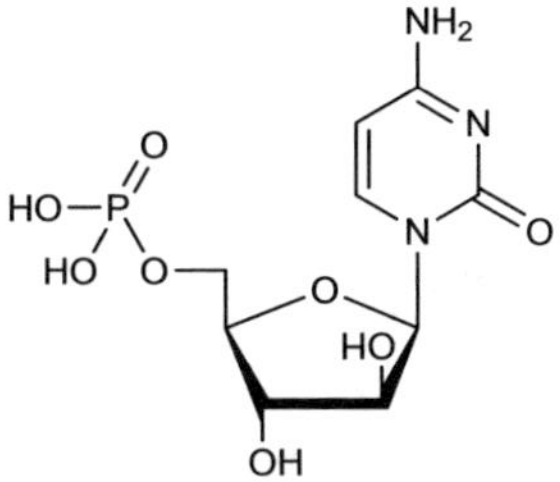 Cytarabine (Ara-C) monophosphate	17	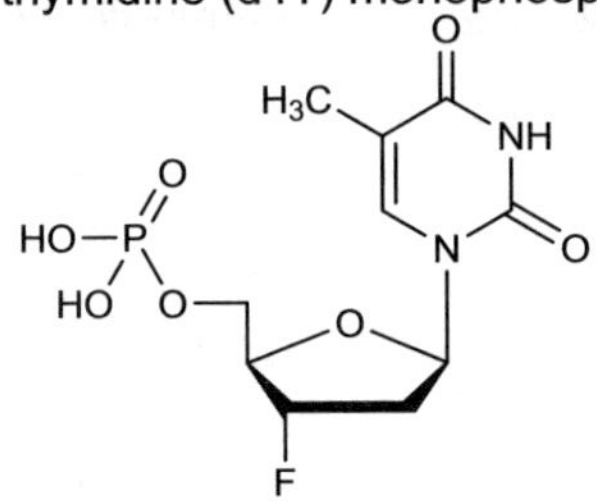 3'-Deoxy-3'-fluoro-thymidine monophosphate

18

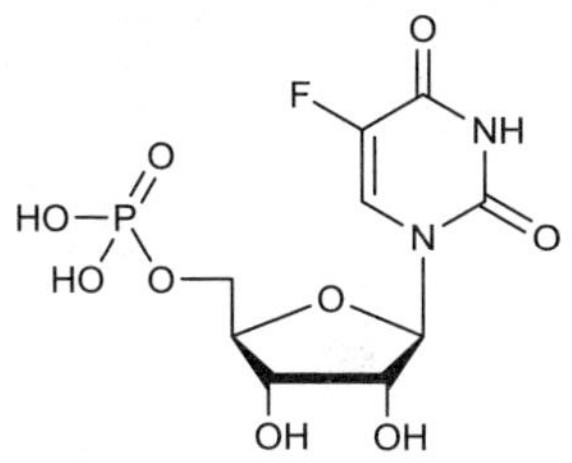

5-Fluoro-uridine monophosphate

19

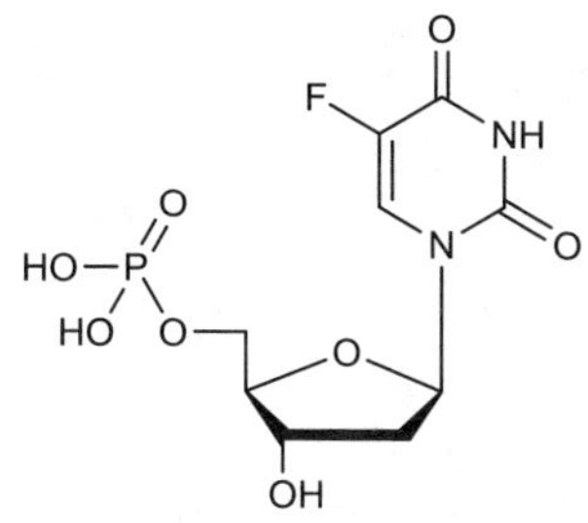

5-Fluoro-2'-deoxy-
uridine (FUdR) monophosphate

20

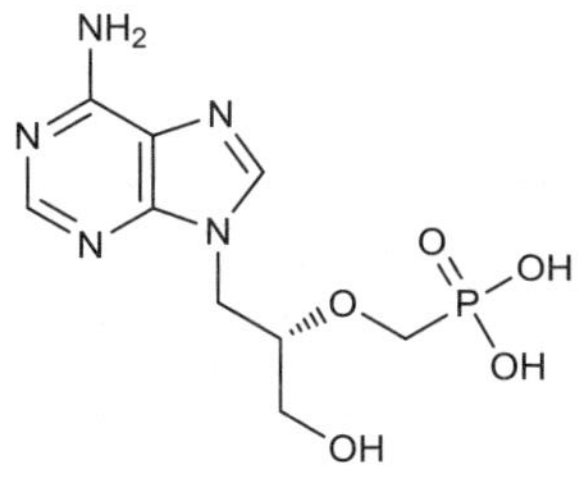

(S)-HPMPA

21

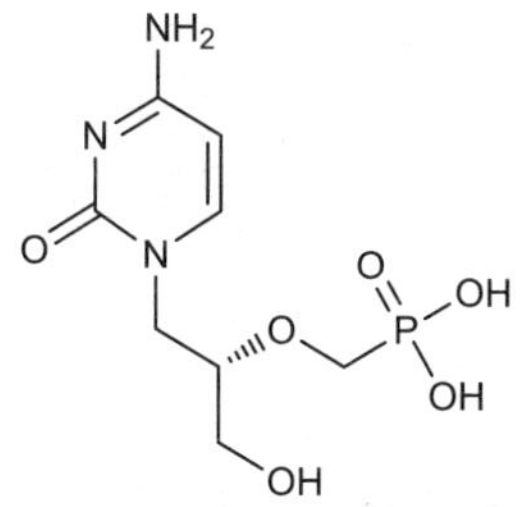

(S)-HPMPC

22

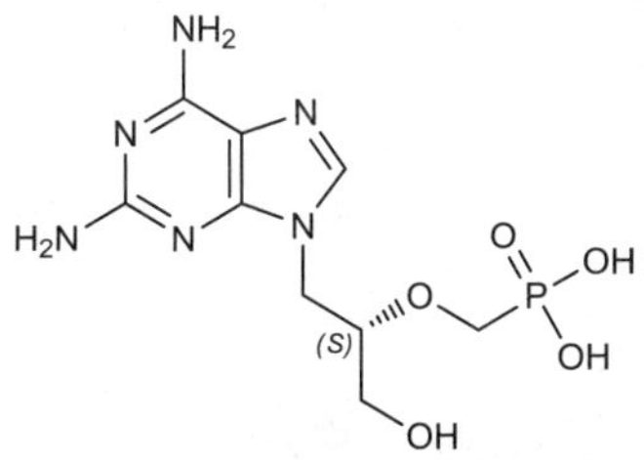

(S)-HPMPDAP

23

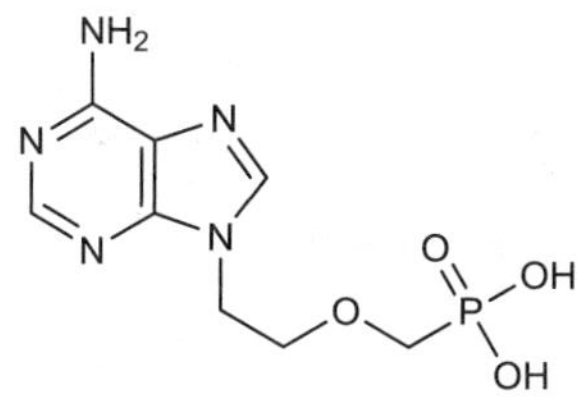

PMEA

(*Continued overleaf*)

TABLE 1 (*Continued*)

No	Structures	No	Structures
24	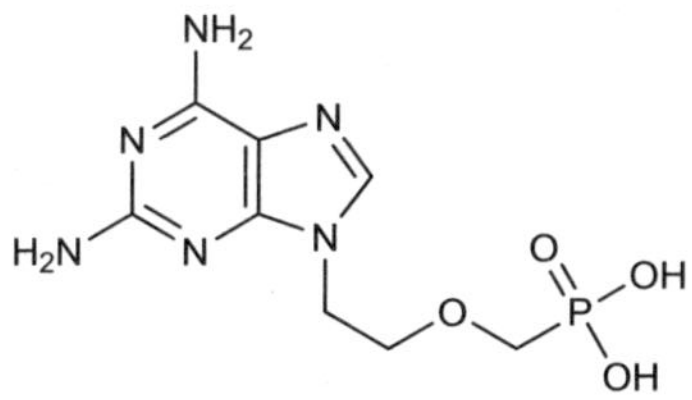 PMEDAP	25	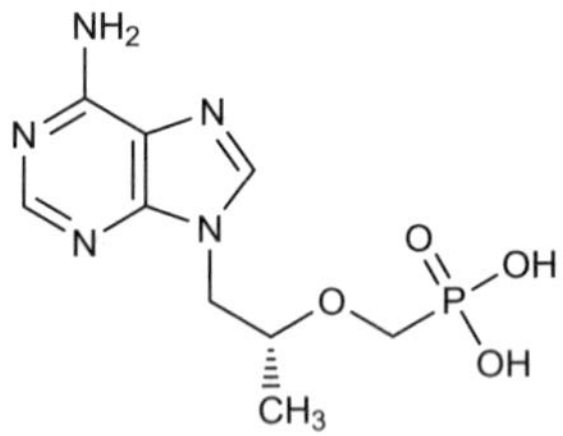 (R)-PMPA
26	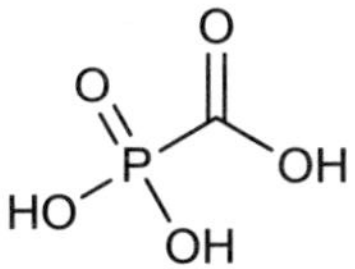 PFA (Foscarnet)		

II. PHOSPHATE GROUP USED FOR SOLUBILITY ENHANCEMENT

Solubility – the property limiting the maximum amount of a drug that can be dissolved in a given volume of a solvent or a solution – is one of the most important physicochemical properties affecting therapeutic potential. Thus, limited aqueous and particularly gastrointestinal (GI) fluid solubility (below 100 μg/ml) of orally administered drugs will reduce their absorption from the GI tract due to the resulting inability to achieve a sufficient concentration gradient between the apical and serosal sides of the intestinal epithelium and, for insoluble compounds, due to the negligible permeability of particulate material over the GI mucosa [16]. Aqueous solubility is also a key factor for parenterally administered drugs [17]. Since nearly 40% of the drug candidates from combinatorial screening programs encounter a solubility problem, various prodrug approaches, as well as formulation techniques, have been developed to address this limitation [18]. The design of prodrugs improves aqueous solubility and can be based on incorporation into the drug's structure of an enzymatically labile, ionizable, or highly polar neutral functional group, such as a phosphate, an amino acid, or a sugar. However, in the case of orally administered prodrugs, along with increased aqueous solubility, introduction of the ionizable or polar group into the hydrophobic molecule will significantly reduce its passive diffusion across intestinal epithelial cell membranes, unless the parent drug is rapidly released from the hydrophilic prodrug at the interface of the intestinal fluid and epithelial cell membrane before or during the absorption. Therefore, to achieve oral bioavailability while avoiding early release leading to precipitation of the parent drug, an enzyme cleaving the ionizable group should be present at the brush border of the enterocytes.

Although the phosphate monoester group is chemically stable under physiological conditions, it is rapidly cleaved by endogenous enzymes, particularly alkaline phosphatases, localized in the intestinal brush border membrane [19]. Phosphate ester prodrugs are especially useful for drug candidates that require administration of a high dose but have limited absorption due to a low dissolution rate. Although a number of prodrug concepts for improved solubility have been developed, only a few orally administered prodrugs have actually reached the clinic. Examples of phosphate ester prodrugs currently used in the clinic include fosamprenavir, estramustine phosphate (EMP), and fosphenytoin (Fig. 2) [18].

Fosamprenavir (Telzir®, Lexiva™; Fig. 2) is a phosphate monoester prodrug of poorly soluble amprenavir (Agenerase®), a protease inhibitor used to treat HIV-1 infection [20,21]. A high dose of amprenavir (1200 mg bid, eight capsules) is needed to achieve a therapeutically useful pharmacological effect, and its very low aqueous solubility (0.04 mg/mL) entails inconvenient administration of a large number of capsules with relatively low drug content. Thus, to increase solubility, the free hydroxyl group of amprenavir was phosphorylated to form the more water-soluble fosamprenavir (Na salt ~ 0.3 mg/mL; Ca salt > 100 mg/mL), which releases the parent drug via alkaline phosphatase-mediated dephosphorylation in the GI tract. The Food

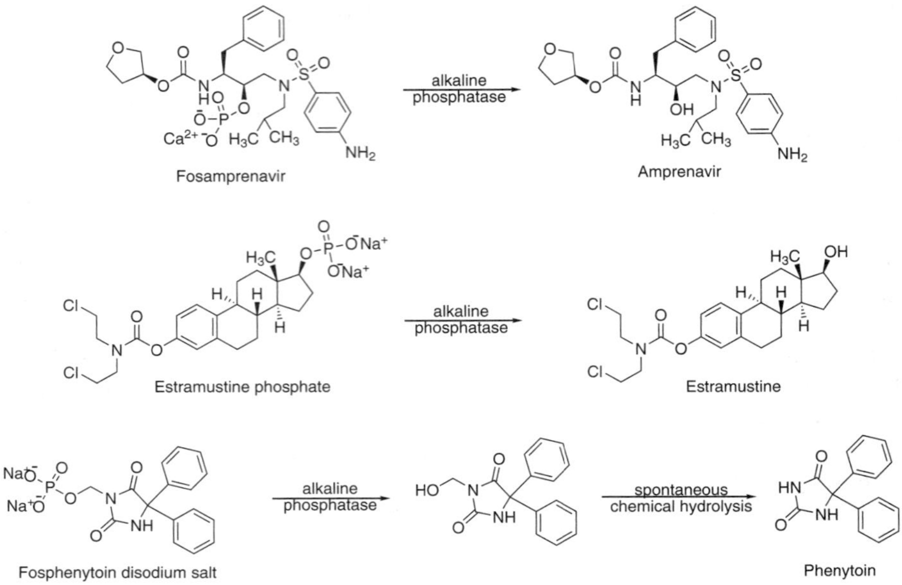

FIGURE 2 Structures and activation pathways of selected phosphate prodrugs.

and Drug Administration (FDA) has approved the calcium salt form of this prodrug, which exhibits greatly increased aqueous solubility and higher oral bioavailability relative to amprenavir, thus permitting more convenient and patient-compliant once-a-day dosing regimen [22].

Another FDA-approved phosphate prodrug is EMP (Emcyt®, Fig. 2), a prodrug of the lipophilic cytotoxic drug estramustine which is used to treat advanced prostatic carcinoma [23,24]. Although the origin of the parent drug cytotoxicity is only partially understood, it was proposed that this nor-nitrogen mustard-carbamate derivative of 17β-estradiol binds to microtubule-associated proteins and/or to tubulin, causing the arrest of cell division in the G_2/M phase of the cell cycle. In order to develop injectable and oral formulations of estramustine, its insufficient water solubility was dramatically increased (200-fold) by the incorporation of an ionizable phosphate group into its structure (via phosphorylation of the free hydroxyl group). EMP rapidly released the parent drug upon enzymatic dephosphorylation; however, PK studies revealed incomplete absorption of EMP from the GI tract (observed oral bioavailability ranges from 44% to 75%), which might be explained by profound first-pass metabolism and undesirable GI interactions of the phosphate prodrug with calcium-containing food and antacids [23].

A further example is fosphenytoin disodium salt (Cerebyx®, Fig. 2), a phosphate prodrug of a sparingly soluble antiepileptic agent, phenytoin, the solubility of which strongly depends on pH of the solution (due to the presence of weakly acidic amine, $pK_a = 8.3$). A high concentration of this drug can be achieved only under very basic conditions ($pH \geq 12$) [25]. A small decrease in pH at the injection site can cause the parent drug to precipitate resulting in local irritation. To increase water solubility and avoid this adverse side effect, the prodrug, fosphenytoin, has been developed. In contrast to the estramustine and amprenavir, phenytoin lacks a free hydroxyl group in its structure, and as a result, the phosphate group was introduced using an oxymethylene linker. Fosphenytoin showed 7000-fold greater aqueous solubility compared to the parent drug and readily released phenytoin after enzymatic dephosphorylation followed by spontaneous dehydroxymethylation of the unstable intermediate. It has been approved by the FDA for intramuscular maintenance administration for periods up to 5 days, intramuscular loading dose administration, and intravenous treatment of status epilepticus [26].

In conclusion, the phosphate group is a valuable prodrug tool to increase the aqueous solubility of lipophilic drugs with a high targeted oral or parenteral dose. Potential problems encountering and implying this approach include insufficient intestinal permeability of the parent drugs, premature dephosphorylation of the prodrugs in the GI tract, precipitation of the released parent drug in the intestinal lumen, and chelation of the prodrug phosphate group with calcium-containing food and antacids [18].

III. PHOSPHATE AND PHOSPHONATE PRODRUGS

Nucleoside and nucleotide analogs constitute an important class of antimetabolites used in anticancer and antiviral chemotherapy. These therapeutic agents mimic metabolism of natural nucleosides and nucleotides but typically contain a modified sugar moiety, in which the 3′-hydroxyl group may be sterically hindered, stereotypically displaced, or absent. After intracellular phosphorylation to the corresponding monophosphate and further to the triphosphate followed by incorporation into newly synthesized DNA or RNA by a targeted nucleic acid polymerase, the nucleoside triphosphate analogs terminate nucleic acid chain elongation [27]. In addition, certain nucleoside and nucleotide analogs may act as inhibitors of the key enzymes involved in phosphorylation and RNA synthesis and may be able to directly activate the caspase cascade. Eventually, all these processes lead to a cell death. Thus, nucleoside analogs are prodrugs, since they require additional inter- or intracellular transformations in order to exhibit therapeutic activity. However, their activation is often inefficient due to a slow first phosphorylation step, which leads to the nucleoside monophosphate and is commonly catalyzed by a nucleoside kinase encoded by the host cell or the virus infecting the cell. In the latter case, mutations may result in resistance by slowing drug activation. Incorporation of the phosphate or phosphonate group into the nucleoside analog not only bypasses the rate-limiting step (first phosphorylation) in the activation pathway but also reduces cell membrane permeability (due to the anionic status of the phosphate and phosphonate group at physiological pH) and, in case of phosphates, decreases stability (due to dephosphorylation by endogenous phosphatases in most biological fluids) [28]. Phosphate and phosphonate groups in the nucleotide analogs have been long recognized as targets (substrates) for modification in order to generate prodrugs with improved ADME properties. Simple dialkyl esters used as a promoiety to mask both negative charges of the phosph(on)ate group fail to release the parent nucleotide [29,30] since chemical hydrolysis of the phosphotriesters is particularly slow under physiological conditions and specific enzymes (e.g., phosphotriesterases) that cleave them have not been identified in mammalian cells [31]. In contrast to alkyl diesters, monoester prodrugs are potential substrates for phosphodiesterases [32,33]. In order to provide efficient intracellular delivery of the nucleotides under physiological conditions, more sophisticated prodrug approaches utilizing intracellular enzymes to trigger the cascade of intramolecular chemical reactions liberating the parent drug from the prodrug are been developed. Several examples will be given.

A. Acyloxyalkyl and Alkyloxycarbonyloxymethyl Approach

In an effort to construct a hydrophobic promoiety exhibiting a good balance between stability and efficient liberation of the parent drug, prodrugs have been proposed that would utilize enzymatically labile acyloxy group connected to a phosph(on)ate group through a linker, usually a hydroxymethyl group [34–38]. Promoiety stability was adjusted by modification of the acyloxy structure. The proposed activation mechanism of the bis(acyloxymethyl) phosph(on)ate prodrugs begins with enzymatic hydrolysis of the acyloxy group catalyzed by esterases (including carboxylesterases, paraoxonases, and cholinesterases) followed by spontaneous elimination of the linker as formaldehyde and release of the mono acyloxyalkyl ester of the parent drug (Fig. 3a) [39]. Cleavage of the second acyloxyalkyl promoiety may proceed via the same mechanism or may be catalyzed by phosphodiesterases (Fig. 3b).

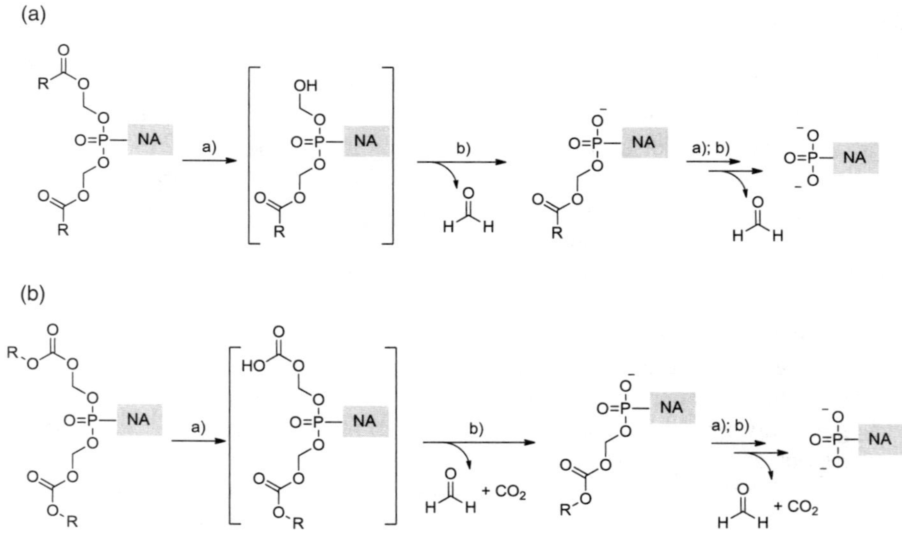

FIGURE 3 General representation of activation pathway of (a) bis(acyloxymethyl) prodrug and (b) bis(alkyloxycarbonyloxymethyl) prodrug to the corresponding nucleoside monophosphate analog: a) enzymatic hydrolysis and b) spontaneous chemical hydrolysis.

Since 1984, when the acyloxyalkyl group was first studied as a promoiety for model phenyl and benzyl phosphates [35], this prodrug approach has been utilized to increase oral bioavailability of numerous nucleotide analogs, including 2′,3′-dideoxyuridine (ddU) monophosphate [40], 5-fluoro-2′-deoxyuridine (FUdR) monophosphate [34], 3′azidothymidine (AZT) monophosphate [37,41], 9-[2-(phosphonomethoxy)-ethoxy] adenine (PMEA) [38], phosphonoformic acid (PFA, foscarnet) [36], and 9-[2-(phosphonomethoxy)propyl]adenine (PMPA) [42]. The bis(acyloxyalkyl) prodrug approach has recently been applied to a non-nucleoside monophosphate, 4-phospho-D-erythronohydroxamic acid (which is 6-phosphogluconate dehydrogenase inhibitor) [43] and also to the novel nucleotide analogs 9-(*S*)-[3-hydroxy-2-(phosphonomethoxy)propyl]-2,6-diaminopurine (HPMPDAP) [44] and thymidine-but-2-enyl phosphonate [45]. An example of successful application of this promoiety design is the bis(pivaloyloxymethyl) prodrug of PMEA (adefovir dipivoxil, Hepsera®; Fig. 4) [38], which has been approved by the FDA and is currently being marketed for antiviral therapy against hepatitis B [46]. This prodrug was selected from a range of acyloxyalkyl prodrugs of PMEA based on its high oral bioavailability in rats [29]. Despite the success of adefovir dipivoxil, application of the pivaloyloxymethyl (POM) promoiety to high-dose drugs used for chronic treatment (e.g., 9-[(*R*)-2-(phosphonomethoxy)propyl]-adenine ((*R*)-PMPA)) could be limited due to concerns about the impact of pivalic acid on carnitine homeostasis, which is released during metabolism of the promoiety [47].

In order to avoid liberation of pivalic acid, an alkyloxycarbonyloxymethyl promoiety containing a carbonate diester group instead of a pivalate ester was utilized to improve the PK properties of (*R*)-PMPA (tenofovir (TFV)), a highly potent anti-HIV and anti-HBV agent [48]. As with cleavage of the acyloxymethyl group, the activation mechanism of alkyloxycarbonyloxymethyl promoiety is believed to proceed via enzymatic hydrolysis of the carboxylic ester but yields a carbonate monoester that releases the nucleoside monophosphate analog by spontaneous decarboxylation and elimination of the hydroxymethyl linker as formaldehyde (Fig. 3b). The generation of the formaldehyde as a result of the hydrolysis of bis(acyloxymethyl) and bis (alkyloxycarbonyloxymethyl) prodrugs could be a source of long-term cellular toxicity arising from the interaction of formaldehyde with cellular nucleophiles, such as DNA, RNA, or proteins [28]. Alkyloxycarbonyloxymethyl prodrugs of TFV were reported to be potent *in vitro* against HIV-1 and showed oral bioavailability ranging 16–30% but exhibited relatively short half-lives in intestinal homogenate ($t_{1/2}$ ≤

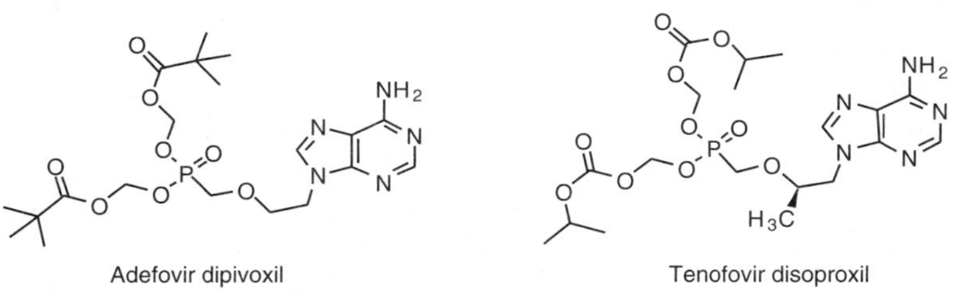

FIGURE 4 Structures of adefovir dipivoxil and tenofovir disoproxil.

5–53 min) [49]. Nonetheless, bis(isopropyl-oxycarbonyloxymethyl) prodrug of (*R*)-PMPA (Fig. 4) containing a more stable isopropyl carbonate ester group ($t_{1/2}$ = 53 min in intestinal homogenate) was selected for further clinical development based on its antiviral potency, water solubility, and oral bioavailability [49]. It was approved by FDA as a fumarate salt (tenofovir disoproxil fumarate (TDF), Viread®) for chronic treatment of HIV in 2001 and for chronic treatment of hepatitis B infection in 2008 [50]. As mentioned, the oral bioavailability of acyloxymethyl and alkyloxycarbonyloxymethyl prodrugs can be limited by premature hydrolysis of the promoiety in the GI tract where esterases, hydrolyzing acyloxy and alkyloxycarbonyl groups, are expressed in numerous tissues including the small intestine [15].

B. Phosphoramidate Approach

The phosphoramidate pronucleotide (ProTide) approach introduced by McGuigan to deliver monophosphorylated nucleosides into cells utilizes amino acid esters (connected via amino group) to mask one negative charge and an alkyl or aryl ester group to mask the remaining negative charge of the nucleotide phosphate group [51]. The mechanism of intracellular release of the nucleoside monophosphate analog from the aryloxy phosphoramidate prodrug involves two enzyme-mediated and two spontaneous steps (Fig. 5), beginning with an enzymatic hydrolysis of the promoiety's carboxylic ester by carboxylesterases, lipases, or proteases. Cleavage of the ester group was identified as a necessary condition for antiviral activity; however, the rate of enzymatic de-esterification did not always correlate with antiviral potency of the prodrug and was found to depend on the structure of both the ester group and the amino acid side chain [52]. Thus, a *tert*-butyl ester-derived phosphoramidate pronucleotide exhibited low potency because *tert*-butyl esters are not good substrates for the cleaving enzymes [53]. Cathepsin A, a major enzyme responsible for the cleavage of the ester group, was unable to hydrolyze phosphoramidates with heavily branched amino acids such as leucine, valine, and isoleucine [54]. The second step of the activation mechanism is proposed to be spontaneous nucleophilic substitution of the aryl moiety by the carboxyl residue formed after ester hydrolysis yielding a five-membered cyclic mixed anhydride. The rate of the cycle formation varies depending on the structures of the nucleoside analog and the amino acid, for example, cyclization of AZT phosphoramidates was more difficult than d4T, and phosphoramidates containing β-alanine failed to cyclize [55]. The transient five-membered cyclic diester is unstable and chemically hydrolyzes via nucleophilic attack of water at the phosphate group, to produce an amino acyl metabolite (phosphoramidate), which finally undergoes phosphoramidase-mediated P-N bond cleavage and releases corresponding nucleoside monophosphate analog.

The phosphoramidate concept has been applied to various antiviral and anticancer nucleoside monophosphate/phosphonate analogs, including 4′-azidoadenosine monophosphate [56], 4′-azidouridine monophosphate [57], AZT monophosphate [58], ddU monophosphate [59,60], 2′,3′-didehydro-2′,3′-dideoxyuridine (d4U) monophosphate [59,60], 2′,3′-didehydro-2′,3′-dideoxythymidine (d4T) monophosphate [61], abacavir (ABC) monophosphate [62], (*E*)-5-(2-bromovinyl)-2′-deoxyuridine (BVDU) monophosphate [63], FUdR monophosphate [64], 2′-C-methylguanosine monophosphate

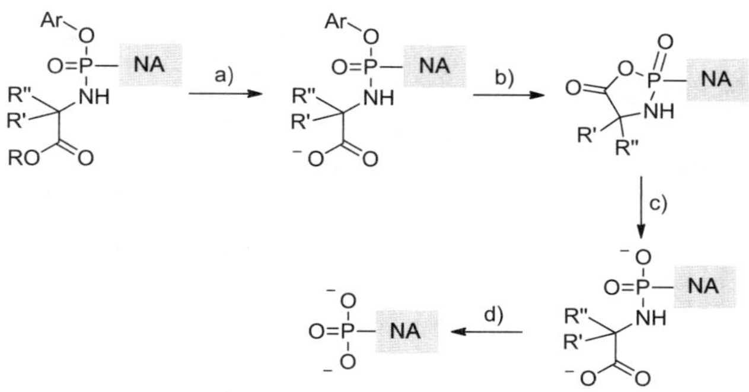

FIGURE 5 General representation of phosphoramidate activation pathway to the corresponding nucleoside monophosphate inside the cell: (a) ester hydrolysis by carboxypeptidase or esterase type enzyme, (b) spontaneous cyclization, (c) aqueous hydrolysis, and (d) enzymatic cleavage of P-N bond by phosphoramidase-type enzyme.

[65], ribavirin monophosphate [66], acyclovir (ACV) monophosphate [67], 9-[2-(phosphonomethoxy)-ethyl]adenine (PMEA) [68], and (*R*)-PMPA (TFV) [68]. Structure–activity relationship (SAR) studies have demonstrated that the biological activity of the pronucleotide depends significantly on the structures of both nucleoside and phosphoramidate promoiety; however (with some exceptions), the most favorable combination of amino acid ester and aryloxy group was found to be L-Ala and a phenyl (or naphthyl) group. In several cases, application of the phosphoramidate approach to inactive nucleosides boosted their potency to a measurable level [56,59,60]. ProTide approach has been recently reviewed [52]. In this chapter, we discuss several phosphoramidate prodrugs that have reached clinical trials.

The phosphoramidate approach has been applied to ameliorate the rate-limiting initial phosphorylation of 2′-C-methylguanosine [65], a nucleoside analog, which undergoes poor phosphorylation but upon conversion into the triphosphate form serves as a potent inhibitor of HCV RNA polymerase [69,70]. In a search for the pronucleotide exhibiting the most favorable antiviral potency and PK properties, various modifications of aryl moiety, ester, and amino acid groups of the phosphoramidate promoiety were evaluated. Alanine ester pronucleotides assayed versus HCV using a replicon assay were found to be 4–50-fold more potent (EC_{50}=0.045–0.21 μM) than the parent drug (EC_{50}=3.5 μM) but exhibited inadequate stability to enzymatic hydrolysis in rodent plasma. The most stable prodrug in the series, L-Ala benzyl ester, showed 40% degradation after 30 min in rodent plasma at 2–4°C. Alteration of the amino acid structure increased stability of the promoiety in plasma but also resulted in decreased pronucleotide antiviral activity [65]. Further development of the 2′-C-methylguanosine pronucleotides included not only variation of the amino acid ester in the phosphoramidate moiety to optimize enzymatic stability but also modification of the purine base at the C-6 position to potentially increase antiviral potency [71,72]. Combination of a naphthyl (L)-Ala neopentyl ester phosphoramidate moiety with a C6-methoxy substituent at the purine base resulted in a dual prodrug (INX-189,

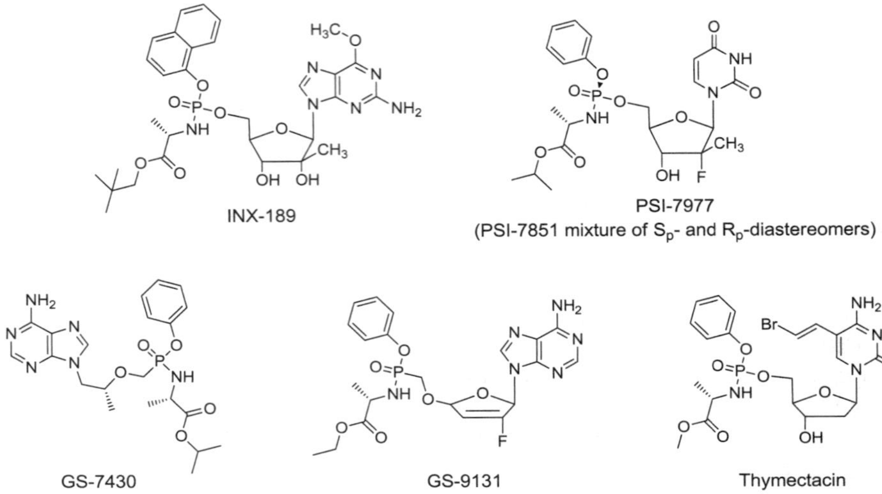

FIGURE 6 Phosphoramidate pronucleotides evaluated in clinical trials.

Fig. 6) of 2′-C-methylguanosine monophosphate, which was found to be 520-fold more active in the genotype 1 HCV replicon assay (EC_{50} = 0.01 μM) than the parent nucleoside analog (EC_{50} = 5.2 μM). Nanomolar activity of INX-189 versus HCV is consistent with intracellular levels of 2′-C-methylguanosine triphosphate in primary human hepatocytes. INX-189 has successfully completed phase 1b clinical trial in HCV-infected patients [72].

Another anti-HCV pronucleotide that has advanced to clinical trials is PSI-7977, a single (S_p)-diastereomer of the phosphoramidate prodrug of β-D-2′-deoxy-2′-fluoro-2′-C-methyluridine-5′-monophosphate (Fig. 6) [73]. The triphosphate derivative of β-D-2′-deoxy-2′-fluoro-2′-C-methyluridine is a potent inhibitor of HCV RNA polymerase (NS5B RNA-directed), whereas the nucleoside analog itself did not exhibit any antiviral activity due to inability to be phosphorylated by cellular nucleoside kinases to its monophosphate form [74]. Application of the phosphoramidate prodrug approach to bypass inefficient first phosphorylation step yielded a number of pronucleotides exhibiting potency in the HCV subgenomic replicon assay (EC_{50} < 1 μM) and producing high levels of β-D-2′-deoxy-2′-fluoro-2′-C-methyluridine-5′-triphosphate in primary hepatocytes and in the livers of rats, dogs, and monkeys when administered *in vivo* [73]. Based on its antiviral, PK, and toxicity profile, the phenyl (L)-alanine isopropyl ester pronucleotide of β-D-2′-deoxy-2′-fluoro-2′-C-methyluridine (PSI-7851) was selected for further development. PSI-7851 is a mixture of (S_p)- and (R_p)-diastereomers, with the (S_p)-diastereomer (PSI-7977) being the more potent (>10-fold) inhibitor of HCV RNA polymerase in a clone A replicon-based assay. Studies of the activation mechanisms of PSI-7851 and PSI-7977 in primary human hepatocytes indicate that (S_p)- and (R_p)-diastereomers produce almost identical amounts of the corresponding triphosphate, suggesting that both diastereomers are equally active in the liver cells, where infection of HCV takes place [75]. Enzymes identified in activation process include human cathepsin A (cat. A) and carboxylesterase 1 (CES1), catalyzing hydrolysis of the carboxyl ester, and histidine triad nucleotide-binding protein 1 (Hint1) cleaving the amino acid moiety prior to release of the free nucleoside monophosphate (Fig. 5) [75]. The (S_p)-diastereomer PSI-7977 has been evaluated a phase II clinical trial for the treatment of HCV infection [73].

Other phosphoramidate prodrugs investigated clinically include GS-7340 and GS-9131, developed by Gilead Sciences to treat HIV/AIDS, and Thymectacin, developed by NewBiotics to treat colon cancer. GS-7340 is the phenyl L-alanine isopropyl ester of TFV (Fig. 6). The use of the FDA-approved bis-isopropoxycarbonyloxymethyl ester of (*R*)-PMPA (TDF, Viread®) is limited due to toxicity, development of resistance, and rapid systemic degradation to the parent nucleotide analog [76–78]. In contrast, GS-7340 has shown 400-fold greater potency against HIV in peripheral blood mononuclear (PBM) cells and 200-fold greater stability in plasma compared to tenofovir disoproxil [79]. GS-7340 has entered a phase 1b clinical trial [80].

GS-9131 is a phenyl-(L)-alanine ethyl ester prodrug of the modified dAMP phosphonate analog 2′-α-fluoro-2′,3′-didehydro-2′,3′-dideoxyadenosine (Fig. 6) [81]. The prodrug moiety was optimized for efficient delivery of GS-9148 together with its diphosphate, active metabolite to lymphoid cells following oral administration. GS-9131 was chosen as a candidate to be developed for treatment of drug-resistant HIV-1 infection based on its (1) anti-HIV potency against HIV-1-resistant virus strains, (2) favorable biological and PK properties, and (3) ability to deliver the parent drug efficiently inside cells [81].

Thymectacin, the phenyl L-alanine methoxy derivative of BVDU, was reported to possess potent anticancer activity in tumor cells with a high expression of thymidylate synthase (TS) and to be non-cross-resistant with other TS-targeted agents [82–84]. Thymectacin was 10-fold more cytotoxic to 5-FU-resistant TS overexpressing colorectal tumor cells than to normal cells [82] and was active in fluoropyrimidine-resistant metastatic colorectal cancer (CRC) during monotherapy trials [85] and has entered undergoing combined phase I/II clinical trials for the treatment of colon cancer [86].

In conclusion, the phosphoramidate prodrug approach can enhance intracellular delivery of nucleotide analogs and has proven to increase activity of a number of antiviral or anticancer nucleosides. Intracellular enzymatic activation of these pronucleotides to liberate nucleoside monophosphates has been studied, but detailed information about all the enzymes involved in the activation process is warranted, given successful progression of a number of phosphoramide triester-based prodrugs (INX-189, PSI-7977, GS-7340, GS-9131, and Thymectacin) to clinical trials.

C. Lipidic Prodrug Approach

Lipid ester prodrug approach aims to improve oral absorption and to reduce adverse side effects of the parent drug by creating drug–lipid conjugates where the drug is covalently bound to a long-chain alkyl group, fatty acid, diglyceride, or phosphoglyceride. Lipid ester prodrugs can exhibit dramatically increased lipophilicity and can cross cell membranes by passive diffusion, membrane flippase activity, or phospholipid uptake. Once inside the cell, they typically require enzymatic activation to release the parent drug. Development of this concept began in the 1970s with lipid-like modification of cytarabine (ara-C), an anticancer agent [87–89]. Incorporation of lipid and phospholipid moieties into ara-C's structure was intended: (1) to prevent metabolic degradation of the drug by cytidine deaminase, (2) to improve the intracellular drug concentration, and (3) to bypass the rate-limiting first phosphorylation step in the ara-C activation pathway via release of ara-C monophosphate from the phospholipid prodrug. This approach was then applied to a variety of antiviral and antitumor agents including 5-fluorouridine monophosphate [90], AZT monophosphate [91,92], 3′-deoxy-3′-fluoro-thymidine (FLT) monophosphate [93], ddC monophosphate [94], ddT monophosphate [95], phosphonoformate (PFA) [96–98], (*S*)-HPMPC (Cidofovir (CDV)) [99], (*S*)-HPMPA [100], PMEA [101], and PMEDAP [101].

An early example of orally administered phospholipid prodrug that reached clinical trials and has been approved in Japan for treatment of leukemia is cytarabine ocfosfate (YNK01, staracid), a stearyl ester of cytarabine monophosphate [102]. YNK01, synthesized by Saneyoshi et al. [103], was resistant to deoxycytidine deaminase and demonstrated high antileukemic activity both *in vitro* and *in vivo* [104]. Mechanism of the parent drug release was initially anticipated to proceed via enzymatic hydrolysis by phosphodiesterases; however, it was later proved to begin with hydroxylation of the side chain at the ω-end of the alkyl chain followed by oxidation to the carboxylic acid and stepwise shortening of the alkyl chain by successive peroxisomal β-oxidation steps (Fig. 7) [105].

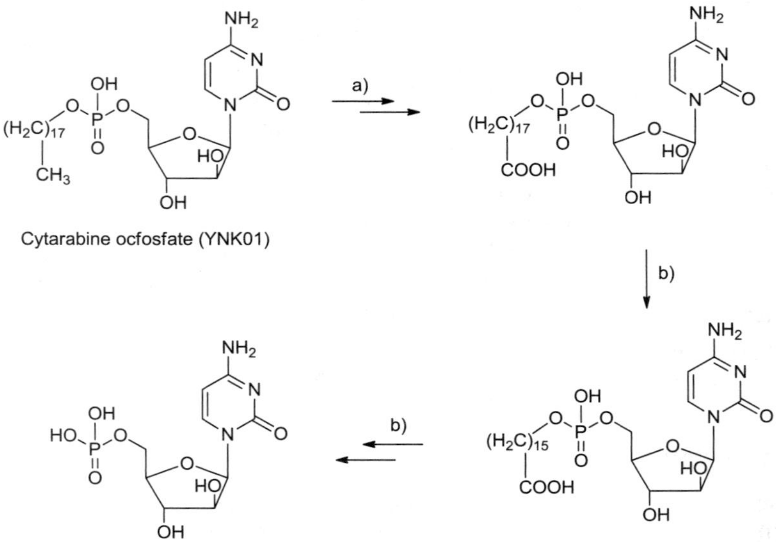

FIGURE 7 Activation pathway of cytarabine ocfosfate (YNK01) to cytarabine monophosphate: (a) ω-hydroxylation and subsequent oxidation to the carboxylic acid and (b) β-oxidation.

The oral bioavailability of cytarabine ocfosfate could not be measured in phase I/II clinical studies due to intravascular hemolysis at the injection site after intravenous administration. In toxicology studies in rodents, metabolic bioavailability of the drug was defined by measuring the renal excretion of ara-C and ara-U taking into account that >90% of YNK01 absorbed after oral administration was metabolized to ara-C and ara-U, which were almost completely excreted. The metabolic bioavailability of YNK01 was estimated to be 15.8% of the total dose [106]. The results of a phase II study of YNK01 in combination with α-interferon used for the treatment of chronic myeloid leukemia have recently been reported [107]. Other phospholipid prodrugs that have advanced to clinical trials include fozivudine tidoxil [108] and fosalvudine tidoxil [93], thioglycerolipid ester prodrugs of AZT and FLT, respectively. The AZT prodrug bypasses rate-limiting phosphorylation and increases concentration of active drug in lymphoid tissues.

In addition to nucleoside analogs, Hostetler has applied a lipidic prodrug approach to various acyclic nucleoside phosphonates (ANPs) [109], including (*S*)-HPMPC, (*S*)-HPMPA, and (*R*)-PMPA [110]. Like nucleoside monophosphates, ANPs bypass the rate-limiting first phosphorylation step during activation, but in contrast to the phosphate derivatives, they are chemically and enzymatically stable due to the non-labile P-C bond [111]. In this approach, a lipid ester promoiety consisting of glycerol linker connected to a long alkyl chain was modeled after lysophosphatidylcholine, which, in contrast to phospholipids, contains only one O-acyl chain and is known to diffuse across cell membranes at a faster rate since their off-rate from phospholipid bilayer (higher disturbance) membranes is higher than that of the diacylphospholipids [112]. Since only about 40% of lysophosphatidylcholine is absorbed intact from the intestine [113,114], the enzymatic and chemical stabilities of the promoiety were increased by replacement of the acyl ester group with an ether linkage (to avoid hydrolysis by lysophospholipase) and by substitution of the hydroxyl group at the *sn*-2 position of the glycerol linker with a hydrogen atom (to prevent reacylation by lysophosphatidylcholine acyltransferases present in small intestinal enterocytes and other tissues) [110]. Liberation of the parent phosphonate drug from the prodrug was proposed to occur by enzymatic cleavage of the P-O-C bond by phospholipase C, an enzyme present in plasma and pancreatic secretions [110].

Structural optimization studies of the promoiety examining variations in the linker and in the alkyl chain length examined combination of a 16-C alkyl group with an oxypropyl linker or 18-C alkyl group with an oxyethyl linker constituting the hexadecyloxypropyl (HDP) or octadecyloxyethyl (ODE) promoieties, respectively (Fig. 8) [115].

Evaluation of the HDP and ODE derivatives of various ANPs for *in vitro* antiviral activity versus dsDNA viruses (cowpox, vaccinia, cytomegalovirus (CMV), variola, HSV-1, VZV) and versus viruses relying on RNA polymerases (HBV and HIV-1) revealed that HDP- and ODE-(*S*)-HPMPC conjugates were most potent against HCMV, exhibiting activity in the low nanomolar range ($EC_{50}=0.0009\,\mu M$) [110]. HDP and ODE prodrugs of (*S*)-HPMPA were potent against HSV-1 and orthopoxviruses but also were the most cytotoxic compounds [110,116]. HDP–PMPA prodrug demonstrated the highest activity against HIV-1 in MT-2 and PBM cells

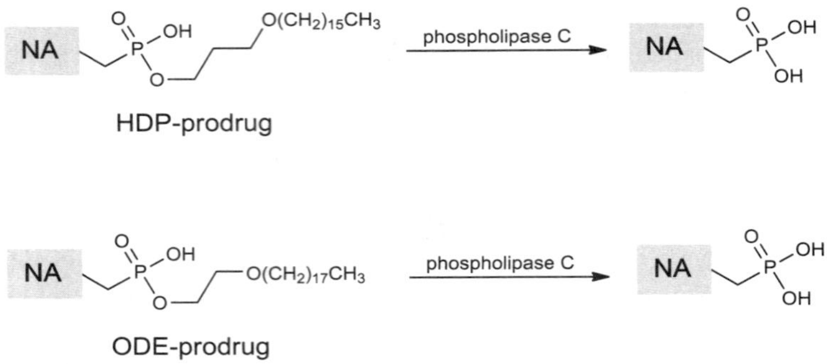

FIGURE 8 Activation pathway of HDP and ODE prodrugs of ANPs.

(EC_{50}<0.00001 and 0.012 μM, respectively) than TFV (EC_{50}=0.65 and 3.20 μM, respectively) [117].

The higher *in vitro* antiviral activities of HDP-(*S*)-HPMPC and HDP-(*S*)-HPMPA relative to the parent drug were rationalized based on their 50–100-fold increased levels of cellular uptake and conversion into the corresponding diphosphates, measured using ^{14}C-labeled HDP-(*S*)-HPMPC, HDP-(*S*)-HPMPA, and corresponding parent drugs in MRC-5 human lung fibroblast cells [118,119]. The relative oral bioavailabilities of the HDP-(*S*)-HPMPC and HDP-(*S*)-HPMPA were estimated to be 88% and 73%, respectively (compared to the <5% and 24% for the parent ANPs), and the half-lives in plasma were found to be 14.9 and 11 h, respectively [120,121]. Both prodrugs exhibited reduced nephrotoxicities compared to the parent (*S*)-HPMPC and (*S*)-HPMPA [110]. Based on these *in vitro* and *in vivo* results, HDP prodrugs of (*S*)-HPMPC (CMX001) and (*R*)-PMPA (CMX157) have advanced into clinical trials; CMX001 is developed as a treatment for CMV, smallpox, and BK viruses, whereas CMX157 is for HIV infection (Fig. 9).

In clinical trials, premature metabolism of HDP-(*S*)-HPMPC has not been reported to be an issue; however, it was responsible for the failure of this prodrug to treat viral infections in cynomolgus monkeys [122]. The proposed mechanism of rapid enzymatic degradation of HDP-CDV includes hydroxylation of the ω or ω-1 carbon atom of the alkyl chain with subsequent oxidation to a carboxylic group, followed by shortening of the alkyl chain via β-oxidation, which is the same mechanism that leads to the release of the nucleoside monophosphate from the phospholipid prodrug, YNK01. Thus, to increase enzymatic stability and prevent oxidation of the long alkyl chains present in HDP and ODE prodrugs, Ruiz et al. synthesized and evaluated a variety of alkoxyalkyl (*S*)-HPMPC and (*S*)-HPMPA prodrugs containing modified alkoxy or alkyl chains [123]. The compounds were evaluated for *in vitro* antiviral potency versus cowpox, vaccinia, and ectromelia viruses, and their metabolic stabilities were estimated in guinea pig, monkey, and human liver S9 fractions. Modifications of the alkyl chain, for example, by introduction of a ω-1 methyl group, a terminal double bond, or a terminal cyclopropyl residue, had only a moderate effect on the metabolic stabilities of the prodrugs, whereas alteration of the alkoxy group by incorporation of a benzyloxy group at the *sn*-2 position increased the metabolic

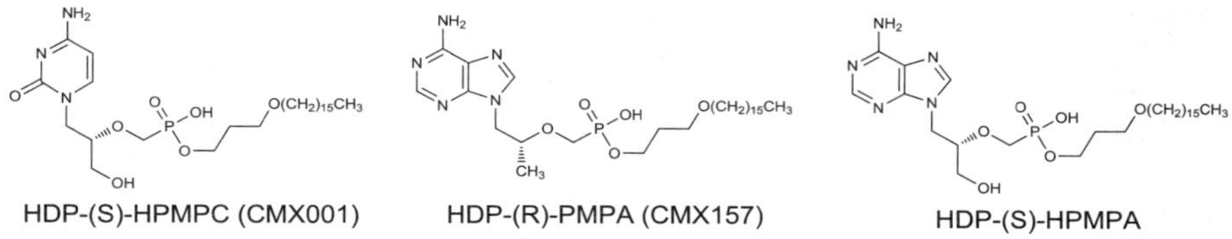

FIGURE 9 Structures of selected lipid ANP prodrugs.

stability of the (*S*)-HPMPC prodrug (88% of modified prodrug was intact after 90 min exposure to monkey liver S9 fractions vs. 16% in case of unmodified prodrug). These structural modifications of the promoieties had little effect on antiviral activities of the prodrugs and all the compounds retained antiviral potency in a submicromolar range [123].

In summary, the lipidic prodrug approach – improvement of parent drug's hydrophobicity via incorporation into its structure of a biolabile lipophilic alkyl chain – has been known for decades and has been applied to a number of nucleotide analogs. To date, only one phospholipid ester prodrug, cytarabine ocfosfate (YNK01) for treatment of leukemia, has been approved for clinical use in Japan. An apparent reason for the relative lack of success of lipid prodrugs in reaching the clinic may stem from elevated cytotoxicity, decreased aqueous solubility, and often insufficient enzymatic stability – properties that they exhibit along with improved hydrophobicity and, in several cases, antiviral activity. Structural modification of the alkyl group *per se* in order to ameliorate those drawbacks by simple adjustment of the chain length may not suffice to achieve an ideal balance between PK properties, cytotoxicity, and antiviral activity of the prodrug. In this respect, incorporation and structural alteration of an easily modifiable linker between the drug and hydrophobic alkyl chain may be advantageous. In this connection, it is worth noting that glycerol linker in two lipidic prodrugs of (*S*)-HPMPC and (*R*)-PMPA, CMX001 and CMX157, did not show significant change in antiviral activity versus HCMV when the ether oxygen was replaced by a carbon atom [115].

D. SATE Approach

The (*S*)-acyl-2-thioethyl (SATE) prodrug approach introduced for intracellular delivery of nucleoside monophosphate/phosphonate analogs [124,125] utilizes enzymatically cleavable SATE protecting groups as promoieties to mask the negative charges of the phosphate/phosphonate group. According to metabolic studies of the radiolabeled SATE pronucleotides in cell extracts and in intact cells [41,126], the mechanism of SATE group cleavage from the pronucleotide and corresponding nucleoside monophosphate release is initiated primarily via intracellular esterase-mediated hydrolysis of the acyl group, yielding an unstable 2-mercaptoethyl ester (Fig. 10, step a). On the next step, the newly formed thiol group intramolecularly attacks the α-carbon of the 2-mercaptoethyl ester group, giving rise to the corresponding phosphate diester and ethylene sulfide (Fig. 10, step b). If the second phosphate ester is also a SATE group, the aforementioned process is repeated, resulting in release of the nucleoside monophosphate analog.

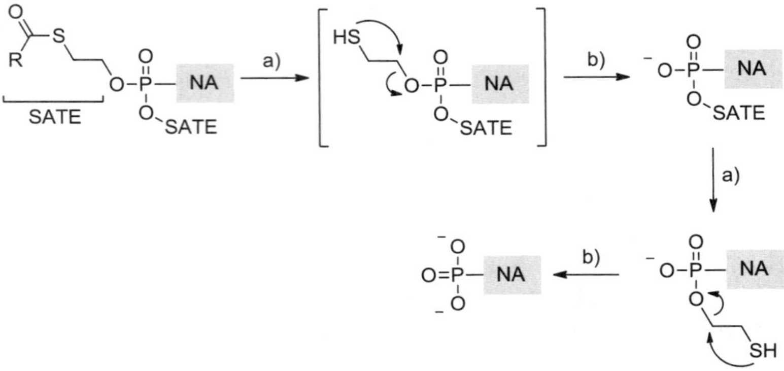

FIGURE 10 General representation of bis(SATE) phosphotriester pronucleotides activation pathway to the corresponding nucleoside monophosphate inside the cell. (a) Esterase-mediated hydrolysis and (b) spontaneous decomposition.

The SATE approach has been applied to a number of nucleoside monophosphate/phosphonate analogs, including 2′,3′-dideoxyadenosine (ddA) monophosphate [127,128], AZT monophosphate [41,129], 2′,3′-dideoxy-2′,3′-didehydrothymidine monophosphate (d4TMP) [130], cytarabine [131], ACV monophosphate [132,133], PMEA [134], 1-[1-(3,4-dimethylcyclopent-3-enylmethoxymethyl)] cytosine phosphonic acid [135], 1-(4-amino-5H-pyrrolo[3,2-d]pyrimidin-7-ylmethyl) cyclobutylmethyl phosphate [136], and 3′,4′-dimethyl-5′-norcarbocyclic adenosine phosphonic acid [137]. In all cases, the bis(SATE) prodrugs exhibited increased *in vitro* antiviral activities compared to the parent nucleoside analogs, for example, bis(MeSATE)-ddA was 450-fold more potent against HIV-1 in PBM cells than the parent (ddA) [127,128], and bis(MeSATE)-d4TMP was 10- to 17-fold more active than d4T versus HIV-1 in wild-type CEM and PBM cells and retained its potency in thymidine kinase (TK)-deficient cells [130].

The toxicity of the SATE promoiety and its metabolites assessed *in vitro* (human bone marrow progenitor cells) and *in vivo* (cynomolgus monkeys) of bis((*S*)-acetyl-2-thioethyl) AZT (bis(MeSATE)-AZT) was comparable to that of the parent nucleoside analog (AZT) [124], indicating that neither the SATE promoiety nor its degradation products (acetic acid and episulfide) induced additional toxicity. The acetate form was proposed to be metabolized by normal pH-homeostatic cellular mechanisms [138], and the episulfide was hypothesized to conjugate with glutathione, the ubiquitous nonprotein thiol that serves as both a nucleophile and a reducing agent in playing a crucial role in detoxification and repair of cell damage [124]. Elevated cytotoxicities observed for hydrophobic SATE acyl prodrugs (e.g., R = *tert*-butyl) were attributed to increased intracellular accumulation of corresponding phosphorylated nucleosides [124,139].

The relative stability of the SATE promoieties in extracellular media (culture medium, human serum) and rapid intracellular liberation of the corresponding nucleoside monophosphate from the SATE prodrug were confirmed in all examined cell lines [124]. The rate of the nucleoside monophosphate release was adjusted by varying the nature of the SATE acyl moiety (R group, Fig. 10). Phosphotriesters incorporating lipophilic bis(*tert*-BuSATE) groups (R = *tert*-butyl; Fig. 10) generally exhibited greater enzymatic stabilities and antiviral activities than compounds with more labile MeSATE promoieties [140]. In contrast to their *in vitro* stability, bis(*tert*-BuSATE) phosphotriester derivatives were unable to survive presystemic metabolism and completely lost their antiviral activity *in vivo*, presumably due to inadequate delivery of the corresponding nucleoside analogs to infected cells and tissues [141]. No intact bis(*tert*-BuSATE)-AZT monophosphate was observed in mouse plasma after oral administration, probably due to extensive esterase-based metabolism in the GI tract or liver [141]. To address presystemic degradation of the bis(*tert*-BuSATE) pronucleotides, two strategies were devised [124,125]. One aimed to improve the enzymatic stability by modifying the SATE promoiety [142–144] on the basis that carboxylesterases, the enzymes most likely involved in the degradation of the SATE group, prefer lipophilic carboxyl esters over polar or charged esters [145]. Incorporation of a polar hydroxyl group into the acyl moiety of the *tert*-BuSATE group (bis(HO-*tert*-BuSATE)) resulted in a decreased rate of enzymatic deacylation in human serum [144]. However, efficient liberation of the active nucleoside monophosphate drug precursor remained elusive because cleavage of the second SATE group in the phosphodiesters proceeded at a much slower rate than hydrolysis of the first SATE moiety in bis(SATE) phosphotriesters [126]. This difference arises because of the presence in the SATE phosphodiesters of a phosphate negative charge, which greatly reduces the rate of enzymatic deacylation of the second SATE promoiety and which also deters the thiol group from attack at the α-carbon of the thioester group, retarding intramolecular nucleophilic substitution and thus preventing release of the nucleoside monophosphate [124,125].

As a result, the prodrug design was revised with the goal of achieving not only effective presystemic metabolism of the first SATE promoiety but accelerated removal of the second promoiety [124,125]. A series of mixed SATE pronucleotides was synthesized wherein one P-OH group was masked with a SATE group and other OH group by a different, enzymatically labile promoiety, such as an aryl ester (removable by phosphodiesterases) [32,146], amino residues (amino acid, alkyl and aryl amine) to form a phosphoramidate cleavable by phosphoramidase [55,147], or a glucosyl residue containing a 2-thioethyl linkage connected to the phosphorus atom of the drug by a P-S bond (isoSGTE group) [125] (Fig. 11).

A series of mixed SATE–AZT pronucleotides was prepared and evaluated for inhibition of HIV-1 replication in a variety of cell lines, including TK-deficient cells where the parent drug (AZT) was inactive at concentrations up to 100 μM [125]. In contrast to AZT, all the mixed SATE prodrugs examined had antiviral potencies in the low micromolar range (EC_{50} = 1–10 μM) in both wild-type and TK-deficient cell lines [125]. SAR studies involving modification of the second promoiety (aryl ester group, amino residue, or glucosyl group) and the SATE group have been undertaken to further optimize pharmacological and PK properties of these pronucleotides.

Modification of the aryl moiety in the mixed SATE pronucleotides was guided by the hypothesis that they are cleaved by 5′-nucleotide phosphodiesterases, a family of

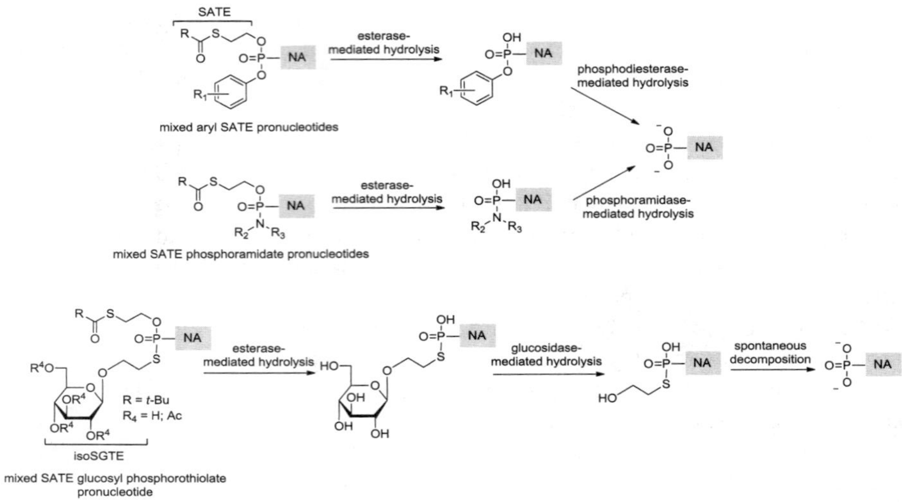

FIGURE 11 General representation of mixed SATE pronucleotide activation pathways.

enzymes catalyzing hydrolysis of a wide range of naturally occurring, as well as synthetic aryl phosphodiesters to the corresponding phosphate monoesters (Fig. 11) [32,146]. The aqueous solubility of the mixed SATE aryl pronucleotides could be increased without significant loss of antiviral activity by utilization of a (L)-tyrosinyl residue instead of a phenyl ester group [148]. However, the (L)-tyrosinyl phosphodiesters released the parent AZT monophosphate in cell extracts at a faster rate than the corresponding phenyl ester derivatives, supposedly because they are better substrates for the phosphodiesterases [124,149]. In another attempt to improve enzymatic stability of the mixed aryl SATE pronucleotide, a hydroxyl group was incorporated into the pivaloyl moiety of the *tert*-BuSATE group, which increased the half-life of the *tert*-BuSATE phenyl AZT phosphotriester in cell extracts 10-fold ($t_{1/2}$ increased from 1.5 to 15 h) with retention of antiviral activity [150]. Replacement of this hydroxyl group with an amino residue drastically decreased both the enzymatic stability (in culture medium and in cell extracts) and the antiviral potency of the pronucleotide in a TK-deficient cell line, relative to the mixed *tert*-BuSATE phenyl phosphotriester [131,150,151]. Introduction of a second hydroxyl group into the pivaloyl residue increased the stability of the SATE group by an additional 12-fold but also led to a total loss of antiviral potency in the TK-deficient cell line [150].

Another approach to enzymatically cleavable promoieties involving mixed SATE pronucleotides has utilized amino residues connected to the phosphorus atom via a P-N bond with the goal of achieving activation by a phosphoramidase to release the nucleoside monophosphate drug (Fig. 11) [124,125]. SAR studies of a library of mixed *tert*-BuSATE phosphoramidates with varying amino group structures revealed that the nitrogen atom basicity principally determines the rate of enzymatic P-N bond hydrolysis to liberate the parent nucleotide analog. Among mixed SATE phosphoramidate pronucleotides of AZT, the isopropylamino derivative showed the greatest potency (EC_{50} = 0.75 μM) against HIV-1 in a TK-deficient cell line, indicating that an amino acid is not required for the pronucleotide to retain antiviral activity [152].

In summary, the SATE approach has been applied to a number of nucleoside monophosphate/phosphonate analogs to create corresponding SATE pronucleotides that exhibit *in vitro* antiviral activity in both wild-type and TK-deficient cell lines. However, rapid *in vivo* metabolism of bis(SATE) phosphotriesters yields chemically and enzymatically stable SATE phosphodiesters, blocking ultimate liberation of the parent nucleoside monophosphate drug. To circumvent this limitation, mixed SATE pronucleotides were developed where one negative charge of the nucleoside monophosphate was masked with a SATE group and the remaining P-OH was derivatized with an enzymatically labile promoiety, such as an aryl ester or amide. Unfortunately, further structural alterations of the SATE group and the second promoiety to improve the PK properties of the mixed pronucleotides increased cytotoxicity without a compensating increase in antiviral activity, thus narrowing their therapeutic windows. Currently, no information is available about SATE pronucleotides undergoing clinical studies, most likely because additional studies need to be conducted to balance their activity, PK properties, and toxicity.

E. *Cyclo*Sal Approach

Originally developed by Meier et al., the *cyclo*Sal (cyclo-saligenyl) concept was formulated as a selective nucleotide delivery system that releases its cargo (nucleotide) exclusively via a chemically induced cascade mechanism [13,153,154]. This approach utilizes salicyl alcohol as a bifunctional masking unit that forms a phenyl and a benzyl ester with the targeted phosphate/phosphonate group of the nucleotide analog. Activation of the pronucleotide is pH dependent and achieved through a sequence of reactions which begins with hydrolytic cleavage of phosphotriester and followed by substitution via S_NP mechanism of the best leaving group – the phenolate – thus releasing corresponding 2-hydroxy-benzyl phosphate diester (Fig. 12). In contrast to unsubstituted benzyl nucleotide diesters that are chemically stable at physiological conditions, newly formed 2-hydroxy-benzyl nucleoside monophosphate ester is activated by the strong electron-donating hydroxyl group in the *ortho* position and therefore is able to spontaneously release the nucleotide. The *cyclo*Sal prodrug approach has been applied to various nucleotide and nucleoside monophosphate analogs, including 2′,3′dideoxyadenosine (ddA) monophosphate [155], 2′,3′-dideoxy-2′,3′-didehydroadenosine(d4A)monophosphate[155],2′-F-*ara*-2′,3′-dideoxyadenosine (F-*ara*-ddA) monophosphate [156], 2′-ribo-F-2′,3′-dideoxyadenosine (F-*ribo*-ddA) monophosphate [156], ABC monophosphate [157], AZT monophosphate [158], d4TMP [159–161], FUdR monophosphate [162,163], 5-[(*E*)-2-bromovinyl]-2′-deoxyuridine (BVdU) monophosphate [13,164], ACV monophosphate [165,166], and PMEA [13,167].

The first generation of *cyclo*Sal pronucleotides did not require enzymatic activity in order to release corresponding nucleoside monophosphate under physiological conditions [153]. Half-lives of these pronucleotides in phosphate buffer at pH 7.3 ranged from 0.25 to 96 h and could be adjusted by introduction into the promoiety's

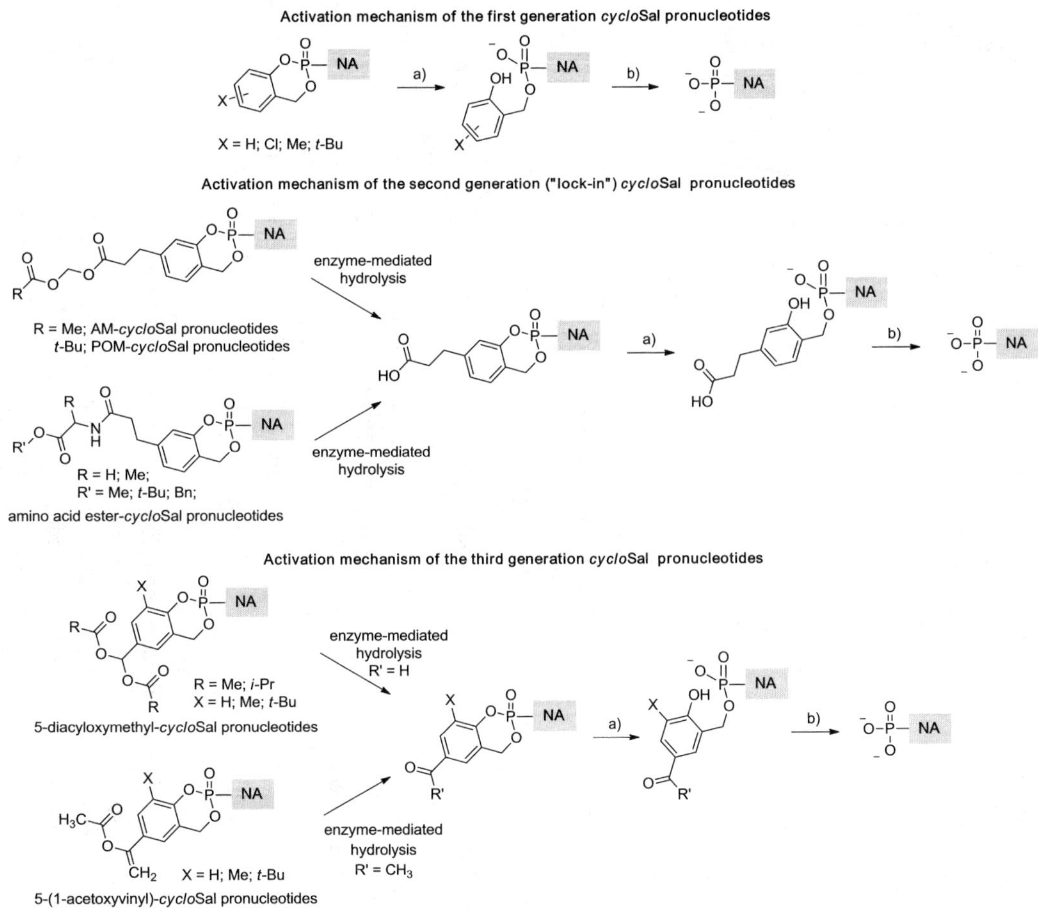

FIGURE 12 General representation of the first-, second-, and third-generation *cyclo*Sal pronucleotides activation pathways. (a) Hydrolysis of phenolate ester via S_NP mechanism and (b) spontaneous cleavage of the benzyl ester group.

aromatic ring of electron-donating or electron-withdrawing substituents to increase or decrease hydrolytic stability [13]. The predominance of chemical activation of the prodrugs over enzymatic activation was demonstrated by comparing the half-lives of the pronucleotides in phosphate buffer (pH 7.3) and in 10% human serum in phosphate buffer (pH 7.3), but significant differences could not be found. The first-generation *cyclo*Sal pronucleotides exhibited similar or better antiviral activities compared to the parent nucleoside analogs, for example, 3-methyl-*cyclo*Sal-ddAMP was 100-fold more potent than ddA against HIV-1 and HIV-2 in a wild-type T-lymphocytic cell line (CEM/0) [168], and 3-methyl-*cyclo*Sal-d4A proved to be 600-fold more potent than d4A [155]. *Cyclo*Sal pronucleotides exist as two diastereomers (S_p and R_p) that showed different PK properties (metabolic stability and antiviral potency). The (R_p)-diastereomers of the first-generation *cyclo*Sal pronucleotides released higher amounts of the parent nucleoside monophosphates, resulting in more elevated intracellular levels of the corresponding nucleotide triphosphates, resulting in higher antiviral potencies than the (S_p)-diastereomers [169,170].

This first generation of *cyclo*Sal pronucleotides, which liberates the parent nucleoside monophosphate analog via a nonenzymatic mechanism, does not depend on enzyme expression differences in tissues, individuals, and species and suffered from several other drawbacks, namely, extracellular release of the parent nucleotide and efflux of the pronucleotide from the cell due to formation of a concentration equilibrium across the cell membrane. To circumvent these limitations, an additional functional group was added to the *cyclo*Sal promoiety to create a second generation of *cyclo*Sal pronucleotides with increased chemical stability in extracellular media and also to promote enzymatic conversion into a more polar group within the cell to retard efflux ("lock-in" concept) [154].

Initially, the second generation of *cyclo*Sal pronucleotides included enzymatically cleavable ester and ether groups connected to the aromatic ring of the *cyclo*Sal promoiety through an ethylene linker [154], to minimize the electronic effect of the ester group on the hydrolysis of the pronucleotide triester. A carboxylic acid group was incorporated to trap the pronucleotide inside the cell, since the free alcohol group released in case of ethers was not polar enough to prevent efflux of the pronucleotides. Simple alkyl ester groups connected to the aromatic ring through an alkyl linker were poor esterase substrates. Acylal systems based on acetoxymethyl (AM) and POM modifications were explored as promoieties to release the carboxylic group (Fig. 12) [13]. Introduction of the AM group resulted in low enzymatic stability ($t_{1/2}$ = 0.25 h) and decreased antiviral activity, whereas POM-modified pronucleotides were only marginally more stable ($t_{1/2}$ = 0.9 h) while suffering from cytotoxic metabolites, such as pivalic acid [13,153]. To avoid the cytotoxic metabolites and to expand structural tunability, the pronucleotides were next modified with uncharged amino acid esters, which upon rapid intracellular hydrolysis by esterases generate the carboxylate group [171]. Among the *cyclo*Sal amino acid ester derivatives of d4TMP studied, the L-alanine benzyl ester derivative provided the most favorable balance between stability of the ester group ($t_{1/2}$ = 0.5 h in CEM cell extracts, pH 6.9) and antiviral potency of the pronucleotide versus HIV-1 and HIV-2 in CEM/0 cells (EC_{50} = 0.3 and 0.4 μM, respectively) [171].

Although effective intracellular trapping was apparently achieved with these second-generation pronucleotides, liberation of the parent drug was slow due to the chemical stability of the *cyclo*Sal moiety [172,173]. To adjust the release rate of the parent drug, a third generation of the *cyclo*Sal pronucleotides was developed in which the aromatic ring of the *cyclo*Sal promoiety was decorated with a 5-diacetoxymethyl [173], 5-diisobutyroxymethyl [159], 3-diacetoxymethyl [159], or 5-(1-acetoxyvinyl) [160] group (Fig. 12). Once these pronucleotides entered cell, the lipophilic, weakly electron-withdrawing acylal substituent would be hydrolyzed by esterases to form a polar, strongly electron-withdrawing aldehyde group, retaining the pronucleotide inside the cell and also facilitating its chemical hydrolysis to a benzyl phosphodiester, which could spontaneously release the parent nucleoside monophosphate. This concept was based on the assumption that the enzymatic conversion occurs mainly inside the cell, since intracellular concentration of esterases is higher than in the extracellular environment [140,174]. 5-Diacetoxymethyl and 5-(1-acetoxyvinyl) derivatives of *cyclo*Sal-d4TMP exhibited decreased half-lives (referring to disappearance of the triesters) in CEM cell extracts at pH 6.9 ($t_{1/2}$ = 0.08 and 0.13 h, respectively) compared to half-lives in phosphate buffer with pH 7.3 ($t_{1/2}$ = 1.2 and 1.4 h, respectively) [160,174] but showed up to 50-fold less in antiviral activity versus HIV-2 in CEM/TK$^-$ cells than versus HIV-1 and HIV-2 in wild-type CEM/0 cells [160,174].

In an effort to maintain antiviral activity, the hydrolytic stability of the third-generation pronucleotides was increased by incorporation of alkyl groups (methyl, *tert*-butyl) into the *cyclo*Sal promoieties at the 3 position of the phenyl ring [160], as such substitution had successfully increased the chemical stability of the first-generation *cyclo*Sal pronucleotides [13]. Introduction of 3-*tert*-butyl group into 5-(1-acetoxyvinyl)-*cyclo*Sal-d4TMP increased its half-life in phosphate buffer (pH 7.3) 10-fold (1.4 to 13.5 h) but slightly decreased the stability of the pronucleotide in CEM cell extracts (pH 6.9; 0.13 to 0.04 h). Most importantly, however, the 3-alkyl derivative of the 5-(1-acetoxyvinyl)-*cyclo*Sal pronucleotide retained their antiviral potency in TK-deficient cells relative to the antiviral activities in wild-type CEM/0 cells [159,160]. Of the two diastereomeric forms of these *cyclo*Sal pronucleotides, the (S_p)-diastereomers were significantly more active [159,160]. This correlation between the phosphorus stereochemistry and the pronucleotide antiviral activity is opposite to that observed for the S_p- and R_p-diastereomers of the first-generation *cyclo*Sal pronucleotides [169,170].

Recently, the *cyclo*Sal approach was applied to a new class of nucleoside cytostatics, 6-heteroaryl-7-deazapurine ribonucleosides; however, the resulting prodrugs had only low to moderate antiviral activity compared to the parent nucleosides [175]. In order to improve the number of equivalents of released nucleotide per masking unit (mask-to-drug ratio) to 1:2, bis-(*cyclo*Sal) pronucleotides of d4TMP were proposed in which two nucleotide molecules are joined to form one "dimeric" *cyclo*Sal masking unit [176–178]. As with first-generation pronucleotides, the activation mechanism of the bis-(*cyclo*Sal) derivatives was nonenzymatic. These pronucleotides showed not only reduced butyrylcholinesterase (BChE) inhibition and marked antiviral activity against HIV-1 and HIV-2 in wild-type CEM cells but were considerably more toxic than the parent nucleoside and, often, partly cost antiviral potency in TK-deficient CEM cells. The bis-(*cyclo*Sal) promoiety (3,3′-bis(hydroxymethyl)-5,5′-di-*tert*-butyldiphenyl-4,4′-diol) itself exhibited an independent antiviral activity against HIV-1 and HIV-2 in CEM/0 cells and against HIV-2 in CEM/TK$^-$ cells only 10-fold inferior to the corresponding pronucleotide itself. The cytotoxicities of this promoiety and the intact prodrug were comparable [178].

In summary, the *cyclo*Sal approach succeeded in improving intracellular delivery of nucleotides with activation solely by a chemical mechanism. Salicyl alcohol was shown to be a bifunctional phosphate/phosphonate masking unit that can also be removed nonenzymatically *in vitro*. Antiviral evaluation showed that some *cyclo*Sal pronucleotides have antiviral activity similar to, or better than, the parent nucleoside, particularly in kinase-deficient cells. Although salicyl alcohol is nontoxic, its nucleotide prodrugs had elevated levels of cytotoxicities. Recognized drawbacks of this approach, such as extracellular release of the nucleotide and excessive efflux of the pronucleotide from the cell, have been addressed in two subsequent generations of the pronucleotides by introduction and further modification of the "lock-in" concept. However, the lack of specific, enzyme-dependent nucleotide release diminishes the benefit of the first-generation pronucleotides, an independence of activation mechanism from an enzyme activity. Currently, there are no examples of *cyclo*Sal pronucleotides in clinical trials.

F. Amino Acid Phosphonate Ester Prodrug Approach

McKenna et al. have explored the use of biologically benign amino acids and peptides as promoieties to increase the oral bioavailability of the broad-spectrum antiviral phosphonate drugs, (*S*)-HPMPC (Cidofovir®) and (*S*)-HPMPA [179–197]. The initial rationale focused on combination of a hydrophobic amino acid such as Ala with Ser in a dipeptide conjugated via an ester linkage between the seryl side-chain OH and the P(O)OH group of (*S*)-HPMPC or (*S*)-HPMPA to target human oligopeptide transporter 1 (hPEPT1) [180]. hPEPT1 is a proton-coupled di- and tripeptide transporter located in the brush border membrane of enterocytes of the small intestine [198]. Broad substrate specificity of hPEPT1 allows it to transport a wide array of chemically diverse compounds, such as β-lactam antibiotics [199–201], angiotensin-converting enzyme inhibitors [202–204], and non-peptidic drugs [205], thus making this transporter an attractive drug delivery target [206]. Prodrugs successfully targeting the hPEPT1 transporter include valacyclovir (Valtrex®) [207], a prodrug of ACV (Zovirax®). ACV is an antiviral agent marketed for the treatment of herpes, varicella zoster, and CMVs [208]. It does not contain a phosphonate group, but has a terminal hydroxyl group that is esterified by (L)-valine to form valacyclovir. Valacyclovir has significantly higher oral bioavailability than the parent drug associated with transport of the prodrug by hPEPT1 [209–211] by a study using radiolabeled valacyclovir in *Xenopus laevis* oocytes expressing the transporter [212]. Although efforts to define the SAR and STR properties of hPEPT1 have been made [213–216], reliably predictive rules for affinity and transport have been elusive [217]. In our approach, various dipeptide and single amino acid prodrugs of (*S*)-HPMPC and (*S*)-HPMPA corresponding to two classes of phosphonate prodrugs, cyclic phosphonate ester prodrugs and acyclic phosphonate diester prodrugs, have been explored (Fig. 13).

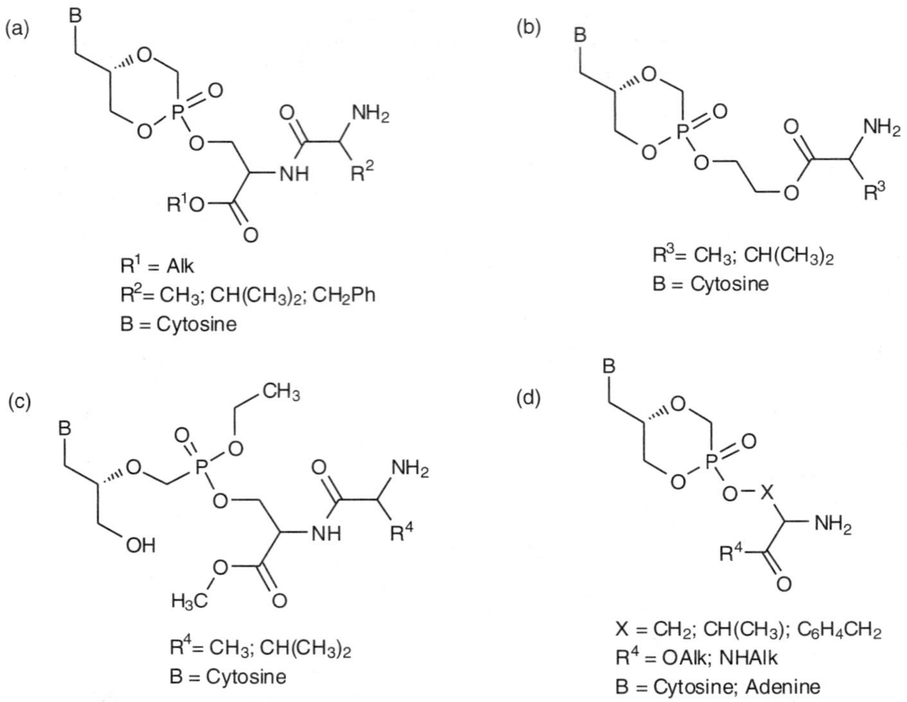

FIGURE 13 General structures of the alcoholic amino acid phosphonate ester prodrugs: (a) cyclic prodrugs incorporating a dipeptide promoiety; (b) cyclic prodrugs incorporating an ethylene glycol linker esterified by the amino acid as a promoiety; (c) acyclic prodrugs incorporating a dipeptide promoiety to mask one phosphonic acid P-OH and an ethyl group to mask the second P-OH group; (d) cyclic prodrugs incorporating a single amino acid promoiety.

In the cyclic phosphonate ester prodrugs, one phosphonic acid OH group of (*S*)-HPMPC or (*S*)-HPMPA was masked by intramolecular cyclization to afford the cyclic forms, cHPMPC and cHPMPA [218]. The remaining P-OH group was esterified with the side-chain hydroxyl group of the alcoholic amino acid ester (e.g., Ser, Thr, or Tyr) serving as a single amino acid promoiety (Fig. 13d) [186,189, 197] or incorporated into a dipeptide promoiety [180,184, 189,197] together with Val, Ala, Phe, and Leu as a second amino acid (Fig. 13a). An ethylene glycol (EG) linker esterified with the amino acid (e.g., (L)-Val or (L)-Phe) was also evaluated as a promoiety for cHPMPC prodrugs (Fig. 13b) [182]. A second group of studied prodrugs, acyclic phosphonate diester prodrugs of (*S*)-HPMPC, incorporated a dipeptide promoiety to mask one phosphonic acid P-OH and an ethyl group to mask the second P-OH group (Fig. 13c) [193].

Although the mechanism of the parent drug release from the alcoholic amino acid prodrug varies depending on the structure of the promoiety, it was hypothesized that in general, cyclic and acyclic phosphonate diester prodrugs undergo chemical hydrolysis

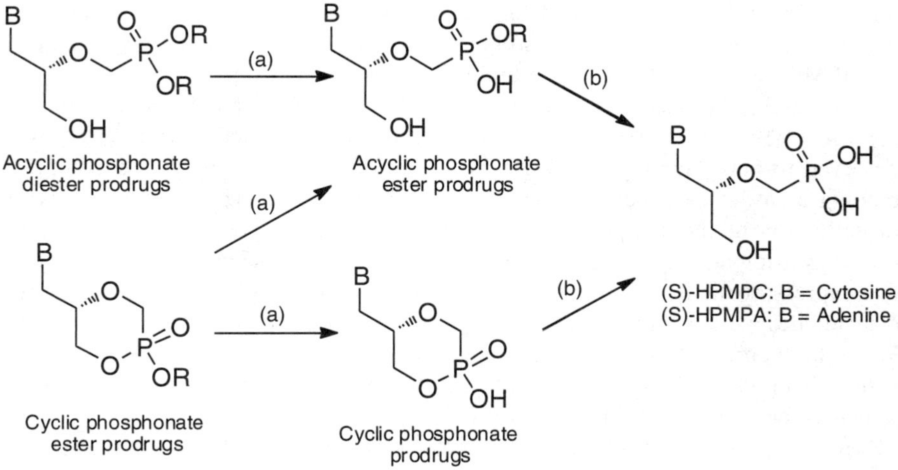

FIGURE 14 General representation of activation pathway of alcoholic amino acid phosphonate ester prodrugs. (a) Chemical hydrolysis and (b) intracellular enzymatic hydrolysis.

first, producing cyclic or acyclic phosphonate monoester prodrugs that liberate (*S*)-HPMPC or (*S*)-HPMPA after enzymatic hydrolysis (Fig. 14). Mendel et al. have shown that intracellular enzymatic conversion of cHPMPC to the parent acyclic (*S*)-HPMPC is catalyzed by a cyclic CMP phosphodiesterase [219].

The prodrugs have been evaluated for *in vitro* antiviral potency and toxicity, for chemical stability in phosphate buffer, and for enzymatic stability in intestinal, liver, and plasma homogenates. Selected prodrugs have also been examined for the ability to be actively transported by hPEPT1 and for oral bioavailability in a murine model. Acyclic phosphonate diester prodrugs of (*S*)-HPMPC exhibited enhanced oral bioavailability; however, the P-OEt monoester metabolite detected in plasma was enzymatically stable and neither released the parent drug in an *in vivo* rat model nor exhibited *in vitro* activity against vaccinia [193].

On the other hand, cyclic phosphonate ester prodrugs of (*S*)-HPMPC incorporating Ser-containing dipeptides and the L-Val-EG-cHPMPC prodrug evaluated for *in vitro* antiviral activity in a plaque reduction assay demonstrated activity against HCMV in the submicromolar range, which was up to 10-fold lower than ganciclovir, the positive control [182,184]. These conjugates exhibited little to no cytotoxicity at concentrations up to 100 μM in human foreskin fibroblast (HFF) and KB cells. In addition to antiviral activity, these prodrugs showed similar chemical and enzymatic stability profiles. They were stable in mildly acidic phosphate buffer (pH = 5) but rapidly metabolized in liver and intestinal homogenates ($t_{1/2}$ = 1–13 min) [182,184]. However, despite the aforementioned similarities, the serine-containing dipeptide conjugates of cHPMPC and the amino acid-EG-cHPMPC prodrug were found to exhibit significantly different oral bioavailabilities defined in a murine model. Thus, the (L)-Val-(L)-Ser(OMe) cHPMPC prodrug [184] demonstrated an eightfold increase in oral bioavailability of total CDV species compared to the parent drug, whereas the L-Val-EG-cHPMPC did not exhibit increased bioavailability compared to cHPMPC after direct injection into the GI tract of rats [182].

Supporting the idea that the observed enhanced oral bioavailability of the (L)-Val-(L)-Ser(OMe) cHPMPC prodrug was mediated by active transport, it was observed to have significant affinity for the hPEPT1 transporter in a competitive binding assay in DC5 cells overexpressing hPEPT1 [180]. However, in a further investigation of these dipeptide and single amino acid cHPMPC and cHPMPA prodrugs using radiotracer uptake and electrophysiologic assays of *Xenopus* oocytes overexpressing the transporter, hPEPT1-mediated transport of the prodrugs was not detected despite the confirmation that these compounds bind to the transporter with low to high affinity [189]. The results suggest that the enhanced oral bioavailability of the dipeptide prodrug of cyclic (*S*)-HPMPC is due to an unknown, alternative transport mechanism, not involving hPEPT1.

The presence of multiple functional groups in the amino acids constructing the promoiety allows its "fine-tuning" during a search for a suitable balance between prodrug stability and efficient conversion to the active parent drug and/or metabolite *in vivo*, for example, N-terminal amino acids with D-stereochemistry were utilized to construct the dipeptide promoieties, generating additional enzymatic stability and thereby increasing the fraction of intact prodrug transported into the plasma [183]. In addition to the modification of dipeptide promoieties, single amino acids (e.g., Ser, Thr, and Tyr) were evaluated as promoieties for cHPMPC and cHPMPA (Fig. 13d) [186,187,197]. Serine and threonine prodrugs of cHPMPC and cHPMPA exhibited short half-lives ($t_{1/2}$ = 20–60 min) but released fewer metabolites in phosphate buffer compared to the dipeptide prodrugs. They liberated mainly (~90%) the cyclic forms of (*S*)-HPMPC and (*S*)-HPMPA. In contrast to Ser- and Thr-containing cHPMPC and cHPMPA prodrugs, Tyr-based cHPMPC and cHPMPA prodrugs demonstrated enhanced chemical stability in phosphate buffer at the physiological pH ($t_{1/2}$ = 150–300 min) and produced only two major metabolites, cHPMPC (or cHPMPA) and the acyclic phosphonate tyrosine ester prodrug which was as active as the parent versus HCMV, vaccinia, and cowpox viruses [197]. A significant difference in stability was also observed between the two diastereomers of the tyrosine-containing prodrugs of cHPMPC and cHPMPA, with the (R_p)-diastereomer being 3- to 10-fold more stable [197,220].

Despite the increased chemical stability, the methyl ester tyrosine prodrugs in phosphate buffer compared to the Ser and Thr derivatives, all the single amino acid prodrugs had similar, rather short, half-lives ($t_{1/2}$ < 30 min) in enzymatic media. The enzymatic stability of the Tyr-based prodrugs was increased and practically to the level of chemical stability when the methyl ester group was changed to an *N*-alkyl amide group [197]. Indeed, among the single amino acid prodrugs of cHPMPC and cHPMPA tested, the tyrosine *N*-alkyl amide derivatives exhibited the most favorable balance between biological stability and efficient conversion to the active parent drug. The tyrosine *iso*-butyl amide prodrug of cHPMPA was evaluated for oral bioavailability in a mouse model and found to have an oral bioavailability of 39%, compared to <5% for the parent drug, (*S*)-HPMPA [197].

The single amino acid prodrugs (*S*)-HPMPC and (*S*)-HPMPA were evaluated *in vitro* for antiviral potency against various viruses, including HCMV, vaccinia, and cowpox viruses and for cytotoxicity in HFF and KB cells [186,187,197]. In contrast to the serine-containing dipeptide prodrugs [184], these compounds demonstrated antiviral activity not only against HCMV but also against vaccinia and cowpox viruses, which was comparable to or better than the activity of the parent drugs, (*S*)-HPMPC and (*S*)-HPMPA [186,187,197]. The antiviral activities of the acyclic phosphonate tyrosine ester prodrug, produced as a stable metabolite from the Tyr-based cyclic (*S*)-HPMPA prodrug against all the viruses, were in the same range as the activities of the corresponding Tyr-based cyclic (*S*)-HPMPA prodrug [197]. These data suggest that this chemically stable metabolite is able to enter the cell and be metabolized to the parent (*S*)-HPMPA. Additionally, none of the single amino acid prodrugs of cHPMPC and cHPMPA showed any significant cytotoxicity toward HFF and KB cells up to a concentration of at least 100 μM [186,187,197].

In summary, an amino acid moiety, strategically linked to (*S*)-HPMPA or (*S*)-HPMPC to mask its anionic properties, significantly enhances intestinal transport, creating the possibility of an orally bioavailable form with low toxicity.

IV. CONCLUSIONS

Prodrug design continues to be one of the most important approaches in the medicinal chemistry armamentarium in improving undesirable PK properties of drugs. Drugs incorporating phosphoric or phosphonic acid group are a focus of modern prodrug strategies. This chapter has surveyed several such strategies with an emphasis on the utility of leveraging specific *in vivo* enzyme activities with the goal of making activation localized, selective, more rapid. At the same time, the promoiety must be carefully tuned to optimize the balance among transport, toxicity, and activation. It should be noted that despite ongoing attempts to create phosphate/phosphonate prodrugs liberating the parent drug solely via chemistry *in vivo*, all phosphate/phosphonate prodrugs currently available on the market or undergoing clinical trials partly or wholly depend on enzymatic activity during their activation process. Further development of pronucleotides that can effectively achieve oral absorption, bypass presystemic metabolism, and deliver nucleotide analogs to exert their therapeutic action is likely to depend on a more detailed knowledge of enzyme-mediated pathways for prodrug activation.

Acknowledgements

We thank Ms. Inah Kang for assistance in preparation of the manuscript and the National Institutes of Health (grant AI091216) for support.

REFERENCES

1. ME Dowty, DM Messing, Y Lai, L Kirkovsky. ADME. In: K Tsaioun, SA Kates, Wiley InterScience, eds. *ADMET for Medicinal Chemists*. Hoboken, NJ: John Wiley & Sons, Inc., 2011, pp. 145–200.
2. DF Veber, SR Johnson, H-Y Cheng, BR Smith, KW Ward, KD Kopple. Molecular properties that influence the oral bioavailability of drug candidates. J Med Chem 45:2615–2623, 2002.
3. HHF Refsgaard, BF Jensen, PB Brockhoff, SB Padkjr, M Guldbrandt, MS Christensen. In silico prediction of membrane permeability from calculated molecular parameters. J Med Chem 48:805–811, 2005.
4. TH Keller, A Pichota, Z Yin. A practical view of druggability. Curr Opin Chem Biol 10:357–361, 2006.
5. CA Lipinski, F Lombardo, BW Dominy, PJ Feeney. Experimental and computational approaches to estimate solubility and permeability in drug discovery and development settings. Adv Drug Deliv Rev 23:3–25, 1997.
6. T Hou, J Wang, W Zhang, X Xu. ADME evaluation in drug discovery. Can oral bioavailability in humans be effectively predicted by simple molecular property-based rules? J Chem Inf Model 47:460–463, 2007.
7. A Albert. Chemical aspects of selective toxicity. Nature 182:421–423, 1958.
8. VJ Stella. Prodrugs as therapeutics. Expert Opin Ther Pat 14:277–280, 2004.
9. Statistical reports were obtained from Web of Knowledge®. http://apps.webofknowledge.com/UA_GeneralSearch_input.do?product=UA&SID=4BcOihD2I5MIk66FB@m&search_mode=GeneralSearch (accessed July 23, 2013).
10. S Jana, S Mandlekar, P Marathe. Prodrug design to improve pharmacokinetic and drug delivery properties: challenges to the discovery scientists. Curr Med Chem 17:3874–3908, 2010.
11. BJ Denny, RT Wheelhouse, MFG Stevens, LLH Tsang, JA Slack. NMR and molecular modeling investigation of the mechanism of activation of the antitumor drug temozolomide and its interaction with DNA. Biochemistry 33:9045–9051, 1994.
12. M Krause, H Stark, W Schunack. Azomethine prodrugs of (R)-α-methylhistamine, a highly potent and selective histamine H3-receptor agonist. Curr Med Chem 8:1329–1340, 2001.
13. C Meier. CycloSal phosphates as chemical Trojan horses for intracellular nucleotide and glycosyl-monophosphate delivery – chemistry meets biology. Eur J Org Chem 1081–1102, 2006.
14. J Rautio, K Laine, L Fang, B Han, K Huttunen, D Sun. Enzyme-activated prodrug strategies for targeted drug action. In: RI Mahato, AS NArang, eds. *Targeted Delivery of Small and Macromolecular Drugs*. Boca Raton, FL: CRC Press/Taylor & Francis, 2010, pp. 311–348.
15. BM Liederer, RT Borchardt. Enzymes involved in the bioconversion of ester-based prodrugs. J Pharm Sci 95:1177–1195, 2006.
16. D Horter, JB Dressman. Influence of physicochemical properties on dissolution of drugs in the gastrointestinal tract. Adv Drug Deliv Rev 46:75–87, 2001.
17. J Broadhead, M Gibson. Parenteral dosage forms. Drugs Pharm Sci 199:325–347, 2009.

18. J Rautio, H Kumpulainen, T Heimbach, R Oliyai, D Oh, T Jaervinen, J Savolainen. Prodrugs: design and clinical applications. Nat Rev Drug Discov 7;255–270, 2008.
19. S Vaishnava, LV Hooper. Alkaline phosphatase: keeping the peace at the gut epithelial surface. Cell Host Microbe 2:365–367, 2007.
20. TM Chapman, GL Plosker, CM Perry. Fosamprenavir: a review of its use in the management of antiretroviral therapy-naive patients with HIV infection. Drugs 64:2101–2124, 2004.
21. ES Furfine, CT Baker, MR Hale, DJ Reynolds, JA Salisbury, AD Searle, SD Studenberg, D Todd, RD Tung, A Spaltenstei. Preclinical pharmacology and pharmacokinetics of GW433908, a water-soluble prodrug of the human immunodeficiency virus protease inhibitor amprenavir. Antimicrob Agents Chemother 48:791–798, 2004.
22. HA Torres, RC Arduino. Fosamprenavir calcium plus ritonavir for HIV infection. Expert Rev Anti-Infect Ther 5:349–363, 2007.
23. AT Bergenheim, R Henriksson. Pharmacokinetics and pharmacodynamics of estramustine phosphate. Clin Pharmacokinet 34:163–172, 1998.
24. T Kitamura. Necessity of re-evaluation of estramustine phosphate sodium (EMP) as a treatment option for first-line monotherapy in advanced prostate cancer. Int J Urol 8:33–36, 2001.
25. VJ Stella. A case for prodrugs: fosphenytoin. Adv Drug Deliv Rev 19:311–330, 1996.
26. TR Browne. Fosphenytoin (Cerebyx). Clin Neuropharmacol 20:1–12, 1997.
27. E De Clercq, J Neyts. Antiviral agents acting as DNA or RNA chain terminators. Handb Exp Pharmacol 189:53–84, 2009.
28. CR Wagner, VV Iyer, EJ McIntee. Pronucleotides: toward the in vivo delivery of antiviral and anticancer nucleotides. Med Res Rev 20:417–451, 2000.
29. JE Starrett, Jr., DR Tortolani, J Russell, MJM Hitchcock, V Whiterock, JC Martin, MM Mansuri. Synthesis, oral bioavailability determination, and in vitro evaluation of prodrugs of the antiviral agent 9-[2-(Phosphonomethoxy)ethyl]adenine (PMEA). J Med Chem 37:1857–1864, 1994.
30. HT Serafinowska, RJ Ashton, S Bailey, MR Harnden, SM Jackson, D Sutton. Synthesis and in vivo evaluation of prodrugs of 9-[2-(Phosphonomethoxy)ethoxy]adenine. J Med Chem 38:1372–1379, 1995.
31. SJ Hecker, MD Erion. Prodrugs of phosphates and phosphonates. J Med Chem 51:2328–2345, 2008.
32. SJ Kelly, DE Dardinger, LG Butler. Hydrolysis of phosphonate esters catalyzed by 5'-nucleotide phosphodiesterase. *Biochemistry* 14:4983–4988, 1975.
33. SJ Kelly, LG Butler. Enzymic hydrolysis of phosphonate esters. Reaction mechanism of intestinal 5'-nucleotide phosphodiesterase. *Biochemistry* 16:1102–1104, 1977.
34. D Farquhar, NJ Kuttesch, MG Wilkerson, T Winkler. Synthesis and biological evaluation of neutral derivatives of 5-fluoro-2′-deoxyuridine 5′-phosphate. J Med Chem 26:1153–1158, 1983.
35. DN Srivastva, D Farquhar. Bioreversible phosphate protective groups: synthesis and stability of model acyloxymethyl phosphates. Bioorg Chem 12:118–129, 1984.
36. RP Iyer, LR Phillips, JA Biddle, DR Thakker, W Egan, S Aoki, H Mitsuya. Synthesis of acyloxyalkyl acylphosphonates as potential prodrugs of the antiviral, trisodium phosphonoformate (foscarnet sodium). Tetrahedron Lett 30:7141–7144, 1989.

37. C McGuigan, C Nickson, J Petrik, A Karpas. Phosphate derivatives of AZT display enhanced selectivity of action against HIV1 by comparison to the parent nucleoside. FEBS Lett 310:171–174, 1992.
38. JE Starrett, Jr., DR Tortolani, MJM Hitchcock, JC Martin, MM Mansuri. Synthesis and in vitro evaluation of a phosphonate prodrug: bis(pivaloyloxymethyl)-9-(2-phosphonylmethoxyethyl)adenine. Antiviral Res 19:267–273, 1992.
39. A Pompon, I Lefebvre, JL Imbach, S Kahn, D Farquhar. Decomposition pathways of the mono- and bis(pivaloyloxymethyl) esters of azidothymidine 5′-monophosphate in cell extract and in tissue culture medium: an application of the "online ISRP-cleaning" HPLC technique. Antivir Chem Chemother 5:91–98, 1994.
40. JK Sastry, PN Nehete, S Khan, BJ Nowak, W Plunkett, RB Arlinghaus, D Farquhar. Membrane-permeable dideoxyuridine 5′-monophosphate analog inhibits human immunodeficiency virus infection. Mol Pharmacol 41:441–445, 1992.
41. I Lefebvre, C Perigaud, A Pompon, AM Aubertin, JL Girardet, A Kirn, G Gosselin, JL Imbach. Mononucleoside phosphotriester derivatives with S-Acyl-2-thioethyl Bioreversible Phosphate-Protecting Groups: intracellular delivery of 3′-azido-2′,3′-dideoxythymidine 5′-monophosphate. J Med Chem 38:3941–3950, 1995.
42. HB Fung, EA Stone, FJ Piacenti. Tenofovir disoproxil fumarate: a nucleotide reverse transcriptase inhibitor for the treatment of HIV infection. Clin Ther 24:1515–1548, 2002.
43. GF Ruda, VP Alibu, C Mitsos, O Bidet, M Kaiser, R Brun, MP Barrett, IH Gilbert. Synthesis and biological evaluation of phosphate prodrugs of 4-phospho-D-erythronohydroxamic acid, an inhibitor of 6-phosphogluconate dehydrogenase. ChemMedChem 2:1169–1180, 2007.
44. M Krecmerova, A Holy, G Andrei, K Pomeisl, T Tichy, P Brehova, M Masojidkova, M Dracinsky, R Pohl, G Laflamme, et al. Synthesis of ester prodrugs of 9-(S)-[3-Hydroxy-2-(phosphonomethoxy)propyl]-2,6-diaminopurine (HPMPDAP) as anti-poxvirus agents. J Med Chem 53:6825–6837, 2010.
45. D Topalis, U Pradere, V Roy, C Caillat, A Azzouzi, J Broggi, R Snoeck, G Andrei, J Lin, S Eriksson, et al. Novel antiviral C5-substituted pyrimidine acyclic nucleoside phosphonates selected as human thymidylate kinase substrates. J Med Chem 54:222–232, 2011.
46. SJ Hadziyannis, GV Papatheodoridis. Adefovir dipivoxil in the treatment of chronic hepatitis B virus infection. Expert Rev Anti Infect Ther 2:475–483, 2004.
47. EP Brass. Pivalate-generating prodrugs and carnitine homeostasis in man. Pharmacol Rev 54:589–598, 2002.
48. MN Arimilli, CU Kim, J Dougherty, A Mulato, R Oliyai, JP Shaw, KC Cundy, N Bischofberger. Synthesis, in vitro biological evaluation and oral bioavailability of 9-[2-(phosphonomethoxy)propyl]adenine (PMPA) prodrugs. Antivir Chem Chemother 8:557–564, 1997.
49. JP Shaw, CM Sueoka, R Oliyai, WA Lee, MN Arimilli, CU Kim, KC Cundy. Metabolism and pharmacokinetics of novel oral prodrugs of 9-[(R)-2-(phosphonomethoxy)propyl] adenine (PMPA) in dogs. Pharm Res 14:1824–1829, 1997.
50. AM Jenh, CL Thio, PA Pham. Tenofovir for the treatment of hepatitis B virus. Pharmacotherapy 29:1212–1227, 2009.
51. D Cahard, C McGuigan, J Balzarini. Aryloxy phosphoramidate triesters as protides. Mini Rev Med Chem 4:371–381, 2004.

52. Y Mehellou, J Balzarini, C McGuigan. Aryloxy phosphoramidate triesters: a technology for delivering monophosphorylated nucleosides and sugars into cells. ChemMedChem 4:1779–1791, 2009.

53. C McGuigan, PW Sutton, D Cahard, K Turner, G O'Leary, Y Wang, M Gumbleton, CE De, J Balzarini. Synthesis, anti-human immunodeficiency virus activity and esterase lability of some novel carboxylic ester-modified phosphoramidate derivatives of stavudine (d4T). Antivir Chem Chemother 9:473–479, 1998.

54. G Birkus, R Wang, X Liu, N Kutty, H MacArthur, T Cihlar, C Gibbs, S Swaminathan, W Lee, M McDermott. Cathepsin A is the major hydrolase catalyzing the intracellular hydrolysis of the antiretroviral nucleotide phosphonoamidate prodrugs GS-7340 and GS-9131. Antimicrob Agents Chemother 51:543–550, 2007.

55. D Saboulard, L Naesens, D Cahard, A Salgado, R Pathirana, S Velazquez, C McGuigan, CE De, J Balzarini. Characterization of the activation pathway of phosphoramidate triester prodrugs of stavudine and zidovudine. Mol Pharmacol 56:693–704, 1999.

56. P Perrone, F Daverio, R Valente, S Rajyaguru, JA Martin, V Leveque, PS Le, I Najera, K Klumpp, DB Smith, et al. First example of phosphoramidate approach applied to a 4′-substituted purine nucleoside (4′-azidoadenosine): conversion of an inactive nucleoside to a submicromolar compound versus hepatitis C virus. J Med Chem 50:5463–5470, 2007.

57. P Perrone, GM Luoni, MR Kelleher, F Daverio, A Angell, S Mulready, C Congiatu, S Rajyaguru, JA Martin, V Leveque, et al. Application of the phosphoramidate ProTide approach to 4′-azidouridine confers sub-micromolar potency versus hepatitis C virus on an inactive nucleoside. J Med Chem 50:1840–1849, 2007.

58. C McGuigan, KG Devine, TJ O'Connor, D Kinchington. Synthesis and anti-HIV activity of some haloalkyl phosphoramidate derivatives of 3′-azido-3′-deoxythymidine (AZT): potent activity of the trichloroethyl methoxyalaninyl compound. Antiviral Res 15:255–263, 1991.

59. Y Mehellou, C McGuigan, A Brancale, J Balzarini. Design, synthesis, and anti-HIV activity of 2′,3′-didehydro-2′,3′-dideoxyuridine (d4U), 2′,3′-dideoxyuridine (ddU) phosphoramidate "ProTide" derivatives. Bioorg Med Chem Lett 17:3666–3669, 2007.

60. Y Mehellou, J Balzarini, C McGuigan. An investigation into the anti-HIV activity of 2′,3′-didehydro-2′,3′-dideoxyuridine (d4U) and 2′,3′-dideoxyuridine (ddU) phosphoramidate "ProTide" derivatives. Org Biomol Chem 7:2548–2553, 2009.

61. C McGuigan, D Cahard, HM Sheeka, CE De, J Balzarini. Phosphoramidate derivatives of d4T with improved anti-HIV efficacy retain full activity in thymidine kinase-deficient cells. Bioorg Med Chem Lett 6:1183–1186, 1996.

62. C McGuigan, SA Harris, SM Daluge, KS Gudmundsson, EW McLean, TC Burnette, TH Marr, R Hazen, LD Condreay, L Johnson, et al. Application of phosphoramidate pronucleotide technology to abacavir leads to a significant enhancement of antiviral potency. J Med Chem 48:3504–3515, 2005.

63. SA Harris, C McGuigan, G Andrei, R Snoeck, CE De, J Balzarini. Synthesis and antiviral evaluation of phosphoramidate derivatives of (E)-5-(2-bromovinyl)-2′-deoxyuridine. Antivir Chem Chemother 12:293–300, 2001.

64. C McGuigan, P Murziani, M Slusarczyk, B Gonczy, VJ Vande, S Liekens, J Balzarini. Phosphoramidate ProTides of the anticancer agent FUDR successfully deliver the preformed bioactive monophosphate in cells and confer advantage over the parent nucleoside. J Med Chem 54:7247–7258, 2011.

65. C McGuigan, A Gilles, K Madela, M Aljarah, S Holl, S Jones, J Vernachio, J Hutchins, B Ames, KD Bryant, et al. Phosphoramidate ProTides of 2′-C-methylguanosine as highly potent inhibitors of hepatitis C virus. Study of their in vitro and in vivo properties. J Med Chem 53:4949–4957, 2010.
66. M Derudas, A Brancale, L Naesens, J Neyts, J Balzarini, C McGuigan. Application of the phosphoramidate ProTide approach to the antiviral drug ribavirin. Bioorg Med Chem 18:2748–2755, 2010.
67. C McGuigan, M Derudas, JJ Bugert, G Andrei, R Snoeck, J Balzarini. Successful kinase bypass with new acyclovir phosphoramidate prodrugs. Bioorg Med Chem Lett 18:4364–4367, 2008.
68. C Ballatore, C McGuigan, CE De, J Balzarini. Synthesis and evaluation of novel amidate prodrugs of PMEA and PMPA. Bioorg Med Chem Lett 11:1053–1056, 2001.
69. AB Eldrup, CR Allerson, CF Bennett, S Bera, B Bhat, N Bhat, MR Bosserman, J Brooks, C Burlein, SS Carroll, et al. Structure-activity relationship of purine ribonucleosides for inhibition of hepatitis C virus RNA-dependent RNA polymerase. J Med Chem 47:2283–2295, 2004.
70. AB Eldrup, M Prhavc, J Brooks, B Bhat, TP Prakash, Q Song, S Bera, N Bhat, P Dande, PD Cook, et al. Structure-activity relationship of heterobase-modified 2′-C-methyl ribonucleosides as inhibitors of hepatitis C virus RNA replication. J Med Chem 47:5284–5297, 2004.
71. C McGuigan, K Madela, M Aljarah, A Gilles, A Brancale, N Zonta, S Chamberlain, J Vernachio, J Hutchins, A Hall, et al. Design, synthesis and evaluation of a novel double prodrug: INX-08189. A new clinical candidate for hepatitis C virus. Bioorg Med Chem Lett 20:4850–4854, 2010.
72. C McGuigan, K Madela, M Aljarah, A Gilles, SK Battina, CVS Ramamurty, RC Srinivas, J Vernachio, J Hutchins, A Hall, et al. Dual pro-drugs of 2′-C-methyl guanosine monophosphate as potent and selective inhibitors of hepatitis C virus. Bioorg Med Chem Lett 21:6007–6012, 2011.
73. MJ Sofia, D Bao, W Chang, J Du, D Nagarathnam, S Rachakonda, PG Reddy, BS Ross, P Wang, HR Zhang, et al. Discovery of a β-D-2′-deoxy-2′-α-fluoro-2′-β-C-methyluridine nucleotide prodrug (PSI-7977) for the treatment of hepatitis C virus. J Med Chem 53:7202–7218, 2010.
74. AM Lam, E Murakami, C Espiritu, HMM Steuer, C Niu, M Keilman, H Bao, V Zennou, N Bourne, JG Julander, et al. PSI-7851, a pronucleotide of β-D-2′-deoxy-2′-fluoro-2′-C-methyluridine monophosphate, is a potent and pan-genotype inhibitor of hepatitis C virus replication. Antimicrob Agents Chemother 54:3187–3196, 2010.
75. E Murakami, T Tolstykh, H Bao, C Niu, HMM Steuer, D Bao, W Chang, C Espiritu, S Bansal, AM Lam, et al. Mechanism of activation of PSI-7851 and its diastereoisomer PSI-7977. J Biol Chem 285:34337–34347, 2010.
76. T Cihlar, G Birkus, DE Greenwalt, MJM Hitchcock. Tenofovir exhibits low cytotoxicity in various human cell types: comparison with other nucleoside reverse transcriptase inhibitors. Antiviral Res 54:37–45, 2002.
77. T Antoniou, LY Park-Wyllie, AL Tseng. Tenofovir: a nucleotide analog for the management of human immunodeficiency virus infection. Pharmacotherapy 23:29–43, 2003.
78. SA Grim, F Romanelli. Tenofovir disoproxil fumarate. Ann Pharmacother 37:849–859, 2003.

79. WA Lee, GX He, E Eisenberg, T Cihlar, S Swaminathan, A Mulato, KC Cundy. Selective intracellular activation of a novel prodrug of the human immunodeficiency virus reverse transcriptase inhibitor tenofovir leads to preferential distribution and accumulation in lymphatic tissue. Antimicrob Agents Chemother 49:1898–1906, 2005.

80. M Markowitz, A Zolopa, P Ruane, K Squires, L Zhong, B Kearney, W Lee. GS-7340 demonstrates greater declines in HIV-1 RNA than tenofovir disoproxil fumarate during 14 days of monotherapy in HIV-1 infected subjects. In *18th Conference on Retroviruses and Opportunistic Infections*, Boston, MA, February 27 – March 3, 2011; Paper # 152LB.

81. RL Mackman, AS Ray, HC Hui, L Zhang, G Birkus, CG Boojamra, MC Desai, JL Douglas, Y Gao, D Grant, et al. Discovery of GS-9131: design, synthesis and optimization of amidate prodrugs of the novel nucleoside phosphonate HIV reverse transcriptase (RT) inhibitor GS-9148. Bioorg Med Chem 18:3606–3617, 2010.

82. DB Lackey, MP Groziak, M Sergeeva, M Beryt, C Boyer, RM Stroud, P Sayre, JW Park, P Johnston, D Slamon, et al. Enzyme-catalyzed therapeutic agent (ECTA) design: activation of the antitumor ECTA compound NB1011 by thymidylate synthase. Biochem Pharmacol 61:179–189, 2001.

83. CR Boyer, PL Karjian, GM Wahl, M Pegram, STC Neuteboom. Nucleoside transport inhibitors, dipyridamole and p-nitrobenzylthioinosine, selectively potentiate the antitumor activity of NB1011. Anticancer Drugs 13:29–36, 2002.

84. STC Neuteboom, PL Karjian, CR Boyer, M Beryt, M Pegram, GM Wahl, HM Shepard. Inhibition of cell growth by NB1011 requires high thymidylate synthase levels and correlates with p53, p21, Bax, and GADD45 induction. Mol Cancer Ther 1:377–384, 2002.

85. M Pegram, CH Yeon, N Ku, C Gottlieb, M Shepard, P Cossum, E John, S Iqbal, A Garcia, HJ Lenz. Enzyme catalyzed therapeutic activation of NB1011 (N) selectively targets thymidylate synthase (TS)-overexpressing tumor cells: Phase I results. J Clin Oncol 22:230S, 2004.

86. RH Wilson. Novel therapeutic developments other than EGFR and VEGF inhibition in colorectal cancer. Oncologist 11:1018–1024, 2006.

87. DT Warner, GL Neil, AJ Taylor, WJ Wechter. Nucleic acids. 13. 3′-0- and 2′-0-esters of 1-β-D-arabinofuranosylcytosine as antileukemic and immunosuppressive agents. J Med Chem 15:790–792, 1972.

88. WJ Wechter, MA Johnson, CM Hall, DT Warner, AE Berger, AH Wenzel, DT Gish, GL Neil. Nucleic acids. 14. ara-cytidine acylates. Use of drug design predictors in structure-activity relation correlation. J Med Chem 18:339–344, 1975.

89. CRH Raetz, MY Chu, SP Srivastava, JG Turcotte. A phospholipid derivative of cytosine arabinoside and its conversion to phosphatidylinositol by animal tissue. Science 196:303–305, 1977.

90. A Sakai, N Mori, S Shuto, T Suzuki. Deacylation-reacylation cycle: a possible absorption mechanism for the novel lymphotropic antitumor agent dipalmitoylphosphatidylfluorouridine in rats. J Pharm Sci 82:575–578, 1993.

91. JM Steim, NC Camaioni, PS Sarin, DK Sun, RK Sehgal, JG Turcotte. Lipid conjugates of antiretroviral agents. I. Azidothymidine-monophosphate-diglyceride: anti-HIV activity, physical properties, and interaction with plasma proteins. Biochem Biophys Res Commun 171:451–457, 1990.

92. KY Hostetler, JR Beadle, GD Kini, MF Gardner, KN Wright, TH Wu, BA Korba. Enhanced oral absorption and antiviral activity of 1-O-octadecyl-sn-glycero-3-phospho-acyclovir and related compounds in hepatitis B virus infection, in vitro. Biochem Pharmacol 53:1815–1822, 1997.

93. AC Venhoff, D Lebrecht, FU Reuss, B Heckl-Oestreicher, R Wehr, UA Walker, N Venhoff. Mitochondrial DNA depletion in rat liver induced by fosalvudine tidoxil, a novel nucleoside reverse transcriptase inhibitor prodrug. Antimicrob Agents Chemother 53:2748–2751, 2009.
94. KY Hostetler, LM Stuhmiller, HBM Lenting, BH Van, DD Richman. Synthesis and antiretroviral activity of phospholipid analogs of azidothymidine and other antiviral nucleosides. J Biol Chem 265:6112–6117, 1990.
95. WGMT Van, KY Hostetler, BH Van. Antiviral nucleoside diphosphate diglycerides: improved synthesis and facilitated purification. J Lipid Res 33:1211–1219, 1992.
96. GD Kini, SE Hostetler, JR Beadle, KA Aldern. Synthesis and antiviral activity of 1-octadecyl-2-0-alkyl-sn-glycero-3-foscarnet conjugates in human cytomegalovirus-infected cells. Antiviral Res 36:115–124, 1997.
97. KY Hostetler, JL Hammond, GD Kini, SE Hostetler, JR Beadle, KA Aldern, TC Chou, DD Richman, JW Mellors. In vitro anti-HIV-1 activity of sn-2-substituted 1-O-octadecyl-sn-glycero-3-phosphonoformate analogues and synergy with zidovudine. Antiviral Chem Chemother 11:213–220, 2000.
98. CE McKenna, ZM Li, XW Liu. Preparation and antiviral activity of pyrophosphate analogs. US 6444837B1, September 3, 2002.
99. JR Beadle, C Hartline, KA Aldern, N Rodriguez, E Harden, ER Kern, KY Hostetler. Alkoxyalkyl esters of cidofovir and cyclic cidofovir exhibit multiple-log enhancement of antiviral activity against cytomegalovirus and herpesvirus replication in vitro. Antimicrob Agents Chemother 46:2381–2386, 2002.
100. JR Beadle, WB Wan, SL Ciesla, KA Keith, C Hartline, ER Kern, KY Hostetler. Synthesis and antiviral evaluation of alkoxyalkyl derivatives of 9-(S)-(3-hydroxy-2-phosphonomethoxypropyl)adenine against cytomegalovirus and orthopoxviruses. J Med Chem 49:2010–2015, 2006.
101. N Valiaeva, JR Beadle, KA Aldern, J Trahan, KY Hostetler. Synthesis and antiviral evaluation of alkoxyalkyl esters of acyclic purine and pyrimidine nucleoside phosphonates against HIV-1 in vitro. Antiviral Res 72:10–19, 2006.
102. P Heussner, R Willemze, A Ganser, A Hanauske, S Amadori, G Heil, E Schleyer, W Hiddemann, J Selbach. YNK01, an oral cytosine arabinoside derivative in acute myeloid leukemia and chronic myeloid leukemia. Haematol Blood Transfus 38:882–885, 1997.
103. M Saneyoshi, M Morozumi, K Kodama, H Machida, A Kuninaka, H Yoshino. Synthetic nucleosides and nucleotides. XVI. Synthesis and biological evaluations of a series of 1-Î²-D-arabinofuranosylcytosine 5′-alkyl or arylphosphates. Chem Pharm Bull 28:2915–2923, 1980.
104. K Kodama, M Morozumi, K Saitoh, A Kuninaka, H Yoshino, M Saneyoshi. Antitumor activity and pharmacology of 1-β-D-arabinofuranosylcytosine-5′-stearylphosphate: an orally active derivative of 1-β-D-arabinofuranosylcytosine. Jpn J Cancer Res 80:679–685, 1989.
105. Y Yoshida, J Yamada, T Watanabe, T Suga, H Takayama. Participation of the peroxisomal β-oxidation system in the chain-shortening of PCA16, a metabolite of the cytosine arabinoside prodrug, YNKO1, in rat liver. Biochem Pharmacol 39:1505–1512, 1990.
106. J Braess, M Freund, A Hanauske, G Heil, C Kaufmann, W Kern, M Schussler, W Hiddemann, E Schleyer. Oral cytarabine ocfosfate in acute myeloid leukemia and non-Hodgkin's lymphoma phase I/II studies and pharmacokinetics. Leukemia 12:1618–1626, 1998.

107. G Rosti, F Bonifazi, E Trabacchi, S Bassi, G Martinelli, N Testoni, D Russo, M Baccarani. A phase II study of Î±-interferon and oral arabinosyl cytosine (YNK01) in chronic myeloid leukemia. *Leukemia* 17:554–559, 2003.

108. PM Girard, PS Pegram, B Diquet, R Anderson, F Raffi, R Tubiana, D Sereni, D Boerner. Phase II placebo-controlled trial of fozivudine tidoxil for HIV infection: pharmacokinetics, tolerability, and efficacy. J Acquir Immune Defic Syndr 23:227–235, 2000.

109. E De Clercq, A Holy. Case history: acyclic nucleoside phosphonates: a key class of antiviral drugs. Nat Rev Drug Discov 4:928–940, 2005.

110. KY Hostetler. Alkoxyalkyl prodrugs of acyclic nucleoside phosphonates enhance oral antiviral activity and reduce toxicity: current state of the art. Antiviral Res. 82:A84–A98, 2009.

111. E De Clercq. The acyclic nucleoside phosphonates from inception to clinical use: historical perspective. Antiviral Res 75:1–13, 2007.

112. RJ Raggers, T Pomorski, JCM Holthuis, N Kalin, MG Van. Lipid traffic: the ABC of transbilayer movement. Traffic (Copenhagen) 1:226–234, 2000.

113. A Nilsson, B Borgstrom. Absorption and metabolism of lecithin and lysolecithin by intestinal slices. Biochim Biophys Acta Lipids Lipid Metab 137:240–254, 1967.

114. RO Scow, Y Stein, O Stein. Incorporation of dietary lecithin and lysolecithin into lymph chylomicrons in the rat. J Biol Chem 242:4919–4924, 1967.

115. WB Wan, JR Beadle, C Hartline, ER Kern, SL Ciesla, N Valiaeva, KY Hostetler. Comparison of the antiviral activities of alkoxyalkyl and alkyl esters of cidofovir against human and murine cytomegalovirus replication in vitro. Antimicrob Agents Chemother 49:656–662, 2005.

116. N Valiaeva, MN Prichard, RM Buller, JR Beadle, CB Hartline, KA Keith, J Schriewer, J Trahan, KY Hostetler. Antiviral evaluation of octadecyloxyethyl esters of (S)-3-hydroxy-2-(phosphonomethoxy)propyl nucleosides against herpesviruses and orthopoxviruses. *Antiviral Res* 84:254–259, 2009.

117. GR Painter, MR Almond, LC Trost, BM Lampert, J Neyts, CE De, BE Korba, KA Aldern, JR Beadle, KY Hostetler. Evaluation of hexadecyloxypropyl-9-R-[2-(phosphonomethoxy) propyl]adenine, CMX157, as a potential treatment for human immunodeficiency virus type 1 and hepatitis B virus infections [Erratum to document cited in CA148:023814]. Antimicrob Agents Chemother 51:4538, 2007.

118. KA Aldern, SL Ciesla, KL Winegarden, KY Hostetler. Increased antiviral activity of 1-O-hexadecyloxypropyl-[2–14C]cidofovir in MRC-5 human lung fibroblasts is explained by unique cellular uptake and metabolism. Mol Pharmacol 63:678–681, 2003.

119. WC Magee, KA Aldern, KY Hostetler, DH Evans. Cidofovir and (S)-9-[3-hydroxy-(2-phosphonomethoxy)propyl]adenine are highly effective inhibitors of vaccinia virus DNA polymerase when incorporated into the template strand. Antimicrob Agents Chemother 52:586–597, 2008.

120. SL Ciesla, J Trahan, WB Wan, JR Beadle, KA Aldern, GR Painter, KY Hostetler. Esterification of cidofovir with alkoxyalkanols increases oral bioavailability and diminishes drug accumulation in kidney. Antiviral Res 59:163–171, 2003.

121. DC Quenelle, DJ Collins, BP Herrod, KA Keith, J Trahan, JR Beadle, KY Hostetler, ER Kern. Effect of oral treatment with hexadecyloxypropyl-[(S)-9-(3-hydroxy-2-phosphonylmethoxypropyl)adenine] [HPD-(S)-HPMPA] or octadecyloxyethyl-(S)-HPMPA on cowpox or vaccinia virus infections in mice. Antimicrob Agents Chemother 51:3940–3947, 2007.

122. R Lanier, L Trost, T Tippin, B Lampert, A Robertson, S Foster, M Rose, W Painter, R O'Mahony, M Almond, et al. Development of CMX001 for the treatment of poxvirus infections. Viruses 2:2740–2762, 2010.

123. J Ruiz, JR Beadle, BR Mark, J Schreiwer, MN Prichard, KA Keith, KC Lewis, KY Hostetler. Synthesis, metabolic stability and antiviral evaluation of various alkoxyalkyl esters of cidofovir and 9-(S)-[3-hydroxy-2-(phosphonomethoxy)propyl]adenine. Bioorg Med Chem 19:2950–2958, 2011.

124. S Peyrottes, D Egron, I Lefebvre, G Gosselin, JL Imbach, C Perigaud. SATE pronucleotide approaches: an overview. Mini Rev Med Chem 4:395–408, 2004.

125. A Jochum, N Schlienger, D Egron, S Peyrottes, C Perigaud. Biolabile constructs for pronucleotide design. J Organomet Chem 690:2614–2625, 2005.

126. L Placidi, A Faraj, AG Loi, C Pierra, D Egron, E Cretton-Scott, G Gosselin, C Perigaud, LT Martin, RF Schinazi, et al. Antiviral activity and intracellular metabolism of Bis(tButylSATE) phosphotriester of Î²-L-2′,3′dideoxyadenosine, a potent inhibitor of HIV and HBV replication. Antiviral Chem Chemother 12:99–108, 2001.

127. S Benzaria, JL Girardet, C Periguad, AM Aubertin, H Pelicano, G Maury, G Goselin, A Kirn, JL Imbach. The SATE pronucleotide derivative of ddA: a more potent HIV inhibitor than AZT. Nucleic Acids Symp Ser 31:129–130, 1994.

128. C Perigaud, AM Aubertin, S Benzaria, H Pelicano, JL Girardet, G Maury, G Gosselin, A Kirn, JL Imbach. Equal inhibition of the replication of human immunodeficiency virus in human T-cell culture by ddA bis(SATE)phosphotriester and 3′-azido-2′,3′-dideoxythymidine. Biochem Pharmacol48:11–14, 1994.

129. J Cinatl, Jr., B Groschel, R Zeghner, J Cinatil, C Perigaud, G Gosselin, JL Imbach, HW Doerr. Human immunodeficiency virus resistance to AZT in MOLT4/8 cells is associated with a lack of AZT phosphorylation and is bypassed by AZT-morphosphate SATE products. Antiviral Chem Chemother 8:343–352, 1997.

130. JL Girardet, C Perigaud, AM Aubertin, G Gosselin, A Kirn, JL Imbach. Increase of the anti-HIV activity of D4T in human T-cell culture by the use of the sate pronucleotide approach. Bioorg Med Chem Lett 5:2981–2984, 1995.

131. MH Gouy, LP Jordheim, I Lefebvre, E Cros, C Dumontet, S Peyrottes, C Perigaud. Special feature of mixed phosphotriester derivatives of cytarabine. Bioorg Med Chem 17:6340–6347, 2009.

132. O Hantz, C Perigaud, C Borel, C Jamard, F Zoulim, C Trepo, JL Imbach, G Gosselin. The SATE pronucleotide approach applied to acyclovir. Part II. Effects of bis(SATE) phosphotriester derivatives of acyclovir on duck hepatitis B virus replication in vitro and in vivo. Antiviral Res 40:179–187, 1999.

133. C Perigaud, G Gosselin, JL Girardet, BE Korba, JL Imbach. The S-acyl-2-thioethyl pronucleotide approach applied to acyclovir. Part I. Synthesis and in vitro anti-hepatitis B virus activity of bis(S-acyl-2-thioethyl)phosphotriester derivatives of acyclovir. Antiviral Res 40:167–178, 1999.

134. S Benzaria, G Gosselin, H Pelicano, G Maury, AM Aubertin, G Obert, A Kirn, JL Imbach. New prodrugs of 9-(2-phosphonomethoxyethyl) adenine [PMEA]: synthesis and stability studies. Nucleosides Nucleotides Nucleic Acids 14:563–565, 1995.

135. H Li, JH Hong. Synthesis and anti-HIV evaluation of new acyclic phosphonate nucleotide analogues and their bis(SATE) derivatives. Nucleosides Nucleotides Nucleic Acids 29:581–590, 2010.

136. LJ Liu, JH Hong. Design and synthesis of novel sate derivatives of acyclic isocytosine and 9-deazaadenine C-nucleosides. Nucleosides Nucleotides Nucleic Acids 29:257–266, 2010.

137. CH Oh, LJ Liu, JH Hong. Design and synthesis of dually branched 5′-norcarbocyclic adenosine phosphonodiester analogue as a new anti-HIV prodrug. Nucleosides Nucleotides Nucleic Acids 29:721–733, 2010.

138. R Tsien, T Pozzan. Measurement of cytosolic free calcium with quin2. Methods Enzymol 172:230–262, 1989.

139. C Perigaud, JL Girardet, I Lefebvre, MY Xie, AM Aubertin, A Kirn, G Gosselin, JL Imbach, JP Sommadossi. Comparison of cytotoxicity of mononucleoside phosphotriester derivatives bearing biolabile phosphate protecting groups in normal human bone marrow progenitor cells. Antiviral Chem Chemother 7:338–345, 1996.

140. C Perigaud, G Gosselin, JL Imbach. Minireview: from the pronucleotide concept to the SATE phosphate protecting groups. Curr Top Med Chem 2:15–29, 1997.

141. X Tan, FD Boudinot, CK Chu, D Egron, C Perigaud, GGosselin, JL Imbach. Pharmacokinetics of bis(t-butyl-SATE)-AZTMP, a bispivaloylthioethyl prodrug for intracellular delivery of zidovudine monophosphate, in mice. Antiviral Chem Chemother 11:203–212, 2000.

142. D Egron, C Perigaud, G Gosselin, AM Aubertin, JL Imbach. Effect of the alkyl chain length on the anti-HIV efficiency of pronucleotides bearing S-acyl-thioalkyl phosphate protecting groups. Bull Soc Chim Belg 106:461–466, 1997.

143. S Peyrottes, C Perigau, AM Aubertin, G Gosselin, JL Imbach. Influence of a Calpha-substitution on the S-pivaloyl-2-thioethyl chain on the anti-HIV activity and stability of the resulting zidovudine mononucleoside phosphotriester. Antivir Chem Chemother 12:223–232, 2001.

144. M Shafiee, S Deferme, AL Villard, D Egron, G Gosselin, JL Imbach, T Lioux, A Pompon, S Varray, AM Aubertin, et al. New bis(SATE) prodrug of AZT 5′-monophosphate: in vitro anti-HIV activity, stability, and potential oral absorption. J Pharm Sci 90:448–463, 2001.

145. T Satoh. Role of carboxylesterases in xenobiotic metabolism. Rev Biochem Toxicol 8:155–181, 1987.

146. M Landt, RA Everard, LG Butler. 5′-Nucleotide phosphodiesterase: features of the substrate binding site as deduced from specificity and kinetics of some novel substrates. Biochemistry 19:138–143, 1980.

147. G Valette, A Pompon, JL Girardet, L Cappellacci, P Franchetti, M Grifantini, CP La, AG Loi, C Perigaud. Decomposition pathways and in vitro HIV inhibitory effects of isoddA pronucleotides: toward a rational approach for intracellular delivery of nucleoside 5′-monophosphates. J Med Chem 39:1981–1990, 1996.

148. N Schlienger, S Peyrottes, T Kassem, JL Imbach, G Gosselin, AM Aubertin, C Perigaud. S-Acyl-2-thioethyl aryl phosphotriester derivatives as mononucleotide prodrugs. J Med Chem 43:4570–4574, 2000.

149. S Peyrottes, G Coussot, I Lefebvre, JL Imbach, G Gosselin, AM Aubertin, C Perigaud. S-Acyl-2-thioethyl aryl phosphotriester derivatives of AZT: synthesis, antiviral activity, and stability study. J Med Chem 46:782–793, 2003.

150. AL Villard, G Coussot, I Lefebvre, P Augustijns, AM Aubertin, G Gosselin, S Peyrottes, C Perigaud. Phenyl phosphotriester derivatives of AZT: variations upon the SATE moiety. Bioorg Med Chem 16:7321–7329, 2008.

151. S Peyrottes, AL Villard, G Coussot, P Augustijns, I Lefebvre, AM Aubertin, G Gosselin, C Perigaud. A step further in the SATE mononucleotide prodrug approach. Nucleic Acids Symp Ser 52:539–540, 2008.
152. D Egron, I Lefebvre, C Perigaud, T Beltran, A Pompon, G Gosselin, AM Aubertin, JL Imbach. Anti-HIV pronucleotides: decomposition pathways and correlation with biological activities. Bioorg Med Chem Lett 8:1045–1050, 1998.
153. C Meier. cycloSal-pronucleotides – design of chemical Trojan horses. Mini Rev Med Chem 2:219–234, 2002.
154. C Meier, C Ducho, H Jessen, D Vukadinovic-Tenter, J Balzarini. Second-generation cycloSal-d4TMP pro-nucleotides bearing esterase-cleavable sites – the "trapping" concept. Eur J Org Chem 2006: 197–206, 2006.
155. C Meier, T Knispel, CE De, J Balzarini. cycloSal-Pronucleotides of 2′,3′-Dideoxyadenosine and 2′,3′-dideoxy-2′,3′-didehydroadenosine: synthesis and antiviral evaluation of a highly efficient nucleotide delivery system. J Med Chem 42:1604–1614, 1999.
156. C Meier, T Knispel, VE Marquez, MA Siddiqui, CE De, J Balzarini. cycloSal-pronucleotides of 2′-fluoro-ara- and 2′-fluoro-ribo-2′,3′-dideoxyadenosine as a strategy to bypass a metabolic blockade. J Med Chem 42:1615–1624, 1999.
157. J Balzarini, F Haller-Meier, CE De, C Meier. Antiviral activity of cyclosaligenyl prodrugs of acyclovir, carbovir, and abacavir. Antiviral Chem Chemother 12:301–306, 2001.
158. C Meier, CE De, J Balzarini. Nucleotide delivery from cycloSaligenyl-3′-azido-3′-deoxythymidine monophosphates (cycloSal-AZTMP). Eur J Org Chem 1998:837–846, 1998.
159. N Gisch, J Balzarini, C Meier. Studies on enzyme-cleavable dialkoxymethyl-cycloSaligenyl-2′,3′-dideoxy-2′,3′-didehydrothymidine monophosphates. J Med Chem 51:6752–6760, 2008.
160. N Gisch, F Pertenbreiter, J Balzarini, C Meier. 5-(1-Acetoxyvinyl)-cycloSaligenyl-2′,3′-dideoxy-2′,3′- didehydrothymidine monophosphates, a second type of new, enzymatically activated cycloSaligenyl pronucleotides. J Med Chem 51:8115–8123, 2008.
161. CA Roman, J Balzarini, C Meier. Diastereoselective synthesis of aryloxy phosphoramidate prodrugs of 3′-Deoxy-2′,3′didehydrothymidine monophosphate. J Med Chem 53:7675–7681, 2010.
162. M Lorey, C Meier, E De Clercq, J Balzarini. Cyclo-saligenyl-5-fluoro-2′-deoxyuridinemonophosphate (cycloSal-FDUMP) – a new prodrug approach for FDUMP. Nucleosides Nucleotides 16:1307–1310, 1997.
163. M Lorey, C Meier, E De Clercq, J Balzarini. New synthesis and antitumor activity of cycloSal-derivatives of 5-fluoro-2′-deoxyuridinemonophosphate. Nucleosides Nucleotides 16:789–792, 1997.
164. C Meier, A Lomp, A Meerbach, P Wutzler. CycloSal-BVDUMP pronucleotides: how to convert an antiviral-inactive nucleoside analogue into a bioactive compound against EBV. J Med Chem 45:5157–5172, 2002.
165. C Meier, L Habel, F Haller-Meier, A Lomp, M Herderich, RKlocking, A Meerbach, P Wutzler. Chemistry and anti-herpes simplex virus type 1 evaluation of cycloSal-nucleotides of acyclic nucleoside analogs. Antiviral Chem Chemother 9:389–402, 1998.
166. A Meerbach, R Klocking, C Meier, A Lomp, B Helbig, P Wutzler. Inhibitory effect of cycloSaligenyl-nucleoside monophosphates (cycloSal-NMP) of acyclic nucleoside analogues on HSV-1 and EBV. Antiviral Res 45:69–77, 2000.

167. C Meier, U Goerbig, C Mueller, J Balzarini. cycloSal-PMEA and cycloAmb-PMEA: potentially new phosphonate prodrugs based on the cycloSal-pro-nucleotide approach. J Med Chem 48:8079–8086, 2005.

168. C Meier, T Knispel, E De Clercq, J Balzarini. ADA-bypass by lipophilic cycloSal-ddAMP pro-nucleotides a second example of the efficiency of the cycloSal-concept. Bioorg Med Chem Lett 7:1577–1582, 1997.

169. C Meier, M Lorey, E De Clercq, J Balzarini. cycloSal-2′,3′-dideoxy-2′,3′-didehydrothymidine monophosphate (cycloSal-d4TMP): synthesis and antiviral evaluation of a new d4TMP delivery system. J Med Chem 41:1417–1427, 1998.

170. J Balzarini, S Aquaro, T Knispel, C Rampazzo, V Bianchi, CF Perno, CE De, C. Meier. Cyclosaligenyl-2′,3′-didehydro-2′,3′-dideoxythymidine monophosphate: efficient intracellular delivery of d4TMP. Mol Pharmacol 58:928–935, 2000.

171. HJ Jessen, J Balzarini, C Meier. Intracellular trapping of cycloSal-pronucleotides: modification of prodrugs with amino acid esters. J Med Chem 51:6592–6598, 2008.

172. C Meier, MFH Ruppel, D Vukadinovic, J Balzarini. Second generation of cycloSal-pronucleotides with esterase-cleavable sites: the “Lock-In”-concept. Nucleosides Nucleotides Nucleic Acids 23:89–115, 2004.

173. N Gisch, J Balzarini, C Meier. Enzymatically activated cycloSal-d4T-monophosphates: the third generation of cycloSal-pronucleotides. J Med Chem 50:1658–1667, 2007.

174. N Gisch, J Balzarini, C Meier. 5-Diacetoxymethyl-cycloSal-d4TMP – a prototype of enzymatically activated cycloSal-pronucleotides. Nucleosides Nucleotides Nucleic Acids 26:861–864, 2007.

175. P Spacilova, P Naus, R Pohl, I Votruba, J Snasel, H Zabranska, I Pichova, R Ameral, G Birkus, T Cihlar, et al. CycloSal-phosphate pronucleotides of cytostatic 6-(Het)aryl-7-deazapurine ribonucleosides: synthesis, cytostatic activity, and inhibition of adenosine kinases. ChemMedChem 5:1386–1396, 2010.

176. C Ducho, U Goerbig, S Jessel, N Gisch, J Balzarini, C Meier. Bis-cycloSal-d4T-monophosphates: drugs that deliver two molecules of bioactive nucleotides. J Med Chem 50:1335–1346, 2007.

177. C Ducho, S Jessel, N Gisch, J Balzarini, C Meier. Novel cycloSal nucleotides with reduced inhibitory potency toward human butyrylcholinesterase. Nucleosides Nucleotides Nucleic Acids 24:519–522, 2005.

178. N Gisch, J Balzarini, C Meier. Doubly loaded cycloSaligenyl-pronucleotides – 5,5′-Bis-(cycloSaligenyl-2′,3′-dideoxy-2′,3′-didehydrothymidine monophosphates). J Med Chem 52:3464–3473, 2009.

179. MS Marma, BA Kashemirov, CE McKenna. Synthesis and stability studies of phosphonoformate-amino acid conjugates: a new class of slowly releasing foscarnet prodrugs. Bioorg Med Chem Lett 14:1787–1790, 2004.

180. CE McKenna, BA Kashemirov, U Eriksson, GL Amidon, PE Kish, S Mitchell, JS Kim, JM Hilfinger. Cidofovir peptide conjugates as prodrugs. J Organomet Chem 690:2673–2678, 2005.

181. U Eriksson, LW Peterson, JS Kim, S Mitchell, P Kijek, JM Hilfinger, JC Drach, BA Kashemirov, CE McKenna. Ethylene glycol-linked amino acid conjugates of cyclic cidofovir: synthesis and biological activity. Antiviral Res 70:A58, 2006.

182. U Eriksson, JM Hilfinger, JS Kim, S Mitchell, P Kijek, KZ Borysko, JM Breitenbach, JC Drach, BA Kashemirov, CE McKenna. Synthesis and biological activation of an ethylene

glycol-linked amino acid conjugate of cyclic cidofovir. Bioorg Med Chem Lett 17:583–586, 2007.

183. LW Peterson, BA Kashemirov, KM Blazewska, J Breitenback, K Borysko, JC Drach, JS Kim, P Kijek, JM Hilfinger, CE McKenna. Synthesis and structure-activity aspects of some cyclic cidofovir peptidomimetic prodrugs. Antiviral Res 74:A33, 2007.

184. U Eriksson, LW Peterson, BA Kashemirov, JM Hilfinger, JC Drach, KZ Borysko, JM Breitenbach, JS Kim, S Mitchell, P Kijek, et al. Serine peptide phosphoester prodrugs of cyclic cidofovir: synthesis, transport, and antiviral activity. Mol Pharm 5:598–609, 2008.

185. LW Peterson, BA Kashemirov, U Eriksson, JS Kim, S Mitchell, P Kijek, KD Lee, JM Hilfinger, CE McKenna. Serine side-chain-linked peptidomimetic prodrugs of cidofovir and cyclic cidofovir: C-ester effects on transport and activation. Antiviral Res 78:A46, 2008.

186. IS Krylov, LW Peterson, BA Kashemirov, J Breitenbach, K Borysco, JC Drach, JS Kim, JM Hilfinger, CE McKenna. In vitro transport, activation and antiviral evaluation of new HPMPA prodrugs synthesized on a solid support. Antiviral Res 82:A75, 2009.

187. CE McKenna, LW Peterson, BA Kashemirov, M Serpi, S Mitchell, JS Kim, JM Hilfinger, JC Drach. New peptidomimetic prodrugs of acyclic and cyclic cidofovir: SAR studies of chemical and enzymatic activation mechanisms. Antiviral Res 82:A75, 2009.

188. CE McKenna, BA Kashemirov, LW Peterson, MF Goodman. Modifications to the dNTP triphosphate moiety: From mechanistic probes for DNA polymerases to antiviral and anti-cancer drug design. Biochim Biophys Acta 1804:1223–1230, 2010.

189. LW Peterson, M Sala-Rabanal, IS Krylov, M Serpi, BA Kashemirov, CE McKenna. Serine side chain-linked peptidomimetic conjugates of cyclic HPMPC and HPMPA: synthesis and interaction with hPEPT1. Mol Pharm 7:2349–2361, 2010.

190. M Serpi, IS Krylov, VM Zakharova, CE McKenna. Synthesis of peptidomimetic conjugates of cyclic nucleoside phosphonates. Curr Protoc Nucleic Acid Chem 43:15.4.1-15.4.13, 2010.

191. U Tehler, CH Nelson, LW Peterson, CJ Provoda, JM Hilfinger, KD Lee, CE McKenna, GL Amidona. Puromycin-sensitive aminopeptidase: an antiviral prodrug activating enzyme. Antiviral Res 85:482–489, 2010.

192. IS Krylov, VM Zakharova, M Serpi, R Haiges, BA Kashemirov, CE McKenna. Structure of cyclic nucleoside phosphonate ester prodrugs: an inquiry. J Org Chem 77(1):684–689, 2012.

193. LW Peterson, JS Kim, P Kijek, S Mitchell, J Hilfinger, J Breitenbach, K Borysko, JC Drach, BA Kashemirov, CE McKenna. Synthesis, transport and antiviral activity of Ala-Ser and Val-Ser prodrugs of cidofovir. Bioorg Med Chem Lett 21:4045–4049, 2011.

194. M Williams, IS Krylov, VM Zakharova, M Serpi, LW Peterson, M Krecmerova, BA Kashemirov, CE McKenna. Cyclic and acyclic phosphonate tyrosine ester prodrugs of acyclic nucleoside phosphonates. *Coll. Symp. Series* 12:167–170, 2011.

195. M Williams, V Zakharova, M Serpi, I Krylov, B Kashemirov, J Breitenbach, K Borysko, JC Drach, JM Hilfinger, CE McKenna. Synthesis and Biological Evaluation of Tyrosine Based Acyclic Nucleoside Phosphonate Prodrugs. In *241st American Chemical Society National Meeting*, Anaheim, CA, March 27–31, 2011; Paper # ORGN-103.

196. VM Zakharova, IS Krylov, M Serpi, BA Kashemirov, CE McKenna. Approaches to tyrosine-linked peptidomimetic prodrugs of (S)-HPMP-based acyclic nucleoside phosphonates. Phosphorus Sulfur Silicon Relat Elem 186:968–969, 2011.

197. VM Zakharova, M Serpi, IS Krylov, LW Peterson, JM Breitenbach, KZ Borysko, JC Drach, M Collins, JM Hilfinger, BA Kashemirov, et al. Tyrosine-based 1-(S)-[3-hydroxy-

2-(phosphonomethoxy)propyl]cytosine and -adenine ((S)-HPMPC and (S)-HPMPA) prodrugs: synthesis, stability, antiviral activity, and in vivo transport studies. J Med Chem 54:5680–5693, 2011.

198. L Charrier, D Merlin. The oligopeptide transporter hPepT1: gateway to the innate immune response. Lab Invest 86:538–546, 2006.

199. ME Ganapathy, M Brandsch, PD Prasad, V Ganapathy, FH Leibach. Differential recognition of Î^2-lactam antibiotics by intestinal and renal peptide transporters, PEPT 1 and PEPT 2. J Biol Chem 270:25672–25677, 1995.

200. U Wenzel, I Gebert, H Weintraut, WM Weber, W Clauss, H Daniel. Transport characteristics of differently charged cephalosporin antibiotics in oocytes expressing the cloned intestinal peptide transporter PepT1 and in human intestinal Caco-2 cells. J Pharmacol Exp Ther 277:831–839, 1996.

201. B Bretschneider, M Brandsch, R Neubert. Intestinal transport of Î^2-lactam antibiotics: analysis of the affinity at the H+/peptide symporter (PEPT1), the uptake into Caco-2 cell monolayers and the transepithelial flux. Pharm Res 16:55–61, 1999.

202. M Boll, D Markovich, WM Weber, H Korte, H Daniel, H Murer. Expression cloning of a cDNA from rabbit small intestine related to proton-coupled transport of peptides, β-lactam antibiotics and ACE-inhibitors. Pfluegers Arch 429:146–149, 1994.

203. CS Temple, CAR Boyd. Proton-coupled oligopeptide transport by rat renal cortical brush border membrane vesicles: a functional analysis using ACE inhibitors to determine the isoform of the transporter. Biochim Biophys Acta 1373:277–281, 1998.

204. T Zhu, XZ Chen, A Steel, MAHediger, DE Smith. Differential recognition of ACE inhibitors in Xenopus laevis oocytes expressing rat PEPT1 and PEPT2. Pharm Res 17:526–532, 2000.

205. F Doring, J Walter, J Will, M Focking, M Boll, S Amasheh, W Clauss, H Daniel. Delta-aminolevulinic acid transport by intestinal and renal peptide transporters and its physiological and clinical implications. J Clin Invest 101:2761–2767, 1998.

206. CU Nielsen, B Brodin. Di/tri-peptide transporters as drug delivery targets: regulation of transport under physiological and patho-physiological conditions. Curr Drug Targets 4:373–388, 2003.

207. MD Antman, OS Gudmundsson. Valacyclovir: a prodrug of acyclovir. Biotechnol Pharm Aspects 5:669–676, 2007.

208. AJ Wagstaff, D Faulds, KL Goa. Aciclovir. A reappraisal of its antiviral activity, pharmacokinetic properties and therapeutic efficacy. Drugs 47:153–205, 1994.

209. ME Ganapathy, W Huang, H Wang, V Ganapathy, FH Leibach. Valacyclovir: a substrate for the intestinal and renal peptide transporters PEPT1 and PEPT2. Biochem Biophys Res Commun 246:470–475, 1998.

210. PJ Sinko, PV Balimane. Carrier-mediated intestinal absorption of valacyclovir, the L-valyl ester prodrug of acyclovir: 1. interactions with peptides, organic anions and organic cations in rats. Biopharm Drug Dispos 19:209–217, 1998.

211. K Sawada, T Terada, H Saito, Y Hashimoto, KI Inui. Recognition of L-amino acid ester compounds by rat peptide transporters PEPT1 and PEPT2. J Pharmacol Exp Ther 291:705–709, 1999.

212. PV Balimane, I Tamai, A Guo, T Nakanishi, H Kitada, FH Leibach, A Tsuji, PJ Sinko. Direct evidence for peptide transporter (PepT1)-mediated uptake of a nonpeptide prodrug, valacyclovir. Biochem Biophys Res Commun 250:246–251, 1998.

213. SB Larsen, FS Jorgensen, L Olsen. QSAR models for the human H+/peptide symporter, hPEPT1: affinity prediction using alignment-independent descriptors. J Chem Inf Model 48:233–241, 2008.

214. A Pedretti, LL De, C Marconi, G Negrisoli, G Aldini, G Vistoli. Modeling of the intestinal peptide transporter hPepT1 and analysis of its transport capacities by docking and pharmacophore mapping. ChemMedChem 3:1913–1921, 2008.

215. S Winiwarter, C Hilgendorf. Modeling of drug-transporter interactions using structural information. Curr Opin Drug Discov Devel 11:95–103, 2008.

216. G Kottra, I Frey, H Daniel. Inhibition of intracellular dipeptide hydrolysis uncovers large outward transport currents of the peptide transporter PEPT1 in Xenopus oocytes. Pfluegers Arch 457:809–820, 2009.

217. D Foley, M Pieri, R Pettecrew, R Price, S Miles, HK Lam, P Bailey, D Meredith. The in vitro transport of model thiodipeptide prodrugs designed to target the intestinal oligopeptide transporter, PepT1. Org Biomol Chem 7:3652–3656, 2009.

218. N Bischofberger, MJM Hitchcock, MS Chen, DB Barkhimer, KC Cundy, KM Kent, SA Lacy, WA Lee, ZH Li. 1-[((S)-2-Hydroxy-2-oxo-1,4,2-dioxaphosphorinan-5-yl)methyl] cytosine, an intracellular prodrug for (S)-1-(3-hydroxy-2-phosphonylmethoxypropyl) cytosine with improved therapeutic index in vivo. Antimicrob Agents Chemother 38:2387–2391, 1994.

219. DB Mendel, T Cihlar, K Moon, MS Chen. Conversion of 1-[((S)-2-hydroxy-2-oxo-1,4,2-dioxaphosphorinan-5-yl)methyl]cytosine to cidofovir by an intracellular cyclic CMP phosphodiesterase. Antimicrob Agents Chemother 41:641–646, 1997.

220. R Oliyai, JP Shaw, CM Sueoka-Lennen, KC Cundy, MN Arimilli, RJ Jones, WA Lee. Aryl ester prodrugs of cyclic HPMPC. I: physicochemical characterization and in vitro biological stability. Pharm Res 16:1687–1693, 1999.

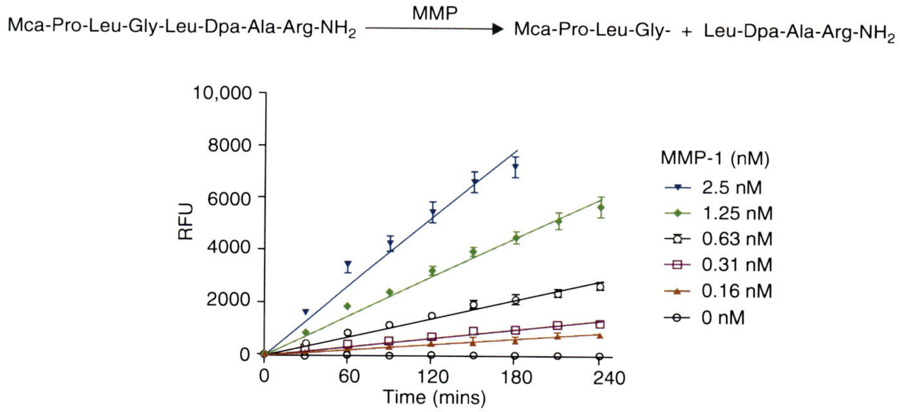

Chapter 1, Figure 8(b) Detection of MMP-1 activity using an MCA/ Dpa FRET peptide as substrate. The graph shows kinetics of increase in fluorescence of hydrolyzed peptide which is monitored at Ex/Em = 340/490 nm. The increase in luminescence after hydrolysis is shown for different concentrations of MMP-1.

Enzyme Technologies: Pluripotent Players in Discovering Therapeutic Agents, First Edition.
Edited by Hsiu-Chiung Yang, Wu-Kuang Yeh, and James R. McCarthy.

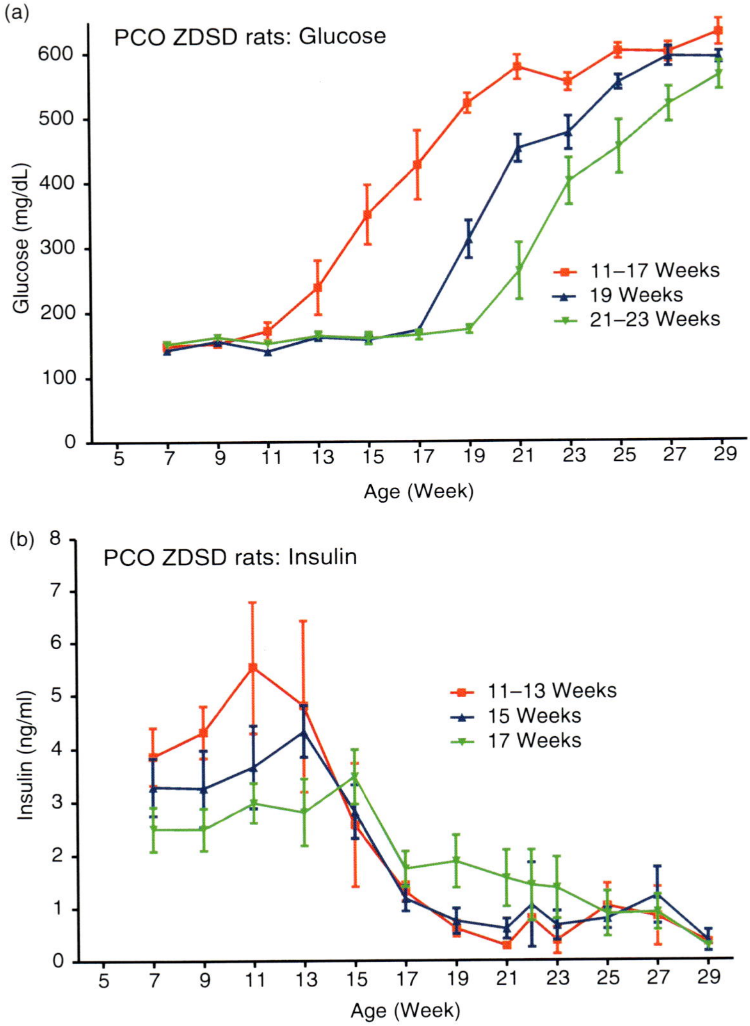

Chapter 4, Figure 1 ZDSD rats: spontaneous diabetic development. Glucose (a) and insulin (b) were monitored for ZDSD male rats that became diabetic, using three different age ranges.

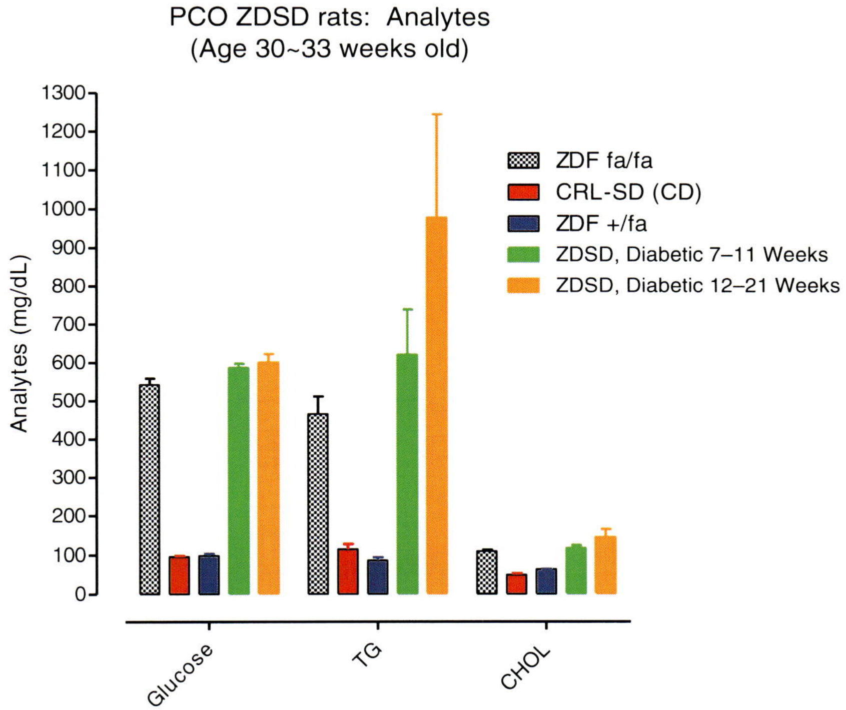

Chapter 4, Figure 2 ZDSD rats: spontaneous diabetic development. Glucose, TG, and total CHOL were determined for five different 30–33-week-old rat models: ZDF fa/fa, CRL-SD (CD), ZDF + /fa, ZDSD diabetic from 7–11 weeks, and ZDSD diabetic from 12–21 weeks. For statistical analysis of this and following studies, as specified, *t*-test from GraphPad Prism 5 was used. ZDF and ZDSD diabetic from 7–11 weeks and from 12–21 weeks were analyzed with CRL-SD (CD) and ZDF + /fa; $P<0.0001$ for glucose. Similarly for TG, P values are <0.0001 with ZDF, $= 0.0104$ and $=0.0073$ with ZDSD diabetic 7–11 weeks, and $=0.0090$ and $=0.0075$ with ZDSD diabetic 12–21 weeks. For total CHOL, P values are <0.0001 with ZDF, <0.0001 and $=0.0003$ with ZDSD diabetic 7–11 weeks, and $=0.0012$ and $=0.0032$ with ZDSD diabetic 12–21 weeks.

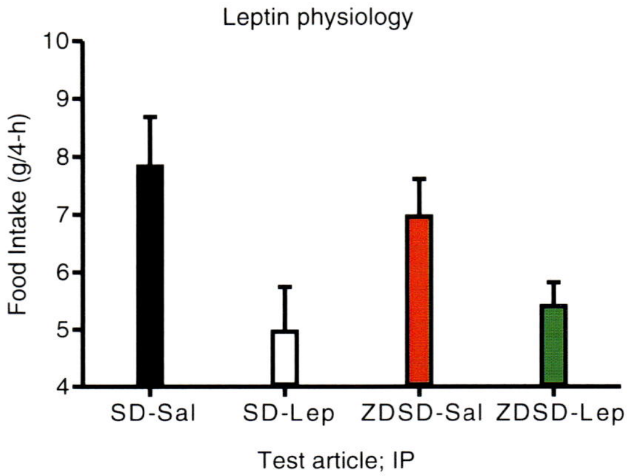

Chapter 4, Figure 3 ZDSD rats: leptin physiology. Assessment of leptin pathway function was determined by feeding response to leptin injection. ZDSD rats of 15 weeks old and SD rats of 17 weeks old, $n = 9$, were dosed with leptin, 1 mg/kg intraperitoneally (IP), just before starting the dark cycle, and food intake was measured for the first 4 dark hours. Sal: saline; Lep: leptin. The food intake of this study was analyzed by *t*-test; in comparison to the saline group, $P = 0.0196$ for SD-Lep and $P = 0.0406$ for ZDSD-Lep.

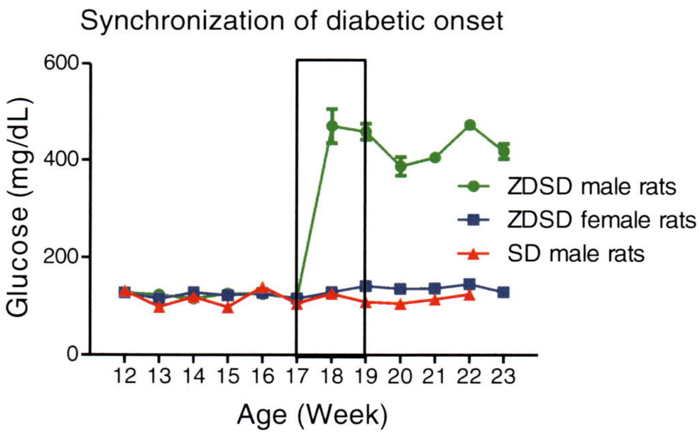

Chapter 4, Figure 4 ZDSD rats: diet-induced diabetic synchronization. The ZDSD male rats, $n = 12$, can be placed on either D12468 (Research Diets) or 5SCA (LabDiet) at 17 weeks of age to synchronize the onset of diabetes. Two-way ANOVA from GraphPad Prism 5 was used in statistical analysis of this and following studies, as specified. Each *P* value is from interaction of column (effect) factor and row (time) factor, unless specified (i.e., from column factor only); in comparison to SD male rats, for analysis of glucose, $P < 0.0001$ for ZDSD male rats.

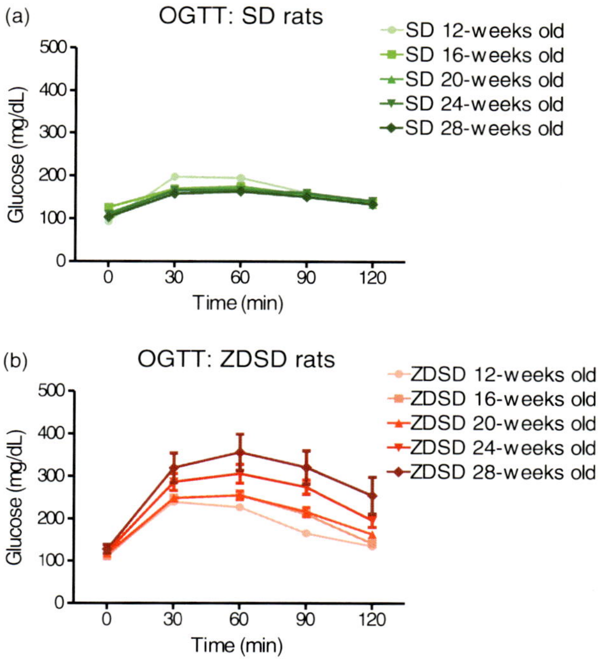

Chapter 4, Figure 6 ZDSD rats: age-dependent impaired glucose tolerance. SD rats (a), $n = 10$, and ZDSD rats (b), $n = 10$, were used in an oral glucose tolerance test (OGTT) starting 12 weeks old. From 12 to 28 weeks old, for every 2 weeks, the rats were dosed with 2 g/kg glucose orally, and the OGTT data are shown for every 4 weeks only. Two-way ANOVA was used in analyzing glucose of this study; in comparison to SD rats, $P < 0.0001$ for ZDSD rats from four age groups (12 to 24 weeks old) and $P = 0.0083$ from 28 weeks old.

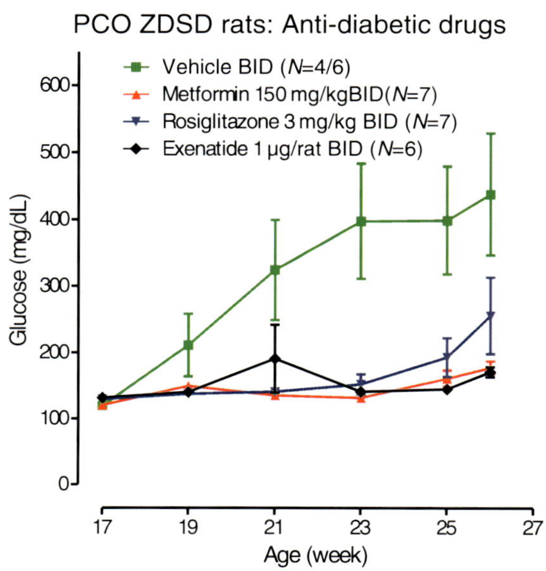

Chapter 4, Figure 7 ZDSD rats: antidiabetic drug treatment with metformin, rosiglitazone, and exenatide. Using two-way ANOVA for glucose of the study, in comparison to the vehicle group, $P < 0.0001$ for metformin, $P = 0.0024$ for rosiglitazone, and $P = 0.0002$ for exenatide.

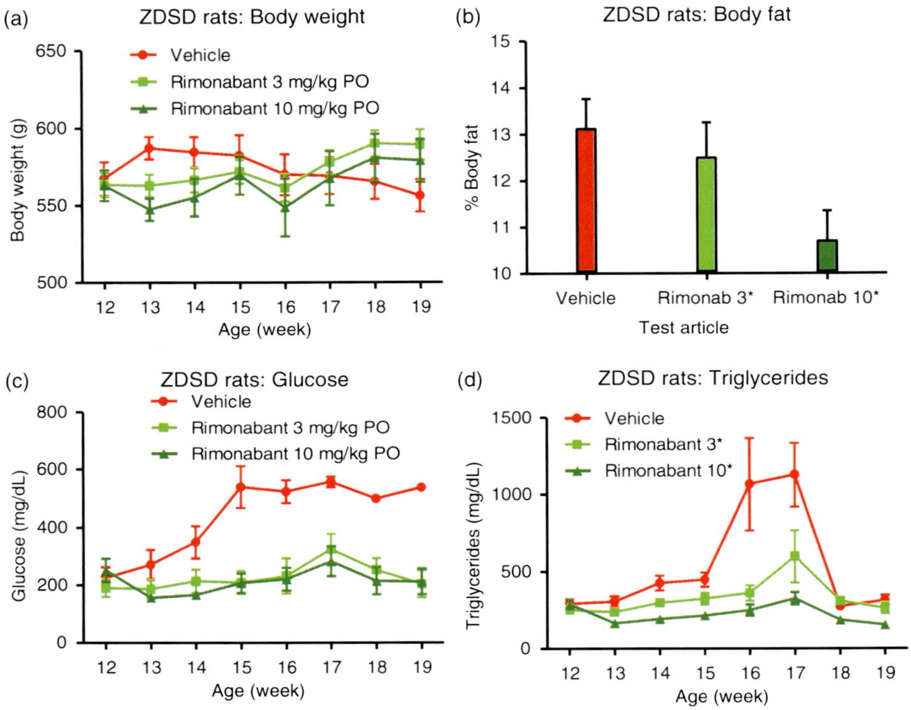

Chapter 4, Figure 8 ZDSD rats: anti-obesity drug treatment with rimonabant. The drug treatment started with 12-week-old rats, $n=8$, that were on a standard diet for 3 weeks, followed by a high-fat diet for 2 weeks (15–17 weeks of age) and returned to the standard diet for 2 weeks. Body weight (a), fed plasma glucose (c), and TG (d) were monitored weekly; 3* is 3 mg/kg PO and 10* is 10 mg/kg PO. Body fat (b) was determined at 27-week-old rats by Quantitative Nuclear Magnetic Resonance (QNMR); Rimonab 3* is rimonabant 3 mg/kg and Rimonab 10* is rimonabant 10 mg/kg. Using two-way ANOVA, in comparison to the vehicle group, $P<0.0001$ for glucose at rimonabant 3 and 10 mg/kg and $P=0.0018$ and $P<0.0001$ for TG at rimonabant 3 and 10 mg/kg, respectively. For 12–15-week-old rats on the standard diet, $P=0.2517$ and $P=0.0087$ for body weight at rimonabant 3 and 10 mg/kg, respectively. Using *t*-test, in comparison to the vehicle group, $P=0.5419$ and $P<0.0202$ for body fat at rimonabant 3 and 10 mg/kg, respectively.

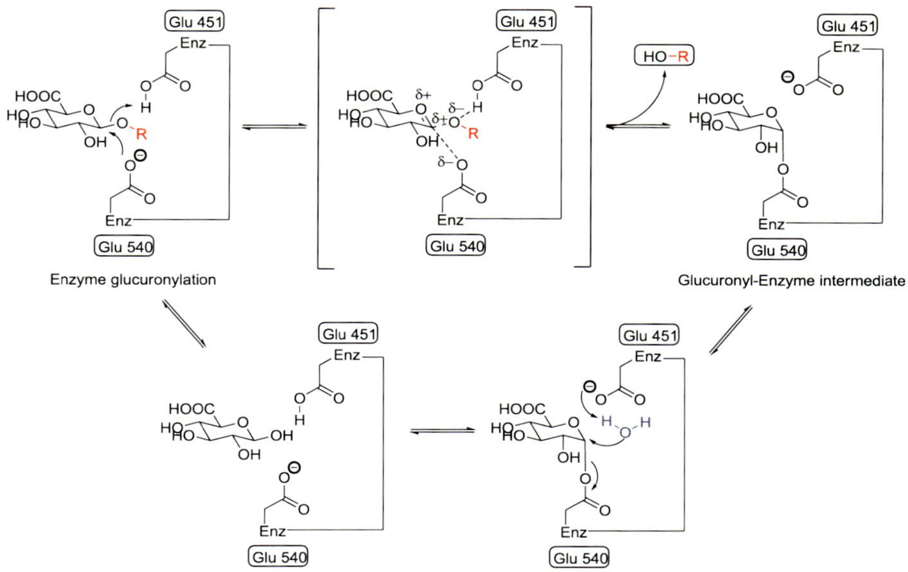

Chapter 5, Figure 22 The catalytic cycle of β-glucuronidase. The R group shown in red is released upon hydrolysis of the glycosidic bond.

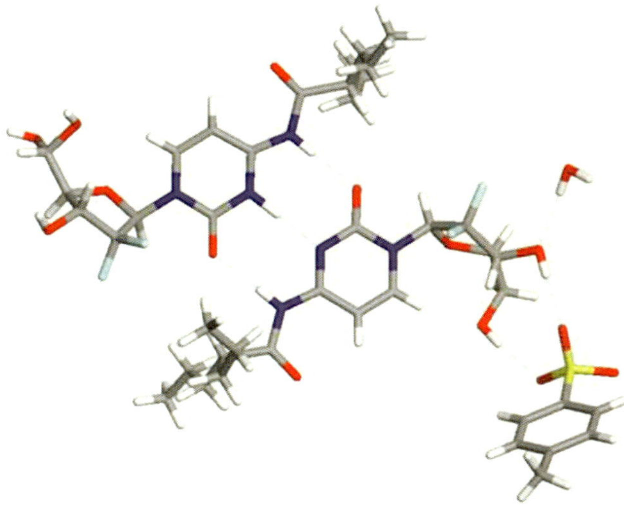

Chapter 6, Figure 5 X-ray structure of the 2:1:1 gemcitabine prodrug hemi-tosylate hemihydrate **31**.

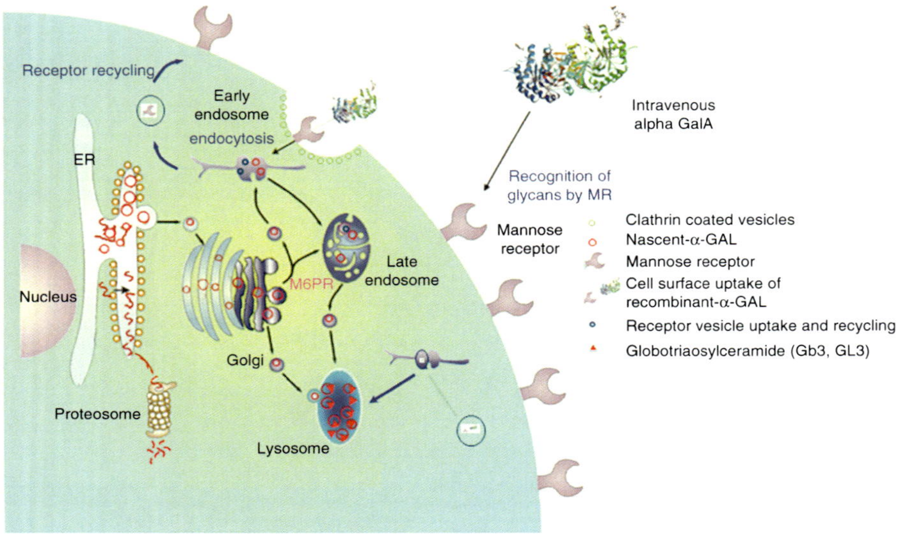

Chapter 9, Figure 1 When a cell is genetically deficient in producing its own α-GalA or if the mutant enzyme cannot reach the lysosome, GL-3 accumulates there leading to Fabry disease. There exists a mechanism for the cell to uptake exogenous enzymes; α-GalA present in the extracellular milieu can bind to the cell surface via mannose receptors (MR) through recognition of the glycan moieties while it is transported into the lysosomal compartment using mannose 6-phosphate receptor (M6PR) system. Once in the lysosomal compartment, the enzyme can carry out its appropriate function. Thus, α-GalA (red circles) is targeted to the lysosomal compartment through interaction with M6PR, which binds to the oligosaccharide moieties of enzymes destined for the lysosomal compartment. The primary enzyme–receptor interaction occurs in the Golgi, where enzymes bound to Golgi M6PRs are segregated for transport to lysosomes. A secondary type of interaction may also occur between extracellular α-GalA and M6PR at the cell surface; α-GalA is transported through the cytoplasm in endocytic vesicles which fuse with late endosomes (multivesicular bodies) and empty their contents into the lysosomes. Receptors are then recycled (blue circles) through the endosomal compartment.

PART C

ENZYMES – POWERFUL WEAPONS FOR CORRECTING NATURE'S ERRORS

8

TREATMENT OPTIONS FOR MUCOPOLYSACCHARIDOSIS TYPE II (HUNTER'S SYNDROME)

MICHAEL BECK
University Medical Center, Children's Hospital, Mainz, Germany

I. INTRODUCTION

Hunter's syndrome (mucopolysaccharidosis type II (MPS II), OMIM 309900) is an X-linked lysosomal storage disorder that belongs to the group of mucopolysaccharidoses (MPS). Mucopolysaccharidoses are lysosomal storage disorders each caused by the deficiency of enzymes that are responsible for the degradation of glycosaminoglycans (GAGs) (previously called mucopolysaccharides). GAGs are complex carbohydrate chains consisting of an uronic acid and an amino sugar. In Hunter's syndrome the deficiency of the enzyme iduronate-2-sulfatase (IDS) leads to the accumulation of the un-degraded GAGs dermatan sulfate (DS) and heparan sulfate (HS) (Fig. 1). Major clinical manifestations include joint contractures, obstructive and restrictive airway disease, cardiac disease, skeletal deformities, and often mental retardation. MPS II shows a broad clinical variability in terms of the extent and rate of progression of organ impairment. Patients with the more attenuated (so-called adult) form usually have a normal intelligence but often have many complications regarding the skeleton, the eye, and peripheral nervous system.

Until recently, no specific treatment was available for Hunter's patients; management mainly consisted of supportive care and treatment of complications. Hematopoietic stem cell transplantation (HSCT) has been performed only in a few

Enzyme Technologies: Pluripotent Players in Discovering Therapeutic Agents, First Edition.
Edited by Hsiu-Chiung Yang, Wu-Kuang Yeh, and James R. McCarthy.

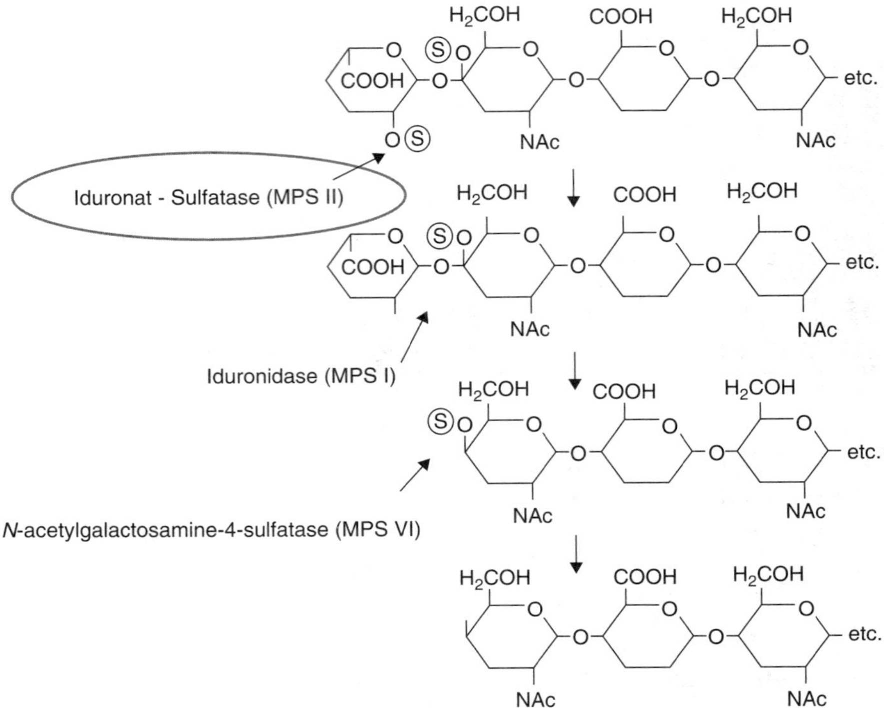

FIGURE 1 Stepwise degradation of DS. In Hunter's syndrome the enzyme IDS is deficient that removes sulfate groups from the 2-position of iduronic acid.

patients with minor success. Enzyme replacement therapy (ERT) with recombinant IDS (idursulfase), however, is now available for patients affected by MPS II. And in a number of clinical studies, it could be demonstrated that weekly intravenous infusions of idursulfase are able to improve many of the symptoms and signs of Hunter's syndrome. The clinical efficacy of ERT is limited by the fact that the intravenously applied enzyme does not reach all target organs and has no effect on the central nervous manifestation as it is not able to enter the blood–brain barrier. This review will present the efficacy and safety data of the enzyme preparation and discuss benefits and limitations of this therapeutic option. In addition, new therapeutic approaches such as substrate deprivation and substrate optimization will be discussed.

II. HISTORY

In 1917, Charles Hunter, a Canadian physician, has described two boys who were affected by a "rare disease" [1]. Characteristic clinical signs of these brothers included small stature, large head, and a dysmorphic face. In addition, enlargement of liver and spleen, joint contractures, and hearing impairment have been reported in this first description of Hunter's syndrome. Apparently, the boys were of normal intelligence. Two years later two unrelated boys with similar clinical findings were reported by Gertrud Hurler, a German pediatrician [2]. In her publication, the coarse facial features, organomegaly, mental impairment, and corneal clouding were very well described, and she gave the skeletal abnormalities, which were seen in these patients, the term *dysostosis multiplex*. During the following years many similar cases were described, and as the coarse facial features and the large tongue resembled the grotesque images of the medieval cathedral gargoyles, the term "gargoylism" was introduced that should not be used anymore. Whereas in most of these patients family recurrence was observed, in some families only boys were affected, suggesting an X-linked form in these cases [3].

The term MPS was the first time used by Brante when he has detected that the liver of Hurler's patients contained increased amounts of chondroitin sulfate (CS)-like material [4]. Based on these results, Dorfman and Lorincz analyzed the urine of an affected individual for the presence of this compound and found excess urinary excretion of mucopolysaccharides, later called GAGs. They concluded that Hurler's syndrome represents a disorder that is caused by an abnormality of mucopolysaccharide metabolism [5]. Eleven years later Fratantoni et al. could prove that the accumulation of GAGs was not caused by increased synthesis, but by a defect in the degradation of these substances [6]. In the same year, Fratantoni et al. also have demonstrated by cell complementation studies that the metabolic defect in fibroblasts cultured from patients with Hurler's syndrome could be corrected by a factor secreted by cells from an individual affected by Hunter's syndrome, and *vice versa* [7]. The protein that was able to correct the metabolic defect in Hunter's fibroblasts was named "Hunter corrective factor" [8]. And by isolating this factor from human urine, it could be identified as a sulfoiduronate sulfatase, now known as an IDS [9]. A cDNA clone coding for the entire sequence of IDS was isolated, and by analyzing the deduced amino acid sequence, it could be demonstrated that IDS consists of 550 amino acids, including a 25-amino acid amino-terminal signal sequence, and has eight potential N-glycosylation sites [10]. In addition, it has been shown that the IDS sequence shows strong sequence homology with other sulfatases such as arylsulfatases and glucosamine 6-sulfatase. In clinically severe Hunter's patients, structural alterations and gross deletions of the IDS gene were detected. Only in 1997 Bondeson et al. described the presence of an IDS pseudogene that is located at the telomeric side of the IDS gene and contains several homologous sequences [11].

III. BIOCHEMISTRY

A. Glycosaminoglycans

Repeating disaccharide units form GAGs that can be divided in four structurally distinct families: HS/heparin, CS/DS, keratan sulfate (KS), and hyaluronan (HA). The disaccharides are composed of an amino sugar (*N*-acetylglucosamine or *N*-acetylgalactosamine) and an uronic acid (glucuronic or iduronic acid). In KS the uronic acid is replaced by galactose. Besides HA, the GAG chains that have a length of about 50–150 disaccharide units undergo extensive modification by N-acetylation, N-sulfation, and N-epimerization. With the exception of hyaluronic acid, GAGs are attached to a core protein through an asparagine or serine residue forming a complex that is called a proteoglycan [12]. After formation proteoglycans are released either to the cell surface or to the extracellular matrix. Free GAGs and proteoglycans play a significant role in the cell biology such as formation of the extracellular matrix (including bone), cell signaling, inflammation, and neuronal function. The degradation of the GAG chains takes place within the lysosomes by the stepwise exoglycosidic cleavage of specific terminal sugars from the nonreducing end of the chain. IDS, the enzyme that is deficient in Hunter's syndrome, is responsible for the removal of sulfate groups from the 2-position of iduronic acid present in DS and HS (Fig. 1). The loss of enzyme activity leads to progressive lysosomal storage of these GAGs in many tissues and organs such as liver, spleen, heart, bone and joints, and upper and lower airways. The mechanisms by which the un-degraded GAGs disturb cellular function and result in a disease are not well understood until now; they include activation of signal transduction by nonphysiological substances, alterations of intracellular calcium homeostasis, impairment of autophagy, and inflammation [13].

B. Synthesis of Iduronate-2-sulfatase

From the IDS gene that encodes a 550-amino acid IDS protein by alternative polyadenylation signals, three distinct transcripts are generated that contain the same open reading frame [14]. By a specific signal the IDS transcripts are directed to translation by ribosomes of the rough endoplasmic reticulum. In the endoplasmic reticulum the posttranslational modification begins after the cleavage of the signal peptide. In the first posttranslational step that is common to all sulfatases, cysteine 59 is converted to C_α-formylglycine (FGly), the catalytic residue of eukaryotic sulfatases. A genetic defect of FGly formation leads to multiple sulfatase deficiency, a severe lysosomal storage disorder in which all sulfatases are lacking [15]. After FGly formation glycosylation commences at selected asparagine residues by transfer of a large preformed oligosaccharide followed by modification of oligosaccharide sugar residues. Thereafter IDS moves by vesicular transport to the Golgi apparatus where it undergoes further posttranslational modifications: At first the oligosaccharides are processed to sialic acid-containing complex-type units. Thereafter IDS receives a mannose-6-phosphate (M6P) residue that all lysosomal enzymes require for targeting to the lysosome. In the formation of the phosphomannosyl residue, two enzymes are involved: a phosphotransferase that transfers *N*-acetylglucosamine 1-phosphate to mannose residues of the enzymes and an *N*-acetylglucosaminidase for removing the *N*-acetylglucosamine residue. The resultant phosphomonoester then leads IDS to the lysosome via the M6P receptor. Only a small amount of the enzyme is secreted from the cell and can be recaptured and delivered to the lysosome via the M6P receptor [16]. This is the prerequisite for the therapy by exogenously administered enzymes.

IV. GENETICS

A. Iduronate-2-sulfatase Gene

The IDS gene that has been mapped to Xq28 consists of nine exons spread over 24kb. A pseudogene, containing copies of exons 2 and 3 and introns 2, 3, and 7, is located about 80kb downstream from the coding gene. To date, more than 350 mutations have been identified in the IDS gene, whereby exonic point mutations encompass half of the mutations. Furthermore, small deletions as altered splicing and small insertions have been described. Major deletions and recombinations between the IDS gene and the pseudogene lead to loss of function of IDS, resulting in severe form of Hunter's disease. Very large deletions of the IDS locus may extend to neighboring genes, and the unusual phenotypes seen in some severely affected patients could possibly be explained by a contiguous syndrome involving the fragile X-genes FMR1 and FMR2 [17].

In European countries the incidence of Hunter's syndrome has been estimated to be about 1.3 per 100.00 male live births [18]. In China, Hunter's disease seems to be the most common MPS. And in an investigation of the IDS gene in 38 unrelated Chinese patients, 30 different mutations could be identified, whereby only 9 were novel [19].

B. Carriers of the IDS gene

In order to characterize the clinical and biochemical features of MPS II carriers, Schwartz et al. evaluated females who were at risk to be heterozygous for the IDS gene. In 52 women extensive investigations have been performed including physical examination, analysis of the IDS gene, measurement of the IDS activity, and quantification of urinary GAG excretion. By pedigree and genetic analysis, 40 from these 52 females could be identified as carriers. IDS activity in plasma and leukocytes of those females was lower than the values of noncarriers, but there was an overlap between the levels of heterozygotes and non-heterozygotes. The carriers did not show any clinical manifestation, and GAG excretion was not increased [20].

To date, only a few females with Hunter's syndrome have been reported. Phenotypic expression of this X-linked disorder in females can be the result of homozygosity for disease-causing mutations, structural abnormalities of the X-chromosome, or preferential X-chromosome inactivation. Preferential X-inactivation of the nonmutant allele seems to be the most common cause of clinical manifestation for MPS II in females who mostly show a severe phenotype [21].

V. CLINICAL MANIFESTATION

Hunter's disease (MPS II) is a clinically heterogeneous disease in terms of the extent and rate of progression of organ impairment in affected individuals. In its most severe form, somatic involvement becomes evident in the first years of life. Leading symptoms at this age are inguinal and umbilical hernia, hepatosplenomegaly, macrocephaly, and coarse facial features [22]. At birth, the patients' birth weight is often at the higher normal range [23]. Even before the second year of life, a gibbus may be seen. Due to joint contractures mobility becomes more and more restricted. Recurrent upper and lower respiratory infections are very common. The skeletal disorder, generally known as dysostosis multiplex, is characterized by chest deformity and short trunk, resulting in disproportionate dwarfism. A delay in developmental milestones such as ability to walk or to speak represents the first sign of cerebral involvement. The mental impairment is progressive, leading to rapid deterioration of social and adaptive skills. As a result of the progressive dementia, patients lose contact with the environment. Obstructive airway disease, cardiac failure due to valvular dysfunction, pulmonary hypertension, and myocardial disease are common causes of death that usually occurs within the second decade of life [24]. A skin abnormality, namely, pebbly, ivory-colored lesions over the back, upper arms, and lateral aspects of the thigh, is unique to Hunter's syndrome; their presence, however, does not correlate with the clinical severity [25].

Patients with the more attenuated (so-called adult) form usually have a normal intelligence but often have many complaints such as progressive loss of vision due to retinal dysfunction, spastic paresis due to myelon compression at the cranio-cervical region, severe hip disease, and cardiac complications [26].

VI. MANAGEMENT AND TREATMENT

A. Supportive Care and Treatment of Complications

As Hunter's syndrome is a complex disorder affecting multiple organs, a multidisciplinary management is required, whereby subspecialties such as otorhinolaryngology, orthopedics, neurosurgery, cardiology, pulmonology, and anesthesiology play an important role [27]. When the diagnosis of Hunter's syndrome has been made in a boy, a comprehensive initial assessment of all organ systems should be undertaken. Genetic counseling and psychological assistance should be offered to all members of the family [28].

As mentioned earlier, besides hernias that may be observed even in a newborn and need early surgical repair, upper and lower airway infections are the most common clinical manifestation in Hunter's syndrome. Tonsillectomy and adenoidectomy are strongly recommended in order to decrease airway obstruction and the frequency of ear infections, and early placement of ventilating tubes is also very helpful. In MPS II the combination of both conductive and sensorineural deafness is commonly seen, and it requires regular audiology assessments and provision of proper hearing aids.

In adult patients, obstructive and restrictive respiratory dysfunction is not rare; for treatment of episodes of hypoxia, particularly during the night, oxygen supply and continuous positive airway pressure devices may become necessary.

As for several reasons Hunter's disease is associated with a major anesthetic risk, the affected patients should undergo general anesthesia only in centers that have longtime experience in this disorder [29]. Bone involvement and joint stiffness are the cause of orthopedic complications. For the treatment of progressive arthropathy, particularly of the hip joints, prosthetic surgery is often required, which in many cases is very difficult to manage because of the deformed acetabulum and pelvis. If patients complain of intolerable hip pain, they may become wheelchair bound. In order to preserve joint function and sustain general mobility, physical therapy should be started at an early age. In Hunter's patients, the heart should regularly be assessed by electrocardiogram and echocardiography. Mitral and aortic regurgitation and/or stenosis makes valve replacement therapy necessary [30]. Cardiomyopathy is observed in some cases that may be associated with an increased risk of cardiac arrhythmia leading to sudden death [31].

Ocular findings in MPS II include papilledema, atrophy of the optic nerve, and retinal dystrophy. In contrast to mucopolysaccharidosis type I (MPS I) and mucopolysaccharidosis type IV (MPS VI), corneal opacity is rare in Hunter's patients and in general does not impair vision [32]. Schumacher et al. have performed ultrasound investigations of the eye of several MPS patients. They found thickening of the optic nerve sheath and sclera, probably due to GAG deposition within the sclera causing pressure on the optic nerve at the interscleral level [33]. Retinal degeneration leading to night blindness is a further complication that can be detected by electroretinography. Regular ophthalmological assessments should include a measure of intraocular pressure, and glasses should be prescribed as appropriate. In severely affected patients tonic–clonic seizures are not uncommon that should be treated by standard anticonvulsant medication. For the management of behavior problems such as hyperactivity and aggression, regular psychological assessment is recommended.

GAG deposition within the dura and transverse ligament results in progressive compression of the spinal cord leading to cervical myelopathy. For detection of spinal compression, MRI scans and neurophysiological investigations should be performed regularly before clinical symptoms such as reduced activity or exercise intolerance arise. And severe neurological complications and irreversible cord damage can be prevented if cervical decompression is performed at an early stage. Carpal tunnel syndrome is a very common complication in patients with the attenuated form of MPS II that requires an early surgical intervention.

Because of the clinical complexity of Hunter's syndrome, the patients should be under the care of a medical center that has extensive experience in treatment and management of this rare disorder.

B. Stem Cell Transplantation

HSCT has been used as treatment for lysosomal storage diseases over 30 years. This therapeutic approach has become the therapy of choice for the severe form of MPS I (Hurler's syndrome). HSCT has led to promising results also in patients with other lysosomal storage disorders, such as α-mannosidosis, metachromatic leukodystrophy, and Krabbe disease, provided the transplantation is performed before signs of involvement of the peripheral and/or central nervous system are present [34].

As until now no controlled clinical trials have been conducted, long-term experience of HSCT in Hunter's syndrome is lacking. An allogenic bone marrow transplantation has been performed in a 29-month-old boy. However, despite of full engraftment, the neurocognitive abilities deteriorated [35]. The effect of bone marrow transplantation was studied in eight boys who received the graft at the ages between 3 and 16 years [36]. The procedure resulted in a reduction of liver and spleen size and an improvement of cardiovascular abnormalities and joint stiffness. Two patients with the attenuated phenotype maintained their normal IQ. The severely affected patients, however, showed a deterioration of their mental status; they developed epileptic seizures and lost the ability to walk and speak. In summary, the use of HSCT in Hunter's syndrome remains controversial in view of the significant morbidity associated with this procedure and the poor results regarding the neurological outcome. Patients and their parents should weigh the risks of HSCT against the potential improvement in the quality of life they may gain from transplantation.

C. Enzyme Replacement Therapy

Synthesis of Idursulfase Idursulfase, the recombinant form of IDS, is produced at Shire Human Genetic Therapies, Inc., Cambridge, MA, by use of a genetically engineered cell line yielding a protein with a glycosylation analogous to that of the native human enzyme [37]. The idursulfase-producing cell line is generated by transfecting the fibrosarcoma HT1080 cells with an expression plasmid encoding the 550 amino acids of human IDS. The produced enzyme is secreted into the culture medium and subsequently purified. As evaluated by several chromatographic and electrophoretic assays, the purified protein is 99.9% pure. The N-linked glycosylation sites contain M6P residues that enable receptor-mediated cell uptake and targeting to the lysosomes via the M6P receptor. In addition, the purified idursulfase contains complex highly sialylated glycans that lead to prolongation of the circulating half-life of the enzyme. About 50% of the protein has received the posttranslational modification of cysteine 59 to formylglycine that is required for enzymatic activity (see preceding text).

Animal Studies First pharmacokinetic studies with idursulfase were performed in normal mice by injecting a single intravenous bolus of idursulfase at 0.5 and 1.5 mg/kg [38]. Serum distribution and elimination were biphasic, showing half-lives of 144 and 90 min, respectively. Two hours after the injection of 1.0 mg/kg of idursulfase, about one-third of the administered enzyme was found in the liver. The spleen had 1.38% of the dose; other organs contained less than 1%. By immunohistological examination of the brain, no enzyme staining could be detected.

In order to examine whether the injection of IDS would be able to reduce the storage material in several organs, a mouse model of MPS II (IDS-knockout mouse (IDS-KO mouse)) was created. The disruption of the IDS gene resulted in a loss of enzymatic activity and GAG accumulation in a number of tissues. The features the animals exhibited are characteristic for MPS II. By 6 weeks of age, urinary GAG levels were significantly elevated compared with wild-type mice. Later on (10 weeks of age), skeletal abnormalities such as swollen hocks, thickened paws, and facial coarsening became evident. Radiographic examinations revealed widened and thickened ribs. GAG accumulation could be seen by Alcian blue staining in the liver, kidney, lung, and heart valves [39]. In this IDS-KO mouse, the effect of purified human recombinant IDS was investigated. After five weekly injections of the enzyme, urinary GAG levels decreased to normal levels. In addition, a reduction in GAG levels in the liver, kidney, heart, spleen, lung, and skeletal muscle was observed [38].

Clinical Trials Based on these animal experiments, the first clinical trial with idursulfase was initiated in patients with Hunter's syndrome. The biochemical and clinical effects of three doses (0.15, 0.5, and 1.5 mg/kg) of idursulfase infused intravenously every 2 weeks were evaluated in this double-blind, placebo-controlled, dosage-range trial. In the treated patients a significant (49%) reduction in urinary GAG excretion could be demonstrated. In addition, a remarkable decrease of liver and spleen volumes and an improvement of the 6-min walk test (6MWT) was observed [40].

Pharmacokinetic studies showed a biphasic serum clearance over a 24-h period. The maximum serum concentration (C_{max}) was reached at the end of the 1-h infusion and was proportional to the dose. As the area under the curve (AUC) was not dose proportional from 0.5 to 1.5 mg/kg, it can be concluded that serum clearance (cellular uptake via the M6P receptor) had become saturated. This serum profile showed how rapidly the administered idursulfase was removed from the circulation and taken up by the target tissues. Pharmacokinetic parameters were not substantially changed after repeat enzyme administration. Anti-idursulfase antibodies had only a minimal effect on serum clearance [41].

Based on the experiences that were gained in the foregoing trial, a phase III, randomized, double-blind, placebo-controlled clinical study was started [42]. In nine centers from four countries (the United States, England, Germany, Brazil), 96 patients between the ages of 5 and 31 years with a diagnosis of MPS II were enrolled. From the study were excluded individuals who were mentally retarded and were not able to perform the pulmonary function testing or the 6MWT. The patients were randomized to placebo infusions, weekly idursulfase (0.5 mg/kg) infusions, or every-other-week (EOW) idursulfase (0.5 mg/kg) infusions. The three groups were well balanced with respect to age, time since diagnosis of MPS II, and baseline disease characteristics.

A so-called two-component composite score was chosen as the primary efficacy endpoint of this study whereby two parameters were used, the forced vital capacity (FVC) and the distance walked in 6 min. FVC was measured by spirometry using standard methods and expressed as percent predicted of normal, which was calculated based on published reference values. The 6MWT was performed by measuring the walking distance that could be achieved within 6 min [43]. Secondary efficacy endpoints included changes in the liver and spleen volumes (measured by abdominal MRI), urinary GAG excretion, and passive joint range of motion. Statistical analysis was performed according to the method described by O'Brien [44].

After 1 year of treatment, a statistically significant ($p=0.0049$) improvement in this composite primary efficacy endpoint was seen in patients who received IDS weekly or EOW compared to placebo. At 1 year of treatment, the weekly dosing group showed a 37-m increase in the 6MWT ($p=0.013$), a 2.7% increase in percentage of FVC ($p=0.065$), and a 160-ml increase in absolute FVC ($p=0.001$) compared to placebo group. Liver volume decreased from baseline by 25.3 ± 1.6% in the idursulfase weekly group and by 24.0 ± 1.7% in the idursulfase EOW group. The change in both treatment groups was statistically significantly greater than the change in the placebo group ($p<0.0001$). The same was true for the spleen volume.

Besides a difference in the improvement of elbow mobility between the weekly idursulfase group compared to placebo, no other significant differences between treatment groups for any joint range of motion were observed.

Patients receiving the enzyme demonstrated a significant reduction in urine GAGs. The effect of ERT on the neurological manifestation of the disorder has not been studied.

As the results of the phase II/III study were positive regarding safety and clinical efficacy, the recombinant human IDS (idursulfase, Elaprase®, Shire Human Genetics Therapies, Inc., Cambridge, MA) received commercial approval in 2006.

In order to evaluate the safety and efficacy of long-term ERT with idursulfase, a multicentered, multinational, open-labeled extension study was initiated [45]. The primary endpoints were identical to those that had been used in the phase III trial, namely, the lung function, measured as FVC and the 6MWT. In this 2-year trial all 94 patients who completed the 53-week double-blinded study were enrolled, and 85 patients completed the end of the study assessments. One patient died of respiratory failure and eight patients were withdrawn from the trial as they received the enzyme at their home and therefore the follow-up evaluations could not be performed anymore.

Data analysis of this extension study showed sustained improvement of absolute FVC and even a 25% increase at the end of the study. For most time points statistically significant increases in 6MWT distances were seen. Liver and spleen size remained reduced. Whereas in almost all joints the range of motion remained stable, an improvement was seen in the shoulder.

The clinical effects of idursulfase infusions have been studied in an open-label clinical trial in 10 Japanese patients aged 21–53 years who had an attenuated form of Hunter's disease [46]. The idursulfase dosage was 0.5 mg/kg per week. After 1 year of treatment, urinary GAG excretion normalized in all patients, and a reduction of liver and spleen volume was noted. There was a significant improvement of the 6MWT, lung function, and left ventricular mass. In some joints an increase of mobility was observed. The enzyme was generally well tolerated; one patient died of causes unrelated to the idursulfase infusion. In summary, these results are comparable to those of earlier studies that enrolled predominantly younger patients.

The Hunter Outcome Survey (HOS) is an international, long-term observational survey created to evaluate safety and efficacy of ERT with idursulfase and to increase the knowledge of the natural history of Hunter's syndrome. Only data that are collected during regular physicians' examinations such as signs and symptoms, laboratory values, and results of functional tests (ECG, echocardiogram, lung function tests, etc.) are entered in this electronic database [22]. In order to assess the safety and efficacy of ERT with idursulfase in Hunter's patients who started treatment before the age of 6 years, data from HOS were analyzed by Muenzer et al. [47]. In this study 124 patients were included from whom at least the result of one follow-up examination in the HOS data bank was available. The mean age at start of idursulfase infusions was 3.6 ± 1.6 years (mean ± SD). The efficacy results after 6 months of treatment were similar to those seen in older patients. In addition to a significant reduction of urine GAG levels, a decrease of liver size was observed.

Enzyme Replacement Therapy in a Female Hunter's Patient ERT was studied in a girl who was affected by the severe form of Hunter's disease. The patient received her first infusion of idursulfase at the age of 8 years. After 2½ years of treatment, an improvement of lung function, joint mobility, and general endurance was observed. Furthermore, upper respiratory tract infections seemed to be reduced. Liver and spleen size normalized. However, more epileptic seizures occurred, requiring a therapeutic adjustment [48].

Safety and Antibodies The regular intravenous application of a protein obviously is associated with a risk of developing antibodies that may lead to allergic reactions and/or inactivate enzyme activity.

The safety profile of idursulfase is similar to that reported for ERT of other lysosomal storage disorders such as MPS I [49], MPS VI [50], or Fabry disease [51]. In all studies the majority of adverse events was consistent with complications that were expected to be seen in untreated MPS II patients such as fever, headache, cough, (upper and lower) respiratory tract infections, abdominal pain, and diarrhea. In the phase I/II trial, no patients receiving the lower dose (0.15 mg/kg) experienced infusion-associated events, defined as an adverse event that occurs within the day after infusion. Six out of eight patients in the higher-dose regimens, however, reported infusion-associated reactions. These moderate-to-severe reactions could be easily managed by increasing the infusion time and premedication with an antihistamine and/or corticosteroid [40].

In the phase III trial, one or more infusion-related adverse events were observed in each treatment group in almost the same number of patients (placebo, 21; EOW, 22; weekly, 22). Most of the infusion-related adverse events were seen between weeks 4 and 12, and the incidence decreased in all three groups thereafter. No patient withdrew from the study because of infusion-related adverse events. Two patients died during the study: one patient in the placebo group and one in the idursulfase weekly group. Neither death was considered related to blinded study medication by the investigator [42].

In the long-term extension study, it could be demonstrated that the incidence of infusion reactions decreased during the course of up to 3 years of treatment [45]. In the study in patients under 6 years of age, 69 infusion-related reactions have been noticed in 33 (26.6%) patients, whereby three serious infusion-related reactions occurred in a single patient [47].

Taking into account the prescribing information for Elaprase® – life-threatening anaphylactic reactions may occur during infusions and also anaphylactic reactions have been observed in some patients—appropriate medical support needs to be available during infusions. For patients who have experienced anaphylactic reactions, prolonged monitoring is required. Infusion reactions may lead to acute exacerbation of respiratory compromise in patients who have a chronic lung disease.

Immune activation represents a concern in therapeutic strategies in which recombinant proteins are used. Antibodies against the infused enzyme have been observed in all lysosomal storage disorders that are treated by ERT [52].

During the phase III of the Hunter's trial, IgG anti-idursulfase antibodies were detected in 15 patients (46.9%) in the idursulfase weekly group and in 15 patients (46, 9%) in the EOW group. The highest levels of IgG antibodies were observed at week 27. The reduction in urinary GAG levels was lower in the antibody-positive patients in comparison with that of the antibody-negative individuals. However, there was no association with the presence of antibodies and adverse events or with clinical assessments (such as 6MWT or FVC measurements).

At the end of the extension study, 23 of 85 patients (27.1%) were IgG positive, and in 19 of these patients, neutralizing antibodies were detected, suggesting that toleration had not occurred. As the increase in absolute FVC in patients who were neutralizing antibody negative was higher than in patients who were neutralizing antibody positive, it can be concluded that neutralizing antibodies have a significant impact in the long-term clinical efficacy. The neutralizing antibody status, however, did not affect the responses on the 6MWT, liver and spleen volume, and urine GAG levels. In none of the clinical studies, IgE antibodies had been found. In summary, the enzyme preparation showed a risk/benefit profile that is similar to that reported for other replacement therapies.

VII. FUTURE PERSPECTIVES

In spite of the encouraging clinical results that have been observed in clinical trials of ERT in Hunter's syndrome, it has become clear that this therapeutic approach has some limitations and is not able to address all aspects of the disorders equally. Several organs such as cartilage, bone, and heart valve are resistant to correction by intravenous ERT. From an observation of siblings affected by MPS type VI (*N*-acetylgalactosamine-4-sulfatase deficiency), it can be concluded that very early treatment may possibly lead to a better clinical response [53]: the younger sibling, who received the first infusion of recombinant human *N*-acetylgalactosamine-4-sulfatase soon after birth, after more than 3 years of treatment did not develop the signs and symptoms such as scoliosis or joint contractures that were seen in the older sibling in whom ERT was initiated at the age of 3½ years.

In lysosomal storage disorders that affect the central nervous system, ERT is limited by the fact that these proteins are not able to cross the blood–brain barrier. One way to overcome this limitation is to inject the enzyme intrathecally in order to get enzyme activity within the brain. This procedure has been performed in a patient affected by MPS I, leading to improvement in the 6MWT and in lung function [54]. A clinical trial of intrathecal idursulfase administration via an intrathecal drug delivery device in pediatric Hunter's patients with involvement of the central nervous system has been initiated by Shire Human Genetic Therapies, Inc., Cambridge, MA [55].

Delivery of an enzyme to the brain may also be achieved by using the human insulin receptor instead of the M6P receptor [56]: IDS was fused to the carboxyl-terminus of the heavy chain of a monoclonal antibody (mAb), which was directed to the human insulin receptor. The antibody–iduronate sulfatase fusion protein was taken up by human fibroblasts and retained specific activity. Treatment of fibroblasts from a Hunter's patient with the fusion protein resulted in a 84% decrease of GAG accumulation. In which degree the bifunctional IgG–sulfatase can be transported into the brain has not yet been tested.

Substrate deprivation represents an alternative approach for treatment of lysosomal storage disorders. The concept of this therapeutic principle is not to increase the activity of the degrading enzyme but to decrease the amount of storage material. The imino sugar *N*-butyldeoxynojirimycin – also named miglustat – is able to inhibit the activity of ceramide glucosyltransferase, the enzyme that synthesizes glycosphingolipid, the storage compound in Gaucher's disease [57]. This drug has been approved for symptomatic Gaucher's patients with mild-to-moderate clinical manifestation for whom ERT is not an option.

Genistein, an isoflavone extract from soybeans, has been shown to be an inhibitor of GAG synthesis [58]. In an open-label study, Piotrowska et al. have analyzed the effect of genistein on urinary GAG excretion, hair morphology, and behavior in 10 patients, affected by either MPS IIIA or MPS IIIB [59]. After oral administration of genistein-rich soy isoflavone extract (5 mg/kg/d) for 1 year, a statistically significant improvement in all these parameters was found. The results of a placebo-controlled clinical trial are still outstanding.

Present substrate reduction therapies such as genistein are directed to enzymes which are involved in the early pathway of substrate synthesis. This strategy can eliminate the substrate completely and may prevent its normal function. In contrast, inhibitors, which modify a substrate to make it more degradable for the lysosomal enzymes, but do not prevent the synthesis, should be able to preserve its normal function. By screening a library of a huge number of compounds, 15 inhibitors of GAG synthesis have been found that fell into three classes: (1) N-sulfation inhibitor, (2) 2-O-sulfation inhibitor, and (3) 6-O-sulfation inhibitor. The substance ZP2345, for example, leads to a reduction in 2-O-sulfation of HS with compensatory increase in 6-O-sulfation. And this modified HS could be degraded by fibroblasts from a patient affected by Hunter's disease much easier than the native HS [60]. This approach, which is called substrate optimization therapy, may have some advantages as it is able to reduce storage in organs that are difficult to reach (heart, muscle tissue, skeleton, central nervous system), has less toxicity associated with reduced GAG synthesis, and – as an oral drug – is more convenient for the patients in comparison with regular intravenous infusions [61].

As each of the treatment modalities discussed in this chapter is targeted to only one or a few aspects of the complex pathophysiology of MPS I, it seems that by combination of several therapeutic regimens, a better outcome may be achieved than by one therapy alone. Some limitations of therapeutic interventions are possibly due to the timing of therapy, and early treatment may be more efficient. For this reason newborn screening programs for lysosomal storage disorder have been developed [62]. However, as for most of these conditions a strict genotype–phenotype correlation does not exist, the screening programs will probably not be generally introduced until the phenotype from the newborn can be exactly predicted [63].

REFERENCES

1. C Hunter. A rare disease in two brothers. Proc R Soc Med 10:104–116, 1917.
2. G Hurler. Über einen Typ multipler Abartungen, vorweigend am Skelettsystem. Z Kinderheilk 24:220–234, 1919.
3. RT Beebe, PF Formel. Gargoylism: sex-linked transmission in nine males. Trans Am Clin Climatol Assoc 66:199–207, 1954.
4. G Brante. Gargoylism; a mucopolysaccharidosis. Scand J Clin Lab Invest 4:43–46, 1952.
5. A Dorfman, AE Lorincz. Occurrence of urinary acid mucopolysaccharides in the Hurler syndrome. Proc Natl Acad Sci USA 43:443–446, 1957.
6. JC Fratantoni, CW Hall, EF Neufeld. The defect in Hurler's and Hunter's syndromes: faulty degradation of mucopolysaccharide. Proc Natl Acad Sci USA 60:699–706, 1968.
7. JC Fratantoni, CW Hall, EF Neufeld. Hurler and Hunter syndromes: mutual correction of the defect in cultured fibroblasts. Science 162:570–572, 1968.
8. M Cantz, A Chrambach, G Bach, EF Neufeld. The Hunter corrective factor. Purification and preliminary characterization. J Biol Chem 247:5456–5462, 1972.
9. G Bach, F Eisenberg, Jr., M Cantz, EF Neufeld. The defect in the Hunter syndrome: deficiency of sulfoiduronate sulfatase. Proc Natl Acad Sci USA 70:2134–2138, 1973.
10. PJ Wilson, CP Morris, DS Anson, T Occhiodoro, J Bielicki, PR Clements, JJ Hopwood. Hunter syndrome: isolation of an iduronate-2-sulfatase cDNA clone and analysis of patient DNA. Proc Natl Acad Sci USA 87:8531–8535, 1990.
11. ML Bondeson, H Malmgren, N Dahl, BM Carlberg, U Pettersson. Presence of an IDS-related locus (IDS2) in Xq28 complicates the mutational analysis of Hunter syndrome. Eur J Hum Genet 3:219–227, 1995.
12. KR Taylor, RL Gallo. Glycosaminoglycans and their proteoglycans: host-associated molecular patterns for initiation and modulation of inflammation. FASEB J 20:9–22, 2006.
13. A Ballabio, V Gieselmann. Lysosomal disorders: from storage to cellular damage. Biochim Biophys Acta 1793:684–696, 2009.
14. S Cudry, R Froissart, O Bouton, I Maire, D Bozon. The 2.1-, 5.4- and 5.7-kb transcripts of the IDS gene are generated by different polyadenylation signals. Biochim Biophys Acta 1447:35–42, 1999.
15. T Dierks, B Schmidt, LV Borissenko, J Peng, A Preusser, M Mariappan, K von Figura. Multiple sulfatase deficiency is caused by mutations in the gene encoding the human C(alpha)-formylglycine generating enzyme. Cell 113:435–444, 2003.
16. S Kornfeld. Trafficking of lysosomal enzymes. FASEB J 1:462–468, 1987.
17. KM Timms, ML Bondeson, MA Ansari-Lari, K Lagerstedt, DM Muzny, SP Dugan-Rocha, DL Nelson, U Pettersson, RA Gibbs. Molecular and phenotypic variation in patients with severe Hunter syndrome. Hum Mol Genet 6:479–486, 1997.
18. F Baehner, C Schmiedeskamp, F Krummenauer, E Miebach, M Bajbouj, C Whybra, A Kohlschutter, C Kampmann, M Beck. Cumulative incidence rates of the mucopolysaccharidoses in Germany. J Inherit Metab Dis 28:1011–1017, 2005.
19. H Zhang, J Li, X Zhang, Y Wang, W Qiu, J Ye, L Han, X Gao, X Gu. Analysis of the IDS gene in 38 patients with Hunter syndrome: the c.879G>A (p.Gln293Gln) synonymous variation in a female create exonic splicing. PloS One 6:e22951, 2011.

20. I Schwartz, L Pinto, G Breda, L Lima, M Ribeiro, J Mota, A Acosta, P Correia, D Horovitz, C Porciuncula, et al. Clinical and biochemical studies in mucopolysaccharidosis type II carriers. J Inherit Metab Dis 32:732–738, 2009.
21. K Tuschl, A Gal, E Paschke, S Kircher, OA Bodamer. Mucopolysaccharidosis type II in females: case report and review of literature. Pediatr Neurol 32:270–272, 2005.
22. JE Wraith, M Beck, R Giugliani, J Clarke, R Martin, J Muenzer. Initial report from the Hunter Outcome Survey. Genet Med 10:508–516, 2008.
23. IV Schwartz, MG Ribeiro, JG Mota, MB Toralles, P Correia, D Horovitz, ES Santos, IL Monlleo, AC Fett-Conte, RP Sobrinho, et al. A clinical study of 77 patients with mucopolysaccharidosis type II. Acta Paediatr Suppl 96:63–70, 2007.
24. SA Jones, Z Almassy, M Beck, K Burt, JT Clarke, R Giugliani, C Hendriksz, T Kroepfl, L Lavery, SP Lin, et al. Mortality and cause of death in mucopolysaccharidosis type II-a historical review based on data from the Hunter Outcome Survey (HOS). J Inherit Metab Dis 32:534–543, 2009.
25. S Sakata, D Orchard, G Mason. Skin rash with the histological absence of metachromatic granules as the presenting feature of Hunter syndrome in a 6-year-old boy. Br J Dermatol 159:249–250, 2008.
26. R Martin, M Beck, C Eng, R Giugliani, P Harmatz, V Munoz, J Muenzer. Recognition and diagnosis of mucopolysaccharidosis II (Hunter syndrome). Pediatrics 121:e377–e386, 2008.
27. J Muenzer, M Beck, CM Eng, ML Escolar, R Giugliani, NH Guffon, P Harmatz, W Kamin, C Kampmann, ST Koseoglu, et al. Multidisciplinary management of Hunter syndrome. Pediatrics 124:e1228–e1239, 2009.
28. I Kuratsubo, Y Suzuki, KO Orii, T Kato, T Orii, N Kondo. Psychological status of patients with mucopolysaccharidosis type II and their parents. Pediatr Int 51:41–47, 2009.
29. L Gaitini, M Fradis, S Vaida, G Collins, M Croitoru, M Somri, Z Borochovitz, A Golz. Failure to control the airway in a patient with Hunter's syndrome. J Laryngol Otol 112:380–382, 1998.
30. CF Wippermann, M Beck, D Schranz, R Huth, I Michel-Behnke, BK Jungst. Mitral and aortic regurgitation in 84 patients with mucopolysaccharidoses. Eur J Pediatr 154:98–101, 1995.
31. T Hishitani, S Wakita, T Isoda, T Katori, A Ishizawa, R Okada. Sudden death in Hunter syndrome caused by complete atrioventricular block. J Pediatr 136:268–269, 2000.
32. JL Ashworth, FE Kruse, B Bachmann, AP Tormene, A Suppiej, R Parini, N Guffon. Ocular manifestations in the mucopolysaccharidoses – a review. Clin Experiment Ophthalmol 38:12–22, 2010.
33. RG Schumacher, R Brzezinska, G Schulze-Frenking, S Pitz. Sonographic ocular findings in patients with mucopolysaccharidoses I, II and VI. Pediatr Radiol 38:543–550, 2008.
34. W Krivit. Allogeneic stem cell transplantation for the treatment of lysosomal and peroxisomal metabolic diseases. Springer Semin Immunopathol 26:119–132, 2004.
35. EJ McKinnis, S Sulzbacher, JC Rutledge, J Sanders, CR Scott. Bone marrow transplantation in Hunter syndrome. J Pediatr 129:145–148, 1996.
36. N Guffon, Y Bertrand, I Forest, A Fouilhoux, R Froissart. Bone marrow transplantation in children with Hunter syndrome: outcome after 7 to 17 years. J Pediatr 154:733–737, 2009.
37. LA Clarke. Idursulfase for the treatment of mucopolysaccharidosis II. Expert Opin Pharmacother 9:311–317, 2008.

38. AR Garcia, JM DaCosta, J Pan, J Muenzer, JC Lamsa. Preclinical dose ranging studies for enzyme replacement therapy with idursulfase in a knock-out mouse model of MPS II. Mol Genet Metab 91:183–190, 2007.
39. AR Garcia, J Pan, JC Lamsa, J Muenzer. The characterization of a murine model of mucopolysaccharidosis II (Hunter syndrome). J Inherit Metab Dis 30:924–934, 2007.
40. J Muenzer, M Gucsavas-Calikoglu, SE McCandless, TJ Schuetz, A Kimura. A phase I/II clinical trial of enzyme replacement therapy in mucopolysaccharidosis II (Hunter syndrome). Mol Genet Metab 90:329–337, 2007.
41. K Loveday, J Lieb, R Gamble, J Ruiz, A Kimura, J Muenzer. Single and repeat-dose pharmacokinetics of idursulfase in patients with MPS II (Hunter Syndrome). *8th International Symposium on Mucopolysaccharide and Related Diseases*. Mainz, June 10–13, 2004.
42. J Muenzer, JE Wraith, M Beck, R Giugliani, P Harmatz, CM Eng, A Vellodi, R Martin, U Ramaswami, M Gucsavas-Calikoglu, et al. A phase II/III clinical study of enzyme replacement therapy with idursulfase in mucopolysaccharidosis II (Hunter syndrome). Genet Med 8:465–473, 2006.
43. ATS Committee on Proficiency Standards for Clinical Pulmonary Function Laboratories. ATS statement: guidelines for the six-minute walk test. Am J Respir Crit Care Med 166:111–117, 2002.
44. PC O'Brien. Procedures for comparing samples with multiple endpoints. Biometrics 40:1079–1087, 1984.
45. J Muenzer, M Beck, CM Eng, R Giugliani, P Harmatz, R Martin, U Ramaswami, A Vellodi, JE Wraith, M Cleary, et al. Long-term, open-labeled extension study of idursulfase in the treatment of Hunter syndrome. Genet Med 13:95–101, 2011.
46. T Okuyama, A Tanaka, Y Suzuki, H Ida, T Tanaka, GF Cox, Y Eto, T Orii. Japan Elaprase Treatment (JET) study: idursulfase enzyme replacement therapy in adult patients with attenuated Hunter syndrome (Mucopolysaccharidosis II, MPS II). Mol Genet Metab 99:18–25, 2010.
47. J Muenzer, M Beck, R Giugliani, Y Suzuki, A Tylki-Szymanska, V Valayannopoulos, A Vellodi, JE Wraith. Idursulfase treatment of Hunter syndrome in children younger than 6 years: results from the Hunter Outcome Survey. Genet Med 13:102–109, 2011.
48. R Manara, A Rampazzo, M Cananzi, L Salviati, R Mardari, P Drigo, R Tomanin, N Gasparotto, E Priante, M Scarpa. Hunter syndrome in an 11-year old girl on enzyme replacement therapy with idursulfase: brain magnetic resonance imaging features and evolution. J Inherit Metab Dis, published online: January 6, 2010. See http://link.springer.com/article/10.1007%2Fs10545-009-9023-8 (accessed July 22, 2013).
49. ED Kakkis, J Muenzer, GE Tiller, L Waber, J John Belmont, M Merry Passage, B Izykowski, J Phillips, R Doroshow, I Irv Walot, et al. Enzyme-replacement therapy in mucopolysaccharidosis I. N Engl J Med 344:182–188, 2001.
50. P Harmatz, R Giugliani, IVD Schwartz, N Guffon, EL Teles, MCS Miranda, JE Wraith, M Beck, L Arash, M Scarpa, et al. Long-term follow-up of endurance and safety outcomes during enzyme replacement therapy for mucopolysaccharidosis VI: final results of three clinical studies of recombinant human *N*-acetylgalactosamine 4-sulfatase. Mol Genet Metabol 94:469–475, 2008.
51. M Beck. Agalsidase alfa for the treatment of Fabry disease: new data on clinical efficacy and safety. Expert Opin Biol Ther 9:255–261, 2009.

52. J Wang, J Lozier, G Johnson, S Kirshner, D Verthelyi, A Pariser, E Shores, A Rosenberg. Neutralizing antibodies to therapeutic enzymes: considerations for testing, prevention and treatment. Nat Biotechnol 26:901–908, 2008.

53. JJ McGill, AC Inwood, DJ Coman, ML Lipke, D de Lore, SJ Swiedler, JJ Hopwood. Enzyme replacement therapy for mucopolysaccharidosis VI from 8 weeks of age – a sibling control study. Clin Genet 77:492–498, 2010.

54. MV Munoz-Rojas, T Vieira, R Costa, S Fagondes, A John, LB Jardim, LM Vedolin, M Raymundo, PI Dickson, E Kakkis, et al. Intrathecal enzyme replacement therapy in a patient with mucopolysaccharidosis type I and symptomatic spinal cord compression. Am J Med Genet A 146A:2538–2544, 2008.

55. See http://clinicaltrials.gov/ct2/show/NCT00920647?term=hunter+syndrome&rank=10 (accessed November 21, 2011).

56. JZ Lu, EK Hui, RJ Boado, WM Pardridge. Genetic engineering of a bifunctional IgG fusion protein with iduronate-2-sulfatase. Bioconjug Chem 21:151–156, 2010.

57. FM Platt, GR Neises, RA Dwek, TD Butters. N-butyldeoxynojirimycin is a novel inhibitor of glycolipid biosynthesis. J Biol Chem 269:8362–8365, 1994.

58. E Piotrowska, J Jakóbkiewicz-Banecka, S Baranska, A Tylki-Szymanska, B Czartoryska, A Wegrzyn, G Wegrzyn. Genistein-mediated inhibition of glycosaminoglycan synthesis as a basis for gene expression-targeted isoflavone therapy for mucopolysaccharidoses. Eur J Hum Genet 14:846–852, 2006.

59. E Piotrowska, J Jakóbkiewicz-Banecka, A Tylki-Szymanska, A Liberek, A Maryniak, M Malinowska, B Czartoryska, E Puk, A Kloska, T Liberek, et al. Genistin-rich soy isoflavone extract in substrate reduction therapy for Sanfilippo syndrome: an open-label, pilot study in 10 pediatric patients. Curr Therap Res 69:166–179, 2008.

60. JR Brown, RS Carroll, S Duron, CA Glass, BE Crawford. Small molecule inhibitors of glycosaminoglycan biosynthesis: substrate optimization therapy for the mucopolysaccharidoses. *WORLD Symposium*. Miami, FL, February 10–12, 2010.

61. JR Brown, BE Crawford, JD Esko. Glycan antagonists and inhibitors: a fount for drug discovery. Crit Rev Biochem Mol Biol 42:481–515, 2007.

62. K Nakamura, K Hattori, F Endo. Newborn screening for lysosomal storage disorders. Am J Med Genet C Semin Med Genet 157:63–71, 2011.

63. RY Wang, OA Bodamer, MS Watson, WR Wilcox. Lysosomal storage diseases: diagnostic confirmation and management of presymptomatic individuals. Genet Med 13:457–484, 2011.

9

ENZYME REPLACEMENT THERAPY FOR FABRY DISEASE

LEY NADINE LACBAWAN

LSD Research and Treatment Unit, Center for Clinical Trials, Fairfax, Virginia

WEI ZHENG

National Center for Advancing Translational Sciences, National Institutes of Health, Bethesda, Maryland

OZLEM GOKER-ALPAN

LSD Research and Treatment Unit, Center for Clinical Trials, Fairfax, Virginia

I. INTRODUCTION

Fabry disease is an X-linked lysosomal storage disorder with an estimated incidence of 1 in 40 000 males [1]. The disorder is caused by a functional deficiency of the lysosomal enzyme alpha-galactosidase A (α-GalA, E.C. 3.2.1.22) that hydrolyzes the terminal alpha-galactosyl moieties from glycolipids and glycoproteins. Most Fabry disease patients have a single point mutation in their *GLA* gene that codes for α-GalA, a homodimeric glycoprotein, and over 400 missense and nonsense mutations have been identified in various patients [2]. Deficiency in α-GalA results in progressive accumulation of globotriaosylceramide (GL-3) in the lysosomes of the vascular endothelial cells of multiple organ systems such as the kidneys, heart, skin, and brain resulting in a microvascular disease. The clinical manifestation of Fabry disease includes angiokeratoma, corneal opacities and whorls, acroparesthesia, sweating abnormalities, and cardiomyopathy with cardiac conduction abnormalities

Enzyme Technologies: Pluripotent Players in Discovering Therapeutic Agents, First Edition.
Edited by Hsiu-Chiung Yang, Wu-Kuang Yeh, and James R. McCarthy.

and Fabry disease primarily affects males, but female heterozygotes can develop symptoms of varying severity [3]. Nephropathy is a dominant feature of advanced Fabry disease that leads to renal function impairment [3,4].

In 1973, Johnson and Brady [5] showed that the infusion of purified α-GalA, derived from human placenta, rapidly cleared GL-3 from systemic circulation in patients with Fabry disease, however, technical complications prevented further development of enzyme replacement therapy (ERT) for Fabry disease until 2001 when recombinant α-GalA became available for therapeutic use. Two recombinant α-GalA preparations are currently available for the treatment of Fabry disease including (1) agalsidase alfa (Replagal, Shire Human Genetic Therapies, Cambridge, MA), produced in human fibroblast cell lines, and (2) agalsidase beta (Fabrazyme, Genzyme Inc., Cambridge, MA), manufactured using Chinese hamster ovary (CHO) cells. Only Fabrazyme is approved by the Food and Drug Administration (FDA) for use in the United States. The recombinant α-GalA enzymes are intravenously injected into patients every 2 weeks. These GLAs are glycoproteins with multiple mannose-6-phosphates (M6P) on the surface, which bind to M6P receptors on the plasma membrane of endothelial cells in many tissues and organs except the liver (which includes hepatocytes and Kupffer cells) [6]. A receptor-mediated endocytosis occurs after the binding of α-GalA to M6P receptor, which delivers recombinant enzymes to its target site lysosomes and restores the lysosome function of α-GalA [7]. The overall experience with ERT has been favorable with therapeutic effects on the kidney and heart during early phases of the disease, further alleviation of pain, and improvement in overall quality of life. Specifically, ERT has also been shown to reduce GL-3 deposits in podocytes, which are the affected cells in the progress of nephropathy [4].

However, the two aforementioned recombinant α-GalA preparations (agalsidase alfa and beta) for Fabry disease treatment are unstable in peripheral blood and not well incorporated into target organs including the kidneys and the heart [7]. Producing recombinant enzymes using mammalian cells can be a disadvantage as mammalian cell culture usually requires bovine serum or a serum-free synthetic medium and thus is very expensive and needs careful monitoring for the potential contamination [6].

In general, ERT needs to be initiated early on in the patient's treatment plan for maximum effectiveness. It usually requires lifetime treatment that is very costly. The long-term clinical benefits of ERT for Fabry disease patients are still unclear and still need to be investigated: ERT has limited effect in patients with cerebrovascular involvement, as the recombinant enzymes cannot penetrate the blood–brain barrier. In addition, the short half-life of the enzyme and the need for repeated administration of large amounts of protein are not the limitations of ERT. For instance, some patients may develop an immune response to the infused recombinant enzyme as a result of repeated injections, thus impeding its therapeutic effects and applications. Specifically, in males with Fabry disease, it has been found that repeated administration of recombinant α-GalA causes infusion reactions, primarily allergic origin which may be due to lack of any endogenous α-GalA in most male Fabry patients.

II. THE STRUCTURE AND CATALYTIC MECHANISMS OF HUMAN ALPHA-GALACTOSIDASE A

Human α-GalA belongs to the group of α-retaining enzymes, in which both substrate and product have anomeric carbons with alpha configurations due to a "double-displacement reaction mechanism," as proposed by Koshland [8]. The nucleophilic attacks on the anomeric carbon result in overall retention of the anomeric configuration [8]. This reaction mechanism requires two carboxylates, with one carboyxl acting as the nucleophile and the other acting as the acid–base [9]. Human α-GalA belongs to the glycoside hydrolase family 27 (GH27), in which identical catalytic mechanism is conserved across glycoside hydrolase clan D, including 1927 amino acid sequences in GH27 [10]. Structurally, human α-GalA is a homodimer, in which each monomer has two domains, an N-terminal (β/α) barrel containing the active site and a C-terminal antiparallel β-domain. The homodimer contains two active sites separated by 43°A. Within the dimer, there is no cooperativity between the active sites on the three-dimensional structure. The two monomers show no significant differences in the asymmetric unit or among the different steps of the catalytic cycle [2].

Guce and her colleagues [2] have recently described the catalytic mechanisms of human α-Gal. There are four different stages of the catalytic cycle of human α-Gal: empty active state; substrate-bound state; covalent intermediate state; and the product-soaked state. The *free enzyme* shows a collection of water molecules at the locations of the galactose ligand. Three waters reflect the position of the oxygen at O2 (0.92°A), O3 (0.27°A), and O6 (0.21°A) of the galactose, and two waters reflect the positions of carbons C1 (0.51°A) and C4 (1.1°A) of the galactose. The wild-type enzyme when *substrate bound* has a *Km* of 8.3 ± 5 mM and a *kcat* of 63.5 ± 0.1 s^{-1}, and the two recombinant α-Gals used in ERT have comparable *Km* and *kcat* [11]. The substrate binds to the active site with the glucose portion extending out, and the galactose portion at the standard ${}^4C_1$ chair conformation. The enzyme makes specific interactions with each galactose portion of the substrate, but only a few interactions with the glucoside portion of the disaccharide.

The *covalent intermediate* is in a 1S3 skew-boat conformation. The electron density shows a clear connection between C1 of the ligand and the Oδ2 catalytic nucleophile, Asp-170. The conformation is distorted from the favored chair conformation of the sugar ring and the ligand is held by the covalent bond between C1 and Oδ2 of Asp-170 at one end and at the other by Tryp-47 and the -4, -5, and 6 carbons on the β-face of the sugar.

In the *product-bound* state, the active site selects α-anomer of the galactose only, which is unique to the animal kingdom enzymes in GH27. The galactose is standard ${}^4C_3$ conformation but after the second nucleophile attack, the ligand converts to a low-energy state. There is a second binding site, comparable in size to the active site, that prefers the β-anomer of the galactose, using the plane of Tyr-329 against the β-face of monosaccharide. Binding to this additional site may stabilize the molecule without competitively inhibiting the enzyme [12](Fig. 1).

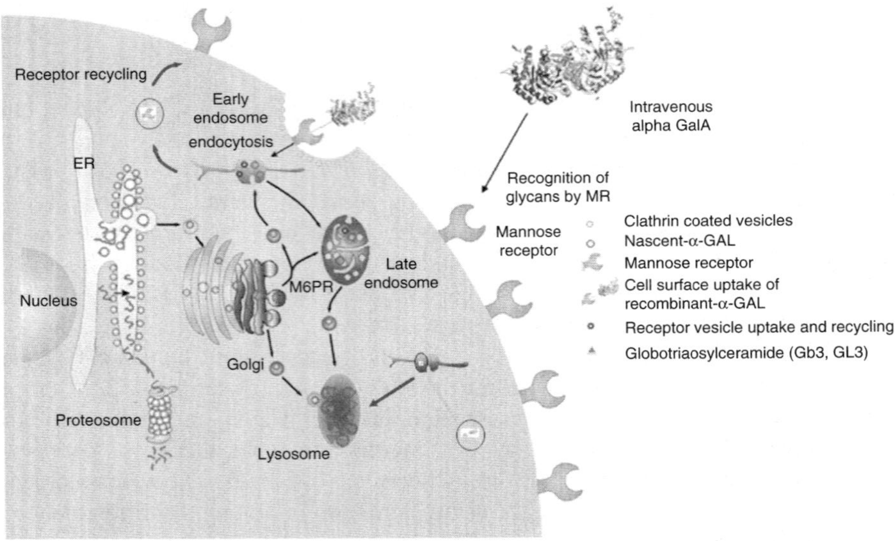

FIGURE 1 When a cell is genetically deficient in producing its own α-GalA or if the mutant enzyme cannot reach the lysosome, GL-3 accumulates causing Fabry disease. There exists a mechanism for the cell to uptake exogenous enzymes; α-GalA present in the extracellular milieu can bind to the cell surface via mannose receptors (MR) through recognition of the glycan moieties while it is transported into the lysosomal compartment using mannose 6-phosphate receptor (M6PR) system. Once in the lysosomal compartment, the enzyme can carry out its appropriate function. Thus, α-GalA (red circles) is targeted to the lysosomal compartment through interaction with M6PR, which binds to the oligosaccharide moieties of enzymes destined for the lysosomal compartment. The primary enzyme–receptor interaction occurs in the Golgi, where enzymes bound to Golgi M6PRs are segregated for transport to lysosomes. A secondary type of interaction may also occur between extracellular α-GalA and M6PR at the cell surface in which α-GalA is transported through the cytoplasm in endocytic vesicles which fuse with late endosomes (multivesicular bodies) and empty their contents into the lysosomes. Receptors are then recycled (blue circles) through the endosomal compartment. (*See insert for color representation of the figure.*)

III. INTRACELLULAR METABOLISM AND LYSOSOMAL TARGETING OF HUMAN α-GALACTOSIDASE A

α-GalA and other lysosomal enzymes are targeted to the lysosomal compartment through interaction with the M6P receptor, which then binds to M6P residues present in the oligosaccharide moieties of enzymes destined for the lysosomal compartment [13]. Primary interaction occurs in the Golgi, where enzymes bound to Golgi M6P receptors are segregated for transport to lysosomes. A secondary type of interaction may also occur between extracellular α-GalA and M6P receptors at the cell surface. Extracellular substances internalized by cells are transported through the cytoplasm in endocytic vesicles, which fuse with primary lysosomes and empty their contents [13].

In this process, cell surface M6P receptors are incorporated into endocytic vesicles and transported to lysosomes.

If any α-GalA with M6P residues is present in the extracellular milieu, they can bind to the cell surface M6P receptors and may be transported into the lysosomal compartment along with the receptors. Once in the lysosomal compartment as a result of this scavenger pathway, the enzyme can carry out its appropriate function. Thus, even if a cell is genetically deficient in producing its own α-Gal or the mutant enzyme cannot reach the lysosome, there exists a mechanism for it to uptake exogenous enzymes, provided that the enzymes are suitably glycosylated and the deficient cell bears M6P receptors.

In Fabry disease, vascular endothelial cells of the kidney and heart display severe histopathologic abnormalities and contribute to the clinical disease pathology. Thus, recombinant α-GalA can be delivered either locally or systemically to the affected cells, by gene therapy (i.e., by genetically modified cells that express and secrete the glycosylated enzyme) or by conventional pharmacologic routes of administration. The degree to which the N-linked oligosaccharides of α-GalA are modified by sialylation is important for delivery to target tissues. In the absence of appropriate sialylation, α-GalA will be cleared from circulation through binding by hepatic asialoglycoprotein receptors, internalization, and degradation in hepatocytes [14]. This hepatocytes elimination reduces the amount of α-GalA available in the circulation for binding to M6P receptors which contribute to the clinical pathology of Fabry disease, already mentioned. Surprisingly, α-GalA secreted by stably transfected cells has glycosylation properties, including the presence of mannose-terminated oligosaccharide side chains, the amount of M6P present on the oligomannose side chains, and the absence of terminal galactose residues that decreases uptake by asialoglycoprotein receptor in the liver, that make Fabry disease suitable for the treatment by conventional intravenous administration of the recombinant proteins [11]. In contrast, the lysosomal delivery of the enzyme, glucocerebrosidase, which is purified from human placenta or secreted from transfected CHO cells to the cells in the patient body, requires complex enzymatic modification to expose the mannose residues on the purified enzymes [15]. Receptor binding studies using surface plasmon resonance to measure interaction of α-GalA with immobilized bovine soluble cation-independent M6P receptor (sCIMPR) and cell uptake studies demonstrates correlation between the level of phosphorylated oligomannose chains and the level of M6P, resulting in improved binding and enhanced uptake in *in vitro* systems [11].

IV. RECOMBINANT HUMAN α-GALS AND PRINCIPLES OF ENZYME REPLACEMENT THERAPY

Primary, secondary, or immortalized human cells can be genetically modified to overexpress human α-GalA that has been used for *in vitro* protein production for ERT. Secondary or immortalized human cells may be a better choice for genetic modification and explanation. After genetic modification, the cells are cultured under conditions permitting overexpression and secretion of α-GalA. The secreted recombinant proteins are then collected from the medium in which the cells are grown and/or from cells followed by the standard protein purification process. One such method involves passing the culture medium, or any sample containing human α-GalA, over a column with hydrophobic interaction resins such as Butyl Sepharose® or a column with the resins that include a functional butyl moiety. This constitutes the first chromatography in the protein purification process. If further purification is required, the α-GalA-containing elutes may then be passed over a second column containing other resins, including immobilized heparin resins such as Heparin Sepharose®, a hydroxyapatite resin, or anion exchange resins such as Q Sepharose®, or a size exclusion resin such as Superdex® 200. Preferably, the purification protocol would include use of each of the aforementioned types of resins. Alternatively, one could use one or more of the latter resins prior to or instead of the hydrophobic interaction resin.

An individual suspected of having a α-GalA deficiency may be treated by administration of pharmaceutical grade, purified human α-GalA by any standard method, including but not limited to intravenous, subcutaneous, or intramuscular injection, or as a solid implant. The purified protein should be formulated in a therapeutic composition consisting of an aqueous solution containing a physiologically acceptable excipient, such as human serum albumin carrier at pH 6.5 or below.

Previous methods for the preparation of α-GalA with relatively high purity were dependent on the use of affinity chromatography requiring a combination of lectin affinity chromatography (concanavalin A (Con A) Sepharose) and affinity chromatography based on binding of α-GalA to the substrate analog (*N*-6-aminohexanoyl-α-D-galactosylamine coupled to a Sepharose matrix) [16]. The use of proteinaceous lectin affinity resins and substrate analog resins is usually associated with the continuous leaching of the affinity agent from the solid support, resulting in contamination of the purified product with the affinity agents either free in solution or bound to eluted protein. Such contaminants make the product unsuitable for use in pharmaceutical preparations. Bound substrate analogs and lectins can also have substantial negative effects on the enzymatic, functional, and structural properties of proteins. Furthermore, such affinity resins are typically expensive to prepare, making the use of such chromatography resins less suitable for production on a commercial scale.

The human α-GalA DNA sequence (either cDNA or genomic DNA) or similar sequences, that differ due to silent codon changes or to codon changes that produce conservative amino acid substitutions, can be used to genetically modify cultured human cells for overexpression of the secreted enzymes. In an α-GalA DNA sequence, certain mutations may encode polypeptides that retain or exhibit improved α-GalA enzymatic activity (made apparent by expressing the mutant DNA molecule in cultured cells, purifying the encoded polypeptide and measuring the catalytic activity) [17].

Production of α-GalA by the cells can be maximized by certain genetic manipulations. For example, the DNA molecule that encodes α-GalA may also contain a signal peptide, such as the signal peptide of human growth hormone (hGH), erythropoietin,

Factor VIII, Factor IX, glucagons, the low-density lipoprotein receptor, or a lysosomal enzyme other than α-GalA. Preferably, the signal peptide is the hGH signal peptide (SEQ ID NO: 21) and is at the N-terminus of the encoded protein. The DNA sequence encoding the signal peptide may contain an intron, such as the first intron of the hGH gene, resulting in a DNA sequence SEQ ID NO: 27. Furthermore, the DNA molecule may also contain a 3′ untranslated sequence (UTS) that is at least six nucleotides in length (in contrast to the α-GalA mRNA found in humans, which has no 3′ UTS, the requisite polyadenylation site being within the coding sequence). The UTS is positioned immediately 3′ to the termination codon of the coding sequence and includes a polyadenylation site. It is preferably at least 6 nucleotides in length, and more preferably at least 30, and in all cases contains the sequence AATAAA, or a related sequence, which serves to promote polyadenylation [17].

The recombinant human α-Gal has been expressed in different systems as a 429-amino acid polypeptide, including CHO cells (Desnick US Patent number 5356804) [18], insect cells (Calhoun US patent number 5179023), and COS cells [19]. The recombinant enzyme expressed in a baculovirus vector and purified using conventional chromatography retains significant catalytic activity after modification with poly(ethylene glycol), a treatment that decreases the immunogenicity and increases the circulation life of many proteins used therapeutically [20]. However, the level of expression was too low to permit economical production. Thus, in a 2000 study, the cDNA of the enzyme was cloned to an improved baculovirus vector and purified to homogeneity using ion exchange (Poros 20-CM, Poros 20-HQ) and hydrophobic chromatography (Toso-ether, Toso-butyl) in the BioCAD HPLC workstation. These chromatographic steps are readily scalable to larger volumes and are appropriate for the purification of the recombinant human α-galactosidase A (r-h-α-GalA) to be used in clinical trials of ERT [21].

V. PHARMACOKINETICS, SAFETY, AND EFFICACY OF RECOMBINANT HUMAN α-GAL

After early trials demonstrated the feasibility of ERT for Fabry disease, a Phase 1 trial was conducted with a single dose of r-h-α-GalA with successful reduction of GL-3 in the urine and liver [22]. A Phase 1/2 open-label dose escalation study at a single center resulted in clearance of GL-3 from plasma and other target organs in 15 male subjects with classic Fabry disease. During this trial, five infusions of recombinant α-Gal A (agalsidase beta) were safely administered and, although the effects were observed in all dosing regimens, there was a dose-related response in which higher doses resulted in better GL-3 clearance [23]. After intravenous administration, r-h-α-GalA was cleared from the plasma by both saturable and concentration-independent pathways, appearing to be biphasic and with the more rapid elimination 1–2 h after the infusion. Both plasma and tissue GL-3 clearance occurred at a dose-dependent manner, biweekly regimens cleared more substrate, and the 1 and 3 mg/kg doses were found to be more effective. The most common reaction after the enzyme infusions was hypersensitivity type and the most common adverse event associated with enzyme infusions was a mild, transient increase in blood pressure. Two serious adverse events occurred in this study; one was an allergic reaction and the other was pulmonary embolism.

In a multicenter, double-blind Phase 3 trial, 58 patients (56 M:2 F) were randomized to r-h-α-GalA or placebo for 20 weeks, followed by open-label treatment protocol for 6 months with agalsidase beta. A dose of 1 mg/kg body weight of agalsidase beta that was administered every other week safely and effectively reversed the accumulation of GL-3 in microvascular endothelium. The primary outcome measure was a reduction of renal deposition of microvascular endothelial GL-3 to normal or near-normal levels. 20 of 29 subjects that received agalsidase beta achieved complete clearance of GL-3 from renal interstitial capillaries ($p<0.001$) compared to none in the placebo group. There was also a reduction of microvascular endothelial GL-3 deposits in the skin and heart in patients receiving r-h-α-GalA ($p<0.001$). Plasma GL-3 levels directly correlated with clearance of microvascular deposits. In general, infusions were well tolerated and all subjects were able to enter the open-label study [24]. During the 36-month extension period, plasma GL-3 levels remained normal, and 39 out 40 subjects sustained endothelial GL-3 clearance in skin biopsy samples. Kidney function, measured with serum creatinine levels and estimated glomerular filtration rate, remained stable [25].

Another double-blind, placebo-controlled Phase 3 study using agalsidase beta, at a dose of 0.2 mg/kg EOW in 26 male subjects, resulted in reduction in the primary outcome measure neuronopathic pain, in 13 subjects, compared to no observed differences in placebo-treated controls. Mesangial widening decreased by 12.5% in kidney biopsies evaluated before and after 24 weeks of agalsidase beta therapy. Further stabilization and improvement of kidney functions was observed in the treatment group. Mean inulin clearance decreased by 6.2 ml/min ($p=0.19$), and mean creatinine clearance was increased by 2.1 ml/min ($p=0.02$) in patients receiving agalsidase beta over placebo group. There was also a 50% decrease in plasma GL-3 levels, improvements in cardiac functions, and increase in body weight. Most of subjects in the treatment arm had mild infusion-related reactions and developed IgG antibodies [22].

In a multicenter Phase 4 trial with agalsidase beta, 82 subjects were randomized in a ratio of 2:1 to either treatment or placebo arms [26]. In this study, the composite primary end point was measured by reaching any of the listed events: end-stage renal disease; a decrease of kidney function as measured with a 33% increase in creatinine clearance; cardiac or central nervous system events; or death. The expected event rate was 40% in placebo and 10% for the treatment group and the study was terminated when one-third of subjects experienced a total of 27 events, mostly decreased kidney function. Two deaths occurred in each group secondary to cardiac arrest, and cardiac arrest after stroke and pulmonary embolism in agalsidase beta-treated group. Most of the adverse events were infusion related, occurring within the first 6 months, and in about 55% of treated subjects versus 23% of the placebo group, and were mild to moderate in intensity. About 68% of the subjects developed antibodies against r-h-α-GalA and one patient had a severe hypotensive episode with positive serum IgE response.

Although both enzyme preparations have demonstrated clinical efficacy in human clinical trials, the different outcome measures and dosing regimens preclude direct comparisons of the relative efficacies of the two different products [27].

VI. IMMUNE RESPONSE AND TOLERANCE

Clinically, the development of an immune response is anticipated in a number of patients treated with any recombinant human proteins [28] and is suggested to be more common, especially when the native protein is deficient or absent as is the case with many male patients with Fabry disease. The immune response that results in the development of antibodies against the infused proteins may affect the clinical outcome by the development of hypersensitivity, anaphylactoid, or febrile reactions or may lead to the development of cytokine release and a generalized inflammatory response or immune complex formation. Furthermore, the mounted immune response may lead to inactivation or degradation of the recombinant enzyme or may change the pharmacokinetic properties of the therapeutic protein. Immunogenicity of r-h-α-GalA was assessed in different clinical trials [22]. Although Phase 3 trials for both agalsidase alfa and beta showed that the majority of patients have developed an antibody response during the course of therapy, there are only rare case reports for the development of IgE antibodies. The seroconversion is mainly due to IgG response, and IgG antibodies may develop in patients treated with either agalsidase alfa or beta and often cross-react [29]. The immunogenicity is suggested to be similar for both agalsidase alfa and agalsidase beta. It has also been suggested that agalsidase alfa produced by human fibroblasts and agalsidase beta produced in CHO cells do not produce different antibodies when administered at the same dose [29]. However, different r-h-α-GalA preparations may elicit different immunogenicity in the same host, as in one case, the patient developed antibody response only to agalsidase beta (Fabrazyme) but not against alfa (Replagal) [30]. The observation that the antibodies exhibit *in vitro* neutralizing capacity may suggest the presence of a single immunogenic epitope of h-Gal that is close to the catalytic site, and the binding of the antibody may prevent proper functioning of the enzyme. Many factors can influence immunogenicity of a protein, including host factors and the properties of the therapeutic protein. Among the properties of the therapeutic agent, structural similarity is one of the key factors, and alterations in the structural confirmation can make a protein more immunogenic [31].

The incidence of lower rates of antibody formation with agalsidase alfa is often attributed to differences in techniques used to measure antibody formation [29]. Other factors that may affect host response are the dose and the infusion frequency. Antibodies are more often produced in patients receiving higher doses of ERT. For r-h-α-GalA the dose at 1.0 mg/kg was associated with higher rates of antibody formation than that at 0.2 mg/kg. During ERT therapy, antibody titers may not decline over time regardless of the types of r-h-α-GalA preparations [32]. However, higher doses of r-h-α-GalA may counter the inhibitory effect of antibody formation as the better urine GL-3 clearance was observed in some patients when the dose switched from 0.2 to 1.0 mg/kg [32].

Clinically, predicting immune response in an individual patient has been difficult, especially when diverse mutations play a role in disease pathology [33]. Various missense mutations may alter either the processing or stability of the enzyme, which may result in either deficient or complete lack of enzymatic activity despite detectable protein levels. Multiple molecular mechanisms, such as recombinant arrangements, frameshift mutations, or splice junction defects, may result in lack of production of enzyme protein. The most common method for detection of endogenous enzyme protein is assessing cross-reactive immunologic material (CRIM) by using an antibody against the enzyme protein. CRIM status has been correlated with development of immune response. Although it has been hypothesized that patients with residual endogenous proteins, thus are CRIM (+), have an attenuated response to recombinant proteins, CRIM status was suggested to not be predictive of antibody development [31].

The variability in the phenotypes and the difference in diseases progression make it difficult to assess the actual impact of antibodies on disease outcomes. During the course of ERT, it was shown that the increase in α-GalA IgG titers correlated with the amount of dermal capillary endothelial GL-3 deposits [34]. Although this may suggest that the increase in antibody titers may have a negative impact on the effect of therapy, there has been no statistically significant correlation between the antibody levels and plasma GL-3 clearance or renal function as measured by indirect methods using estimated glomerular filtration rate.

VII. TYPES OF DIFFERENT ENZYMATIC PREPARATIONS AND OTHER ENZYMES ON THE HORIZON

Two α-GalA preparations, agalsidase alfa (Replagal) and agalsidase beta (Fabrazyme), were developed by TKT (Transkaryotic Therapeutics), subsequently acquired by Shire Human Genetic Therapies, and Genzyme, respectively. Both drugs were approved by the European Agency for the Evaluation of Medicinal products in 2001, albeit at different doses. Agalsidase alfa (Replagal) is administered at 0.2 mg/kg every other week, and agalsidase beta (Fabrazyme), at 1 mg/kg every other week. According to the orphan drug policy of the United States in which only one drug formation can be approved for 7 years of marketing exclusivity (US FDA CDER: Accelerated Development Review and Definition. www.fda.gov/cder/handbook), Fabrazyme® (Genzyme) was approved by FDA for the treatment of Fabry disease with an orphan drug status in 2003.

Both agalsidase alfa and agalsidase beta are derived from the same human DNA sequence and have similar amino acid sequences with the exception of C-terminal residues. Both proteins display C-terminal heterogeneity, most likely due to proteolytic processing of the mature full-length protein [11]. Although the removal of C-terminal residues was suggested to increase the specific activity of the enzyme [35], more recent studies suggest that C-terminal differences may not result in different enzymatic properties. Any differences affecting biological activity and availability of r-h-α-GalA were suggested to result from posttranslational modifications. The major species in each preparation has a monomeric molecular weight of 51.2 kDa, but the main difference is the pattern of glycosylation. Although both have similar type and location of oligosaccharide attachments, the primary difference in glycosylation between the two preparations is the number of sialic acid residues. There is a higher level of fucose, *N*-acetylglucosamine, and galactose in agalsidase alfa (Replagal). Although the number of sialic acid residues is similar on a mole/mole protein, agalsidase alfa (Replagal) contains higher amounts of complex chains and thus lower sialic acid/galactose residues compared to agalsidase beta (Fabrazyme). Although the three glycosylation sites at Asn-108, Asn-161, and Asn-184 are similar, there is a higher degree of sialylation and/or phosphorylation in Fabrazyme [11,36]. In addition, Fabrazyme contains a higher amount of M6P on oligomannose side chains, resulting in a higher level of M6P in molar amount than Replagal.

Protein glycosylation impacts not only the therapeutic efficacy but also immunogenicity of protein therapeutics. Multiple factors, including cell culture conditions and expression systems used to produce protein, can affect glycosylation patterns [37,38]. Thus, the different expression systems used, CHO cells for Fabrazyme and human fibroblast cell line for Replagal, were suggested to result in the glycosylation differences between the two enzyme preparations.

In a study by Tajima and colleagues [7] looking at structure-based design of modified lysosomal enzymes, the substrate specificity of α-*N*-acetylgalactosaminidase (NAGA, E.C. 3.2.1.49) was altered after a structure-based amino acid substitution in the active site. The modified enzyme, α-galactosidase B, has similar protein structure to NAGA but displays catalytic activity toward GL-3 and was developed as a new recombinant enzyme for the treatment of Fabry disease [7]. NAGA catalyzes the hydrolysis of the terminal α-*N*-acetylgalactosamine of glycoconjugates, including polysaccharides, glycolipids, and glycoproteins. Because the overall structure of the modified NAGA is not altered, despite moderate changes in the substrate-recognition region, the administration of the modified NAGA to Fabry patients (who usually have endogenous NAGA protein) should not cause an allergic reaction. The modified NAGA has many advantages because of its high stability and high M6P content and the low probability to induce an allergic reaction. These characteristics should be confirmed in future clinical studies.

In a study of searching for the alternatives to replace bovine serum used in the production of recombinant enzymes, scientists created a recombinant human *GLA* with a high content of phosphorylated *N-glycans* using a methylotrophic yeast, *Ogataea minuta*, in which the *OCH1* gene was disrupted and the *MNN4* gene was overexpressed. The enzyme was well incorporated into the kidneys easily and cleaved the accumulated glycosphingolipids. As an enzyme production system involving methylotrophic yeast is economical and does not require bovine serum in the procedure, it should be useful for the developments of a new ERT for Fabry disease [6].

VIII. CONCLUSIONS

The rationale for ERT for lysosomal storage diseases such as Fabry disease is quite intriguing. While the lack of a crucial α-GalA enzyme in lysosomes is indicated in the pathophysiology in Fabry disease, replacing the deficient enzyme exogenously can be expected to treat some or most of the disease manifestations, if administered in quantities adequate for equal distribution of affected organ systems. Any disease that affects less than 200 000 is subject to the designation of orphan disease. Because of the difficulty of recruiting a large number of patients with an orphan disease, the FDA permits clinical trials for orphan drugs to have fewer subjects for enrollment. After the Biologics License Application (BLA) is filed, the FDA also provides accelerated review for orphan disease. For any slowly progressive disorder such as Fabry disease, the usual clinical trial durations of 6–12 months are not sufficient to demonstrate clinical efficacy. Thus, the evidence of clinical efficacy is sought using surrogate markers. However, using surrogate markers such as GL-3 has significant limitations, especially in some patient populations include females and children. In Fabry disease, almost a decade of information with ERT acquired through different post-marketing studies, such as Fabry Registry (Genzyme) and FOS (Fabry Outcome Survey, Shire), has provided us the most valuable insight into the spectrum of disease pathology in Fabry disease. Further understanding the multiple pathways that are beyond the biological effects of enzymatic deficiency and storage of abnormal lipid substrate and into the role of lysosomal dysfunction, protein misfolding, and the inflammatory processes in Fabry disease are essential to devise a comprehensive therapeutic approach. Soon, there may be various ERT alternatives to treat Fabry disease that are equally effective, more safer, and more economical to produce than those currently available.

REFERENCES

1. RJ Desnick, R Brady, J Barranger, AJ Collins, DP Germain, M Goldman, G Grabowski, S Packman, WR Wilcox. Fabry disease, an under-recognized multisystemic disorder: expert recommendations for diagnosis, management, and enzyme replacement therapy. Ann Intern Med 138:338–346, 2003.
2. AI Guce, NE Clark, EN Salgado, DR Ivanen, AA Kulminskaya, H Brumer, SC Garman. Catalytic mechanism of human α–galactosidase. J Biol Chem 285(6):3625–3632, 2010.
3. BL Thurburg, H Rennke, RB Colvin, S Dikman, RE Gordon, AB Collins, RJ Desnick, M O'Callaghan. Globotriaosylceramide accumulation in the Fabry kidney is cleared from multiple cell types after enzyme replacement therapy. Kidney Int 62:1933–1946, 2002.
4. T Prabakaran, R Nielsen, JV Larsen, SS Serensen, U Feldt-Rasmussen, MA Saleem, CM Petersen, PJ Verroust, EI Christensen. Receptor-mediated endocytosis of α–galactosidase A in human podocytes in Fabry disease. PLoS One 6(9):1–9, 2011.
5. RO Brady. Enzyme replacement for lysosomal diseases. Annu Rev Med 57:283–296, 2006.
6. T Tsukimura, I Kawashima, T Togawa, T Kodama, T Suzuki, T Watanabe, Y Chiba, Y Jigami, T Fukushige, T Kanekura, et al. Efficient uptake of recombinant α–galactosidase A produced with a gene-manipulated yeast by Fabry mice kidneys. Mol Med 18(1):76–82, 2012.
7. Y Tajima, I Kawashima, T Tsukimura, K Sugawara, M Kuroda, T Suzuki, T Togawa, Y Chiba, Y Jigami, K Ohno, et al. Use of a modified α–*N*-acetylgalactosaminidase in the development of enzyme replacement therapy for Fabry disease. Am J Hum Genet 85:569–580, 2009.

8. DE Koshland. Stereochemistry and the mechanism of enzymatic reactions. Biol Rev Cambridge Philos Soc 28:416–436, 1953.
9. DJ Vocadlo, GJ Davies. Mechanistic insights into glycosidase chemistry. Curr Opin Chem Biol 12:539–555, 2008.
10. BL Cantarel, PM Coutinho, C Rancurel, T Bernard, V Lombard, B Henrissat. The Carbohydrate-Active EnZymes database (CAZy): an expert resource for Glycogenomics. Nucl Acid Res 37:D233–D238, 2009.
11. K Lee, X Jin, K Zhang, L Copertino, L Andrews, J Baker-Malcolm, L Geagan, H Qiu, K Seiger, D Barngover, et al. A biochemical and pharmacological comparison of enzyme replacement therapies for the glycolipid storage disorder Fabry disease. Glycobiology 13(4):305–313, 2003.
12. N Asano. Glycosidase inhibitors: update and perspectives on practical use. Glycobiology 13(10):93R–104R, 2003.
13. S Kornfeld, I Mellman. The biogenesis of lysosomes. Annu Rev Cell Biol 5:483–525, 1989.
14. G Ashwell, J Harford. Carbohydrate specific receptors of the liver. Annu Rev Biochem 51:531–554, 1982.
15. E Beutler. Gaucher's disease. New Engl J Med 325:1354–1360, 1991.
16. DF Bishop, RJ Resnick. Affinity purification of alpha-galactosidase A from human spleen, placenta, and plasma with elimination of pyrogen contamination. Properties of the purified splenic enzyme compared to other forms. J Biol Chem 256:1307–1316, 1981.
17. M Yasuda, J Shabbeer, M Osawa, RJ Desnick. Fabry disease: novel α-galactosidase A 3′-terminal mutations result in multiple transcripts due to aberrant 3′-end formation. Am J Hum Genet 73(1):162–173, 2003.
18. A Ioannou, DF Bishop, RJ Resnick. Overexpression of human alpha-galactosidase A results in its intracellular aggregation, crystallization in lysosomes, and selective secretion. J Cell Biol 119:1137–1150, 1992.
19. S Tsuji, BM Martin, DC Kaslow, BR Migeon, PV Choudary, BK Stubbleflied, JA Mayor, GJ Murray, JA Barranger, et al. Signal sequence and DNA-mediated expression of human lysosomal alpha-galactosidase A. Eur J Biochem 165:275–280, 1987.
20. G Coppola, Y Yan, P Hantzopoulos, E Segura, JG Stroh, DH Calhoun. Characterization of glycosylated and catalytically active recombinant human alpha-galactosidase A using a baculovirus vector. Gene 144(2):197–203, 1994.
21. Y Chen, M Jin, L Goodrich, G Smith, G Coppola, DH Calhoun. Purification and characterization of human alpha-galactosidase A expressed in insect cells using a baculovirus vector. Protein Expr Purif 20(2):228–236, 2000.
22. R. Schiffmann, GJ Murray, D Treco, P Daniel, M Sellos-Moura, M Myers, JM Quirk, GC Zirzow, M Borowski, K Loveday, et al. Infusion of a-galactosidase A reduces tissue globotriaosylceramide storage in patients with Fabry disease. Proc Natl Acad Sci USA 97:365–370, 2000.
23. CM Eng, M Banikazemi, RE Gordon, M Goldman, R Phelps, L Kim, A Gass, J Winston, S Dikman, JT Fallon, et al. A phase 1/2 clinical trial of enzyme replacement in Fabry disease: pharmacokinetic, substrate clearance, and safety studies. Am J Hum Genet 68:711–722, 2001.
24. CM Eng, N Guffon, WR Wilcox, DP Germain, P Lee, S Waldek, L Caplan, GE Linthorst, RJ Desnick. Safety and efficacy of recombinant human alpha-galactosidase A – replacement therapy in Fabry's disease. N Engl J Med 345:9–16, 2001.

25. WR Wilcox, M Banikazemi, N Guffon, S Waldek, P Lee, GE Linthorst, RJ Desnick, DP Germain. Long-term safety and efficacy of enzyme replacement therapy for Fabry disease. Am J Hum Genet 75:65–74, 2004.
26. M Banikazemi, J Bultas, S Waldek, WR Wilcox, CB Whitley, M McDonald, R Finkel, S Packman, DG Bichet, DG Warnock, et al. Fabry Disease Clinical Trial Study Group. Agalsidase-beta therapy for advanced Fabry disease: a randomized trial. Ann Intern Med 146(2):77–86, 2007.
27. GM Pastores, R Tadhani. Advances in the management of Anderson-Fabry disease: enzyme replacement therapy. Expert Opin Biol Ther 2(3):325–333, 2002.
28. S. Porter. Human immune response to recombinant human proteins. J Pharm Sci 90(1): 1–11, 2001.
29. GE Linthorst, CEM Hollak, WE Donker-Koopman, A Strijland, JMFG Aerts. Enzyme therapy for Fabry disease: neutralizing antibodies toward agalsidase alpha and beta. Kidney Int 66(4):1589–1595, 2004.
30. A Tanaka, T Takeda, T Hoshina, K Fukai, T Yamano. Enzyme replacement therapy in a patient with Fabry disease and the development of IgE antibodies against agalsidase beta but not agalsidase alpha. J Inherit Metab Dis, published online: June 22, 2010. DOI: 10.1007/s10545-010-9136-0.
31. SM Richards. Immunologic considerations for enzyme replacement therapy in the treatment of lysosomal storage disorders. Clin Appl Immunol Rev 2:241–253, 2002.
32. AC Vedder, F Breunig, WE Donker-Koopman, K Mills, E Young, B Winchester, IJM Ten Berge, JEM Groener, JMFG Aerts, C Wanner, et al. Treatment of Fabry disease with different dosing regimens of agalsidase: effects on antibody formation and GL-3. Mol Genet Metab 94:319–325, 2008.
33. CM Eng, LA Resnick-Silverman, DJ Niehaus, KH Astrin, RJ Desnick. Nature and frequency of mutations in α–galactosidase A gene that cause Fabry disease. Am J Hum Genet 53:1186–1197, 1993.
34. B Benichou, S Goyal, C Sung, AM Norfleet, F O'Brien. A retrospective analysis of the potential impact of IgG antibodies to agalsidase beta on efficacy during enzyme replacement therapy for Fabry disease. Mol Genet and Metab 96(1):4–12, 2009.
35. N Miyamura, E Araki, K Matsuda, R Yoshimura, N Furukawa, K Tsuruzoe, T Shirotani, H Kishikawa, K Yamaguchi, M Shichiri. A carboxy-terminal truncation of human alpha-galactosidase A in a heterozygous female with Fabry disease and modification of the enzymatic activity by the carboxy-terminal domain. Increased, reduced or absent enzyme activity depending on number of amino acid residues deleted. J Clin Invest 98(8):1809–1817, 1996.
36. F Matsuura, M Ohata, YA Ioannou, RJ Desnick. Human α–galactosidase A: characterization of the N-linked oligosaccharides on the intracellular and secreted glycoforms overexpressed by Chinese hamster ovary cells. Glycobiology 8:329–339, 1998.
37. CF Goochee. Bioprocess factors affecting glycoprotein oligosaccharide structure. Dev Biol Stand 76:95–104, 1992.
38. N Jenkins, RB Parekh, DC James. Getting the glycosylation right: implications for the biotechnology industry. Nature Biotechnol 14:975–981, 1996.

10

METHODS AND PRINCIPLES OF PANCREATIC FUNCTION TESTS

HENRIKE VON SCHASSEN, JUTTA KELLER, AND PETER LAYER
Department of Internal Medicine, Israelitic Hospital in Hamburg, Hamburg, Germany

I. INTRODUCTION

Pancreatic exocrine insufficiency is defined as a reduced secretory capacity of pancreatic enzymes and bicarbonate with the main symptoms as weight loss, steatorrhea (often associated or confused with diarrhea), signs of reduced resorption of lipid-soluble vitamins, and nutrient malabsorption. Exocrine insufficiency is mainly caused by diseases affecting directly the pancreatic tissue, such as chronic or acute pancreatitis, the hereditary disease cystic fibrosis, and pancreatic cancer. However, other entities such as gastric or small bowel resection and probably also inflammatory diseases of the gastrointestinal tract can induce regulatory disturbances and can thereby indirectly impair pancreatic exocrine function [1]. Since pancreatic exocrine dysfunction may occur in several diseases, its recognition is of great clinical interest.

In patients with chronic pancreatitis, exocrine function decreases slowly over several years. Symptoms like steatorrhea occur, when enzyme synthesis and secretion (mainly that of lipase) are reduced to less than 10% of normal [1,2]. In patients with alcoholic chronic pancreatitis, it usually takes 10–20 years until symptoms of malabsorption occur. Steatorrhea precedes overt malabsorption of carbohydrates and proteins [1].

There are several direct and indirect tests for evaluation of pancreatic exocrine function. The direct tests measure secreted pancreatic enzymes in chyme or stool, and the indirect tests investigate effects caused by lack of pancreatic enzymes [1,3–5]. Pancreatic function tests ideally meet all of the following requirements: They should

Enzyme Technologies: Pluripotent Players in Discovering Therapeutic Agents, First Edition.
Edited by Hsiu-Chiung Yang, Wu-Kuang Yeh, and James R. McCarthy.

be specific for pancreatic insufficiency, be sensitive enough to detect early stages of exocrine pancreatic dysfunction, be noninvasive and not very stressful for the patient, and be cheap and easy to perform. None of the available pancreatic function tests satisfies all these aspects [6]; especially the diagnosis of mild forms of pancreatic exocrine insufficiency is often difficult. In particular, most indirect tests cannot assess mild stages of pancreatic exocrine dysfunction accurately [3].

II. TESTING OF PANCREATIC EXOCRINE FUNCTION

A. Direct Tests

Stool Tests The fecal excretion of pancreatic enzymes correlates with duodenal enzyme secretion [7]. However, pancreatic enzymes are inactivated to different degrees during gastrointestinal transit. Chymotrypsin and elastase-1 are more stable enzymes and are therefore suitable for stool testing [8].

The activity of chymotrypsin in stool can be tested photometrically [9]. To improve the sensitivity of the test, three different stool samples are necessary. Since a differentiation between human and substituted chymotrypsin is not possible, the test results are influenced by pancreatin supplementation. Thus, it is important that these drugs are discontinued at least 5 days before the examination. On the other hand, the chymotrypsin test can also be used to monitor a patient's compliance with enzyme supplementation in refractory cases. The normal chymotrypsin activity in stool is >6 U/g. The main drawbacks of this test are its low sensitivity and specificity in patients with only mild or moderate exocrine pancreatic insufficiency [8–12].

Currently, measurement of fecal elastase-1 concentration in a single stool sample is the preferred and best available pancreatic function test. The concentration of elastase-1 is investigated by an ELISA test using a specific antibody against the human enzyme, so that pancreatin supplements have no influence on the results and there is no need to discontinue them. Normal stool concentration of elastase-1 exceeds 200 μg/g stool (depending on the method), and a concentration less than 100 μg/g stool usually means severe pancreatic exocrine insufficiency. Measurement of fecal elastase-1 is more sensitive and specific than chymotrypsin testing [12]. However, sensitivity for detection of mild to moderate pancreatic exocrine insufficiency is still limited. Moreover, in the differential diagnosis of diarrhea, specificity of the test is rather low since increased stool-water content leads to false-positive results [1].

Secretin Test The secretin (or secretin–pancreozymin) test is an invasive test that requires placement of a duodenal tube. It is regarded as the reference method for evaluation of pancreatic exocrine function and can also detect mild and moderate cases but has several disadvantages including invasiveness, high costs, necessity of special equipment and trained personnel [1], and lack of standardization between different centers.

In order to achieve reliable test results, pancreatin preparations have to be discontinued several days in advance. Nicotine and drugs with sedative or anticholinergic effects have to be discontinued at least 24 h before the secretin test is performed, and the patient has to be fasting for at least 12 h. In patients with acute pancreatitis, it is necessary to wait for 8 to 12 weeks after the acute episode.

For the measurement, the tip of a double-lumen nasoduodenal tube is placed near the ligament of Treitz. Via the lumen placed in the gastric antrum, gastric secretions are continuously aspirated and discarded. Duodenal contents are aspirated via the second lumen of the tube for 30 min under basal conditions followed by a 60 min collection period with intravenous application of secretin. Subsequently, secretion volume, bicarbonate concentration, and activity of pancreatic enzymes (trypsin, chymotrypsin, lipase, and amylase) need to be determined in duodenal juice samples. Secretin stimulation leads to maximal bicarbonate output but induces only moderate stimulation of pancreatic enzyme secretion. This is why a second stimulation period using a combination of secretin and cholecystokinin (CCK) or the CCK analog cerulein was usually performed. However, these substances are currently not available in many countries.

To compensate for incomplete aspiration of duodenal contents, a dilution marker can be added, but this addition further complicates the procedure. Moreover, endoscopy-based modifications are used by some specialized centers with intermittent aspiration of duodenal chyme over a prolonged period of time [13].

Lundh Test The Lundh test [14] also requires intestinal intubation for direct measurement of enzyme output in duodenal juice. However, in contrast to the secretin test, pancreatic exocrine secretion is stimulated by a standardized test meal. Thus, release of regulatory mediators from the intestinal mucosa is needed for stimulation of pancreatic secretion. As a consequence, false-positive results may occur in intestinal diseases such as celiac sprue.

B. Indirect Tests

Fluorescein Dilaurate and NBT-PABA Test Neither the fluorescein dilaurate (pancreolauryl test (PLT)) nor the *N*-benzoyl-L-tyrosyl-*p*-aminobenzoic acid test (NBT-PABA test) is commercially available in many countries at present and, therefore, these tests will only be discussed briefly. For both tests, the patient ingests a substrate that is metabolized into two or more products by pancreatic enzymes. At least one of the metabolites is absorbed from the gut, conjugated, and excreted in urine, where it can be measured. In subjects with pancreatic exocrine insufficiency, increased fecal excretion of the unsplit molecule and decreased absorption, blood levels, and urinary excretion of the metabolite will occur. Because of interindividual variability of intermediary steps such as intestinal absorption and renal function, the fluorescein dilaurate test includes application of the absorbable metabolite (fluorescein) on a second day, and the results of the test are expressed as the ratio of excreted fluorescein on the test and the control day in percent. A ratio of less than 20% is abnormal, 20–30% is equivocal, and more than 30% is considered normal. A modified serum test eliminates the need for a second test day but does not increase sensitivity and specificity [12,15,16].

^{13}C-Breath Tests Several breath tests using ^{13}C-labeled substrates for measurement of pancreatic function have been developed during the recent years. Of these, tests using

^{13}C-labeled lipids are most promising because lipase synthesis and secretion tend to be impaired earlier than those of other pancreatic enzymes in chronic pancreatitis [17]. The labeled lipids are ingested orally together with a test meal and need to be digested to monoglycerides and free fatty acids by pancreatic lipase prior to absorption. Hepatic metabolism of the absorbed lipids leads to production of $^{13}CO_2$ which is transported to the lung and exhaled. Thus, the ratio of breath $^{13}CO_2/^{12}CO_2$ over time reflects intestinal lipolysis by pancreatic lipase as the rate-limiting step of lipid absorption. Available substrates include 1,3-distearyl-2[^{13}C]-octanoate (^{13}C-mixed triglyceride) which is most commonly used, but also uniformly labeled Hiolein® (mixture of long-chain triglycerides) and cholesteryl-^{13}C-octanoate [18–21]. Sensitivity and specificity of certain test modifications have been reported to exceed 90% [22]. Moreover, a modified version of the ^{13}C-mixed triglyceride breath test has been shown to also detect mild to moderate pancreatic exocrine insufficiency [23].

III. SUMMARY

Pancreatic exocrine insufficiency defined as a reduced synthesis and secretion of bicarbonate and pancreatic enzymes is caused by several disorders affecting the pancreas directly (e.g., chronic pancreatitis) or indirectly (e.g., gastric or small bowel resection). Pancreatic function tests are mandatory tools for objective diagnosis of pancreatic exocrine insufficiency in numerous clinical settings, and a variety of different test principles have been developed. Invasive tests like the secretin or secretin–cerulein test that directly measure bicarbonate and enzyme output are still the gold standard with highest sensitivity and specificity. However, availability of these tests is limited to centers due to invasiveness and high costs. Determination of fecal elastase-1 is convenient and widely available, but sensitivity and specificity of this test are low, particularly in patients with mild to moderate forms of exocrine insufficiency. ^{13}C-breath tests have been shown to accurately diagnose pancreatic exocrine insufficiency and are becoming more important. Unfortunately, they are still not widely established.

REFERENCES

1. J Keller, AA Aghdassi, MM Lerch, JV Mayerle, P Layer. Tests of pancreatic exocrine function – clinical significance in pancreatic and non-pancreatic disorders. Best Pract Res Clin Gastroenterol 23:425–439, 2009.
2. P Layer, H Yamamoto, L Kalthoff, JE Clain, LJ Bakken, EP Di Magno. The different courses of early and late onset idiopathic and alcoholic chronic pancreatitis. Gastroenterology 107:1481–1487, 1994.
3. RS Chowdhury, CE Forsmark. Review article: pancreatic function testing. Aliment Pharmacol Ther 17:733–750, 2003.
4. C Niederau, JH Grendell. Diagnosis of chronic pancreatitis. Gastroenterology 88:1973–1995, 1985.
5. K Ochi, T Mizushima, H Harada, S Matsumoto, N Matsumura, T Seno. Chronic pancreatitis: functional testing. Pancreas 16:343–348, 1998.

6. M Kahl, J Keller, P Layer. Functional testing, radiologic work up of chronic pancreatitis. In: HG Beger, S Matsuno, JL Cameron, eds. *Diseases of the Pancreas: Current Surgical Therapy*. Berlin/New York: Springer Verlag, 2008, pp. 319–330.
7. M Katschinski, J Schirra, A Bross, B Göke, R Arnold. Duodenal secretion and fecal excretion of pancreatic elastase-1 in healthy humans and patients with chronic pancreatitis. Pancreas 15:191–200, 1997.
8. J Keller, C Jahr, V Andresen, U Rosien, P Layer. Funktionsdiagnostik von Sekretion, Absorption und Motilität. In: P Layer, U Rosien, eds. *Praktische Gastroenterologie*, 4th ed. München: Urban und Fischer Verlag, 2010, pp. 76–77, 79, 81.
9. RW Ammann, A Akovbiantz, W Häcki, F Largiadèr, M Schmid. Diagnostic value of the fecal chymotrypsin test in pancreatic insufficiency, particularly chronic pancreatitis: correlation with the pancreozymin-secretin test, fecal fat excretion and final clinical diagnosis. Digestion 21:281–289, 1981.
10. A Bonin, CC Roy, R Lasalle, A Weber, CL Morin. Fecal chymotrypsin: a reliable index of exocrine pancreatic function in children. J Pediatr 83:594–600, 1973.
11. PG Lankisch, A Schreiber, J Otto. Pancreolauryl test. Evaluation of a tubeless pancreatic function test in comparison with other indirect and direct tests for exocrine pancreatic function. Dig Dis Sci 28:490–493, 1983.
12. E Siegmund, JM Löhr, P Schuff-Werner. Die diagnostische Validität nichtinvasiver Pankreasfunktionstests – Eine Metaanalyse. Z Gastroenterol 42:1117–1128, 2004.
13. DL Conwell, G Zuccaro, Jr., JJ Vargo, JA Dumot, F VanLente, F Khandwala, PA Trolli, C O'Laughlin. Comparison of the secretin stimulated endoscopic pancreatic function test to retrograde pancreatogram. Dig Dis Sci 52:1076–1081, 2007.
14. O James. The lundh Test. Gut 14:582–591, 1973.
15. G Lock, R Kadow, H Messmann, H Zirngibl, J Schölmerich, A Holstege. Modified serum pancreolauryl test in chronic pancreatitis: evaluation in comparison to endoscopic retrograde pancreatography. Hepatogastroenterology 44:1110–1116, 1997.
16. P Malfertheiner, M Buchler, A Müller, H Ditschuneit. Fluorescein dilaurate serum test: a rapid tubeless pancreatic function test. Pancreas 2:53–60, 1987.
17. EP DiMagno, JR Malagelada, VL Go. Relationship between alcoholism and pancreatic insufficiency. Ann NY Acad Sci 252:200–207, 1975.
18. C Loser, C Brauer, S Aygen, O Hennemann, UR Fölsch. Comparative clinical evaluation of the 13C-mixed triglyceride breath test as an indirect pancreatic function test [see comments]. Scand J Gastroenterol 33:327–334, 1998.
19. B Lembcke, B Braden, WF Caspary. Exocrine pancreatic insufficiency: accuracy and clinical value of the uniformly labeled 13C-Hiolein breath test. Gut 39:668–674, 1996.
20. B Lembcke. Current role of breath tests in gastroenterology. Z Gastroenterol 34 (Suppl. 4):46–53, 1996.
21. KD Wutzke, M Radke, K Breuel, S Gurk, JD Lafrenz, WE Heine. Triglyceride oxidation in cystic fibrosis: a comparison between different 13C-labeled tracer substances. J Pediatr Gastroenterol Nutr 29:148–154, 1999.
22. JE Dominguez-Munoz, ed. Pancreatic function tests for diagnosis and staging of chronic pancreatitis, cystic fibrosis, and exocrine pancreatic insufficiency of other etiologies. In: *Clinical Pancreatology*, 1st ed. Malden, MA: Wiley-Blackwell, 2005, pp. 259–266.
23. J Keller, S Brückel, C Jahr, P Layer. A modified 13C-mixed triglyceride breath test detects moderate pancreatic exocrine insufficiency. Pancreas 40:1201–1205, 2011.

INDEX

Enzyme Technologies: Pluripotent Players in Discovering Therapeutic Agents, First Edition.
Edited by Hsiu-Chiung Yang, Wu-Kuang Yeh, and James R. McCarthy.